Finite Element Analysis of Composite Materials Using Abaqus®

Developed from the author's course on advanced mechanics of composite materials, *Finite Element Analysis of Composite Materials Using Abaqus®* shows how powerful finite element tools tackle practical problems in the structural analysis of composites. This second edition includes two new chapters on fatigue and Abaqus programmable features, a major update of Chapter 10 on delaminations, and significant updates throughout the remaining chapters. Furthermore, the book updates all examples, sample code, and problems to Abaqus 2020.

Unlike other texts, this one takes theory to a hands-on level by actually solving problems. It explains the concepts involved in the detailed analysis of composites, the mechanics needed to translate those concepts into a mathematical representation of the physical reality, and the solution of the resulting boundary value problems using Abaqus. The reader can follow a process to recreate every example using Abaqus graphical user interface (CAE) by following step-by-step directions in the form of pseudo-code or watching the solutions on YouTube.

The first seven chapters provide material ideal for a one-semester course. Along with offering an introduction to finite element analysis for readers without prior knowledge of the finite element method, these chapters cover the elasticity and strength of laminates, buckling analysis, free edge stresses, computational micromechanics, and viscoelastic models for composites. Emphasizing hereditary phenomena, the book goes on to discuss continuum and discrete damage mechanics as well as delaminations and fatigue. The text also shows readers how to extend the capabilities of Abaqus via user subroutines and Python scripting.

Aimed at advanced students and professional engineers, this textbook features 62 fully developed examples interspersed with the theory, 82 end-of-chapter exercises, and 50+ separate pieces of Abaqus pseudo-code that illustrate the solution of example problems. The author's website offers the relevant Abaqus and MATLAB® model files available for download, enabling readers to easily reproduce the examples and complete the exercises. Video recordings of solutions to examples are available on YouTube with multilingual captions.

Composite Materials: Analysis and Design

Series Editor
Ever J. Barbero

Finite Element Analysis of Composite Materials Using Abaqus•
Ever J. Barbero

Finite Element Analysis of Composite Materials Using ANSYS®, Second Edition
Ever J. Barbero

Strengthening Design of Reinforced Concrete with FRP
Hayder A. Rasheed

Natural Fiber Composites
R.D.S.G. Campilho

Structural Health Monitoring Technologies and Next-Generation Smart Composite Structures
Jayantha Ananda Epaarachchi and Gayan Chanaka Kahandawa

FRP Deck and Steel Girder Bridge Systems: Analysis and Design
Julio F. Davalos, An Chen, and Pizhong Qiao

Introduction to Composite Materials Design, Third Edition
Ever J. Barbero

Smart Composites: Mechanics and Design
Rani Elhajjar, Valeria La Saponara, and Anastasia Muliana

Braided Structures and Composites: Production, Properties, Mechanics, and Technical Applications
Sohel Rana and Raul Fangueiro

Finite Element Analysis of Composite Materials Using Abaqus®, Second Edition
Ever J. Barbero

For more information about this series, please visit:
https://www.routledge.com/Composite-Materials/book-series/CRCCOMMATDES

Finite Element Analysis of Composite Materials Using Abaqus®
Second Edition

Ever J. Barbero

CRC Press
Taylor & Francis Group
Boca Raton London New York

CRC Press is an imprint of the
Taylor & Francis Group, an **informa** business

Second edition published 2023
by CRC Press
6000 Broken Sound Parkway NW, Suite 300, Boca Raton, FL 33487-2742

and by CRC Press
4 Park Square, Milton Park, Abingdon, Oxon, OX14 4RN

CRC Press is an imprint of Taylor & Francis Group, LLC

[LCCN # 2022948138]

ISBN: 978-0-367-62145-2 (hbk)
ISBN: 978-0-367-62149-0 (pbk)
ISBN: 978-1-003-10815-3 (ebk)

DOI: 10.1201/9781003108153

Typeset in CMR10 font
by KnowledgeWorks Global Ltd.

Dedicated to my graduate students, who taught me as much as I taught them.

Contents

Preface

This edition of *Finite Element Analysis of Composite Materials Using Abaqus*® updates 62 examples and 82 end-of-chapter problems using Abaqus 2020 from the first edition (2013) that used Abaqus 6.10 (2010). All chapters are updated and expanded. For example, Chapter 2 has a new section on Interactions and Constraints, Chapter 10 has a new section on Determination of CZM Parameters from experimental data, and so on. Furthermore, Chapter 11 (Fatigue) and Chapter 12 (Programmable Features) are new to this second edition.

UMAT and UGENS subroutines used in Chapters 3, 7, 8, 9, 10, and 11 are updated to Abaqus Release 2020 and thoroughly explained in the new Chapter 12 (Programmable Features). The Appendix dealing with software setup, which is necessary for using programmable features with Abaqus 2020, is updated as well.

The reader will be able to follow a process to recreate every example using the Abaqus graphical user interface (CAE) by following step-by-step directions in the form of pseudo-code. Also, screen shots of the CAE interface are included in earlier chapters to train the reader on this process and to supplement the pseudo-code in later chapters. Furthermore, narrated video recordings of most examples are available in [1]. Finally, the most complex examples are also presented in the form of Python® scripts that can be interpreted by Abaqus CAE. A fair amount of explanation is included on both how these scripts work and how they can be harvested from CAE sessions.

The textbook deals with the analysis of structures made of composite materials, also called *composites*. The analysis of composites treated in this textbook includes the analysis of the material itself, at the micro-level, and the analysis of structures made of composite materials. This textbook evolved from the class notes of *MAE 646 Advanced Mechanics of Composite Materials* that I teach as a graduate course at West Virginia University. Although this is a textbook on advanced mechanics of composite materials, the use of the finite element method is essential for the solution of the complex boundary value problems encountered in the advanced analysis of composites, and thus the title of the book.

There are a number of good textbooks on advanced mechanics of composite materials, but none carries the theory to a practical level by actually solving problems, as it is done in this textbook. Some books devoted exclusively to finite element analysis include some examples about modeling composites but fall quite short of dealing with the actual analysis and design issues of composite materials and composite structures.

This textbook includes an explanation of the concepts involved in the detailed analysis of composites, a sound explanation of the mechanics needed to translate those concepts into a mathematical representation of the physical reality, and a detailed explanation of the solution of the resulting boundary value problems by using commercial Finite Element Analysis software such as Abaqus. Furthermore, this textbook includes 62 fully developed examples interspersed with the theory, as well as 82 exercises at the end of chapters.

The reader will be able to reproduce the examples and complete the exercises. When a finite element analysis is called for, the reader will be able to do it with commercially or otherwise available software. A website is set up with links to download the necessary software unless it is easily available from Finite Element Analysis software vendors. Use of Abaqus and MATLAB® is explained with numerous examples, and the relevant code can be downloaded from the website. Furthermore, the reader will be able to extend the capabilities of Abaqus by use of user material subroutines and Python scripting, as demonstrated in the examples included in this textbook.

Chapters 1 through 7 can be covered in a one-semester graduate course. Chapter 2 (*Introduction to Finite Element Analysis*) contains a brief introduction intended for those readers who have not had a formal course or prior knowledge about the finite element method. Chapter 4 (*Buckling*) is not referenced in the remainder of the textbook and thus it could be omitted in favor of more exhaustive coverage of content in later chapters. Chapters 7 (*Viscoelasticity*), 8 (*Continuum Damage Mechanics*), 9 (*Discrete Damage Mechanics*), and 11 (*Fatigue*) emphasize hereditary phenomena. Chapter 10 (*Delaminations*) emphasizes cohesive behavior and Chapter 12 (*Programmable Features*) discusses user-programmable features such as user materials (UMAT) and user general sections (UGENS) that are used to expand the capabilities of Abaqus to deal with issues germane to composites. Any of Chapters 7 to 12 can be skipped to fit the content into a one-semester course.

The *inductive method* is applied as much as possible in this textbook. That is, topics are introduced with examples of increasing complexity, until sufficient physical understanding is reached to introduce the general theory without difficulty. This method sometimes requires that at earlier stages of the presentation, certain facts, models, and relationships be accepted as fact, until they are completely proven later on. For example, in Chapter 7, viscoelastic models are introduced early to aid the reader in gaining an appreciation for the response of viscoelastic materials. This is done simultaneously with a cursory introduction to the superposition principle and the Laplace transform, which are formally introduced only later in the chapter. For those readers accustomed to the *deductive method*, this may seem odd, but many years of teaching have convinced me that students acquire and retain knowledge more efficiently in this way.

It is assumed that the reader is familiar with basic mechanics of composites as covered in introductory textbooks such as my previous textbook, *Introduction to Composite Material Design–Third Edition (2018)*. Furthermore, it is assumed that the reader masters a body of knowledge that is commonly acquired as part of a bachelor of science degree in Aerospace, Mechanical, Civil, or related disciplines.

References to books and to other sections in this textbook, as well as footnotes, are used to assist the reader in refreshing those concepts and to clarify the notation used. Prior knowledge of continuum mechanics, tensor analysis, and the finite element method would enhance the learning experience but are not necessary for studying with this textbook.

The finite element method is used as a tool to solve practical problems. For the most part, Abaqus is used throughout the book. Fortran, Python, MATLAB, and Scilab® are used for coding material models and post-processing algorithms. Basic knowledge of these programming languages is useful but not essential.

Only three software packages are used throughout the book. Abaqus is needed for finite element solutions of numerous examples and suggested problems. MATLAB is needed for both symbolic and numerical solutions of examples and suggested problems. Additionally, BMI3©, which is available free of charge on the book's website, is used in Chapter 4. Other software such as ANSYS Mechanical®, LSDYNA®, MSCMarc®, and SolidWorks™ are cited, but not used in the examples. Relevant code used in the examples is available on the book's website [2].

Composite materials are now ubiquitous in the marketplace, including extensive applications in aerospace, automotive, civil infrastructure, sporting goods, and so on. Their design is especially challenging because, unlike conventional materials such as metals, the composite material itself is designed concurrently with the composite structure. Preliminary design of composites is based on the assumption of a state of plane stress in the laminate. Furthermore, rough approximations are made about the geometry of the part, as well as the loading and support conditions. In this way, relatively simple analysis methods exist and computations can be carried out simply using algebra. However, preliminary analysis methods have a number of shortcomings that are remedied with advanced mechanics and finite element analysis, as explained in this textbook.

Recent advances in commercial finite element analysis packages, with user-friendly pre- and post-processing, as well as powerful user-programmable features, have made detailed analysis of composites quite accessible to the designer. This textbook bridges the gap between powerful finite element tools and practical problems in structural analysis of composites. I expect that many graduate students, practicing engineers, and instructors will find this to be a useful and practical textbook on finite element analysis of composite materials based on sound understanding of advanced mechanics of composite materials.

Ever J. Barbero, 2023

Author Bio

Ever J. Barbero, BSME/BSEE UNRC (1983), Ph.D. Va. Tech (1989), ASME and SAMPE Fellow, Professor of Mechanical and Aerospace Engineering at West Virginia University, Honorary Professor at Universidad Nacional de Trujillo (Peru), Universidad Carlos III de Madrid (Spain), and Universidad de Puerto Rico at Mayaguez (PR). He is the author of *Introduction to Composite Materials Design-Third Ed.* CRC (2018), *Finite Element Analysis of Composite Materials Using Abaqus* CRC (2013), and *Finite Element Analysis of Composite Materials Using ANSYS-Second Ed.* CRC (2014), *Multifunctional Composites* (2016), and *Workbook for Introduction to Composite Materials Design* (2018), as well as nine book chapters and 165+ journal publications. He holds US Patents #6,455,131 (2002) and #6,544,624 (2003) and has mentored over 30 MS and 20 Ph.D. graduates. He is a member of the editorial board of *Annals of Solid and Structural Mechanics*, *Journal of Applied and Computational Mechanics*, and *Mechanics of Advanced Materials and Structures*. Recognized internationally for his work on material models for advanced materials such as Aerogels, Composites, Inflatable Structures, Self-healing Composites, Superalloys, and Thermoelectric materials. He received the AE Alumni Academy Award for Outstanding Teaching and numerous research awards. H-index: 55. Citations: 12,700+. See: `http://barbero.cadec-online.com/publications_list.pdf`

Acknowledgments

I wish to thank Elio Sacco, Raimondo Luciano, and Fritz Campo for their contributions to Chapter 6 and to Tom Damiani, Joan Andreu Mayugo, and Xavier Martinez, who taught the course in 2004, 2006, and 2009, making many corrections and additions to the course notes on which this textbook is based. Also, thanks to Eduardo Sosa for helping set up the imperfection analysis in Chapter 4 and to Fabrizio Greco for reviewing Chapter 10. Thank you to Dario Barulich, Dick Rotelli, Ivan Podkolzin, and Fritz A. Campo for their suggestions of state-of-the-art topics to be added to this second edition. Special thanks to Michail Tzimas, who recorded videos for all the examples in this textbook and spotted the changes needed for the pseudo-code to reflect the updates in Abaqus CAE from release 2010 (used in the first edition) to release 2020 (used in this second edition).

Acknowledgment is due also to those who reviewed the manuscript including Enrique Barbero, Guillermo Creus, Luis Godoy, Paolo Lonetti, Severino Marques, Pizhong Qiao, Timothy Norman, Sonia Sanchez, and Eduardo Sosa. Furthermore, recognition is due to those who helped me compile the solutions manual, including Hermann Alcazar, John Sandro Rivas, and Rajiv Dastane. Also, I wish to thank John Sandro Rivas for manually checking that the pseudo-code of the examples translates into viable Abaqus CAE keystrokes and mouse clicks that accurately represent the problem at hand. Also, many thanks to my colleagues and students for their valuable suggestions and contributions to this textbook. Lastly, my gratitude goes to my wife, Ana Maria, and to my children, Margaret and Daniel, who gave up many opportunities to bond with their dad so that I might write this book.

List of Symbols and Acronyms

Symbols Related to Mechanics of Orthotropic Materials

α_1	Longitudinal thermal expansion coefficient
α_2	Transverse thermal expansion coefficient
α_f	Isotropic fiber modulus
α_m	Matrix thermal expansion coefficient
α_A	Fiber axial thermal expansion coefficient
α_T	Fiber transverse thermal expansion coefficient
$\boldsymbol{\epsilon}$	Strain tensor
ε_{ij}	Strain components in tensor notation
ϵ_α	Strain components in contracted notation
ϵ_α^e	Elastic strain
ϵ_α^p	Plastic strain
λ	Lame constant
ν	Poisson's ratio
ν_{12}	In-plane Poisson's ratio
ν_{13}, ν_{23}	Interlaminar Poisson's ratios
ν_f	Isotropic fiber Poisson's ratio
ν_m	Matrix Poisson's ratio
ν_{xy}	Apparent laminate Poisson's ratio x-y
ν_A	Fiber axial Poisson's ratio
ν_T	Fiber transverse Poisson's ratio
$\boldsymbol{\sigma}$	Stress tensor
σ_{ij}	Stress components in tensor notation
σ_α	Stress components in contracted notation
$[a]$	Transformation matrix for vectors
e_i	Unit vector components in global coordinates
e_i'	Unit vector components in materials coordinates
f_i, f_{ij}	Tsai-Wu coefficients
K	Bulk modulus
l, m, n	Direction cosines
$\widetilde{u}(\varepsilon_{ij})$	Strain energy per unit volume
u_i	Displacement vector components
x_i	Global directions or axes
x_i'	Materials directions or axes
$\mathbf{C}$	Stiffness tensor

C_{ijkl} Stiffness in index notation
$C_{\alpha,\beta}$ Stiffness in contracted notation
E Young's modulus
E_1 Longitudinal modulus
E_2 Transverse modulus
E_f Isotropic fiber modulus
E_m Matrix modulus
E_x Apparent laminate modulus in the global x-direction
E_A Fiber axial modulus
E_T Fiber transverse modulus
$G = \mu$ Shear modulus
G_{12} In-plane shear modulus
G_{13}, G_{23} Interlaminar shear moduli
G_f Isotropic fiber shear modulus
G_m Matrix shear modulus
G_{xy} Apparent laminate shear modulus x-y
G_A Fiber axial shear modulus
G_T Fiber transverse shear moduli
I_{ij} Second-order identity tensor
I_{ijkl} Fourth-order identity tensor
Q'_{ij} Lamina stiffness components in lamina coordinates
$[R]$ Reuter matrix
$\mathbf{S}$ Compliance tensor
S_{ijkl} Compliance in index notation
$S_{\alpha,\beta}$ Compliance in contracted notation
$[T]$ Coordinate transformation matrix for stress
$[\overline{T}]$ Coordinate transformation matrix for strain

Symbols Related to Finite Element Analysis

$\underline{\partial}$ Strain-displacement equations in matrix form
$\underline{\epsilon}$ Six-element array of strain components
$\theta_x, \theta_y, \theta_z$ Rotation angles following the right-hand rule (Figure 2.19)
$\underline{\sigma}$ Six-element array of stress components
ϕ_x, ϕ_y Rotation angles used in plate and shell theory
$\underline{a}$ Nodal displacement array
u_j^e Unknown parameters in the discretization
$\underline{\underline{B}}$ Strain-displacement matrix
$\underline{\underline{C}}$ Stiffness matrix
$\underline{\underline{K}}$ Assembled global stiffness matrix
$\underline{\underline{K}}^e$ Element stiffness matrix
$\underline{N}$ Interpolation function array
N_j^e Interpolation functions in the discretization
$\underline{P}^e$ Element force array
$\underline{P}$ Assembled global force array

Symbols Related to Elasticity and Strength of Laminates

γ_{xy}^0	In-plane shear strain
γ_{4u}	Ultimate interlaminar shear strain in the 2-3 plane
γ_{5u}	Ultimate interlaminar shear strain in the 1-3 plane
γ_{6u}	Ultimate in-plane shear strain
$\epsilon_x^0, \epsilon_y^0$	In-plane strains
ϵ_{1t}	Ultimate longitudinal tensile strain
ϵ_{2t}	Ultimate transverse tensile strain
ϵ_{3t}	Ultimate transverse-thickness tensile strain
ϵ_{1c}	Ultimate longitudinal compressive strain
ϵ_{2c}	Ultimate transverse compressive strain
ϵ_{3c}	Ultimate transverse-thickness compressive strain
κ_x, κ_y	Bending curvatures
κ_{xy}	Twisting curvature
ϕ_x, ϕ_y	Rotations of the middle surface of the shell (Figure 2.19)
c_4, c_5, c_6	Tsai-Wu coupling coefficients
t_k	Lamina thickness
u_0, v_0, w_0	Displacements of the middle surface of the shell
z	Distance from the middle surface of the shell
A_{ij}	Components of the extensional stiffness matrix $[A]$
B_{ij}	Components of the bending-extension coupling matrix $[B]$
D_{ij}	Components of the bending stiffness matrix $[D]$
$[E_0]$	Extensional stiffness matrix $[A]$, in ANSYS notation
$[E_1]$	Bending-extension matrix $[B]$, in ANSYS notation
$[E_2]$	Bending stiffness matrix $[D]$, in ANSYS notation
F_{1t}	Longitudinal tensile strength
F_{2t}	Transverse tensile strength
F_{3t}	Transverse-thickness tensile strength
F_{1c}	Longitudinal compressive strength
F_{2c}	Transverse compressive strength
F_{3c}	Transverse-thickness compressive strength
F_4	Interlaminar shear strength in the 2-3 plane
F_5	Interlaminar shear strength in the 1-3 plane
F_6	In-plane shear strength
H_{ij}	Components of the interlaminar shear matrix $[H]$
I_F	Failure index
M_x, M_y, M_{xy}	Moments per unit length (Figure 3.3)
$\widehat{M_n}$	Applied bending moment per unit length
N_x, N_y, N_{xy}	In-plane forces per unit length (Figure 3.3)
$\widehat{N_n}$	Applied in-plane force per unit length, normal to the edge
$\widehat{N_{ns}}$	Applied in-plane shear force per unit length, tangential
$\left(\overline{Q}_{ij}\right)_k$	Lamina stiffness components in laminate coordinates, layer k
V_x, V_y	Shear forces per unit length (Figure 3.3)

Symbols Related to Buckling

λ, λ_i	Eigenvalues
s	Perturbation parameter
Λ	Load multiplier
$\Lambda^{(cr)}$	Bifurcation multiplier or critical load multiplier
$\Lambda^{(1)}$	Slope of the post-critical path
$\Lambda^{(2)}$	Curvature of the post-critical path
v	Eigenvectors (buckling modes)
$[K]$	Stiffness matrix
$[K_s]$	Stress stiffness matrix
P_{CR}	Critical load

Symbols Related to Free Edge Stresses

$\eta_{xy,x}, \eta_{xy,y}$	Coefficients of mutual influence
$\eta_{x,xy}, \eta_{y,xy}$	Alternate coefficients of mutual influence
F_{yz}	Interlaminar shear force y-z
F_{xz}	Interlaminar shear force x-z
M_z	Interlaminar moment

Symbols Related to Micromechanics

$\overline{\epsilon}_\alpha$	Average engineering strain components
$\overline{\varepsilon}_{ij}$	Average tensor strain components
$\epsilon_\alpha^0, \varepsilon_{ij}^0$	Far-field applied strain components
$\overline{\sigma}_\alpha$	Average stress components
$\mathbf{A}^i$	Strain concentration tensor, i-th phase, contracted notation
$2a_1, 2a_2, 2a_3$	Dimensions of the RVE
A_{ijkl}	Components of the strain concentration tensor
$\mathbf{B}^i$	Stress concentration tensor, i-th phase, contracted notation
B_{ijkl}	Components of the stress concentration tensor
I	6×6 identity matrix
P_{ijkl}	Eshelby tensor
V_f	Fiber volume fraction
V_m	Matrix volume fraction

Symbols Related to Viscoelasticity

$\dot{\varepsilon}$	Stress rate
η	Viscosity
θ	Age or aging time

$\dot{\sigma}$	Stress rate
τ	Time constant of the material or system
Γ	Gamma function
s	Laplace variable
t	Time
$C_{\alpha,\beta}(t)$	Stiffness tensor in the time domain
$C_{\alpha,\beta}(s)$	Stiffness tensor in the Laplace domain
$\widehat{C}_{\alpha,\beta}(s)$	Stiffness tensor in the Carson domain
$D(t)$	Compliance
$D_0, (D_i)_0$	Initial compliance values
$D_c(t)$	Creep component of the total compliance $D(t)$
D', D''	Storage and loss compliances
$E_0, (E_i)_0$	Initial moduli
E_∞	Equilibrium modulus
E, E_0, E_1, E_2	Parameters in the viscoelastic models (Figure 7.1)
$E(t)$	Relaxation
E', E''	Storage and loss moduli
$F[]$	Fourier transform
$(G_{ij})_0$	Initial shear moduli
$H(t - t_0)$	Heaviside step function
$H(\theta)$	Relaxation spectrum
$L[]$	Laplace transform
$L[]^{-1}$	Inverse Laplace transform

Symbols Related to Damage

α	Laminate CTE
$\alpha^{(k)}$	CTE of lamina k
α_{cr}	Critical misalignment angle at longitudinal compression failure
α_σ	Standard deviation of fiber misalignment
$\gamma(\delta)$	Damage hardening function
γ_0	Damage threshold
δ_{ij}	Kronecker delta
δ	Damage hardening variable
ε	Effective strain
$\bar{\varepsilon}$	Undamaged strain
ε^p	Plastic strain
$\dot{\gamma}$	Heat dissipation rate per unit volume
$\dot{\gamma}_s$	Internal entropy production rate
λ	Crack density
λ_{lim}	Saturation crack density
$\dot{\lambda}, \dot{\lambda}^d$	Damage multiplier
$\dot{\lambda}^p$	Yield multiplier
ρ	Density
σ	Effective stress

$\overline{\sigma}$	Undamaged stress
τ_{13}, τ_{23}	Intralaminar shear stress components
φ, φ^*	Strain energy density, and complementary SED
χ	Gibbs energy density
ψ	Helmholtz free energy density
ΔT	Change in temperature
$\boldsymbol{\Omega} = \Omega_{ij}$	Integrity tensor
$2a_0$	Representative crack size
d_i	Eigenvalues of the damage tensor
f^d	Damage flow surface
f^p	Yield flow surface
$f(x), F(x)$	Probability density, and its cumulative probability
g	Damage activation function
g^d	Damage surface
g^p	Yield surface
h	Laminate thickness
h_k	Thickness of lamina k
m	Weibull modulus
p	Yield hardening variable
$\widehat{p}$	Thickness average of quantity p
$\widetilde{p}$	Virgin value of quantity p
$\overline{p}$	Volume average of quantity p
$\mathbf{q}$	Shear flow vector per unit area
r	Radiation heat per unit mass
s	Specific entropy
$u(\varepsilon_{ij})$	Internal energy density
A	Crack area
$[A]$	Laminate in-plane stiffness matrix
A_{ijkl}	Tension-compression damage constitutive tensor
B_{ijkl}	Shear damage constitutive tensor
B_a	Dimensionless number
$\overline{C}_{\alpha,\beta}$	Stiffness matrix in the undamaged configuration
$\mathbf{C}^{ed}$	Tangent stiffness tensor
D_{ij}	Damage tensor
D_{1t}^{cr}	Critical damage at longitudinal tensile failure
D_{1c}^{cr}	Critical damage at longitudinal compression failure
D_{2t}^{cr}	Critical damage at transverse tensile failure
D_2, D_6	Damage variables
$E(D)$	Effective modulus
$\overline{E}$	Undamaged (virgin) modulus
G_c	Critical energy release rate
G_m	Mixed-mode energy release rate
G_m^c	Mixed mode critical energy release rate
$G_I^c = G_n^c$	Critical energy release rate modes I (opening)
G_{II}^c, G_s^c	Critical energy release rate mode II (shear)

G_{III}^c, G_t^c	Critical energy release rate mode III (tearing shear)
J_{ijkl}	Normal damage constitutive tensor
M_{ijkl}	Damage effect tensor
N	Number of laminas in the laminate
$\{N\}$	Membrane stress resultant array
Q	Degraded 3x3 stiffness matrix of the laminate
$R(p)$	Yield hardening function
R_0	Yield threshold
S	Entropy or laminate compliance matrix, depending on context
T	Temperature
U	Strain energy
V	Volume of the RVE
Y_{ij}	Thermodynamic force tensor

Symbols Related to Delaminations

α	Power law crack propagation criterion
δ	CZM separation of the interface
δ^0	CZM separation at damage onset
δ^c	CZM final separation at fracture
δ_m	CZM mixed-mode (effective) separation
η	BK exponent
γ_c	Surface energy
a_0	Initial crack length
m	2D mixed-mode ratio in the BK criterion
m_n, m_s, m_t	CZM mixed-mode ratios
t^0	CZM in-situ strength at damage onset
t_n, t_s, t_t	CZM traction (stress) for modes n, s, t
D_n, D_s, D_t	CZM damage variables for modes n, s, t
D_m	CZM mixed-mode (effective) damage
G	Energy release rate (ERR) or fracture energy
G_n, G_s, G_t	CZM fracture energy of modes n, s, t
G_c	Critical energy release rate (ERR) or critical fracture energy
G_n^c, G_s^c	Critical fracture energy modes n, s
K	Penalty stiffness
$\tilde{K}$	Virgin penalty stiffness
U	Internal energy
W	Work done by the body on its surroundings
$W_{closure}$	Crack closure work

Symbols Related to Fatigue

β_I, β_{II}	Defect nucleation rates, modes I, II
λ	Crack density
$f(N)$	Defect nucleation function

G	Energy release rate (ERR), also fracture energy
G_C	Critical energy release rate (ERR), also fracture toughness
G_I	Energy release rate mode I
$G_{IC} = G_I^c$	Quasi-static fracture toughness mode I
G'_{IC}	Fatigue fracture toughness mode I
$G_{IIC} = G_{II}^c$	Quasi-static fracture toughness mode II
Q	Laminate stiffness
R	Thermal ratio
T_g	Glass transition temperature

Acronyms

BC	Boundary conditions
c.s.	Coordinate system
CAE	Complete analysis environment
CDF	Cumulative distribution function
CDM	Continuum damage mechanics
CE	Constraint equations
CLT	Classical lamination theory
COD	Crack opening displacement
CS	Continuum shell, also crack saturation
CTE	Coefficient of thermal expansion
CZM	Cohesive zone model
DCB	Double cantilever beam
DDM	Discrete damage mechanics
DI	Damage initiation
DOF	Degrees of freedom
DOS	Precursor to Windows OS
ENF	End notch flexure
FC	Failure criteria/on
FEA	Finite element analysis
FEM	Finite element model
FF	Fiber failure
FSDT	First order shear deformation theory
GUI	Graphical user interface
HFC	Hashin failure criterion
HFE	Helmholtz free energy
IVOL	Element volume
LEFM	Linear elastic fracture mechanics
LHS	Left hand side
LSS	Laminate stacking sequence
MDB	Model database (.cae)
MF	Matrix failure
MMD	Micromechanics of damage
MMF	Mixed-mode fracture

MPC	Multi-point constraints
MPL	Master Paris Law
MSTRN	Maximum strain failure criterion
MSTRS	Maximum stress failure criterion
ODB	Object database (.obj)
ODE	Ordinary differential equations
PBC	Periodic boundary conditions
PDA	Progressive damage analysis
PMC	Polymer matrix composite
PMM	Periodic microstructure micromechanics
PVW	Principle of virtual work
RBM	Rigid body motion
RHS	Right hand side
RMA	Return mapping algorithm
RP	Reference point
RTA	Room temperature ambient
SDV	State variables
SFT	Stress free temperature
SMD	Synergistic damage mechanics
TME	Thermomechanical equivalence
TMS	Truncated maximum strain failure criterion
TTSP	Time-temperature superposition principle
UD	Unidirectional
UGENS	User general section
UMAT	User material subroutine
VCCT	Virtual crack closure technique
WS	Workspace
XFEM	Extended finite element method

List of Examples

Errata

For the most current errata, go to
https://barbero.cadec-online.com/feacm-abaqus/ErrataAbaqus.pdf

Chapter 1

Mechanics of Orthotropic Materials

This chapter provides the foundation for the rest of the book. Basic concepts of mechanics, tailored for composite materials, are presented, including coordinate transformations, constitutive equations, and so on. Continuum mechanics is used to describe deformation and stress in an orthotropic material. The basic equations are reviewed in Sections 1.2 to 1.9. Tensor operations are reviewed in Section 1.10 because they are used in the rest of the chapter. Coordinate transformations are required to express quantities such as stress, strain, and stiffness in lamina coordinates, laminate coordinates, and so on. They are reviewed in Sections 1.10 to 1.11. This chapter is heavily referenced in the rest of the book, and thus readers who are already versed in continuum mechanics may choose to come back to review this material as needed.

1.1 Lamina Coordinate System

A single lamina of fiber-reinforced reinforced composite behaves as an orthotropic material. That is, the material has three mutually perpendicular planes of symmetry. The intersection of these three planes defines three axes that coincide with the fiber direction x_1', the thickness coordinate x_3', and a third direction $x_2' = x_3' \times x_1'$ perpendicular to the other two[1] [3, Figure 5.1].

1.2 Displacements

Under the action of forces, every point in a body may translate and rotate as a rigid body as well as deform to occupy a new region. The displacements u_i of any point P in the body (Figure 1.1) are defined in terms of the three components of the vector u_i (in a rectangular Cartesian coordinate system) as $u_i = (u_1, u_2, u_3)$. An

---

[1] $\times$ denotes vector cross product.

1

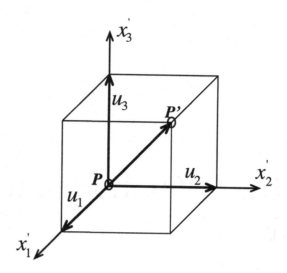

Fig. 1.1: Notation for displacement components.

alternate notation for displacements is $u_i = (u, v, w)$. Displacement is a vector or first-order tensor quantity

$$\mathbf{u} = u_i = (u_1, u_2, u_3) \; ; \; i = 1...3 \tag{1.1}$$

where boldface (e.g., $\mathbf{u}$) indicates a *tensor* written in tensor notation, in this case a vector (or first-order tensor). In this book, all tensors are boldfaced (e.g., $\boldsymbol{\sigma}$), but their components are not (e.g., σ_{ij}). The order of the tensor (i.e., first, second, fourth, etc.) must be inferred from context, or as in (1.1), by looking at the number of subscripts of the same entity written in index notation (e.g., u_i).

1.3   Strain

For geometric nonlinear analysis, the components of the Lagrangian strain tensor are [4]

$$L_{ij} = \frac{1}{2}(u_{i,j} + u_{j,i} + u_{r,i}u_{r,j}) \tag{1.2}$$

where

$$u_{i,j} = \frac{\partial u_i}{\partial x_j} \tag{1.3}$$

If the gradients of the displacements are so small that products of partial derivatives of u_i are negligible compared with linear (first-order) derivative terms, then the (infinitesimal) strain tensor ε_{ij} is given by [4]

$$\boldsymbol{\varepsilon} = \varepsilon_{ij} = \frac{1}{2}(u_{i,j} + u_{j,i}) \tag{1.4}$$

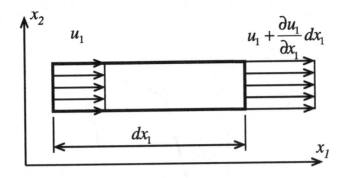

Fig. 1.2: Normal strain.

Again, boldface indicates a tensor, the order of which is implied from the context. For example, ε is a 1D strain, and $\boldsymbol{\varepsilon}$ is the second-order tensor of strain. Index notation (e.g., $= \varepsilon_{ij}$) is used most of the time, and the tensor character of variables (scalar, vector, second-order, and so on) is easily understood from context.

From the definition (1.4), strain is a second-order, symmetric tensor (i.e., $\varepsilon_{ij} = \varepsilon_{ji}$). In expanded form, the strains are defined by

$$\varepsilon_{11} = \frac{\partial u_1}{\partial x_1} = \epsilon_1 \; ; \quad 2\varepsilon_{12} = 2\varepsilon_{21} = \left(\frac{\partial u_1}{\partial x_2} + \frac{\partial u_2}{\partial x_1} \right) = \gamma_6 = \epsilon_6$$

$$\varepsilon_{22} = \frac{\partial u_2}{\partial x_2} = \epsilon_2 \; ; \quad 2\varepsilon_{13} = 2\varepsilon_{31} = \left(\frac{\partial u_1}{\partial x_3} + \frac{\partial u_3}{\partial x_1} \right) = \gamma_5 = \epsilon_5$$

$$\varepsilon_{33} = \frac{\partial u_3}{\partial x_3} = \epsilon_3 \; ; \quad 2\varepsilon_{23} = 2\varepsilon_{32} = \left(\frac{\partial u_2}{\partial x_3} + \frac{\partial u_3}{\partial x_2} \right) = \gamma_4 = \epsilon_4 \qquad (1.5)$$

where ϵ_α with $\alpha = 1..6$ are defined in Section 1.5. The normal components of strain $(i = j)$ represent the change in length per unit length (Figure 1.2). The shear components of strain $(i \neq j)$ represent one-half the change in an original right angle (Figure 1.3). The engineering shear strain $\gamma_\alpha = 2\varepsilon_{ij}$, for $i \neq j$, is often used instead of the tensor shear strain because the shear modulus G is defined by $\tau = G\gamma$ in mechanics of materials [5]. The strain tensor, being of second order, can be displayed as a matrix

$$[\varepsilon] = \begin{bmatrix} \varepsilon_{11} & \varepsilon_{12} & \varepsilon_{13} \\ \varepsilon_{12} & \varepsilon_{22} & \varepsilon_{23} \\ \varepsilon_{13} & \varepsilon_{23} & \varepsilon_{33} \end{bmatrix} = \begin{bmatrix} \epsilon_1 & \epsilon_6/2 & \epsilon_5/2 \\ \epsilon_6/2 & \epsilon_2 & \epsilon_4/2 \\ \epsilon_5/2 & \epsilon_4/2 & \epsilon_3 \end{bmatrix} \qquad (1.6)$$

where [ ] is used to denote matrices.

1.4 Stress

The stress vector associated to a plane passing through a point is the force per unit area acting on the plane passing through the point. A second-order tensor, called stress tensor, completely describes the state of stress at a point. The stress tensor

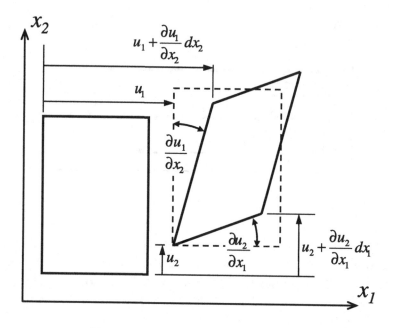

Fig. 1.3: Engineering shear strain.

can be expressed in terms of the components acting on three mutually perpendicular planes aligned with the orthogonal coordinate directions as indicated in Figure 1.4. The tensor notation for stress is σ_{ij} with $(i, j = 1, 2, 3)$, where the first subscript corresponds to the direction of the normal to the plane of interest, and the second subscript corresponds to the direction of the stress. Tensile normal stresses $(i = j)$ are defined to be positive when the normal to the plane and the stress component directions are either both positive or both negative. All components of stress depicted in Figure 1.4 have a positive sense. Force and moment equilibrium of the element in Figure 1.4 requires that the stress tensor be symmetric (i.e., $\sigma_{ij} = \sigma_{ji}$) [5]. The stress tensor, being of second order, can be displayed as a matrix

$$[\sigma] = \begin{bmatrix} \sigma_{11} & \sigma_{12} & \sigma_{13} \\ \sigma_{12} & \sigma_{22} & \sigma_{23} \\ \sigma_{13} & \sigma_{23} & \sigma_{33} \end{bmatrix} = \begin{bmatrix} \sigma_1 & \sigma_6 & \sigma_5 \\ \sigma_6 & \sigma_2 & \sigma_4 \\ \sigma_5 & \sigma_4 & \sigma_3 \end{bmatrix} \tag{1.7}$$

1.5   Contracted Notation

Since the stress is symmetric, it can be written in *Voigt* contracted notation as

$$\sigma_\alpha = \sigma_{ij} = \sigma_{ji} \tag{1.8}$$

with the contraction rule defined as follows

$$\begin{aligned} \alpha = i \qquad & if \quad i = j \\ \alpha = 9 - i - j \quad & if \quad i \neq j \end{aligned} \tag{1.9}$$

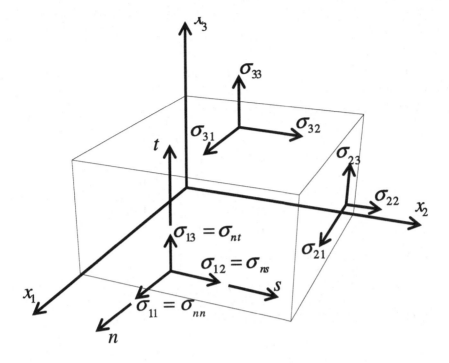

Fig. 1.4: Stress components.

Table 1.1: Contracted notation convention used by various FEA software packages.

Standard Convention	Abaqus/Standard	LS-DYNA and Abaqus/Explicit	ANSYS/Mechanical
$11 \longrightarrow 1$	$11 \longrightarrow 1$	$11 \longrightarrow 1$	$11 \longrightarrow 1$
$22 \longrightarrow 2$	$22 \longrightarrow 2$	$22 \longrightarrow 2$	$22 \longrightarrow 2$
$33 \longrightarrow 3$	$33 \longrightarrow 3$	$33 \longrightarrow 3$	$33 \longrightarrow 3$
$23 \longrightarrow 4$	$12 \longrightarrow 4$	$12 \longrightarrow 4$	$12 \longrightarrow 4$
$13 \longrightarrow 5$	$13 \longrightarrow 5 \longrightarrow V_x$	$23 \longrightarrow 5$	$23 \longrightarrow 5$
$12 \longrightarrow 6$	$23 \longrightarrow 6 \longrightarrow V_y$	$13 \longrightarrow 6$	$13 \longrightarrow 6$

resulting in the contracted version of stress components shown in (1.7). The same applies to the strain tensor, resulting in the contracted version of strain shown in (1.6). Note that the six components of stress σ_α with $\alpha = 1 \ldots 6$ can be arranged into a column array, denoted by curly brackets {} as in (1.10), but $\{\sigma\}$ is not a vector, but just a convenient way to arrange the six unique components of a symmetric second-order tensor.

1.5.1 Alternate Contracted Notation

Some FEA software packages use different contracted notations, as shown in Table 1.1. For example, to transform stresses or strains from standard notation to Abaqus notation, a transformation matrix can be used, as follows

$$\{\sigma_A\} = [T]\{\sigma\} \tag{1.10}$$

where the subscript $()_A$ denotes a quantity in Abaqus notation. Also note that $\{\ \}$ denotes a column array, in this case of six elements, and $[\ ]$ denotes a matrix, in this case the 6×6 rotation matrix given by

$$[T] = \begin{bmatrix} 1 & 0 & 0 & 0 & 0 & 0 \\ 0 & 1 & 0 & 0 & 0 & 0 \\ 0 & 0 & 1 & 0 & 0 & 0 \\ 0 & 0 & 0 & 0 & 0 & 1 \\ 0 & 0 & 0 & 0 & 1 & 0 \\ 0 & 0 & 0 & 1 & 0 & 0 \end{bmatrix} \tag{1.11}$$

The stiffness matrix transforms as follows

$$[C_A] = [T]^T [C][T] \tag{1.12}$$

For LS-DYNA and ANSYS, the transformation matrix is

$$[T] = \begin{bmatrix} 1 & 0 & 0 & 0 & 0 & 0 \\ 0 & 1 & 0 & 0 & 0 & 0 \\ 0 & 0 & 1 & 0 & 0 & 0 \\ 0 & 0 & 0 & 0 & 0 & 1 \\ 0 & 0 & 0 & 1 & 0 & 0 \\ 0 & 0 & 0 & 0 & 1 & 0 \end{bmatrix} \tag{1.13}$$

Note that in Table 1.1, V_x and V_y are included to help establish the relationship between H_{ij} in (3.9) and Abaqus variables K_{11}, K_{22}, K_{12}. See explanation after (3.9) and usage in Example 3.1.

1.6   Equilibrium and Virtual Work

The three equations of equilibrium at every point in a body are written in tensor notation as

$$\sigma_{ij,j} + f_i = 0 \tag{1.14}$$

where f_i is the body force per unit volume and $(\ )_{,j} = \dfrac{\partial}{\partial x_j}$. When body forces are negligible, the expanded form of the equilibrium equations, written in the laminate coordinate system x-y-z, is

$$\frac{\partial \sigma_{xx}}{\partial x} + \frac{\partial \sigma_{xy}}{\partial y} + \frac{\partial \sigma_{xz}}{\partial z} = 0$$

$$\frac{\partial \sigma_{xy}}{\partial x} + \frac{\partial \sigma_{yy}}{\partial y} + \frac{\partial \sigma_{yz}}{\partial z} = 0$$

$$\frac{\partial \sigma_{xz}}{\partial x} + \frac{\partial \sigma_{yz}}{\partial y} + \frac{\partial \sigma_{zz}}{\partial z} = 0 \tag{1.15}$$

The principle of virtual work (PVW) provides an alternative to the equations of equilibrium [6]. Since the PVW is an integral expression, it is more convenient than (1.14) for finite element formulation. The PVW reads

$$\int_V \sigma_{ij} \delta\epsilon_{ij} dV - \int_S t_i \delta u_i dS - \int_V f_i \delta u_i dV = 0 \tag{1.16}$$

where t_i are the surface tractions per unit area acting on the surface S. The negative sign means that work is done by external forces (t_i, f_i) on the body. The forces and the displacements follow the same sign convention; that is, a component is positive when it points in the positive direction of the respective axis. The first term in (1.16) is the virtual work performed by the internal stresses, and it is positive, following the same sign convention.

Example 1.1 *Find the displacement function $u(x)$ for a slender rod of cross-sectional area A, length L, modulus E, and density ρ, hanging from the top end and subjected to its own weight. Use a coordinate x pointing downward with origin at the top end.*

Solution to Example 1.1 *We assume a quadratic displacement function*

$$u(x) = C_0 + C_1 x + C_2 x^2$$

Using the boundary condition at the top yields $C_0 = 0$. The PVW (1.16) simplifies because the only non-zero strain is ϵ_x and there is no surface tractions. Using Hooke's law

$$\int_0^L E\epsilon_x \delta\epsilon_x A dx - \int_0^L \rho g \delta u A dx = 0$$

From the assumed displacement

$$\delta u = x\delta C_1 + x^2 \delta C_2$$

$$\epsilon_x = \frac{du}{dx} = C_1 + 2xC_2$$

$$\delta\epsilon_x = \delta C_1 + 2x\delta C_2$$

Substituting

$$EA \int_o^L (C_1 + 2xC_2)(\delta C_1 + 2x\delta C_2)dx - \rho g A \int_0^L (x\delta C_1 + x^2 \delta C_2)dx = 0$$

Integrating and collecting terms in δC_1 and δC_2 separately

$$(EC_2L^2 + EC_1L - \frac{\rho g L^2}{2})\delta C_1 + (\frac{4}{3}EC_2L^3 + EC_1L^2 - \frac{\rho g L^3}{3})\delta C_2 = 0$$

Since δC_1 and δC_2 have arbitrary (virtual) values, two equations in two unknowns are obtained, one inside each parenthesis. Solving them we get

$$C_1 = \frac{L\rho g}{E} \ ; \ C_2 = -\frac{\rho g}{2E}$$

Substituting back into $u(x)$

$$u(x) = \frac{\rho g}{2E}(2L - x)x$$

which coincides with the exact solution from mechanics of materials.

1.7 Boundary Conditions

1.7.1 Traction Boundary Conditions

The solution of problems in solid mechanics requires that boundary conditions be specified. The boundary conditions may be specified in terms of components of displacement, stress, or a combination of both. For any point on an arbitrary surface, the traction T_i is defined as the vector consisting of the three components of stress acting on the surface at the point of interest. As indicated in Figure 1.4 the traction vector consists of one component of normal stress, σ_{nn}, and two components of shear stress, σ_{nt} and σ_{ns}. The traction vector can be written using Cauchy's law

$$T_i = \sigma_{ji} n_j = \sum_{j}^{3} \sigma_{ji} n_j \qquad (1.17)$$

where n_j is the unit normal to the surface at the point under consideration.[2] For a plane perpendicular to the x_1 axis, $n_i = (1, 0, 0)$, and the components of the traction are $T_1 = \sigma_{11}$, $T_2 = \sigma_{12}$, and $T_3 = \sigma_{13}$.

1.7.2 Free Surface Boundary Conditions

The condition that a surface be free of stress is equivalent to all components of traction being zero, i.e., $T_n = \sigma_{nn} = 0$, $T_t = \sigma_{nt} = 0$, and $T_s = \sigma_{ns} = 0$. It is possible that only selected components of the traction be zero while others are non-zero. For example, pure pressure loading corresponds to non-zero normal stress and zero shear stresses.

---

[2]Einstein's summation convention can be introduced with (1.17) as an example. Any pair of repeated indices implies a summation over all the values of the index in question. Furthermore, each pair of repeated indices represents a *contraction*. That is, the order of resulting tensor, in this case order one for T_i, is two less than the sum of the orders of the tensors involved in the operation. The resulting tensor keeps only the *free* indices that are not involved in the contraction–in this case only i remains.

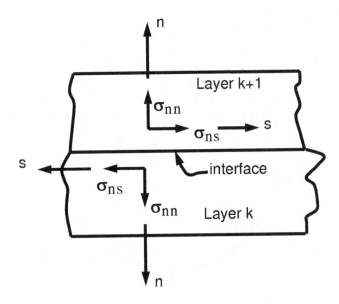

Fig. 1.5: Traction continuity across an interface.

1.8   Continuity Conditions

1.8.1   Traction Continuity

Equilibrium (action and reaction) requires that the traction components T_i must be continuous across any surface. Mathematically this is stated as $T_i^+ - T_i^- = 0$. Using (1.17), $T_i^+ = \sigma_{ji}^+ n_j$. Since $n_j^+ = -n_j^-$, we have $\sigma_{ji}^+ = \sigma_{ji}^-$. In terms of individual stress components, $\sigma_{nn}^+ = \sigma_{nn}^-$, $\sigma_{nt}^+ = \sigma_{nt}^-$, and $\sigma_{ns}^+ = \sigma_{ns}^-$ (Figure 1.5). Thus, the normal and shear components of stress acting on a surface must be continuous across that surface. There are no continuity requirements on the other three components of stress. That is, it is possible that $\sigma_{tt}^+ \neq \sigma_{tt}^-$, $\sigma_{ss}^+ \neq \sigma_{ss}^-$, and $\sigma_{ts}^+ \neq \sigma_{ts}^-$. Lack of continuity of the two normal and one shear components of stress is very common because the material properties are discontinuous across the interface between laminas.

1.8.2   Displacement Continuity

Certain conditions on displacements must be satisfied along any surface in a perfectly bonded continuum. Consider for example, buckling of a cylinder under external pressure (Figure 1.6). The displacements associated with the material from either side of the line A-A must be identical, $u_i^+ = u_i^-$. The continuity conditions must be satisfied at every point in a perfectly bonded continuum. However, continuity is not required in the presence of debonding or sliding between regions or phases of a material. For the example shown, continuity of slope $(\frac{\partial w^+}{\partial \theta} = \frac{\partial w^-}{\partial \theta})$, must be satisfied, where w is the radial displacement.

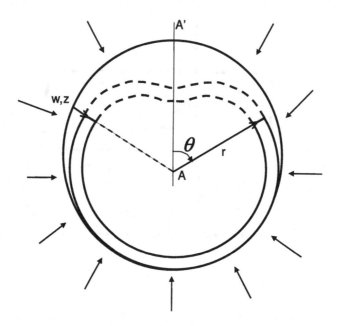

Fig. 1.6: Buckling of an encased cylindrical pipe under external pressure.

1.9 Compatibility

The strain displacement equations (1.5) provide six equations for only three un-known displacements u_i. Thus, integration of equations (1.5) to determine the unknown displacements will not have a single-valued solution unless the strains ε_{ij} satisfy certain conditions. Arbitrary specification of the ε_{ij} could result in discon-tinuities in the material, including gaps and/or overlapping regions.

The necessary conditions for single-valued displacements are the *compatibility conditions*. Although these six equations are available [4], they are not used here because the displacement method, which is used throughout this book, does not require them. That is, in solving problems, the form of displacements u_i is always assumed a priori. Then, the strains are computed with (1.5), and the stress with (1.46). Finally, equilibrium is enforced by using the PVW (1.16).

1.10 Coordinate Transformations

The coordinates of point P in the prime coordinate system can be found from its coordinates in the unprimed system. From Figure 1.7, the coordinates of point P are

$$
\begin{aligned}
x_1' &= x_1 \cos\theta + x_2 \sin\theta \\
x_2' &= -x_1 \sin\theta + x_2 \cos\theta \\
x_3' &= x_3
\end{aligned}
$$

(1.18)

or

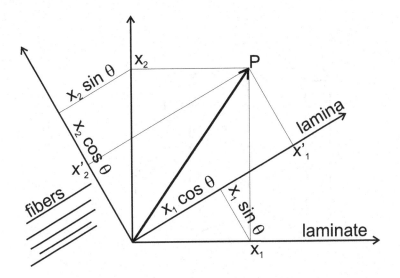

Fig. 1.7: Coordinate transformation.

$$x_i' = a_{ij}x_j \qquad (1.19)$$

or in matrix notation

$$\{x'\} = [a]\{x\} \qquad (1.20)$$

where a_{ij} are the components of the unit vectors of the primed system e_i' on the unprimed system e_j, by rows [4]

$$a_{ij} = \cos(e_i', e_j) = \begin{array}{|c|c|c|c|} \hline & e_1 & e_2 & e_3 \\ \hline e_1' & a_{11} & a_{12} & a_{13} \\ \hline e_2' & a_{21} & a_{22} & a_{23} \\ \hline e_3' & a_{31} & a_{32} & a_{33} \\ \hline \end{array} \qquad (1.21)$$

If primed coordinates denote the lamina coordinates and unprimed denote the laminate coordinates, then (1.19) transforms vectors from laminate to lamina coordinates. The inverse transformation simply uses the transpose matrix

$$\{x\} = [a]^T\{x'\} \qquad (1.22)$$

Example 1.2 *A composite lamina has fiber orientation $\theta = 30°$. Construct the [a] matrix by calculating the direction cosines of the lamina system, i.e., the components of the unit vectors of the lamina system x_i' on the laminate system x_j.*

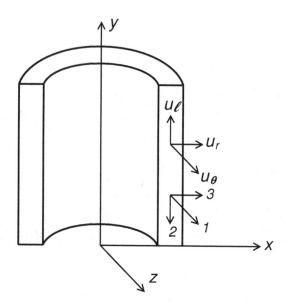

Fig. 1.8: Coordinate transformation for axial-symmetric analysis.

Solution to Example 1.2 *From Figure 1.7 and (1.19) we have*

$$a_{11} = \cos\theta = \frac{\sqrt{3}}{2}$$

$$a_{12} = \sin\theta = \frac{1}{2}$$

$$a_{13} = 0$$

$$a_{21} = -\sin\theta = -\frac{1}{2}$$

$$a_{22} = \cos\theta = \frac{\sqrt{3}}{2}$$

$$a_{23} = 0$$

$$a_{31} = 0$$

$$a_{32} = 0$$

$$a_{33} = 1$$

Example 1.3 *A fiber-reinforced composite tube is wound in the hoop direction (1-direction). Formulas for the stiffness values E_1, E_2, etc., are given in that system. However, when analyzing the cross-section of this material with generalized plane strain elements (CAX4 in Abaqus), the model is typically constructed in the structural X, Y, Z system. It is therefore necessary to provide the stiffness values in the structural system as E_x, E_y, etc. Construct the transformation matrix $[a]^T$ to go from lamina coordinates $1, 2, 3$, to structural coordinates in Figure 1.8.*

Solution to Example 1.3 *First, construct $[a]$ using the definition (1.21). Taking each unit vector $1, 2, 3$, at a time, we construct the matrix $[a]$ by rows. The i-th row contains the components of $i = 1, 2, 3$ along X, Y, Z.*

$[a]$	X	Y	Z
1	0	0	1
2	0	-1	0
3	1	0	0

The required transformation is just the transpose of the matrix above.

1.10.1   Stress Transformation

A second-order tensor σ_{pq} can be thought as the (uncontracted) outer product[3] of two vectors, V_p and V_q

$$\sigma_{pq} = V_p \otimes V_q \qquad (1.23)$$

each of which transforms as (1.19)

$$\sigma'_{ij} = a_{ip}V_p \otimes a_{jq}V_q \qquad (1.24)$$

Therefore,

$$\sigma'_{ij} = a_{ip}a_{jq}\sigma_{pq} \qquad (1.25)$$

or, in matrix notation

$$\{\sigma'\} = [a]\{\sigma\}[a]^T \qquad (1.26)$$

For example, expand σ'_{11} in contracted notation

$$\sigma'_1 = a_{11}^2\sigma_1 + a_{12}^2\sigma_2 + a_{13}^2\sigma_3 + 2a_{11}a_{12}\sigma_6 + 2a_{11}a_{13}\sigma_5 + 2a_{12}a_{13}\sigma_4 \qquad (1.27)$$

Expanding σ'_{12} in contracted notation yields

$$\sigma'_6 = a_{11}a_{21}\sigma_1 + a_{12}a_{22}\sigma_2 + a_{13}a_{23}\sigma_3 + (a_{11}a_{22} + a_{12}a_{21})\sigma_6 \qquad (1.28)$$
$$+ (a_{11}a_{23} + a_{13}a_{21})\sigma_5 + (a_{12}a_{23} + a_{13}a_{22})\sigma_4$$

The following algorithm is used to obtain a 6 × 6 coordinate transformation matrix [T] such that (1.25) is rewritten in contracted notation as

$$\sigma'_\alpha = T_{\alpha\beta}\sigma_\beta \qquad (1.29)$$

If $\alpha \leq 3$ and $\beta \leq 3$ then $i = j$ and $p = q$, so

$$T_{\alpha\beta} = a_{ip}a_{ip} = a_{ip}^2 \quad \text{no sum on } i, p \qquad (1.30)$$

If $\alpha \leq 3$ and $\beta > 3$ then $i = j$ but $p \neq q$, and taking into account that switching p by q yields the same value of $\beta = 9 - p - q$ as per (1.9), we have

$$T_{\alpha\beta} = a_{ip}a_{iq} + a_{iq}a_{ip} = 2a_{ip}a_{iq} \quad \text{no sum on } i, p \qquad (1.31)$$

---

[3]The outer product preserves all indices of the entities involved, thus creating a tensor of order equal to the sum of the order of the entities involved.

If $\alpha > 3$, then $i \neq j$, but we want only one stress, say σ_{ij}, not σ_{ji} because they are numerically equal. In fact $\sigma_\alpha = \sigma_{ij} = \sigma_{ji}$ with $\alpha = 9 - i - j$. If in addition $\beta \leq 3$ then $p = q$ and we get

$$T_{\alpha\beta} = a_{ip}a_{jp} \quad \text{no sum on } i, p \tag{1.32}$$

When $\alpha > 3$ and $\beta > 3$, $i \neq j$ and $p \neq q$ so we get

$$T_{\alpha\beta} = a_{ip}a_{jq} + a_{iq}a_{jp} \tag{1.33}$$

which completes the derivation of $T_{\alpha\beta}$. Expanding (1.30–1.33) and using (1.21) we get

$$[T] = \begin{bmatrix} a_{11}^2 & a_{12}^2 & a_{13}^2 & 2\,a_{12}\,a_{13} & 2\,a_{11}\,a_{13} & 2\,a_{11}\,a_{12} \\ a_{21}^2 & a_{22}^2 & a_{23}^2 & 2\,a_{22}\,a_{23} & 2\,a_{21}\,a_{23} & 2\,a_{21}\,a_{22} \\ a_{31}^2 & a_{32}^2 & a_{33}^2 & 2\,a_{32}\,a_{33} & 2\,a_{31}\,a_{33} & 2\,a_{31}\,a_{32} \\ a_{21}\,a_{31} & a_{22}\,a_{32} & a_{23}\,a_{33} & a_{22}\,a_{33} + a_{23}\,a_{32} & a_{21}\,a_{33} + a_{23}\,a_{31} & a_{21}\,a_{32} + a_{22}\,a_{31} \\ a_{11}\,a_{31} & a_{12}\,a_{32} & a_{13}\,a_{33} & a_{12}\,a_{33} + a_{13}\,a_{32} & a_{11}\,a_{33} + a_{13}\,a_{31} & a_{11}\,a_{32} + a_{12}\,a_{31} \\ a_{11}\,a_{21} & a_{12}\,a_{22} & a_{13}\,a_{23} & a_{12}\,a_{23} + a_{13}\,a_{22} & a_{11}\,a_{23} + a_{13}\,a_{21} & a_{11}\,a_{22} + a_{12}\,a_{21} \end{bmatrix} \tag{1.34}$$

A MATLAB program that can be used to generate (1.34) is shown next (also available in [2]).

```
% Derivation of the transformation matrix [T]
clear all;
syms T alpha R
syms a a11 a12 a13 a21 a22 a23 a31 a32 a33
a = [a11,a12,a13;
     a21,a22,a23;
     a31,a32,a33];
T(1:6,1:6) = 0;
for i=1:1:3
for j=1:1:3
 if i==j; alpha = j; else alpha = 9-i-j; end
 for p=1:1:3
 for q=1:1:3
  if p==q beta = p; else beta = 9-p-q; end
  T(alpha,beta) = 0;
  if alpha<=3 & beta<= 3; T(alpha,beta)=a(i,p)*a(i,p); end
  if alpha> 3 & beta<= 3; T(alpha,beta)=a(i,p)*a(j,p); end
  if alpha<=3 & beta>3; T(alpha,beta)=a(i,q)*a(i,p)+a(i,p)*a(i,q);end
  if alpha>3 & beta>3; T(alpha,beta)=a(i,p)*a(j,q)+a(i,q)*a(j,p);end
 end
 end
end
end
T
R = eye(6,6); R(4,4)=2; R(5,5)=2; R(6,6)=2; % Reuter matrix
Tbar = R*T*R^(-1)
```

1.10.2 Strain Transformation

The tensor components of strain ε_{ij} transform in the same way as the stress components

$$\varepsilon'_{ij} = a_{ip}a_{jq}\varepsilon_{pq} \tag{1.35}$$

or

$$\varepsilon'_\alpha = T_{\alpha\beta}\varepsilon_\beta \tag{1.36}$$

with $T_{\alpha\beta}$ given by (1.34). However, the three engineering shear strains $\gamma_{xz}, \gamma_{yz}, \gamma_{xy}$ are normally used instead of tensor shear strains $\varepsilon_{xz}, \varepsilon_{yz}, \varepsilon_{xy}$. The engineering strains (ϵ instead of ε) are defined in (1.5). They can be obtained from the tensor components by the following relationship

$$\epsilon_\delta = R_{\delta\gamma}\varepsilon_\gamma \tag{1.37}$$

with the Reuter matrix given by

$$[R] = \begin{bmatrix} 1 & 0 & 0 & 0 & 0 & 0 \\ 0 & 1 & 0 & 0 & 0 & 0 \\ 0 & 0 & 1 & 0 & 0 & 0 \\ 0 & 0 & 0 & 2 & 0 & 0 \\ 0 & 0 & 0 & 0 & 2 & 0 \\ 0 & 0 & 0 & 0 & 0 & 2 \end{bmatrix} \tag{1.38}$$

Then, the coordinate transformation of engineering strain results from (1.36) and (1.37) as

$$\epsilon'_\alpha = \overline{T}_{\alpha\beta}\epsilon_\beta \tag{1.39}$$

with

$$[\,\overline{T}\,] = [R][T][R]^{-1} \tag{1.40}$$

used only to transform engineering strains. Explicitly we have

$$[\,\overline{T}\,] =$$
$$\begin{bmatrix} a_{11}^2 & a_{12}^2 & a_{13}^2 & a_{12}\,a_{13} & a_{11}\,a_{13} & a_{11}\,a_{12} \\ a_{21}^2 & a_{22}^2 & a_{23}^2 & a_{22}\,a_{23} & a_{21}\,a_{23} & a_{21}\,a_{22} \\ a_{31}^2 & a_{32}^2 & a_{33}^2 & a_{32}\,a_{33} & a_{31}\,a_{33} & a_{31}\,a_{32} \\ 2\,a_{21}\,a_{31} & 2\,a_{22}\,a_{32} & 2\,a_{23}\,a_{33} & a_{22}\,a_{33} + a_{23}\,a_{32} & a_{21}\,a_{33} + a_{23}\,a_{31} & a_{21}\,a_{32} + a_{22}\,a_{31} \\ 2\,a_{11}\,a_{31} & 2\,a_{12}\,a_{32} & 2\,a_{13}\,a_{33} & a_{12}\,a_{33} + a_{13}\,a_{32} & a_{11}\,a_{33} + a_{13}\,a_{31} & a_{11}\,a_{32} + a_{12}\,a_{31} \\ 2\,a_{11}\,a_{21} & 2\,a_{12}\,a_{22} & 2\,a_{13}\,a_{23} & a_{12}\,a_{23} + a_{13}\,a_{22} & a_{11}\,a_{23} + a_{13}\,a_{21} & a_{11}\,a_{22} + a_{12}\,a_{21} \end{bmatrix}$$
$$\tag{1.41}$$

1.11   Transformation of Constitutive Equations

The constitutive equations that relate stress $\boldsymbol{\sigma}$ to strain $\boldsymbol{\varepsilon}$ are defined using tensor strains (ε, not ϵ), as

$$\boldsymbol{\sigma}' = \mathbf{C}' : \boldsymbol{\varepsilon}'$$
$$\sigma'_{ij} = C'_{ijkl}\varepsilon'_{kl} \tag{1.42}$$

where both *tensor* and *index* notations have been used.[4]

For simplicity, consider an orthotropic material (Section 1.12.3). Then, it is possible to write σ'_{11} and σ'_{12} as

$$\sigma'_{11} = C'_{1111}\varepsilon'_{11} + C'_{1122}\varepsilon'_{22} + C'_{1133}\varepsilon'_{33}$$
$$\sigma'_{12} = C'_{1212}\varepsilon'_{12} + C'_{1221}\varepsilon'_{21} = 2C'_{1212}\varepsilon'_{12} \tag{1.43}$$

Rewriting (1.43) in contracted notation, it is clear that in contracted notation all the shear strains appear twice, as follows

$$\sigma'_1 = C'_{11}\varepsilon'_1 + C'_{12}\varepsilon'_2 + C'_{13}\varepsilon'_3 \tag{1.44}$$
$$\sigma'_6 = 2C'_{66}\varepsilon'_6$$

The factor 2 in front of the tensor shear strains is caused by two facts, the minor symmetry of the tensors C and ε (see (1.5,1.55-1.56)) and the contraction of the last two indices of C_{ijkl} with the strain ε_{kl} in (1.43). Therefore, *any double contraction of tensors with minor symmetry needs to be corrected by a Reuter matrix (1.38) when written in contracted notation*. Next, (1.42) can be written as

$$\sigma'_\alpha = C'_{\alpha\beta}R_{\beta\delta}\varepsilon'_\delta \tag{1.45}$$

Note that the Reuter matrix in (1.45) can be combined with the tensor strains using (1.37), to write

$$\sigma'_\alpha = C'_{\alpha\beta}\epsilon'_\beta \tag{1.46}$$

in terms of engineering strains. To obtain the stiffness matrix $[C]$ in the laminate coordinate system, introduce (1.29) and (1.39) into (1.46) so that

$$T_{\alpha\delta}\sigma_\delta = C'_{\alpha\beta}\overline{T}_{\beta\gamma}\epsilon_\gamma \tag{1.47}$$

It can be shown that

$$[T]^{-1} = [\overline{T}]^T \tag{1.48}$$

Therefore

$$\{\sigma\} = [C]\{\epsilon\} \tag{1.49}$$

with

---

[4]A double contraction involves contraction of two indices, in this case k and l, and it is denoted by : in tensor notation. Also note the use of boldface to indicate tensors in tensor notation.

$$[C] = [\overline{T}]^T [C'][\overline{T}] \tag{1.50}$$

and

$$[C'] = [\overline{T}]^{-T} [C][\overline{T}]^{-1} = [T][C][T]^T \tag{1.51}$$

The compliance matrix is the inverse of the stiffness matrix, not the inverse of the fourth-order tensor C_{ijkl}. Therefore,

$$[S'] = [C']^{-1} \tag{1.52}$$

Taking into account (1.48) and (1.50), the compliance matrix transforms as

$$[S] = [T]^T [S'][T] \tag{1.53}$$
$$[S'] = [T]^{-T} [S][T]^{-1} = [\overline{T}][S][\overline{T}]^T \tag{1.54}$$

1.12   3D Constitutive Equations

Hooke's law in three dimensions (3D) takes the form of (1.42). The 3D stiffness tensor C_{ijkl} is a fourth-order tensor with 81 components. For anisotropic materials, only 21 components are independent. That is, the remaining 60 components can be written in terms of the other 21. The 1D case, studied in mechanics of materials, is recovered when all the stress components are zero except σ_{11}. Only for the 1D case, $\sigma_{11} = \sigma, \varepsilon_{11} = \epsilon, C_{1111} = E$, and $\sigma = E\epsilon$. All the derivations in this section are carried out in lamina coordinates, but for simplicity, *the prime symbol (') is omitted, in this section only.*

In (1.42), exchanging the dummy indexes i by j, and k by l, we have

$$\sigma_{ji} = C_{jilk}\varepsilon_{lk} \tag{1.55}$$

Since the stress and strain tensors are symmetric, i.e, $\sigma_{ij} = \sigma_{ji}$ and $\varepsilon_{kl} = \varepsilon_{lk}$, it follows that

$$C_{ijkl} = C_{jikl} = C_{ijlk} = C_{jilk} \tag{1.56}$$

which effectively reduces the number of independent components from 81 to 36. For example, $C_{1213} = C_{2131}$ and so on. Then, the 36 independent components can be written as a 6×6 matrix.

Furthermore, an elastic material does not dissipate energy. All elastic energy stored during loading is recovered during unloading. Therefore, the elastic energy at any point on the stress-strain curve is independent on the path that was followed to arrive at that point. A path-independent function is called a potential function. In this case, the potential is the strain energy density $\widetilde{u}(\varepsilon_{ij})$. Expanding the strain energy density in a Taylor power series

$$\widetilde{u} = \widetilde{u}_0 + \left.\frac{\partial \widetilde{u}}{\partial \varepsilon_{ij}}\right|_0 \varepsilon_{ij} + \frac{1}{2} \left.\frac{\partial^2 \widetilde{u}}{\partial \varepsilon_{ij} \partial \varepsilon_{kl}}\right|_0 \varepsilon_{ij}\, \varepsilon_{kl} + \dots \tag{1.57}$$

Now take a derivative with respect to ε_{ij}

$$\frac{\partial \widetilde{u}}{\partial \varepsilon_{ij}} = 0 + \beta_{ij} + \frac{1}{2}\left(\alpha_{ijkl}\, \varepsilon_{kl} + \alpha_{klij}\, \varepsilon_{ij}\right) \tag{1.58}$$

where β_{ij} and α_{ijkl} are constants. From here, one can write

$$\sigma_{ij} - \sigma_{ij}^0 = C_{ijkl}\, \varepsilon_{kl} \tag{1.59}$$

where $\sigma_{ij}^0 = \beta_{ij}$ is the residual stress and $\alpha_{ijkl} = 1/2(C_{ijkl} + C_{klik}) = C_{ijkl}$ is the symmetric stiffness tensor (see (1.56)). Equation (1.59) is a generalization of (1.55) including residual stresses.

Using contracted notation, the generalized Hooke's law becomes

$$\begin{Bmatrix} \sigma_1 \\ \sigma_2 \\ \sigma_3 \\ \sigma_4 \\ \sigma_5 \\ \sigma_6 \end{Bmatrix} = \begin{bmatrix} C_{11} & C_{12} & C_{13} & C_{14} & C_{15} & C_{16} \\ C_{12} & C_{22} & C_{23} & C_{24} & C_{25} & C_{26} \\ C_{13} & C_{23} & C_{33} & C_{34} & C_{35} & C_{36} \\ C_{14} & C_{24} & C_{34} & C_{44} & C_{45} & C_{46} \\ C_{15} & C_{25} & C_{35} & C_{45} & C_{55} & C_{56} \\ C_{16} & C_{26} & C_{36} & C_{46} & C_{56} & C_{66} \end{bmatrix} \begin{Bmatrix} \epsilon_1 \\ \epsilon_2 \\ \epsilon_3 \\ \gamma_4 \\ \gamma_5 \\ \gamma_6 \end{Bmatrix} \tag{1.60}$$

Once again, the 1D case is covered when $\sigma_\alpha = 0$ if $\alpha \neq 1$. Then, $\sigma_1 = \sigma, \epsilon_1 = \epsilon, C_{11} = E$.

1.12.1   Anisotropic Material

Equation (1.60) represents a fully anisotropic material. Such a material has properties that change with the orientation. For example, the material body depicted in Figure 1.9 deforms differently in the directions P, T, and Q, even if the forces applied along the directions P, T, and Q are equal. The number of constants required to describe anisotropic materials is 21.

The inverse of the stiffness matrix is the compliance matrix $[S] = [C]^{-1}$. The constitutive equation (3D Hooke's law) is written in terms of compliances, as follows

$$\begin{Bmatrix} \epsilon_1 \\ \epsilon_2 \\ \epsilon_3 \\ \gamma_4 \\ \gamma_5 \\ \gamma_6 \end{Bmatrix} = \begin{bmatrix} S_{11} & S_{12} & S_{13} & S_{14} & S_{15} & S_{16} \\ S_{12} & S_{22} & S_{23} & S_{24} & S_{25} & S_{26} \\ S_{13} & S_{23} & S_{33} & S_{34} & S_{35} & S_{36} \\ S_{14} & S_{24} & S_{34} & S_{44} & S_{45} & S_{46} \\ S_{15} & S_{25} & S_{35} & S_{45} & S_{55} & S_{56} \\ S_{16} & S_{26} & S_{36} & S_{46} & S_{56} & S_{66} \end{bmatrix} \begin{Bmatrix} \sigma_1 \\ \sigma_2 \\ \sigma_3 \\ \sigma_4 \\ \sigma_5 \\ \sigma_6 \end{Bmatrix} \tag{1.61}$$

The [S] matrix is also symmetric and it has 21 independent constants. For the 1D case, $\sigma = 0$ if $p \neq 1$. Then, $\sigma_1 = \sigma, \epsilon_1 = \epsilon, S_{11} = 1/E$.

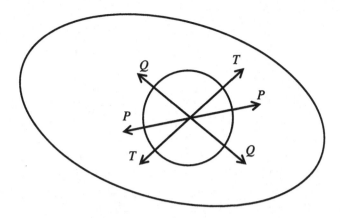

Fig. 1.9: Anisotropic material.

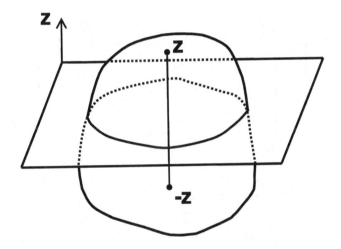

Fig. 1.10: Monoclinic material.

1.12.2 Monoclinic Material

If a material has one plane of symmetry (Figure 1.10), it is called monoclinic and 13 constants are required to describe it. One plane of symmetry means that the properties are the same at symmetric points (z and -z as in Figure 1.10).

When the material is symmetric about the 1-2 plane, the material properties are identical upon reflection with respect to the 1-2 plane. For such reflection, the a-matrix (1.21) is

$$
\begin{array}{c}
 \begin{array}{ccc} x_1 & x_2 & x_3 \end{array} \\
\begin{array}{c} e_1'' \\ e_2'' \\ e_3'' \end{array}
\begin{bmatrix}
1 & 0 & 0 \\
0 & 1 & 0 \\
0 & 0 & -1
\end{bmatrix}
\end{array}
\tag{1.62}
$$

where $''$ has been used to avoid confusion with the lamina coordinate system that is denoted without $()'$ in this section but with $()'$ elsewhere in this book. From (1.40) we get

$$[T] = \begin{bmatrix} 1 & 0 & 0 & 0 & 0 & 0 \\ 0 & 1 & 0 & 0 & 0 & 0 \\ 0 & 0 & 1 & 0 & 0 & 0 \\ 0 & 0 & 0 & -1 & 0 & 0 \\ 0 & 0 & 0 & 0 & -1 & 0 \\ 0 & 0 & 0 & 0 & 0 & 1 \end{bmatrix} \tag{1.63}$$

The effect of $[T]$ is to multiply rows and columns 4 and 5 in $[C]$ by -1. The diagonal terms C_{44} and C_{55} remain positive because they are multiplied twice. Therefore, $C_{i4}'' = -C_{i4}$ with $i \neq 4, 5$, $C_{i5}'' = -C_{i5}$ with $i \neq 4, 5$, with everything else unchanged. Since the material properties in a monoclinic material cannot change by a reflection, it must be $C_{4i} = C_{i4} = 0$ with $i \neq 4, 5$, $C_{5i} = C_{i5} = 0$ with $i \neq 4, 5$. That is, 3D Hooke's law reduces to

$$\begin{Bmatrix} \sigma_1 \\ \sigma_2 \\ \sigma_3 \\ \sigma_4 \\ \sigma_5 \\ \sigma_6 \end{Bmatrix} = \begin{bmatrix} C_{11} & C_{12} & C_{13} & 0 & 0 & C_{16} \\ C_{12} & C_{22} & C_{23} & 0 & 0 & C_{26} \\ C_{13} & C_{23} & C_{33} & 0 & 0 & C_{36} \\ 0 & 0 & 0 & C_{44} & C_{45} & 0 \\ 0 & 0 & 0 & C_{45} & C_{55} & 0 \\ C_{16} & C_{26} & C_{36} & 0 & 0 & C_{66} \end{bmatrix} \begin{Bmatrix} \epsilon_1 \\ \epsilon_2 \\ \epsilon_3 \\ \gamma_4 \\ \gamma_5 \\ \gamma_6 \end{Bmatrix} \tag{1.64}$$

and in terms of the compliances to

$$\begin{Bmatrix} \epsilon_1 \\ \epsilon_2 \\ \epsilon_3 \\ \gamma_4 \\ \gamma_5 \\ \gamma_6 \end{Bmatrix} = \begin{bmatrix} S_{11} & S_{12} & S_{13} & 0 & 0 & S_{16} \\ S_{12} & S_{22} & S_{23} & 0 & 0 & S_{26} \\ S_{13} & S_{23} & S_{33} & 0 & 0 & S_{36} \\ 0 & 0 & 0 & S_{44} & S_{45} & 0 \\ 0 & 0 & 0 & S_{45} & S_{55} & 0 \\ S_{16} & S_{26} & S_{36} & 0 & 0 & S_{66} \end{bmatrix} \begin{Bmatrix} \sigma_1 \\ \sigma_2 \\ \sigma_3 \\ \sigma_4 \\ \sigma_5 \\ \sigma_6 \end{Bmatrix} \tag{1.65}$$

1.12.3   Orthotropic Material

An orthotropic material has three planes of symmetry that coincide with the co-ordinate planes. It can be shown that if two orthogonal planes of symmetry exist, there is always a third orthogonal plane of symmetry. Nine constants are required to describe this type of material.

The symmetry planes can be Cartesian, as depicted in Figure 1.11, or they may correspond to any other coordinate representation (cylindrical, spherical, etc.). For example, the trunk of a tree has cylindrical orthotropy because of the growth rings. However, most practical materials exhibit Cartesian orthotropy. A unidirectional fiber-reinforced composite may be considered to be orthotropic. One plane of symmetry is perpendicular to the fiber direction, another parallel to the fiber direction, and the third perpendicular to the other two.

In addition to the reflection about the 1-2 plane discussed in Section 1.12.2, a second reflection about the 1-3 plane should not affect the properties of the orthotropic materials. In this case the a-matrix is

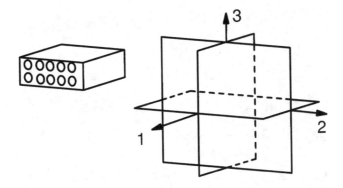

Fig. 1.11: Orthotropic material.

$$[a] = \begin{bmatrix} 1 & 0 & 0 \\ 0 & -1 & 0 \\ 0 & 0 & 1 \end{bmatrix} \tag{1.66}$$

The $\overline{T}$-matrix from (1.40) is

$$[T] = \begin{bmatrix} 1 & 0 & 0 & 0 & 0 & 0 \\ 0 & 1 & 0 & 0 & 0 & 0 \\ 0 & 0 & 1 & 0 & 0 & 0 \\ 0 & 0 & 0 & -1 & 0 & 0 \\ 0 & 0 & 0 & 0 & 1 & 0 \\ 0 & 0 & 0 & 0 & 0 & -1 \end{bmatrix} \tag{1.67}$$

This will make $C_{i6} = -C_{i6}$, $i \neq 4, 6$ and $C_{i4} = -C_{i4}$, $i \neq 4, 6$. Since the material has symmetry about the 1-3 plane, this means that $C_{i6} = C_{6i} = 0$, $i \neq 6$. In this case, 3D Hooke's law reduces to

$$\begin{Bmatrix} \sigma_1 \\ \sigma_2 \\ \sigma_3 \\ \sigma_4 \\ \sigma_5 \\ \sigma_6 \end{Bmatrix} = \begin{bmatrix} C_{11} & C_{12} & C_{13} & 0 & 0 & 0 \\ C_{12} & C_{22} & C_{23} & 0 & 0 & 0 \\ C_{13} & C_{23} & C_{33} & 0 & 0 & 0 \\ 0 & 0 & 0 & C_{44} & 0 & 0 \\ 0 & 0 & 0 & 0 & C_{55} & 0 \\ 0 & 0 & 0 & 0 & 0 & C_{66} \end{bmatrix} \begin{Bmatrix} \epsilon_1 \\ \epsilon_2 \\ \epsilon_3 \\ \gamma_4 \\ \gamma_5 \\ \gamma_6 \end{Bmatrix} \tag{1.68}$$

and in terms of the compliances to

$$\begin{Bmatrix} \epsilon_1 \\ \epsilon_2 \\ \epsilon_3 \\ \gamma_4 \\ \gamma_5 \\ \gamma_6 \end{Bmatrix} = \begin{bmatrix} S_{11} & S_{12} & S_{13} & 0 & 0 & 0 \\ S_{12} & S_{22} & S_{23} & 0 & 0 & 0 \\ S_{13} & S_{23} & S_{33} & 0 & 0 & 0 \\ 0 & 0 & 0 & S_{44} & 0 & 0 \\ 0 & 0 & 0 & 0 & S_{55} & 0 \\ 0 & 0 & 0 & 0 & 0 & S_{66} \end{bmatrix} \begin{Bmatrix} \sigma_1 \\ \sigma_2 \\ \sigma_3 \\ \sigma_4 \\ \sigma_5 \\ \sigma_6 \end{Bmatrix} \tag{1.69}$$

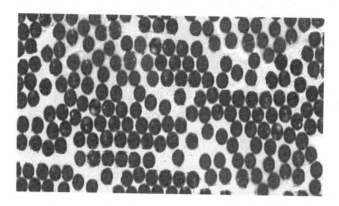

Fig. 1.12: Randomly distributed E-glass fibers with 200X magnification.

Note that if the material has two planes of symmetry, it automatically has three because applying the procedure once more for a third plane (the 2-3 plane) will not change the outcome (1.68-1.69).

1.12.4 Transversely Isotropic Material

A transversely isotropic material has one axis of symmetry. For example, the fiber direction of a unidirectional fiber-reinforced composite can be considered an axis of symmetry if the fibers are randomly distributed in the cross-section (Figure 1.12). In this case, any plane containing the fiber direction is a plane of symmetry. A transversely isotropic material is described by five constants. When the axis of symmetry is the fiber direction (1-direction), 3D Hooke's law reduces to

$$
\begin{Bmatrix} \sigma_1 \\ \sigma_2 \\ \sigma_3 \\ \sigma_4 \\ \sigma_5 \\ \sigma_6 \end{Bmatrix}
=
\begin{bmatrix}
C_{11} & C_{12} & C_{12} & 0 & 0 & 0 \\
C_{12} & C_{22} & C_{23} & 0 & 0 & 0 \\
C_{12} & C_{23} & C_{22} & 0 & 0 & 0 \\
0 & 0 & 0 & (C_{22}-C_{23})/2 & 0 & 0 \\
0 & 0 & 0 & 0 & C_{66} & 0 \\
0 & 0 & 0 & 0 & 0 & C_{66}
\end{bmatrix}
\begin{Bmatrix} \epsilon_1 \\ \epsilon_2 \\ \epsilon_3 \\ \gamma_4 \\ \gamma_5 \\ \gamma_6 \end{Bmatrix}
\tag{1.70}
$$

and in terms of the compliances to

$$
\begin{Bmatrix} \epsilon_1 \\ \epsilon_2 \\ \epsilon_3 \\ \gamma_4 \\ \gamma_5 \\ \gamma_6 \end{Bmatrix}
=
\begin{bmatrix}
S_{11} & S_{12} & S_{12} & 0 & 0 & 0 \\
S_{12} & S_{22} & S_{23} & 0 & 0 & 0 \\
S_{12} & S_{23} & S_{22} & 0 & 0 & 0 \\
0 & 0 & 0 & 2(S_{22}-S_{23}) & 0 & 0 \\
0 & 0 & 0 & 0 & S_{66} & 0 \\
0 & 0 & 0 & 0 & 0 & S_{66}
\end{bmatrix}
\begin{Bmatrix} \sigma_1 \\ \sigma_2 \\ \sigma_3 \\ \sigma_4 \\ \sigma_5 \\ \sigma_6 \end{Bmatrix}
\tag{1.71}
$$

Note the equations would be different if the axis of symmetry was not the 1-direction. In terms of engineering properties (Section 1.13), and taking into account that the directions 2 and 3 are indistinguishable, the following relations apply for a transversely isotropic material

$$E_2 = E_3$$
$$\nu_{12} = \nu_{13} \tag{1.72}$$
$$G_{12} = G_{13}$$

In addition, any two perpendicular directions on the plane 2-3 can be taken as axes. In other words, the 2-3 plane is isotropic. Therefore, the following holds in the 2-3 plane

$$G_{23} = \frac{E_2}{2(1 + \nu_{23})} \tag{1.73}$$

just as it holds for isotropic materials (see Problem 1.14).

1.12.5 Isotropic Material

The most common materials of industrial use are isotropic, like aluminum, steel, etc. Isotropic materials have an infinite number of planes of symmetry, meaning that the properties are independent of the orientation. Only two constants are needed to represent the elastic properties. These two properties can be the Young's modulus E and the Poisson's ratio ν, but several other pairs of constants are used whenever it is convenient. However, any pair of properties has to be related to any other pair. For example, you could describe isotropic materials by E and G, but the shear modulus of isotropic materials is related to E and ν by

$$G = \frac{E}{2(1 + \nu)} \tag{1.74}$$

Also, the Lamé constants are sometimes used for convenience. In this case the two constants are

$$\lambda = \frac{E\nu}{(1 + \nu)(1 - 2\nu)} \tag{1.75}$$
$$\mu = G$$

To form yet another pair, any of the above properties could be substituted by the bulk modulus k, as follows

$$k = \frac{E}{3(1 - 2\nu)} \tag{1.76}$$

which relates the hydrostatic pressure p to the volumetric strain as

$$p = k(\epsilon_1 + \epsilon_2 + \epsilon_3) \tag{1.77}$$

For isotropic materials, the 3D Hooke's law is written in terms of only two constants C_{11} and C_{12} as

$$
\begin{Bmatrix} \sigma_1 \\ \sigma_2 \\ \sigma_3 \\ \sigma_4 \\ \sigma_5 \\ \sigma_6 \end{Bmatrix} =
\begin{bmatrix}
C_{11} & C_{12} & C_{12} & 0 & 0 & 0 \\
C_{12} & C_{11} & C_{12} & 0 & 0 & 0 \\
C_{12} & C_{12} & C_{11} & 0 & 0 & 0 \\
0 & 0 & 0 & \frac{(C_{11}-C_{12})}{2} & 0 & 0 \\
0 & 0 & 0 & 0 & \frac{(C_{11}-C_{12})}{2} & 0 \\
0 & 0 & 0 & 0 & 0 & \frac{(C_{11}-C_{12})}{2}
\end{bmatrix}
\begin{Bmatrix} \epsilon_1 \\ \epsilon_2 \\ \epsilon_3 \\ \gamma_4 \\ \gamma_5 \\ \gamma_6 \end{Bmatrix}
$$

$$(1.78)$$

In terms of compliances, once again, two constants, S_{11} and S_{12}, are used, as follows

$$
\begin{Bmatrix} \epsilon_1 \\ \epsilon_2 \\ \epsilon_3 \\ \gamma_4 \\ \gamma_5 \\ \gamma_6 \end{Bmatrix} =
\begin{bmatrix}
S_{11} & S_{12} & S_{12} & 0 & 0 & 0 \\
S_{12} & S_{11} & S_{12} & 0 & 0 & 0 \\
S_{12} & S_{12} & S_{11} & 0 & 0 & 0 \\
0 & 0 & 0 & 2s & 0 & 0 \\
0 & 0 & 0 & 0 & 2s & 0 \\
0 & 0 & 0 & 0 & 0 & 2s
\end{bmatrix}
\begin{Bmatrix} \sigma_1 \\ \sigma_2 \\ \sigma_3 \\ \sigma_4 \\ \sigma_5 \\ \sigma_6 \end{Bmatrix}
$$

$$(1.79)$$

$$s = S_{11} - S_{12}$$

Not only are the various constants related in pairs, but also certain restrictions apply on the values that these constants may have for real materials. Since the Young and shear moduli must always be positive, the Poisson's ratio must be $\nu > -1$. Furthermore, since the bulk modulus must be positive, we have $\nu < \frac{1}{2}$. Finally, the Poisson's ratio of isotropic materials is constrained by $-1 < \nu < \frac{1}{2}$.

1.13   Engineering Constants

Please note from here forward $()'$ *denotes the lamina coordinate system.* Our next task is to write the components of the stiffness and compliance matrices in terms of engineering constants for orthotropic materials. For this purpose it is easier to work with the compliance matrix, which is defined as the inverse of the stiffness matrix. In *lamina coordinates* $[S'] = [C']^{-1}$. The compliance matrix is used to write the relationship between strains and stresses in (1.69) for an orthotropic material. Let's rewrite the first of (1.69), which corresponds to the strain in the 1-direction (fiber direction)

$$\epsilon_1' = S_{11}'\sigma_1' + S_{12}'\sigma_2' + S_{13}'\sigma_3' \tag{1.80}$$

and let's perform a thought experiment. Note that $[S']$ is used to emphasize the fact that we are working in the lamina coordinate system. First, apply a tensile stress along the 1-direction (fiber direction) as in Figure 1.13, with all the other stresses equal to zero, and compute the strain produced in the 1-direction, which is

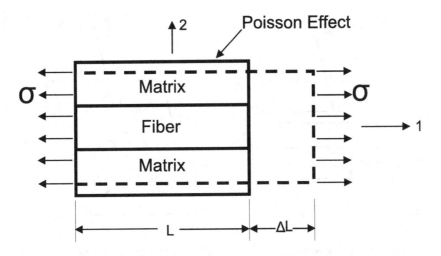

Fig. 1.13: Longitudinal loading.

$$\epsilon'_1 = \frac{\sigma_1'}{E_1} \tag{1.81}$$

Then, apply a stress in the 2-direction only, and compute the strain in the 1-direction using the appropriate Poisson's ratio [3, (5.16)]

$$\epsilon'_1 = -\nu_{21}\frac{\sigma'_2}{E_2} \tag{1.82}$$

Now, apply a stress in the 3-direction only, and compute the strain in the 1-direction using the appropriate Poisson's ratio,

$$\epsilon'_1 = -\nu_{31}\frac{\sigma'_3}{E_3} \tag{1.83}$$

The total strain ϵ'_1 is the sum of equations (1.81), (1.82), and (1.83)

$$\epsilon'_1 = \frac{1}{E_1}\sigma'_1 - \frac{\nu_{21}}{E_2}\sigma'_2 - \frac{\nu_{31}}{E_3}\sigma'_3 \tag{1.84}$$

Comparing (1.84) with (1.80) we conclude that

$$S'_{11} = \frac{1}{E_1}; S'_{12} = -\frac{\nu_{21}}{E_2}; S'_{13} = -\frac{\nu_{31}}{E_3} \tag{1.85}$$

Repeat the same procedure for the equations corresponding to ϵ'_2 and ϵ'_3 to obtain the coefficients in the second and third rows of the compliance matrix (1.69).

For the shear terms, use the 4th, 5th, and 6th rows of the compliance matrix (1.69). For example, from Figure 1.14 we write

$$\sigma'_6 = \epsilon'_6 G_{12} = 2\varepsilon'_6 G_{12} \tag{1.86}$$

which, compared to the 6th row of (1.69), leads to $S_{66} = 1/G_{12}$.

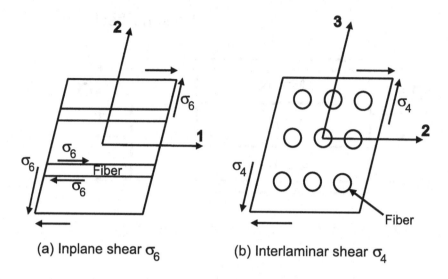

(a) Inplane shear σ_6 (b) Interlaminar shear σ_4

Fig. 1.14: Shear loading.

For an orthotropic material, the compliance matrix $[S']$ is defined in the lamina coordinate system as

$$[S'] = \begin{bmatrix} \dfrac{1}{E_1} & \dfrac{-\nu_{21}}{E_2} & \dfrac{-\nu_{31}}{E_3} & 0 & 0 & 0 \\[2mm] \dfrac{-\nu_{12}}{E_1} & \dfrac{1}{E_2} & \dfrac{-\nu_{32}}{E_3} & 0 & 0 & 0 \\[2mm] \dfrac{-\nu_{13}}{E_1} & \dfrac{-\nu_{23}}{E_2} & \dfrac{1}{E_3} & 0 & 0 & 0 \\[2mm] 0 & 0 & 0 & \dfrac{1}{G_{23}} & 0 & 0 \\[2mm] 0 & 0 & 0 & 0 & \dfrac{1}{G_{13}} & 0 \\[2mm] 0 & 0 & 0 & 0 & 0 & \dfrac{1}{G_{12}} \end{bmatrix} \tag{1.87}$$

where E_i, G_{ij}, and ν_{ij} are the elastic moduli, shear moduli, and Poisson's ratios, respectively. Furthermore, the subscripts indicate lamina coordinates, i.e.,

$$\nu_{ij} = \nu_{x_i' x_j'} \text{ and } E_{ii} = E_{x_i'} \tag{1.88}$$

Since $[S']$ is symmetric, the following must be satisfied

$$\frac{\nu_{ij}}{E_{ii}} = \frac{\nu_{ji}}{E_{jj}} \ , \ i,j = 1\ldots 3 \tag{1.89}$$

Furthermore, Poisson's ratios are defined so that the lateral strain is given by

$$\nu_j = -\nu_{ij}\epsilon_i \tag{1.90}$$

In ANSYS, the Poisson's ratios are defined differently than in this textbook. In fact, $\nu_{xy}, \nu_{xz}, \nu_{yz}$ are denoted PRXY, PRXZ, and PRYZ, while $\nu_{yx}, \nu_{zx}, \nu_{zy}$ are denoted by NUXY, NUXZ, and NUYZ. On the contrary, Abaqus uses the standard notation also used in this textbook. That is, the symbols NU12, NU13, NU23 follow the convention described by (1.90).

After computing S_{ij}, the components of stress are obtained by using (1.46) or (1.49). This formulation predicts realistic behavior for finite displacement and rotations as long as the strains are small. This formulation is expensive to use since it needs 18 state variables: 12 components of the strain displacement matrix computed in the initial configuration ($u_{i,j}$ and $u_{r,i}u_{r,j}$) plus 6 direction cosines $[a]$ to account for finite rotations.

However, in (1.87) only nine constants are independent because the matrix $[S']$ must be symmetric (see 1.93), so

$$[S'] = \begin{bmatrix} \dfrac{1}{E_1} & -\dfrac{\nu_{12}}{E_1} & -\dfrac{\nu_{13}}{E_1} & 0 & 0 & 0 \\ -\dfrac{\nu_{12}}{E_1} & \dfrac{1}{E_2} & -\dfrac{\nu_{23}}{E_2} & 0 & 0 & 0 \\ -\dfrac{\nu_{13}}{E_1} & -\dfrac{\nu_{23}}{E_2} & \dfrac{1}{E_3} & 0 & 0 & 0 \\ 0 & 0 & 0 & \dfrac{1}{G_{23}} & 0 & 0 \\ 0 & 0 & 0 & 0 & \dfrac{1}{G_{13}} & 0 \\ 0 & 0 & 0 & 0 & 0 & \dfrac{1}{G_{12}} \end{bmatrix} \tag{1.91}$$

The stiffness matrix can be computed also in terms of engineering constants by inverting the above equation so that $[C'] = [S']^{-1}$, with components given in terms of engineering constants as

$$C'_{11} = \frac{1 - \nu_{23}\nu_{32}}{E_2 E_3 \Delta}$$

$$C'_{12} = \frac{\nu_{21} + \nu_{31}\nu_{23}}{E_2 E_3 \Delta} = \frac{\nu_{12} + \nu_{32}\nu_{13}}{E_1 E_3 \Delta}$$

$$C'_{13} = \frac{\nu_{31} + \nu_{21}\nu_{32}}{E_2 E_3 \Delta} = \frac{\nu_{13} + \nu_{12}\nu_{23}}{E_1 E_2 \Delta}$$

$$C'_{22} = \frac{1 - \nu_{13}\nu_{31}}{E_1 E_3 \Delta}$$

$$C'_{23} = \frac{\nu_{32} + \nu_{12}\nu_{31}}{E_1 E_3 \Delta} = \frac{\nu_{23} + \nu_{21}\nu_{13}}{E_1 E_2 \Delta}$$

$$C'_{33} = \frac{1 - \nu_{12}\nu_{21}}{E_1 E_2 \Delta}$$

$$C'_{44} = G_{23}$$

$$C'_{55} = G_{13}$$

$$C'_{66} = G_{12} \tag{1.92}$$

$$\Delta = \frac{1 - \nu_{12}\nu_{21} - \nu_{23}\nu_{32} - \nu_{31}\nu_{13} - 2\nu_{21}\nu_{32}\nu_{13}}{E_1 E_2 E_3}$$

So far both $[S']$ and $[C']$ are 6×6 matrices with nine independent constants for the case of orthotropic materials. If the material is transversely isotropic $G_{13} = G_{12}, \nu_{13} = \nu_{12}, E_3 = E_2$.

1.13.1   Restrictions on Engineering Constants

It is important to note that because of the symmetry of the compliance matrix (1.91), the following restrictions on engineering constants apply

$$\frac{\nu_{ij}}{E_i} = \frac{\nu_{ji}}{E_j}; \quad i, j = 1 \ldots 3; \quad i \neq j \tag{1.93}$$

Further restrictions on the values of the elastic constants can be derived from the fact that all diagonal terms in both the compliance and stiffness matrices must be positive. Since all the engineering elastic constants must be positive $(E_1, E_2, E_3, G_{12}, G_{23}, G_{31} > 0)$, all the diagonal terms of the stiffness matrix (1.92) will be positive if the following two conditions are met. The first condition is that $(1 - \nu_{ij}\nu_{ji}) > 0$ for $i, j = 1..3$ and $i \neq j$, which leads to the following restriction on the values of the engineering constants

$$0 < \nu_{ij} < \sqrt{\frac{E_i}{E_j}}; \quad i, j = 1 \ldots 3; \quad i \neq j \tag{1.94}$$

The second condition is that

$$\Delta = 1 - \nu_{12}\nu_{21} - \nu_{23}\nu_{32} - \nu_{31}\nu_{13} - 2\nu_{21}\nu_{32}\nu_{13} > 0 \tag{1.95}$$

These restrictions can be used to check experimental data. For example, consider an experimental program in which E_1 and ν_{12} are measured in a longitudinal test (fibers in the direction of loading) by using two strain gages, one longitudinal and one transverse, and E_2 and ν_{21} are measured in the transverse tensile tests (fibers perpendicular to loading). For the test procedure to be valid, all four data values, E_1, E_2, ν_{12}, and ν_{21}, must conform to (1.93-1.95) within the margin allowed by experimental errors.

Example 1.4 *Sonti et al. [7] performed a series of tests on pultruded glass-fiber-reinforced composites. From tensile tests along the longitudinal axis, the average of eight tests gives $E_1 = 19.981 \ GPa$ and $\nu_{12} = 0.274$. The average of eight tests in the transverse direction gives $E_2 = 11.389 \ GPa$ and $\nu_{21} = 0.192$. Do these data fall within the constraints on elastic constants?*

Solution to Example 1.4 *First compute both sides of (1.93) for $i, j = 1, 2$ as*

$$\frac{E_1}{\nu_{12}} = \frac{19.981}{0.274} = 72.9 \ GPa$$

$$\frac{E_2}{\nu_{21}} = \frac{11.389}{0.192} = 59.3 \ GPa$$

The transverse result is 23% lower than expected. Either E_2 measured is too low or ν_{21} measured is 23% higher than what it should be. In any case a 23% difference deserves some scrutiny.

Next check (1.94)

$$abs(\nu_{12}) < \sqrt{\frac{E_1}{E_2}}$$

$$0.274 < 1.32$$

$$abs(\nu_{21}) < \sqrt{\frac{E_2}{E_1}}$$

$$0.192 < 0.75$$

Finally, there are insufficient data to evaluate the last of the restrictions on elastic constants from (1.95).

1.14   From 3D to Plane Stress Equations

Setting $\sigma_3 = 0$ in the compliance equations (1.69) of an orthotropic material implies that the third row and column of the compliance matrix are not used

$$
\begin{Bmatrix} \epsilon_1' \\ \epsilon_2' \\ \epsilon_3' \\ \gamma_4' \\ \gamma_5' \\ \gamma_6' \end{Bmatrix} =
\begin{bmatrix}
S_{11}' & S_{12}' & S_{13}' & 0 & 0 & 0 \\
S_{12}' & S_{22}' & S_{23}' & 0 & 0 & 0 \\
S_{13}' & S_{23}' & S_{33}' & 0 & 0 & 0 \\
0 & 0 & 0 & S_{44}' & 0 & 0 \\
0 & 0 & 0 & 0 & S_{55}' & 0 \\
0 & 0 & 0 & 0 & 0 & S_{66}'
\end{bmatrix}
\begin{Bmatrix} \sigma_1' \\ \sigma_2' \\ \sigma_3'=0 \\ \sigma_4' \\ \sigma_5' \\ \sigma_6' \end{Bmatrix}
\tag{1.96}
$$

So, the first two equations plus the last one can be written separately of the remaining, in terms of a 3×3 reduced compliance matrix $[S]$ and using $\gamma = 2\epsilon$, we have

$$
\begin{Bmatrix} \epsilon_1' \\ \epsilon_2' \\ \gamma_6' \end{Bmatrix} =
\begin{bmatrix}
S_{11}' & S_{12}' & 0 \\
S_{12}' & S_{22}' & 0 \\
0 & 0 & S_{66}'
\end{bmatrix}
\begin{Bmatrix} \sigma_1' \\ \sigma_2' \\ \sigma_6' \end{Bmatrix}
\tag{1.97}
$$

The third equation is seldom used

$$\epsilon_3' = S_{13}'\sigma_1' + S_{23}'\sigma_2' \tag{1.98}$$

and the remaining two equations can be written separately as

$$
\begin{Bmatrix} \gamma_4' \\ \gamma_5' \end{Bmatrix} =
\begin{bmatrix}
S_{44}' & 0 \\
0 & S_{55}'
\end{bmatrix}
\begin{Bmatrix} \sigma_4' \\ \sigma_5' \end{Bmatrix}
\tag{1.99}
$$

To compute stress components from strains, (1.97) can be inverted to get $\{\sigma\} = [Q]\{\epsilon\}$ or

$$
\begin{Bmatrix} \sigma_1' \\ \sigma_2' \\ \sigma_6' \end{Bmatrix} =
\begin{bmatrix}
Q_{11}' & Q_{12}' & 0 \\
Q_{12}' & Q_{22}' & 0 \\
0 & 0 & Q_{66}'
\end{bmatrix}
\begin{Bmatrix} \epsilon_1' \\ \epsilon_2' \\ \gamma_6' \end{Bmatrix}
\tag{1.100}
$$

where the matrix $[Q'] = [S'_{3\times3}]^{-1}$ is the reduced stiffness matrix for plane stress. Note that while the components of the reduced compliance matrix $[S'_{3\times3}]$ are numerically identical to the corresponding entries in the 6×6 compliance matrix, the components of the reduced stiffness matrix $[Q']$ are not numerically equal to the corresponding entries on the 6×6 stiffness matrix $[C']$, thus the change in name. This is because the inverse of a 3×3 matrix produces different values than the inverse of a 6×6 matrix. The set of equations is completed by writing

$$\left\{ \begin{array}{c} \sigma'_4 \\ \sigma'_5 \end{array} \right\} = \left[\begin{array}{cc} C'_{44} & 0 \\ 0 & C'_{55} \end{array} \right] \left\{ \begin{array}{c} \gamma'_4 \\ \gamma'_6 \end{array} \right\} \tag{1.101}$$

where coefficients C'_{44} and C'_{55} are numerically equal to the corresponding entries in the 6×6 stiffness matrix because the 2×2 matrix in (1.101) is diagonal.

Example 1.5 *Show that the change in the thickness $t\epsilon_3$ of a plate is negligible when compared to the in-plane elongations $a\epsilon_1$ and $b\epsilon_2$. Use the data from a composite plate with thickness $t = 0.635$ mm, and dimensions $a = 279$ mm and $b = 203$ mm. Take $E_1 = 19.981$ GPa, $E_2 = 11.389$ GPa, $\nu_{12} = 0.274$.*

Solution to Example 1.5 *Assuming that the 0.635 mm thick glass-reinforced Polyester plate is transversely isotropic, take $E_3 = E_2 = 11.389$ GPa, $\nu_{13} = \nu_{12} = 0.274$, $G_{31} = G_{12}$. Sonti et al. [7] report the average of eight torsion tests as $G_{12} = 3.789$ GPa. Lacking experimental data, assume $\nu_{23} \approx \nu_m = 0.3$, $G_{23} \approx G_m = 0.385$ GPa, with the properties of the Polyester matrix taken from [3, Tables 2.13–2.14]. The remaining properties in (1.91) can be obtained, using (1.93), as*

$$\nu_{21} = \nu_{12}\frac{E_2}{E_1} = 0.274 \left(\frac{11.389}{19.981} \right) = 0.156$$

$$\nu_{31} = \nu_{13}\frac{E_3}{E_1} = 0.274 \left(\frac{11.389}{19.981} \right) = 0.156$$

$$\nu_{32} = \nu_{23}\frac{E_3}{E_2} = 0.3 \left(\frac{11.389}{11.389} \right) = 0.3$$

Because of transverse isotropy, $G_{13} = G_{12} = 3.789$ GPa. Now, assume a state of stress $\sigma'_1 = \sigma'_2 = 0.1$ GPa, $\sigma'_4 = \sigma'_5 = \sigma'_6 = 0$, and $\sigma'_3 = 0$ because of the assumption of plane stress. Using (1.97) we get

$$\epsilon'_1 = S'_{11}\sigma'_1 + S'_{12}\sigma'_2 = \frac{0.1}{19.981} - \frac{0.1(0.156)}{11.389} = 3.635 \ 10^{-3}$$

$$\epsilon'_2 = S'_{12}\sigma'_1 + S'_{22}\sigma'_2 = -\frac{0.1(0.156)}{11.389} + \frac{0.1}{11.389} = 7.411 \ 10^{-3}$$

$$\epsilon'_3 = S'_{13}\sigma'_1 + S'_{23}\sigma'_2 = -\frac{0.274(0.1)}{19.981} - \frac{0.3(0.1)}{11.389} = -4.005 \ 10^{-3}$$

Finally

$$t\epsilon'_3 = -0.635(4.005 \ 10^{-3}) = -2.543 \ 10^{-3} \ mm$$

$$a\epsilon'_1 = 279(3.635 \ 10^{-3}) = 1.014 \ mm$$

$$b\epsilon'_2 = 203(7.411 \ 10^{-3}) = 1.504 \ mm$$

Since the elongation in the transverse direction is so small, it is neglected in the deriva-
tion of the plate equations in [3, Section 6.1].

1.15 Apparent Laminate Properties

A laminate is called balanced if the total thickness of laminas oriented with respect
to the laminate direction at $+\theta$ and $-\theta$ are the same. The stiffness matrix $[C]$
of a balanced, symmetric laminate with N laminas is built by adding the lamina
matrices in the laminate coordinate system multiplied by the thickness ratio t_k/t of
each lamina, where t is the laminate thickness and t_k denotes the thickness of the
k-th lamina

$$[C] = \sum_{k=1}^{N} \frac{t_k}{t}[C_k] \tag{1.102}$$

with C_k calculated by (1.50) applied to (1.92) for $k = 1 \ldots N$.

Note that compliances cannot be added nor averaged. The laminate compliance
is obtained by inverting the 6×6 stiffness matrix, as

$$[S] = [C]^{-1} \tag{1.103}$$

A balanced laminate has orthotropic stiffness $[C]$ and compliance $[S]$. In terms
of the apparent engineering properties of the laminate, the compliance is

$$[S] = \begin{bmatrix} \dfrac{1}{E_x} & -\dfrac{\nu_{yx}}{E_y} & -\dfrac{\nu_{zx}}{E_z} & 0 & 0 & 0 \\ -\dfrac{\nu_{xy}}{E_x} & \dfrac{1}{E_y} & -\dfrac{\nu_{zy}}{E_z} & 0 & 0 & 0 \\ -\dfrac{\nu_{xz}}{E_x} & -\dfrac{\nu_{yz}}{E_y} & \dfrac{1}{E_z} & 0 & 0 & 0 \\ 0 & 0 & 0 & \dfrac{1}{G_{yz}} & 0 & 0 \\ 0 & 0 & 0 & 0 & \dfrac{1}{G_{xz}} & 0 \\ 0 & 0 & 0 & 0 & 0 & \dfrac{1}{G_{xy}} \end{bmatrix} \tag{1.104}$$

Since the compliance must be symmetric, it must satisfy (1.93) with $i, j =$
x, y, z. Therefore, it is possible to compute the apparent engineering properties of
a laminate in terms of the laminate compliance, as follows

$$E_x = 1/S_{11} \qquad \nu_{xy} = -S_{21}/S_{11}$$
$$E_y = 1/S_{22} \qquad \nu_{xz} = -S_{31}/S_{11}$$
$$E_z = 1/S_{33} \qquad \nu_{yz} = -S_{32}/S_{22}$$
$$G_{yz} = 1/S_{44}$$
$$G_{xz} = 1/S_{55}$$
$$G_{xy} = 1/S_{66} \tag{1.105}$$

Example 1.6 *Compute the laminate properties of $[0/90/\pm 30]_S$ with $t_k = 1.5$ mm, $E_f = 241$ GPa, $\nu_f = 0.2$, $E_m = 3.12$ GPa, $\nu_m = 0.38$, fiber volume fraction $V_f = 0.6$, where f,m, denote fiber and matrix, respectively.*

Solution to Example 1.6 *First use periodic microstructure micromechanics (6.8) to obtain the lamina properties (in MPa).*

$$E_1 = 145,880 \quad G_{12} = 4,386 \quad \nu_{12} = \nu_{13} = 0.263$$
$$E_2 = 13,312 \quad G_{23} = 4,528 \qquad \nu_{23} = 0.470$$

Then, compute the compliance matrix $[S']$ using (1.91), the rotation matrix $[T]$ using (1.34), the compliance $[S]$ in laminate coordinate system using (1.53), and the stiffness $[C] = [S]^{-1}$ in laminate coordinate system for each lamina. Then, average them using (1.102), invert the average, and finally using (1.105) get

$$E_x = 78,901 \quad G_{xy} = 17,114 \quad \nu_{xy} = 0.320$$
$$E_y = 47,604 \quad G_{yz} = 4,475 \quad \nu_{yz} = 0.364$$
$$E_z = 16,023 \quad G_{xz} = 4,439 \quad \nu_{xz} = 0.280$$

Suggested Problems

Problem 1.1 *Using the principle of virtual work, find a quadratic displacement function $u(x)$ in $0 < x < L$ of a tapered slender rod of length L, fixed at the origin and loaded axially in tension at the free end. The cross-section area changes lineally and the areas are $A_1 > A_2$ at the fixed and free ends, respectively. The material is homogeneous and isotropic with modulus E.*

Problem 1.2 *Using the principle of virtual work, find a quadratic rotation angle function $\theta(x)$ in $0 < x < L$ of a tapered slender shaft of circular cross-section and length L, fixed at the origin and loaded by a torque T at the free end. The cross-section area changes lineally and the areas are $A_1 > A_2$ at the fixed and free ends, respectively. The material is homogeneous and isotropic with shear modulus G.*

Problem 1.3 *Construct a rotation matrix $[a]$ resulting from three consecutive reflections about (a) the x-y plane, (b) the x-z plane, (c) the y-z plane. The resulting system does not follow the right-hand rule.*

Problem 1.4 *Construct three rotation matrices $[a]$ for rotations $\theta = \pi$ about (a) the x-axis, (b) the y-axis, (c) the z-axis.*

Problem 1.5 *Using*

$$\sigma = \begin{bmatrix} 10 & 2 & 1 \\ 2 & 5 & 1 \\ 1 & 1 & 3 \end{bmatrix}$$

and [a] of Ex. 1.2, verify that (1.29) yields the same result as (1.26).

Problem 1.6 *Write a computer program to evaluate the compliance and stiffness matrices in terms of engineering properties. Take the input from a file and the output to another file. Validate the program with your own examples. You may use material properties from [3, Tables 1.1–1.4] and assume the material is transversely isotropic as per Section 1.12.4. Show all work in a report.*

Problem 1.7 *Write a computer program to transform the stiffness and compliance matrix from lamina coordinates C', S', to another coordinate system C, S, by a rotation $-\theta$ around the z-axis (Figure 1.7). The data C', S', θ, should be read from a file. The output C, S should be written to another file. Validate your program with your own examples. You may use material properties from [3, Tables 1.1–1.4] and assume the material is transversely isotropic as per Section 1.12.4. Show all work in a report.*

Problem 1.8 *Verify numerically (1.92) against $[S]^{-1}$ for the material of your choice. You may use material properties from [3, Tables 1.1–1.4] and assume the material is transversely isotropic as per Section 1.12.4.*

Problem 1.9 *The following data has been obtained experimentally for a composite based on a unidirectional carbon-epoxy pre-preg (MR50 carbon fiber at 63% by volume in LTM25 epoxy). Determine if the restrictions on elastic constants are satisfied.*

$$E_1 = 156.403 \; GPa, \qquad E_2 = 7.786 \; GPa$$
$$\nu_{12} = 0.352, \qquad \nu_{21} = 0.016$$
$$G_{12} = 3.762 \; GPa$$
$$\sigma_{1t}^u = 1.826 \; GPa, \qquad \sigma_{1c}^u = 1.134 \; GPa$$
$$\sigma_{2t}^u = 19 \; MPa, \qquad \sigma_{2c}^u = 131 \; MPa$$
$$\sigma_6^u = 75 \; MPa$$
$$\epsilon_{1t}^u = 11,900 \; 10^{-6}, \qquad \epsilon_{1c}^u = 8,180 \; 10^{-6}$$
$$\epsilon_{2t}^u = 2,480 \; 10^{-6}, \qquad \epsilon_{2c}^u = 22,100 \; 10^{-6}$$
$$\gamma_{12}^u = 20,000 \; 10^{-6}$$

Problem 1.10 *Explain contracted notation for stresses and strains.*

Problem 1.11 *What is an orthotropic material and how many constants are needed to describe it?*

Problem 1.12 *What is a transversely isotropic material and how many constants are needed to describe it?*

Problem 1.13 *Use the three rotation matrices in Problem 1.4 to verify (1.48) numerically.*

Problem 1.14 *Prove (1.73) using (1.71) and (1.91).*

Problem 1.15 *Demonstrate that a material having two perpendicular planes of symmetry also has a third. Apply a reflection about the 2-3 plane to (1.68) using the procedure in Section 1.12.3.*

Problem 1.16 *What is a plane stress assumption?*

Problem 1.17 *Write a computer program to evaluate the laminate engineering properties for symmetric balanced laminates. All laminas are of the same material. Input data consists of all the engineering constants for a transversely isotropic material, number of laminas N, thickness and angle for all the laminas t_k, θ_k with $k = 1...N$. Use Section 1.15, 1.12.4, and 1.13.*

Chapter 2

Introduction to Finite Element Analysis

In this textbook, the finite element method (FEM) is used as a tool to solve practical problems. For the most part, commercial packages, mainly Abaqus, are used in the examples. Computer programming is limited to implementing material models and post processing algorithms. When commercial codes lack needed features, other codes are used, which are provided in [2]. A basic understanding of the finite element method is necessary for effective use of any finite element (FE) software. Therefore, this chapter contains a brief introduction intended for those readers who have not had a formal course or prior knowledge about FEM. Furthermore, an introduction to the Abaqus CAE graphical user interface (GUI) is presented to familiarize the reader with typical procedures used for finite element modeling using any commercial software.

2.1 Basic FEM Procedure

Consider the axial deformation of a rod. The ordinary differential equation (ODE) describing the deformation of a rod is

$$-\frac{d}{dx}\left(EA\frac{du}{dx}\right) - f = 0 \quad ; \quad 0 \le x \le L \tag{2.1}$$

where E and A are the modulus and cross-section area of the rod, respectively, and f is the distributed force. The boundary conditions for the case illustrated in Figure 2.1 are

$$u(0) = 0$$

$$\left[\left(EA\frac{du}{dx}\right)\right]_{x=L} = P \tag{2.2}$$

As it is customary in mechanics of materials textbooks, the real rod shown in Figure 2.1(a) is mathematically modeled as a line in Figure 2.1(b). The rod occupies the domain $[0, L]$ along the real axis x.

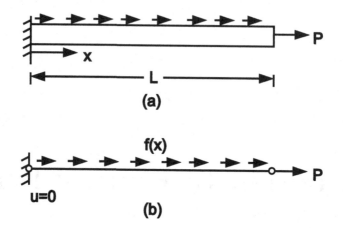

Fig. 2.1: Physical and mathematical (idealization) model.

2.1.1   Discretization

The next step is to divide the domain into discrete elements, as shown in Figure 2.2.

2.1.2   Element Equations

To derive the element equations, an integral form of the ordinary differential equation is used, which is obtained by integrating the product of the ODE times a weight function v, as follows

$$0 = \int_{x_A}^{x_B} v \left[-\frac{d}{dx} \left(EA \frac{du}{dx} \right) - f \right] dx \qquad (2.3)$$

This is called a weak form because the solution $u(x)$ does not have to satisfy the ODE (2.1) for all and every one of the infinite values of x in $[0, L]$, in a strong sense. Instead, the solution $u(x)$ only has to satisfy the ODE (2.3) in a weighted average sense. It is therefore easier to find a weak solution than a strong one. Although for the case of a rod, the strong (exact) solution is known, most problems of composite mechanics do not have exact solutions. The governing equation is obtained by integrating (2.3) by parts, as follows

$$0 = \int_{x_A}^{x_B} EA \frac{dv}{dx} \frac{du}{dx} dx - \int_{x_A}^{x_B} vf dx - \left[v \left(EA \frac{du}{dx} \right) \right]_{x_A}^{x_B} \qquad (2.4)$$

where $v(x)$ is a weight function, which is usually set equal to the primary variable $u(x)$. From the boundary term, it is concluded that

– specifying $v(x)$ at x_A or x_B is an *essential* boundary condition.

– specifying $\left(EA \frac{du}{dx} \right)$ at either end is a *natural* boundary condition.

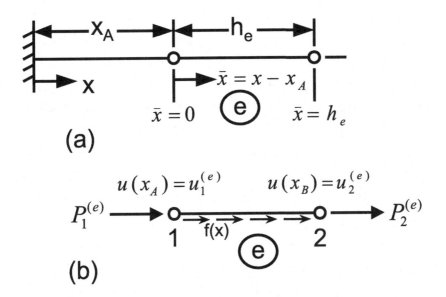

Fig. 2.2: Discretization into elements.

While $u(x)$ is the *primary variable*, $\left(EA\frac{du}{dx}\right) = EA\epsilon_x = A\sigma_x$ is the *secondary variable*. Let

$$u(x_A) = u_1^e$$
$$u(x_B) = u_2^e$$
$$-\left[\left(EA\frac{du}{dx}\right)\right]_{x_A} = P_1^e$$
$$\left[\left(EA\frac{du}{dx}\right)\right]_{x_B} = P_2^e \qquad (2.5)$$

Then, the governing equation becomes

$$0 = \int_{x_A}^{x_B}\left(EA\frac{dv}{dx}\frac{du}{dx} - vf\right)dx - P_1^e v(x_A) - P_2^e v(x_B) = B(v,u) - l(v) \qquad (2.6)$$

with

$$B(u,v) = \int_{x_A}^{x_B} EA\frac{dv}{dx}\frac{du}{dx}dx$$
$$l(v) = \int_{x_A}^{x_B} vfdx + P_1^e v(x_A) + P_2^e v(x_B) \qquad (2.7)$$

2.1.3 Approximation over an Element

Now, the unknown $u(x)$ is approximated as a linear combination (series expansion) of known functions $N_i^e(x)$ and unknown coefficients a_j^e, as

$$u_e(x) = \sum_{j=1}^{n} a_j^e N_j^e(x)$$

where a_j^e are the coefficients to be found and $N_j^e(x)$ are the interpolation functions. For the weight function $v(x)$, the Ritz method can be used [6], in which $v(x) = N_j^e(x)$. Substituting in the governing equation (2.6) we get

$$\sum_{j=1}^{n} \left(\int_{x_A}^{x_B} EA \frac{dN_i^e}{dx} \frac{dN_j^e}{dx} dx \right) a_j^e = \int_{x_A}^{x_B} N_i^e f dx + P_1^e N_i^e(x_A) + P_2^e N_i^e(x_B) \quad (2.8)$$

which can be written as

$$\sum_{j=1}^{n} K_{ij}^e a_j^e = F_i^e \quad (2.9)$$

or in matrix form

$$[K^e]\{a^e\} = \{F^e\} \quad (2.10)$$

where $[K^e]$ is the element stiffness matrix, $\{F^e\}$ is the element vector equivalent force, and $\{a^e\}$ are the element unknown parameters.

2.1.4   Interpolation Functions

Although any complete set of linearly independent functions could be used as interpolation functions, it is convenient to choose the function in such a way that the unknown coefficients represent the nodal displacements, that is $a_i = u_i$. For a two-node element spanning the interval $x_e \leq x \leq x_{e+1}$, the following linear interpolation functions (Figure 2.3) can be used

$$N_1^e = \frac{x_{e+1} - x}{h_e}$$

$$N_2^e = \frac{x - x_e}{h_e} \quad (2.11)$$

where $h_e = x_{e+1} - x_e$ is the element length. These interpolation functions satisfy the following conditions

$$N_i^e(x_j) = \begin{cases} 0 & if \quad i \neq j \\ 1 & if \quad i = j \end{cases} \quad (2.12)$$

$$\sum_{i=1}^{2} N_i^e(x) = 1 \quad (2.13)$$

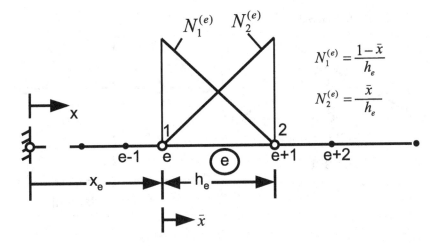

Fig. 2.3: Linear interpolation functions for a two-node element rod.

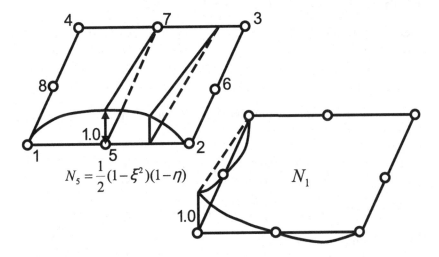

Fig. 2.4: 2D interpolation functions.

which guarantees that the unknown coefficients represent the nodal displacements, i.e., $a_i = u_i$.

Many other interpolation functions can be used, each one with some advantages and disadvantages. The interpolation functions are intimately related to the number of nodes in the element. Figure 2.4 illustrates the shape of interpolation functions N_1 and N_5 (corresponding to nodes 1 and 5) in an eight-node shell element.

Broadly speaking, more nodes per element imply more accuracy and less need for a fine mesh, but also imply higher cost in terms of computer time. Figure 2.5 illustrates how the approximate solution converges to the exact one as the number of elements increases from 2 to 4 or as the number of nodes in the element increases from 2 for the linear element to 3 for the quadratic element.

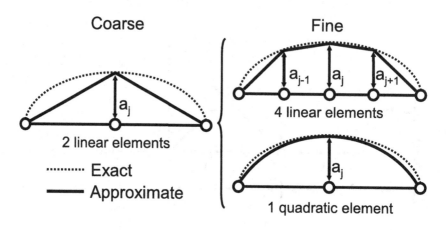

Fig. 2.5: Discretization error.

2.1.5   Element Equations for a Specific Problem

With interpolation functions that satisfy the conditions in (2.12-2.13), it is possible to rewrite (2.10) as

$$[K^e]\{u^e\} = \{F^e\} \tag{2.14}$$

where $\{u^e\}$ are the nodal displacements, $[K^e]$ is the element stiffness matrix given by

$$[K^e] = \begin{bmatrix} \int_{x_A}^{x_B} EA \dfrac{dN_1^e}{dx} \dfrac{dN_1^e}{dx} dx & \int_{x_A}^{x_B} EA \dfrac{dN_1^e}{dx} \dfrac{dN_2^e}{dx} dx \\ \int_{x_A}^{x_B} EA \dfrac{dN_2^e}{dx} \dfrac{dN_1^e}{dx} dx & \int_{x_A}^{x_B} EA \dfrac{dN_2^e}{dx} \dfrac{dN_2^e}{dx} dx \end{bmatrix} \tag{2.15}$$

and $\{F^e\}$ is the element force vector

$$\{F_i^e\} = \left\{ \begin{array}{c} \int_{x_A}^{x_B} N_1^e f dx + P_1^e \\ \int_{x_A}^{x_B} N_2^e f dx + P_2^e \end{array} \right\} \tag{2.16}$$

For a two-node rod element number e, the constant cross-section area A_e, the element length h_e, and the modulus E are fixed. These values define the tensile-compression element stiffness as

$$k^e = \frac{EA_e}{h_e} \tag{2.17}$$

The external loads on the element are the distributed force f_e, the force at end number 1, P_1^e, and the force at end number 2, P_2^e. Using these values, the linear interpolation functions (2.11), as well as (2.15) and (2.16), the element matrix stiffness and the equivalent nodal forces become

$$[K^e] = \begin{bmatrix} k^e & -k^e \\ -k^e & k^e \end{bmatrix} = \frac{EA_e}{h_e} \begin{bmatrix} 1 & -1 \\ -1 & 1 \end{bmatrix} \tag{2.18}$$

Fig. 2.6: Connectivity between three two-node elements.

$$\{F^e\} = \frac{f_e h_e}{2} \left\{ \begin{array}{c} 1 \\ 1 \end{array} \right\} + \left\{ \begin{array}{c} P_1^e \\ P_2^e \end{array} \right\} \tag{2.19}$$

2.1.6 Assembly of Element Equations

The element unknown parameters correspond to displacements at the element nodes. Since a node must have the same displacement on both adjacent elements, the value is unique. For example, using the connectivity of elements shown in Figure 2.6, unique labels are assigned to the displacements, using capital letters. While a superscript denotes element number, a subscript indicates node number, as follows

$$u_1^1 = U_1$$
$$u_2^1 = U_2 = u_1^2$$
$$u_1^2 = U_3 = u_1^3$$
$$u_2^3 = U_4 \tag{2.20}$$

Now, the element equations can be assembled into the global system. First, the contribution of element #1 is

$$\begin{bmatrix} k^1 & -k^1 & 0 & 0 \\ -k^1 & k^1 & 0 & 0 \\ 0 & 0 & 0 & 0 \\ 0 & 0 & 0 & 0 \end{bmatrix} \left\{ \begin{array}{c} U_1 \\ U_2 \\ U_3 \\ U_4 \end{array} \right\} = \left\{ \begin{array}{c} f_1 h_1/2 \\ f_1 h_1/2 \\ 0 \\ 0 \end{array} \right\} + \left\{ \begin{array}{c} P_1^1 \\ P_2^1 \\ 0 \\ 0 \end{array} \right\} \tag{2.21}$$

Add the contribution of element #2, as follows

$$\begin{bmatrix} k^1 & -k^1 & 0 & 0 \\ -k^1 & k^1+k^2 & -k^2 & 0 \\ 0 & -k^2 & k^2 & 0 \\ 0 & 0 & 0 & 0 \end{bmatrix} \left\{ \begin{array}{c} U_1 \\ U_2 \\ U_3 \\ U_4 \end{array} \right\} = \left\{ \begin{array}{c} f_1 h_1/2 \\ f_1 h_1/2 + f_2 h_2/2 \\ f_2 h_2/2 \\ 0 \end{array} \right\} + \left\{ \begin{array}{c} P_1^1 \\ P_2^1 + P_1^2 \\ P_2^2 \\ 0 \end{array} \right\} \tag{2.22}$$

Finally, add element #3 to obtain the fully assembled system, as follows

$$\begin{bmatrix} k^1 & -k^1 & 0 & 0 \\ -k^1 & k^1+k^2 & -k^2 & 0 \\ 0 & -k^2 & k^2+k^3 & -k^3 \\ 0 & 0 & -k^3 & k^3 \end{bmatrix} \left\{ \begin{array}{c} U_1 \\ U_2 \\ U_3 \\ U_4 \end{array} \right\} = \frac{1}{2} \left\{ \begin{array}{c} f_1 h_1 \\ f_1 h_1 + f_2 h_2 \\ f_2 h_2 + f_3 h_3 \\ f_3 h_3 \end{array} \right\} + \left\{ \begin{array}{c} P_1^1 \\ P_2^1 + P_1^2 \\ P_2^2 + P_1^3 \\ P_2^3 \end{array} \right\} \tag{2.23}$$

2.1.7   Boundary Conditions

By equilibrium (see Figure 2.2), the internal loads cancel whenever two elements share a node, or

$$P_2^1 + P_1^2 = 0$$
$$P_2^2 + P_1^3 = 0 \tag{2.24}$$

The remaining P_1^1 and P_3^2 are the forces at the end of the rod. If either end of the rod is fixed, then the displacement must be set to zero at that end. Say the end at $x = 0$ is fixed, then $U_1 = 0$. If the end at $x = L$ is free, then P_3^2 must be specified, since $U_4 \neq 0$. If it is not specified, then it is assumed that the force is zero.

2.1.8   Solution of the Equations

Since $U_1 = 0$, eliminating the first row and column of the stiffness matrix, a 3×3 system of algebraic equations is obtained, and solved for three unknowns: U_2, U_3, U_4. Once a solution for U_2 is found, the reaction P_1^1 is computed from the first equation of (2.23), as follows

$$-k^1 U_2 = \frac{f_1 h_1}{2} + P_1^1 \tag{2.25}$$

2.1.9   Solution Inside the Elements

Now, the solution U_i at four points along the rod is available. Next, the solution at any location x can be computed by interpolating with the interpolation functions, as follows

$$U^e(x) = \sum_{j=1}^{2} U_j^e N_j^e(x) \tag{2.26}$$

or

$$u(x) = \begin{cases} U_1 N_1^1(x) + U_2 N_2^1(x) & if \quad 0 \leq x \leq h_1 \\ U_2 N_1^2(x) + U_3 N_2^2(x) & if \quad h_1 \leq x \leq h_1 + h_2 \\ U_3 N_1^3(x) + U_4 N_2^3(x) & if \quad h_1 + h_2 \leq x \leq h_1 + h_2 + h_3 \end{cases} \tag{2.27}$$

2.1.10   Derived Results

Strains

Strains are computed using (1.5) directly from the known displacements inside the element. For example,

$$\epsilon_x = \frac{du}{dx} = \sum_{j=1}^{2} U_j^e \frac{dN_j^e}{dx} \tag{2.28}$$

Note that if $N_j^e(x)$ are linear functions, the strains are constant over the element. In general, the quality of strains is one order of magnitude poorer than the primary variable (displacements).

Stresses

Stress values are usually computed from strains through the constitutive equations. In this example, with 1D stress-strain behavior

$$\sigma_x = E\ \epsilon_x \tag{2.29}$$

Note that the quality of stresses is the same as that of the strains.

2.2   General Finite Element Procedure

The derivation of element equations, assembly, and solution for any type of elements is similar to that of the 1D rod element described in Section 2.1, with the exception that the principle of virtual work (PVW, 1.16) is used instead of the governing equation (2.1). The PVW provides a weak form similar to that in (2.4). Expanding (1.16) for a full 3D state of deformation, the internal virtual work is

$$\begin{aligned}
\delta W_I &= \int (\sigma_{xx}\delta\epsilon_{xx} + \sigma_{yy}\delta\epsilon_{yy} + \sigma_{zz}\delta\epsilon_{zz} + \sigma_{yz}\delta\gamma_{yz} + \sigma_{xz}\delta\gamma_{xz} + \sigma_{xy}\delta\gamma_{xy})\,dV \\
&= \int_V \underline{\sigma}^T \underline{\delta\epsilon}\,dV
\end{aligned} \tag{2.30}$$

where

$$\begin{aligned}
\underline{\sigma}^T &= \{\sigma_{xx}, \sigma_{yy}, \sigma_{zz}, \sigma_{yz}, \sigma_{xz}, \sigma_{xy}\} \\
\underline{\delta\epsilon}^T &= \{\delta\epsilon_{xx}, \delta\epsilon_{yy}, \delta\epsilon_{zz}, \delta\gamma_{yz}, \delta\gamma_{xz}, \delta\gamma_{xy}\}
\end{aligned} \tag{2.31}$$

Next, the external work is

$$\delta W_E = \int_V \underline{f}^T \underline{\delta u}\,dV + \int_S \underline{t}^T \underline{\delta u}\,dS \tag{2.32}$$

where the volume forces per unit volume and surface forces per unit area are

$$\begin{aligned}
\underline{f}^T &= \{f_x, f_y, f_z\} \\
\underline{t}^T &= \{t_x, t_y, t_z\}
\end{aligned} \tag{2.33}$$

Here, the underlines (_) denote a 1D array, not necessarily a vector. For example, $\underline{u}$ is a vector but $\underline{\sigma}$ are the six components of stress arranged in a six-element array. The virtual strains are those that would be produced by virtual displacements $\underline{\delta u}(\underline{x})$.

Therefore, virtual strains are computed from virtual displacements using the strain-displacement equations (1.5). In matrix notation

$$\underline{\epsilon} = \underline{\underline{\partial}}\, \underline{u}$$
$$\underline{\delta\epsilon} = \underline{\underline{\partial}}\, \underline{\delta u} \tag{2.34}$$

where

$$\underline{\underline{\partial}} = \begin{bmatrix} \dfrac{\partial}{\partial x} & 0 & 0 & \dfrac{\partial}{\partial y} & 0 & \dfrac{\partial}{\partial z} \\[2mm] 0 & \dfrac{\partial}{\partial y} & 0 & \dfrac{\partial}{\partial x} & \dfrac{\partial}{\partial z} & 0 \\[2mm] 0 & 0 & \dfrac{\partial}{\partial z} & 0 & \dfrac{\partial}{\partial y} & \dfrac{\partial}{\partial x} \end{bmatrix}^{T} \tag{2.35}$$

Then, the PVW is written in matrix notation as

$$\int_V \underline{\sigma}^T \underline{\underline{\partial}}\, \underline{\delta u}\, dV = \int_V \underline{f}^T \underline{\delta u}\, dV + \int_S \underline{t}^T \underline{\delta u}\, dS \tag{2.36}$$

The integrals over volume V and surface S of the body can be broken element by element over m elements, as

$$\sum_{e=1}^{m} \left[\int_{V_e} \underline{\sigma}^T \underline{\underline{\partial}}\, \underline{\delta u}\, dV \right] = \sum_{e=1}^{m} \left[\int_{V_e} \underline{f}^T \underline{\delta u}\, dV + \int_{S_e} \underline{t}^T \underline{\delta u}\, dS \right] \tag{2.37}$$

Whenever two elements share a surface, the contributions of the second integral cancel out, just as the internal loads canceled in Section 2.1.7. The stress components are given by the constitutive equations. For a linear material

$$\underline{\sigma} = \underline{\underline{C}}\, \underline{\epsilon} \tag{2.38}$$

with $\underline{\underline{C}}$ given by (1.68). The internal virtual work over each element becomes

$$\delta W_I^e = \int_{V_e} \underline{\sigma}^T \underline{\delta\epsilon}\, dV = \int_{V_e} \underline{\epsilon}^T \underline{\underline{C}}\, \underline{\delta\epsilon}\, dV \tag{2.39}$$

The expansion of the displacements can be written in matrix form as

$$\underline{u} = \underline{\underline{N}}\, \underline{a} \tag{2.40}$$

where $\underline{\underline{N}}$ contains the element interpolation functions and $\underline{a}$ the nodal displacements of the element, just as in Section 2.1.4. Therefore, the strains are

$$\underline{\epsilon} = \underline{\underline{\partial}}\, \underline{u} = \underline{\underline{\partial}}\, \underline{\underline{N}}\, \underline{a} = \underline{\underline{B}}\, \underline{a} \tag{2.41}$$

where $\underline{\underline{B}} = \underline{\underline{\partial}}\, \underline{\underline{N}}$ is the strain-displacement matrix. Now, the discretized form of the internal virtual work over an element can be computed as

$$\delta W_I^e = \int_{V_e} \underline{a}^T \underline{\underline{B}}^T \underline{\underline{C}} \, \underline{\underline{B}} \, \delta \underline{a} \, dV = \underline{a}^T \int_{V_e} \underline{\underline{B}}^T \underline{\underline{C}} \, \underline{\underline{B}} \, dV \delta \underline{a} = \underline{a}^T \underline{\underline{K}}^e \, \delta \underline{a} \tag{2.42}$$

where the element stiffness matrix K^e is

$$\underline{\underline{K}}^e = \int_{V_e} \underline{\underline{B}}^T \underline{\underline{C}} \, \underline{\underline{B}} \, dV \tag{2.43}$$

The external virtual work becomes

$$\delta W_E^e = \int_{V_e} \underline{f}^T \delta \underline{u} \, dV + \int_{S_e} \underline{t}^T \delta \underline{u} \, dS$$

$$= \left(\int_{V_e} \underline{f}^T \underline{N} \, dV + \int_{S_e} \underline{t}^T \underline{N} \, dS \right) \delta \underline{a} = (\underline{P}^e)^T \delta \underline{a} \tag{2.44}$$

where the element force vector is

$$\underline{P}^e = \int_{V_e} \underline{N}^T \underline{f} \, dV + \int_{S_e} \underline{N}^T \underline{t} \, dS \tag{2.45}$$

The integrals over the element volume V_e and element surface S_e are usually evaluated numerically by the Gauss integration procedure. For the volume integral, such a procedure needs evaluation of the integrand at a few points inside the volume. Such points, which are called Gauss points, are important for two reasons. First, the constitutive matrix C is evaluated at those locations. Second, the most accurate values of strains (and stresses) are obtained at those locations too.

The assembly of element equations δW_I^e and δW_E^e into the PVW for the whole body is done similarly to the process in Section 2.1.6. Obviously the process is more complicated than for rod elements. The details of such a process and its computer programming are part of finite element technology, which is outside the scope of this textbook. Eventually all the element stiffness matrices K^e and element force vectors P^e are assembled into a global system for the whole body

$$\underline{\underline{K}} \, \underline{a} = \underline{P} \tag{2.46}$$

Next, boundary conditions are applied on the system (2.46) in a systematic way resembling the procedure in Section 2.1.6. Next, the algebraic system of equations (2.46) is solved to find the nodal displacement array $\underline{a}$ over the whole body. Since the nodal displacements results for every element can be found somewhere in $\underline{a}$, it is possible to go back to (2.34) and to (2.38) to compute the strains and stresses anywhere inside the elements.

Example 2.1 *Compute the element stiffness matrix (2.43) and the equivalent force vector (2.45) of a rod discretized with one element. Use linear interpolation functions such as (2.11). Compare the result with (2.18-2.19).*

Solution to Example 2.1 *Let A_e be the transverse area of the rod and h_e the element length, with $x_e = 0$ and $x_{e+1} = h_e$. Substituting these values in the linear interpolation functions from equation (2.11), the interpolation functions arrays are obtained, as follows*

$$\underline{N}^T = \begin{bmatrix} N_1^e \\ N_2^e \end{bmatrix} = \begin{bmatrix} \dfrac{x_{e+1} - x}{h_e} \\ \dfrac{x - x_e}{h_e} \end{bmatrix} = \begin{bmatrix} 1 - x/h_e \\ x/h_e \end{bmatrix}$$

The strain-displacement array is obtained as

$$\underline{\underline{B}}^T = \underline{\partial}\ \underline{N}^T = \begin{bmatrix} \partial N_1^e/\partial x \\ \partial N_2^e/\partial x \end{bmatrix} = \begin{bmatrix} -1/h_e \\ 1/h_e \end{bmatrix}$$

The rod element has a 1D strain-stress state with linear elastic behavior. Therefore

$$\underline{C} = E$$

Then, using equation (2.43) we can write

$$\underline{\underline{K}}^e = \int_{V_e} \underline{\underline{B}}^T \underline{\underline{C}}\ \underline{\underline{B}}\ dV = \int_0^{h_e} \begin{bmatrix} -1/h_e & 1/h_e \end{bmatrix} E \begin{bmatrix} -1/h_e \\ 1/h_e \end{bmatrix} A_e dx$$

The element stiffness matrix is obtained by integration

$$[K^e] = \frac{EA_e}{h_e} \begin{bmatrix} 1 & -1 \\ -1 & 1 \end{bmatrix}$$

To calculate the equivalent vector force, f_e is defined as the distributed force on element, P_1^e is the force at end $x = 0$, and P_2^e is the force at end $x = h_e$. Substituting into equation (2.45) we obtain

$$\underline{P}^e = \int_{V_e} \underline{N}^T \underline{f}\ dV + \int_{S_e} \underline{N}^T \underline{t}\ dS = \int_0^{h_e} \begin{bmatrix} 1 - x/h_e \\ x/h_e \end{bmatrix} f_e dx + \begin{bmatrix} P_1^e \\ P_2^e \end{bmatrix}$$

The element equivalent force vector is obtained by integration

$$\underline{P}^e = \frac{f_e h_e}{2} \begin{bmatrix} 1 \\ 1 \end{bmatrix} + \begin{bmatrix} P_1^e \\ P_2^e \end{bmatrix}$$

2.3 Solid Modeling, Analysis, and Visualization

Many commercial programs exist with finite element analysis (FEA) capabilities for different engineering disciplines. They help solve a variety of problems from simple linear static analysis to nonlinear transient analysis. A few of these codes, such as ANSYS and Abaqus, have special capabilities to analyze composite materials and they accept custom, user-programmed constitutive equations and element formulations. Since these software packages not only provide analysis tools, geometric modeling, and visualization of results, but can also be integrated into the larger design, production, and product life-cycle process, they are often called complete analysis environments or computer aided engineering (CAE) systems.

Modern FEA software are commonly organized into three blocks: the preprocessor, the processor, and the post-processor. In the preprocessor, the model is built defining the geometry, material properties, and element type. Also, loads and boundary conditions are entered in the preprocessor, but they may also be entered during the solution phase. With this information, the processor can compute the stiffness matrix and the force vector. Next, the algebraic equations (2.46) are solved and the solution is obtained in the form of displacement values. In the last block–the post-processor–derived results, such as stress, strain, and failure ratios, are computed. The solution can be reviewed using graphic tools.

In the remainder of this chapter, a general description of procedures and specific steps for basic FEA are presented using examples executed with Abaqus [8]. Although the emphasis of this textbook is on mechanics of composite materials, concepts are illustrated with examples that are solved using Abaqus. Solutions to similar examples are available in [9], using ANSYS [10].

The first requirement of the model is the geometry. Then, material properties are given for the various parts that make up the geometry. Next, loads and boundary conditions are applied on the geometry. Next, the geometry is discretized into elements, which are defined in terms of the nodes and element connectivity. The element type is chosen to represent the type of problem to be solved. Next, the model is solved. Finally, derived results are computed and visualized.

2.3.1 Interacting with Abaqus

There are three ways for the analyst to interact with Abaqus: the input file, the CAE graphical user interface (GUI), and Python scripting. For high-end modeling applications, it is important that the analyst understand all three of them.

The Abaqus solver only understands the input file (*.inp), which is a text file that can be written and edited by the analyst, but nowadays it is almost exclusively written by CAE.

CAE is a powerful GUI used to generate the model to be solved by the Abaqus solver. In MS-Windows, we can find an App called **Abaqus CAE**.[1] Using Figure 2.7, locate **Module** (third menu bar from the top) and set it to **Job**. Then locate **Job** on the top menu bar (which is dynamic), and from **Job** select **Job Manager**. The following CAE mouse clicks illustrate the procedure

```
Module: Job
Menu:  Job Manager
    Create, Continue, OK # creates a new Job
    Write Input, Yes, Dismiss # writes the .imp file
```

The procedure we just completed generates the following *empty* input file

```
*Heading
** Job name: Job-1 Model name: Model-1
** Generated by: Abaqus CAE 2020
*Preprint, echo=NO, model=NO, history=NO, contact=NO
**
```

---

[1]Click Windows Start, then "Dassault Systemes SIMULIA Established Products 2020".

```
** PARTS
**
**
** ASSEMBLY
**
*Assembly, name=Assembly
*End Assembly
```

In addition to the `Job-1.inp` file, the file `abaqus1.rec` is also saved. Copy `abaqus1.rec` as `abaqus1.py`. Now, the same input file can be created in your *Work Directory* by using the following Python script, `abaqus1.py`

```
from part import *
from material import *
from section import *
from optimization import *
from assembly import *
from step import *
from interaction import *
from load import *
from mesh import *
from job import *
from sketch import *
from visualization import *
from connectorBehavior import *
mdb.Job(activateLoadBalancing=False, atTime=None, contactPrint=OFF,
    description='', echoPrint=OFF, explicitPrecision=SINGLE,
    getMemoryFromAnalysis=True, historyPrint=OFF, memory=90, memoryUnits=
    PERCENTAGE, model='Model-1', modelPrint=OFF, multiprocessingMode=DEFAULT,
    name='Job-1', nodalOutputPrecision=SINGLE, numCpus=1, numDomains=1,
    numGPUs=0, parallelizationMethodExplicit=DOMAIN, queue=None, resultsFormat=
    ODB, scratch='', type=ANALYSIS, userSubroutine='', waitHours=0,
    waitMinutes=0)
mdb.jobs['Job-1'].writeInput(consistencyChecking=OFF)
```

To run the script, open CAE, then

```
Menu: File
    Run Script [abaqus1.py]
```

In this chapter, you will become acquainted with CAE, and in later chapters with scripting and the input file. Extensive use of scripting is presented in Chapter 12.

2.3.2   Model Geometry

The model geometry is obtained by specifying all nodes, their position, and the element connectivity. The connectivity information allows the program to assemble the element stiffness matrix and the element equivalent force vector to obtain the global equilibrium equations, as shown in Section 2.1.6.

There are two ways to generate the model. The first is to manually create a mesh (Section 2.3.3). The second is to use solid modeling (Section 2.3.4), and then mesh the solid to get the node and element distribution.

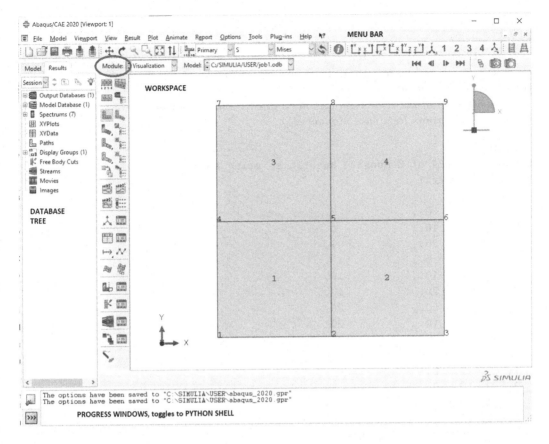

Fig. 2.7: Square mesh in Example 2.2.

2.3.3 Manual Meshing

Manual mesh generation was the only method available before solid modeling became widespread among commercial packages. It is still the only option with some older and custom software, although in those cases it is always possible to use a general purpose, solid modeling preprocessor to generate the mesh. With manual meshing, the user creates nodes, then connects the nodes into elements. Afterward, the user applies boundary conditions and loads directly on nodes and/or elements. Manual meshing is used in Example 2.2.

Example 2.2 *Use manual meshing to set up a finite element model of a 2D square plate, in a state of plane stress, with dimensions 20×20 mm $\times 4$ mm and mesh it with a 2×2 mesh of continuum, plane stress, square, CPS4R elements in Abaqus. The left edge (nodes 1, 4, 7 in Figure 2.7) should be horizontally restrained. Also, node 1 should be vertically restrained. A uniform surface load of -9.5 N/mm^2 should be applied to the right edge (nodes 3, 6, 9). The material is steel with $E = 210,000$ MPa, $\nu = 0.3$. Run the model from the DOS shell window. Then, use Abaqus CAE to visualize the results.*

Solution to Example 2.2 *The commands listed below define the model geometry by using manual meshing, and they are written into what is called an input file (Ex_2.2.inp). They are available on the website [2, file name: Ex_2.2.inp].*

The *Node *command line is followed by a list of nodes, each line containing the node number, plus x, y, and possibly z coordinates of each node.*

The *Element *command line is followed by a list of elements, each line containing the element number and a sequence of node numbers going counterclockwise (ccw) around the boundary of the element (see Figure 2.7).*

The meaning of the remaining command lines is either obvious or can be researched in the Abaqus Keywords Reference Manual [11].

```
*Heading
Example 2.2 FEA of Composite Materials: using Abaqus
*Part, name=Part1
*Node
   1,          0.,          0.
   2,         10.,          0.
   3,         20.,          0.
   4,          0.,         10.
   5,         10.,         10.
   6,         20.,         10.
   7,          0.,         20.
   8,         10.,         20.
   9,         20.,         20.
*Element, type=CPS4R
   1,   1,   2,   5,   4
   2,   2,   3,   6,   5
   3,   4,   5,   8,   7
   4,   5,   6,   9,   8
*Nset, nset=NamedSet, generate
   1,   9,   1
*Elset, elset=NamedSet, generate
   1,   4,   1
** Section: Section-1
*Solid Section, elset=NamedSet, material=Material-1
4.,
*End Part
*Assembly, name=Assembly
*Instance, name=Part1-1, part=Part1
*End Instance
*Nset, nset=Nodeset1, internal, instance=Part1-1
 1
*Nset, nset=Nodeset2, internal, instance=Part1-1
 4, 7
*Elset, elset=Elset1, internal, instance=Part1-1
   2,   4
*Surface, type=ELEMENT, name=Surface1, internal
Elset1, S2
*End Assembly
*Material, name=Material-1
*Elastic
210000., 0.3
*Step, name=Step-1
*Static
```

```
1., 1., 1e-05, 1.
*Boundary
Nodeset1, 1, 2
Nodeset2, 1, 1
*Dsload
Surface1, P, -9.5
*Restart, write, frequency=0
*Output, field, variable=PRESELECT
*Output, history, variable=PRESELECT
*End Step
```

To run Abaqus on a command line, first you have to get a DOS *shell, as follows: On Windows 10, click* Start *(lower left corner), then type* cmd, *then click* Run *or* Run as Administrator, *then inside the* DOS *shell, do this:*

```
>cd c:\SIMULIA
>mkdir User
>cd User
```

where cd *stands for change directory, and* mkdir *creates a new folder (mkdir stands for 'make directory').*[2]

Next, copy the file Ex_2.2.inp *from [2] to the folder* c:\SIMULIA\User, *or type the commands listed above using Notepad or your favorite text editor.*[3] *The command line to run Abaqus is:*

```
>Abaqus job=job1 input=Ex_2.2
```

Now to start CAE on Windows 10, click Start, *then "Dassault Systemes SIMULIA Established Products 2020", then* Abaqus CAE. *Once CAE starts, close the pop-up window. Then, on the Menu bar at the top of the screen*[4] *click* File, Open, *navigate to the* c:\Simulia\User\ *folder, change the File Filter to* *.odb, *and select* Job1.odb. *At this point you should see the mesh, as shown in Figure 2.7.*

To display element and node numbers do this: On the top Menu, Options, Common, Labels *(see Figure 2.8), checkmark* Show element labels *and* Show node labels. *This works only in Module: Visualization. Also in the same window, to increase the font of the labels do this:* Set Font. *For future reference, the instructions given in this paragraph are summarized below (see notation in Table 2.1).*

```
Menu: File,
    Open [c:\Simulia\User\], File Filter [*.odb], File Name [Job1.odb]
Menu: Options,
    Common, Labels,
        # checkmark [Show element labels]
        # checkmark [Show node labels]
        Set Font for All Model Labels, Size [24], OK
    OK
```

---

[2]If your system administrator has restricted write access to c:\SIMULIA then make your Work Directory at some other unrestricted location such as c:\TEMP or see Appendix E for additional details.

[3]We recommend Notepad++ available free at http://notepad-plus-plus.org/.

[4]Also called display, monitor, or window.

Table 2.1: Pseudo-code notation used to list the acronyms and commands[a] used to describe the user interaction with Abaqus CAE.

=	equal
==	identical
(+)	Expand item on left menu tree
(−)	Collapse item on left menu tree
[]	user input such as file name, value, etc.
OK	name of button used often to complete a series of commands
X	name of button used often to cancel a series of commands
Cont	Continue, is the name of a button which means OK
Dis	Dismiss, is the name of a button that means Cancel
Enter	is the Enter key on the keyboard
#	comment line
# pick	pick object as requested in the window at the bottom of the WS
>	DOS prompt
>>>	Python prompt
Esc	Escape key
ctrl-z	undo
ctrl-y	redo
F6	auto fit to workspace
crtl-click	unselect from selection
shift-click	add to selection
mouse wheel	zoom in/out
c.s.	coordinate system
BC	boundary conditions
WD	work directory
WS	workspace, all graphics shown here
Info	Progress Window, below the WS, toggles to >>>
>>>	Python shell, toggles to `Info`

a *Commands* are actions, instructions, steps, mouse clicks, selections, input data, and all other operations performed by the user on the GUI.

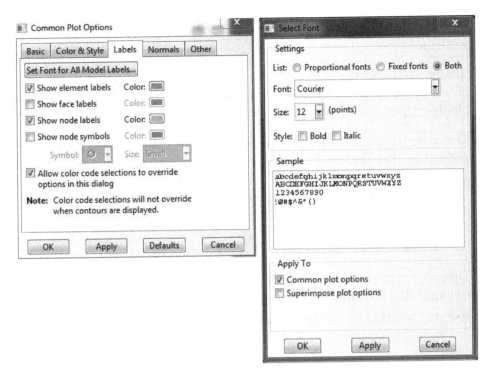

Fig. 2.8: To display element and node numbers in a larger font size.

2.3.4 Solid Modeling

With solid modeling, the user creates a geometric representation of the geometry using solid model constructs, such as volumes, areas, lines, and points. Boundary conditions, loads, and material properties can be assigned to parts of the solid model before meshing. In this way, re-meshing can be done without losing, or having to remove, the loads and boundary conditions. The models are meshed just prior to the solution. Solid modeling is used in Example 2.3.

Abaqus CAE has all the capabilities to construct the solid model, mesh, solve the problem, and visualize the results within the CAE environment. Other solid modeling software, such as SolidWorks [12], can be used to generate the solid model, then import into CAE for adding the boundary conditions and loads, invoking the solver, and finally visualizing the results. A brief description of the CAE GUI is described next. The notation used to describe the steps necessary to construct a model in CAE are listed in Table 2.1.

2.3.5 The CAE Window

The CAE window[5] (see Figure 2.7) is divided in regions, as follows

- Menu bar: The top menu bar (horizontally across the top of the window) is dynamic, i.e., it changes depending on what we are doing. Specifically,

---

[5]At this time, the user should have Abaqus CAE open on a computer.

it changes according to the selection in the drop-down list `Module`, which is explained below.

– Toolbars: Under the menu bar, we have dynamic toolbars. You can undock them, one at a time, to learn their names. They are dynamic because they change depending on what `Module` is selected.

– Database Trees: On the left you see two tabs, Model and Results; one for each of the two database trees. All the items in the trees can be right-clicked to access actions to modify the information stored in each item. The trees correspond to the database files .mdb and .odb, as follows

  – Model: .mdb
  – Results: .odb

– Drop-down menus: To the right of Model and Results tabs, we have several dynamic drop-down menus, such as Module, Model, and Part.[6]

`Module:` contains a number of selections, as follows

  – Part
  – Property
  – Assembly
  – Step
  – Interaction
  – Load
  – Mesh
  – Job
  – Visualization
  – Sketch

– Workspace: The workspace covers most of the screen. It is located under the drop-down menus, to the right of the Model/Results tabs.

– On a column between the Model/Result trees and the Workspace we have a series of icons, dynamically linked to the drop-down menus above. Icons have names that show up if you hover the cursor on them for a while. The icons are shortcuts to commands on the menu bar.

– Below the workspace there is a Progress Window that toggles to a Python shell[7] by clicking on the tabs on the left (shown in Figure 2.7). The Python Shell is recognized by the prompt `>>>`

---

[6]Since the drop-down menus are dynamic, only Module: and Model: are shown in Figure 2.7.

[7]A *shell* is a process window that accepts literal (typed) commands. A Python shell accepts Python code, like a DOS shell accepts DOS script, such as `cd`, and so on.

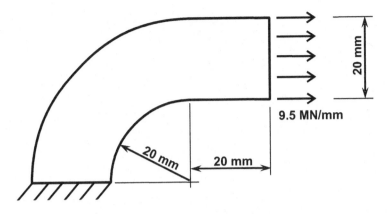

Fig. 2.9: Curved beam in Example 2.3.

Example 2.3 *Generate the geometry shown in Figure 2.9 using solid modeling techniques in Abaqus CAE. Add boundary conditions and loads. Solve the problem and visualize the results. The thickness of the part is 4.0 mm. The material properties are $E = 195,000\ MPa$, $\nu = 0.3$.*

Solution to Example 2.3 *Watch the video in [1].*
*The solution to this example is **distributed among various subsections** that explain the various steps involved in the modeling, solution, and visualization (see text in **italics** within pp. 55 to 76). Although it is possible to follow the example with the aid of the figures, it is best to have Abaqus CAE running on a computer and to execute all the instructions while reading this section.*

To start Abaqus CAE on Windows 10, click Start, *"Dassault Systemes SIMULIA Established Products 2020",* Abaqus CAE.[8] *Once CAE starts, close the pop-up window. Your screen should look more or less like Figure 2.7 without the mesh.*
The first thing you want to do is to make sure the files for this example are saved to a separate folder, so that they don't get mixed up with other examples. The easiest way to do that is to save this example right now, even though it is empty, to a new folder, as follows. On the Menu bar at the top of the screen click File, Save As, *navigate to a new folder of your choice, or create one if you need to; then in the field* File name *enter* Ex_2.3, OK. *These steps are summarized below.*

Menu: File, Save as, New directory [name of new directory],
         Select new directory, [name of file], OK

The first modeling task is to create one or more Parts, which represent the geometry of the model. Since this example is very simple, a single part will suffice, but more complex models will require several parts. There are many ways to define the geometry of a part. In this first example we use Points connected by Arcs and Lines. To work with parts, the Module has to be set to Part. Remember that

---

[8]Productivity tip: Pin the shortcut for Abaqus CAE to the Windows Taskbar. To do this, drag the shortcut to the taskbar.

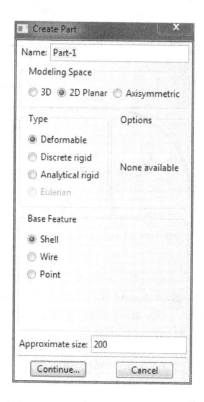

Fig. 2.10: The Create Part dialog box.

- *Module:* refers to the drop-down box immediately left of the Model/Result tabs (see Figure 2.7), and

- *Menu:* refers to the top menu bar running across the top of the CAE window.

In this way, the selections described below will result in opening the dialog box shown in Figure 2.10:

```
Module: Part
    Menu: Part, Create,
    Name[Part-1], 2D, Deformable, Shell, Approx size [200], Cont
    # sketch mode is now active
```

Next, we define Points and join them with Arcs, as summarized below. See Figure 2.11 in conjunction with the step-by-step instructions below.

```
Menu: Add, Point,
    [0,0] # type in the input window below the WS, then [Enter]
    [-20,0]
    [-40,0]
    [0,20]
    [0,40]
    [20,20]
```

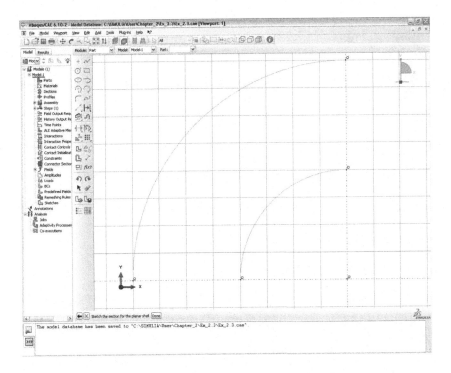

Fig. 2.11: The two arcs are defined.

```
[20,40]
F6    # zoom to fit
Menu: Add, Arc, Center/Endpoints,
    # pick center point (see Figure 2.11)
    # pick point left of center point
    # pick point above of center point, the arc is done
    # again pick center point (see Figure 2.11)
    # pick point left-most of center point
    # pick point above-most of center point, the outer arc is done
    X # to finalize the Add feature (do not click Done)
```

At this point, the workspace should look like Figure 2.11.

Sketch mode *is similar to SolidWorks [12]. Arcs, points, and other geometrical constructs are called* entities. *When adding/editing entities, you must follow the instructions displayed immediately below the WS. The* **X** *situated to the left of the instructions line allows you to finalize the current operation, i.e., the* Add *feature in the set of instructions above.*

Still referring to Figure 2.11, add lines to close the boundary of the figure, and once the figure is closed, the part is created:

```
Menu: Add, Line, Connected,
    # pick two nodes to close bottom left of the part
    X   # to finalize the Line drawing
```

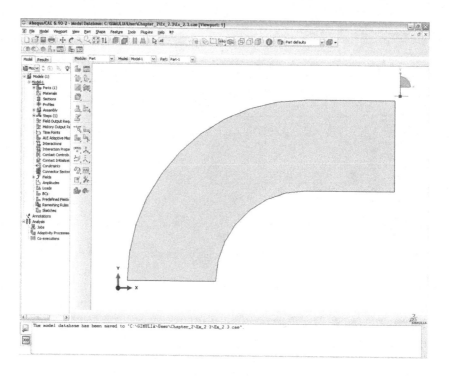

Fig. 2.12: The Part is finalized.

```
Menu: Add, Lines, Connected,
    # pick 4 nodes in sequence to close rectangle at top right
    X   # to finalize the Line drawing
Done    # to finish the sketch, thus creating the part
```

At this point, the workspace should look like Figure 2.12.

2.3.6   Material and Section Properties

Parts must be associated to materials. Depending on the analysis, material properties can be linear (linear elastic analysis) or nonlinear (e.g. damage mechanics analysis), isotropic or orthotropic, constant or temperature-dependent. Entering the correct materials properties is one of the most important aspects of a successful analysis of composite materials. A great deal of attention is devoted to material properties in the rest of the textbook. For now it will suffice to illustrate the process using a linear elastic, isotropic material. For structural analysis, elastic properties must be defined according to Section (1.12). Other mechanical properties, such as strength, density, and thermal expansion coefficients are optional and their definition depends on the objectives of the analysis.

In Abaqus CAE, materials are entered in Module: Property, *Menu:* Material. *The instructions below will bring out a pop-up window as shown in Figure 2.13, where the materials properties are entered, as follows*

```
Module: Property
    Menu: Material, Create,
        Name [Material-1], Mechanical, Elasticity, Elastic,
        Type, Isotropic [195000, 0.3], OK
```

All elements need material properties, but *structural* elements need additional parameters that vary with the type of element. These parameters result from analytical integration of the 3D governing equations while formulating the element. For example, the cross-section area A appears in (2.1) because the 3D partial differential equations have been integrated over the cross-section of the rod to arrive at the ordinary differential equation (2.1).

Beam elements require cross-section area and moments of area. *Laminated shell* elements require the laminate stacking sequence (LSS, see Figure 3.4, p. 102).

Continuum elements are 3D solid elements (see Table 2.2) that do not require additional parameters, only material properties, because the geometry is described by the mesh. However, laminated continuum elements still need the LSS.

Conventional shells and beams are typical *structural* elements. The word *conventional* is used by Abaqus to emphasize shell elements that are not continuum.

Continuum shells are continuum elements with kinematic constraints introduced in order to represent shell behavior [13]. See also, Sections 3.2.1, 3.2.5, and Examples 3.4, 3.7.

Therefore, besides a `Material`, one needs to create a `Section` to provide the additional parameters required by structural elements. In Abaqus, the additional parameters are called *section parameters*; in ANSYS they are called *section constants*.[9] Continuum elements do not require section parameters, but a `Section` has to be created anyway.

In this simple example, section parameters are used to define the state of plane stress for the part shown in Figure 2.9. A complex model may have several different sections. Sections are created in the same Module: Property, as follows

```
Menu: Section, Create,
    # plane stress/strain is considered solid even though
    # pstres needs a thickness
    # pstran does not
    # shells are not solids, they need a thickness
    # beams are not solids, they need area and moments of area
    Name [Section-1], Solid, Homogeneous, Cont
    Material: Material-1,
    Plane stress/strain thickness [4.0], OK
```

The Create Section and Edit Section windows, shown in Figure 2.14, appear in sequence. They are used to define the section.

Finally, we need to tell the software what the section properties are for each part. This is done by assigning Sections to Parts. In this simple model there is only

---

[9]Section constants were called "real constants" in older versions of ANSYS.

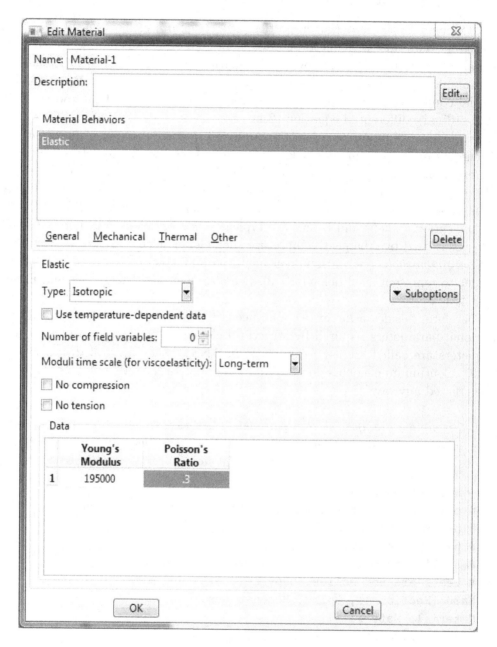

Fig. 2.13: Material properties window.

Table 2.2: Some of the elements available in Abaqus and ANSYS.

Abaqus	ANSYS	Nodes	DOF*	Element Description
Structural elements				
T2D2	LINK180	2	u_X u_Y	line bar/truss, 2D space
T3D2	LINK180	2	u_X u_Y u_Z	line bar/truss, 3D space
-	COMBIN14	2	u_X u_Y u_Z	spring/damper, 3D space
B21	BEAM188	2	u_X u_Y θ_X θ_Y	line beam in 2D space
B31	BEAM188	2	u_X u_Y u_Z θ_X θ_Y θ_Z	line beam in 3D space
CPE4R	PLANE182	4	u_X u_Y	solid quadrilateral in 2D space
CPE8R	PLANE183	8	u_X u_Y	solid quadrilateral in 2D space
S4R	SHELL181	4	u_X u_Y u_Z θ_X θ_Y θ_Z	shell quadrilateral in 3D space (conventional)
S8R	SHELL281	8	u_X u_Y u_Z θ_X θ_Y θ_Z	shell quadrilateral in 3D space (conventional)
S8R5	-	4	u_X u_Y u_Z θ_X θ_Y	thin shell quadrilateral in 3D space (conventional)
Continuum elements				
C3D8	SOLID185	8	u_X u_Y u_Z	solid hexahedra in 3D space
C3D20	SOLID186	20	u_X u_Y u_Z	solid hexahedra in 3D space
Continuum shell elements [13]				
SC6R		6	u_X u_Y u_Z	shell tetra/hexahedra in 3D space (continuum)
SC8R	SOLSH190	8		

(*) DOF: degrees of freedom.

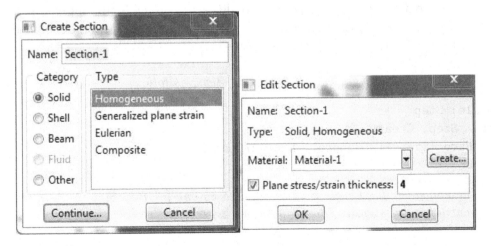

Fig. 2.14: The Create Section and Edit Section windows define the Section.

one part and only one section, so the assignment is trivial, but it has to be done anyway. Note that the assignment is done in the same module, i.e., the Property module. The procedure is illustrated in Figure 2.15, and summarized as follows

```
Menu: Assign, Section
    # pick all (click on part), Done
    OK
```

2.3.7   Assembly

If more than one part exists, assembly is necessary to put the parts together into what is called an *assembly*, which represents the physical object you are trying to analyze.

During assembly it is possible to specify how the mesh is related to the parts. That is, the mesh can be dependent or independent of the part. A dependent mesh is tied to the part. So, if the part is used (i.e., instanced) multiple times in an assembly, all the instances of the part will be meshed identically. An independent mesh means that each instance of the part will have to be meshed independently. The latter provides flexibility to refine the mesh for some instances of the part at the expense of more work when it comes the time to mesh. Since this example has only one part, the assembly process is trivial, as follows (see also Figure 2.16)

```
Module: Assembly
Menu: Instance, Create,
    Independent, OK
```

2.3.8   Solution Steps

Next, the analysis process is normally broken down into several *steps*, each representing different loading and constraint conditions. The minimum number of steps is two: an initial step and at least one additional step. No loads can be applied on the initial step, only boundary conditions.

The current example uses just those two steps, as follows

```
Module: Step
Menu: Step, Create,
    Name [Step-1], General, Static/General, Cont
    OK
```

The dialog box for Create Step is followed by one for Edit Step to complete the data entry for this step, as shown in Figure 2.17.

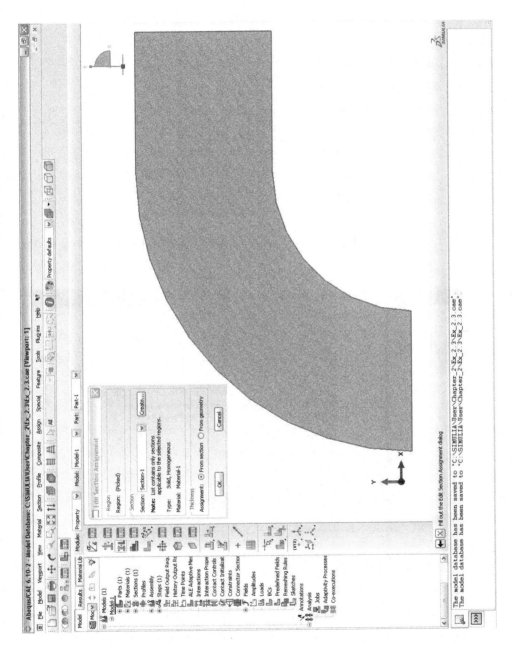

Fig. 2.15: Assigning a section to the part.

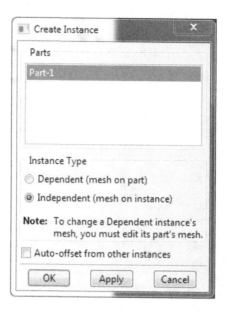

Fig. 2.16: Dialog box to create an instance of a part.

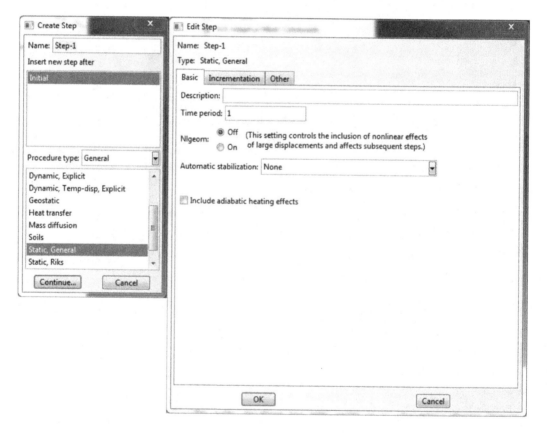

Fig. 2.17: Dialog box to create a Step.

2.3.9 Loads

In structural analysis, loads are defined by forces, pressures, inertial forces (as gravity), and specified displacements, all applied to the model. Specification of different kinds of loads for the FE model is explained in the following sections. The reactions obtained by fixing nodal degrees of freedom (DOF), such as displacements and rotations, are discussed also.

Loads can be applied on nodes by means of concentrated forces and moments, as shown in Example 2.4, p. 76. Also, loads can be distributed over the elements as: surface loads, body loads, inertia loads, or other coupled-field loads (for example, thermal strains). Surface loads are used in Example 2.5, p. 78.

A surface load is a distributed load applied over a surface, for example a pressure. A body load is a volumetric load, for example expansion of material by temperature increase in structural analysis. Inertia loads are those attributable to the inertia of a body, such as gravitational acceleration, angular velocity, and acceleration.

A concentrated load applied on a node is directly added to the force vector (RHS in (2.23)). However, element interpolation functions are used to compute the equivalent forces vector due to distributed loads.

In Abaqus CAE, loads and boundary (support) conditions for the structure are specified with the Loads module, as follows

```
Module: Load
Menu: Load, Create,
    Name [load-1], Mechanical, Pressure, Cont
    # pick the vertical edge on the right, Done
    Magnitude [-9.5], OK
```

The dialog box for Create Load (top left in Figure 2.18) is followed by the selection of the surfaces over which the loads are to be applied (note dialog box below the x-y-z triad in Figure 2.18, then followed by the Edit Load box to enter the values for the load (top right in Figure 2.18).

2.3.10 Boundary Conditions

The boundary conditions (BC) are the known values of the degrees of freedom (DOF) on the boundary. In structural analysis, the DOF are displacements and rotations. With this information, the software knows which values of $\underline{a}$ in (2.46) are known or unknown.

Restrained Displacements and Rotations

In general, a node can have more than one DOF. For example, if the FE model uses beam elements in 2D space, there are three DOF: the horizontal displacement, the vertical displacement, and the rotation around an axis perpendicular to the plane. Restraining different sets of DOF results in different boundary conditions being applied. See Section 3.2.5. In the 2D beam element case, restraining only the

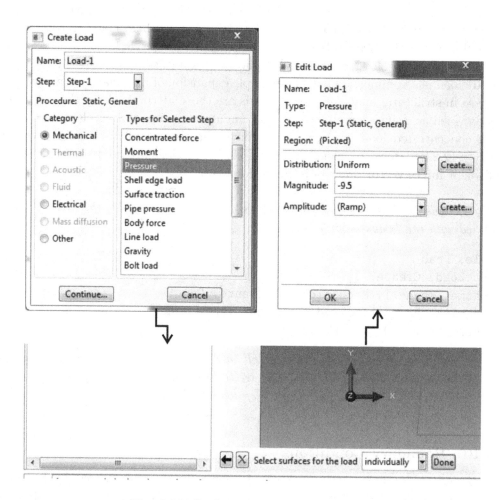

Fig. 2.18: Dialog box to create a Load.

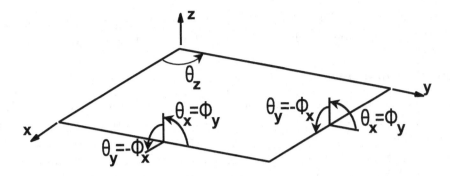

Fig. 2.19: Convention for rotations of a plate or shell.

horizontal and vertical displacements results in a simple support, but restraining all the DOF results in a clamped condition.

Symmetry Conditions

Symmetry conditions can be applied to reduce the size of the model without loss of accuracy. Four types of symmetry must exist concurrently: symmetry of geometry, boundary conditions, material, and loads. Under these conditions the solution will also be symmetric. For example, symmetry with respect to the $y - z$ plane means that the nodes on the symmetry plane have the following constraints

$$u_x = 0 \quad ; \quad \theta_y = 0 \quad ; \quad \theta_z = 0 \qquad (2.47)$$

where u_x is the displacement along the x-direction, θ_y and θ_z are the rotations around the y- and z-axis, respectively (Figure 2.19). Note that the definition of rotations used in shell theory (ϕ_i, see Section 3.1) is different than the usual definition of rotations θ_i that follows the right-hand rule. All rotations in Abaqus[10] are described using right-hand-rule rotations θ_i. Symmetry boundary conditions on nodes in the symmetry plane involve the restriction of DOF translations out-of-plane with respect to the symmetry plane and restriction of the DOF rotations in-plane with respect to the symmetry plane. Symmetry boundary conditions are used in Example 2.5, p. 78.

Antisymmetry Conditions

Antisymmetry conditions are similar to the symmetry conditions. They can be applied when the model exhibits antisymmetry of loads but otherwise the model exhibits symmetry of geometry, symmetry of boundary conditions, and symmetry of material. Antisymmetry boundary conditions on nodes in the antisymmetry plane

---

[10]Quote "Except for axisymmetric elements, the DOF are always referred to as follows1: x-displacement 2: y-displacement 3: z-displacement 4: Rotation about the x-axis, in radians 5: Rotation about the y-axis, in radians 6: Rotation about the z-axis, in radians ... Here the x-, y-, and z-directions coincide with the global X-, Y-, and Z-directions, respectively; however, if a local transformation is defined at a node (see Transformed coordinate systems [8, Section 2.1.5]), they coincide with the local directions defined by the transformation." End quote [8].

involve restriction of the DOF translations in the antisymmetry plane and restriction of the DOF rotations out-of-plane with respect to the antisymmetry plane.

Periodicity Conditions

When the material, load, boundary conditions, and geometry are periodic with period $(x, y, z) = (2a_i, 2b_i, 2c_i)$, only a portion of the structure needs to be modeled, with dimensions $(2a_i, 2b_i, 2c_i)$. The fact that the structure repeats itself periodically means that the solution will also be periodic. Periodicity conditions can be imposed by different means. One possibility involves using constraint equations (CE) between DOF (see Section 6.2) or using Lagrange multipliers.

Interactions

Interactions between regions of a model are explained in Section 2.4.

Boundary Conditions in Abaqus CAE

In Abaqus CAE, the dialog box for Create a Boundary Condition is followed by the Edit Boundary Condition dialog, as shown in Figure 2.20. The dialog box is reached as follows

```
Module: Load
Menu: BC, Create,
    Name [BC-1], Step: Initial, Mechanical, Displacement/Rot, Cont
    # pick the horizontal edge on bottom, Done
    # checkmark U1, U2, UR3, OK
    Step: Step-1  # to see the load and BC together
```

To see the loads together with the BC, one must change the visualized step to Step-1, because on Step: Initial, only the BC are present. To change the visualized step, look for the **Step** *drop-down box to the right of the* **Module** *drop-down menu. Once the change is made, the WS should look like Figure 2.21.*

2.3.11   Meshing and Element Type

Next, the assembly needs to be meshed. There are many ways to mesh a model. For this example we use a very simple approach, called global seeding, as follows

```
Module: Mesh
Menu:   Seed, Instance,
        Approx global size [5.0], Apply, OK
Menu:   Mesh, Controls, Quad, Structured, OK
```

Global seeding, as shown in the left dialog box in Figure 2.22, is perhaps the easiest approach to specify mesh density. It is best to click Apply before OK. This

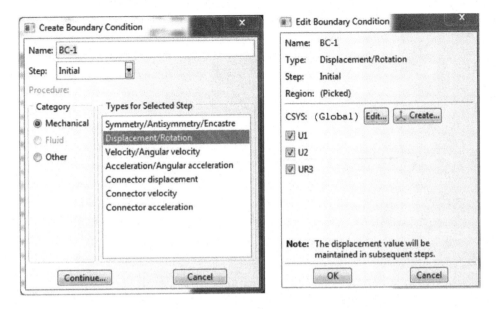

Fig. 2.20: Dialog box to create boundary conditions.

provides an opportunity to see the seeded part, and thus to modify it, before committing to a given seed pattern. Of course everything can be edited, but that is for later.

Then, on the right dialog box in Figure 2.22, Mesh Control is used to specify the element characteristics.

Finite element analysis (FEA) programs have an *element library* that contains many different element types. The element type determines the element formulation used. For example, the degrees of freedom set, the interpolation functions, whether the element is for 2D or 3D space, etc. The element type identifies the element category: bar/rod tensile-compression, beam bending, solid, shell, laminate shell, etc. Each commercial code identifies element formulations with different labels. Identification labels and basic characteristics of a few element formulations are shown in Table 2.2. Also, each element type has different options. For example, on a planar solid element, an option allows one to choose between plane strain and plane stress analysis.

The element type is selected in the Element Type dialog box shown in Figure 2.23. The last step is to finalize (instance) the mesh. The procedure is summarized as follows

```
Menu: Mesh, Element library, Standard, Linear, Plane stress, OK
Menu: Mesh, Instance, Yes
```

Finally, your mesh should look like the one displayed in Figure 2.24.

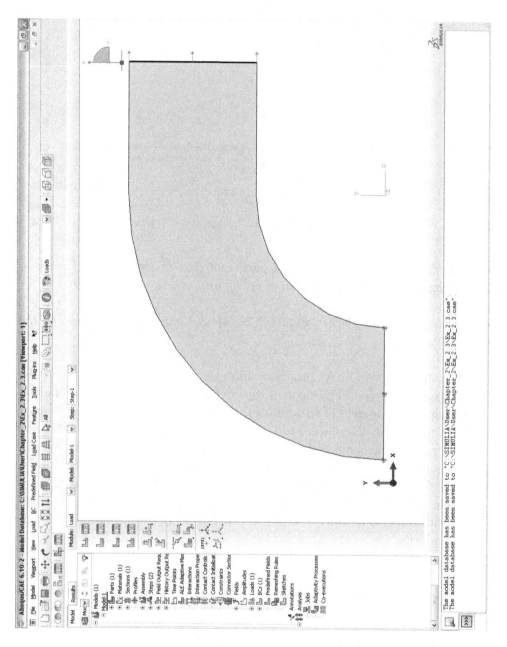

Fig. 2.21: Displaying the loads and boundary conditions.

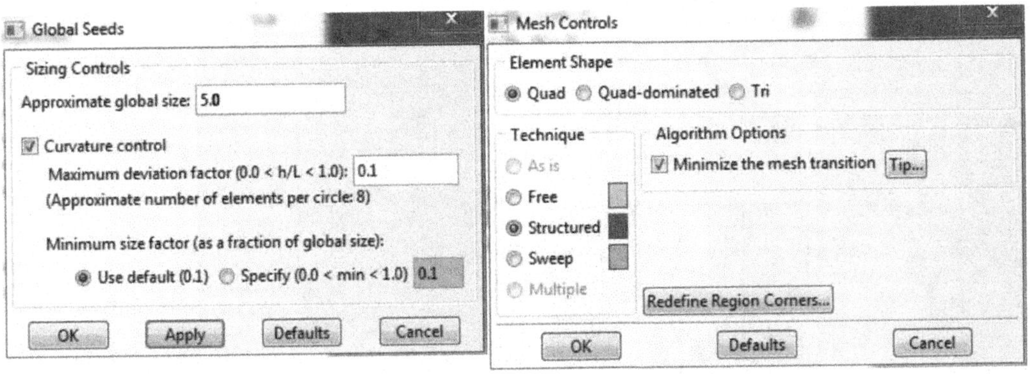

·Fig. 2.22: Using Global Seeds (left dialog box) and Mesh Control (right dialog box).

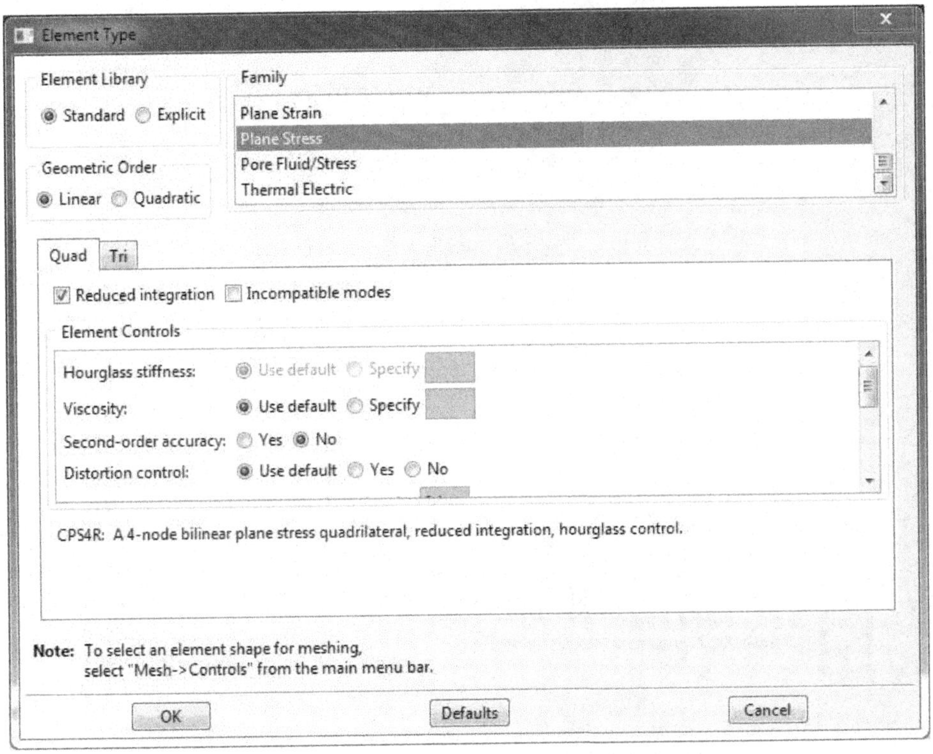

Fig. 2.23: Element Type.

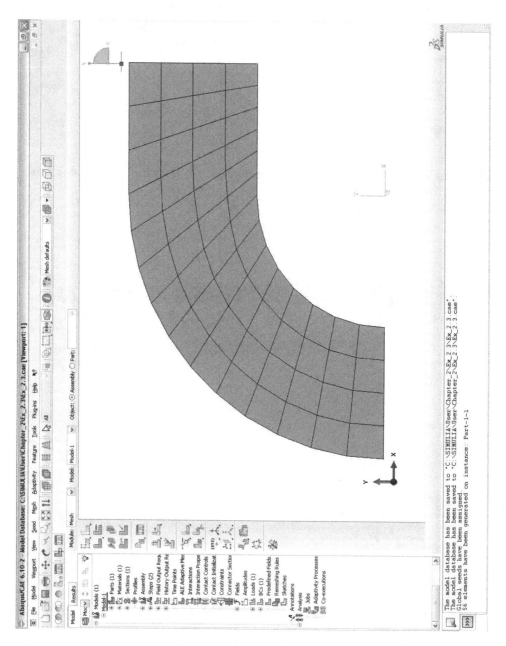

Fig. 2.24: The part is meshed.

2.3.12   Solution Phase

In the solution phase of the analysis, the solver subroutine included in the finite element program solves the simultaneous set of equations (2.46) that the finite element method (FEM) generates. Usually, the primary solution is obtained by solving for the nodal degree of freedom values, i.e., displacements and rotations. Then, derived results, such as stresses and strains, are calculated at the integration points. Primary results are called *nodal solutions* and derived results are called *element solutions.*

Several methods of solving the system of simultaneous equations are available. Some methods are better for larger models, others are faster for nonlinear analysis, others allow one to distribute the solution by parallel computation. Commercial finite element programs solve these equations in batch mode. The frontal direct solution method is commonly used because it is rather efficient for FEA. When the analysis is nonlinear, the equations must be solved repeatedly, thus increasing the computational time significantly.

To solve the model, one must create a Job. The job is then submitted from within CAE for execution by the Abaqus processor. The CAE submit process is simply a way to submit a job as in Example 2.2 but within CAE, thus obviating the use of a DOS shell. Note however that execution is a process done outside CAE. CAE maintains a model database (.mdb), and on **Submit***, it generates an input file (.inp) that is read by the Abaqus processor. Although the processor checks the file for errors, it cannot find all possible modeling errors until actual execution. Since, the error check capability of the processor is much less intensive than the solution itself, a* **Data Check** *is normally requested from the Job Manager before requesting a solution. The response is displayed in the Progress Window located at the bottom of the CAE window. If the data check passes, one can submit the job. Again, progress is shown in the Progress Window. Once the solution is completed, hopefully without errors, the results are written by the processor in the output database (.odb). At this point one can invoke the Visualization module directly from the Job window by clicking* **Results***. This last step will change the module to Visualization, read the .odb, and display a preliminary visualization. The process is summarized as follows*

```
Module: Job
Menu:   Job, Manager,    # opens the Job Manager window
        Create, Name [Job-1], Model: [Model-1],
            Description: [User's heading goes here], Cont, OK
        Data check      # watch the execution window below the WS
        Submit          # watch the execution window below the WS
        Results
```

The Job Manager dialog and the Create Job dialog boxes are shown in Figure 2.25. On the Job Manager, note the buttons for Write Input, Data Check, Submit, Continue, Monitor, Results, and Kill. These allow the user to control the execution

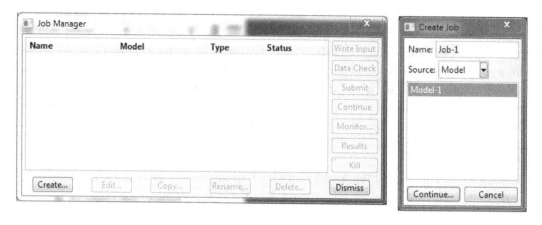

Fig. 2.25: The Job Manager and Create Job dialog boxes.

of the Abaqus processor, and with Results, to jump directly to the visualization mode after the execution is done.

The `Description` *field allows the user to enter a heading that will display on the visualization module provided* `Viewport Annotation Options, General, Show title box` *is checked. The heading appears in the* `.inp` *file following the keyword* `*Heading`*, as shown in Example 2.2, p. 49.*

2.3.13 Post-processing and Visualization

Once the solution has been calculated, the post-processor can be used to review and to analyze the results. Results can be reviewed graphically or by listing the values numerically. Since a model usually contains a considerable amount of results, it may be better to use graphical tools. Post-processors of commercial codes produce contour plots of stress and strain distributions, deformed shapes, etc. The software usually includes derived calculations such as error estimation, load case combinations, or path operations.

The visualization module can be reached either from the Job Manager (button Results in Figure 2.25) or by changing the Module to Visualization using the drop-down Module menu, but the latter requires the user to further instruct CAE to read the *.odb by executing* `File, Open, File Filter, [.odb], File Name, [job1.odb]`*. Assuming you reached the visualization module by clicking Results in the Job Manager, a basic visualization can be done as follows*

```
Module: Visualization
Menu: Plot, Contours, On Deformed Shape
```

This will produce a contour plot of von Mises stress on the deformed shape, as shown in Figure 2.26. The maximum value of stress is 65.78 MPa. Finally, before exiting the program, save the various files that contain the status of Abaqus CAE at this point in the session, as follows

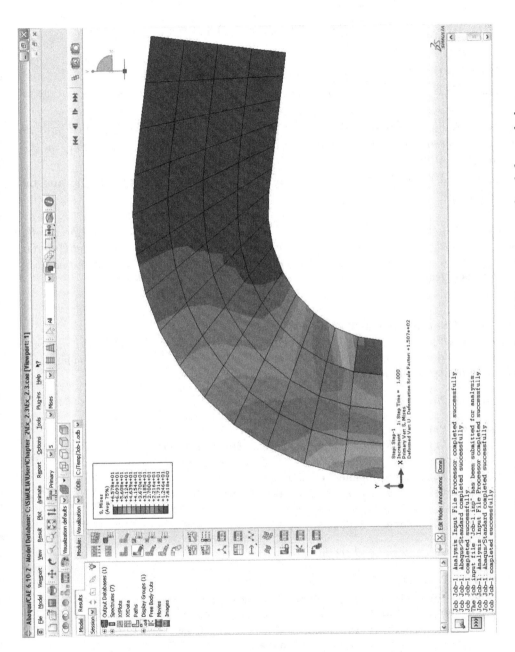

Fig. 2.26: Visualize a contour plot of von Mises stress on the deformed shape.

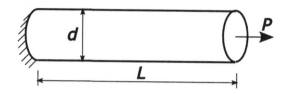

Fig. 2.27: Example 2.4

```
Menu: File, Save    # save the .cae and .mdb files
Menu: File, Exit    # exit CAE
```

The .mdb *file contains all the model information. The file* .cae *allows you to resume the session at a later time. A copy of* **Ex_2.3.cae** *is available in the website [2].*

Examples 2.4 and 2.5 include commands to review the results by listing and by graphic output, respectively.

Example 2.4 *Use Abaqus to find the axial displacement, stress, and strain, at the axially loaded end of a rod clamped at the other end (Figure 2.27). The rod is made of steel $E = 200$ GPa, diameter $d = 9$ mm, length $L = 750$ mm, and load $P = 100$ kN. Use three two-node (linear)* **truss** *elements.*

Solution to Example 2.4 *Watch the video in [1].*
The Abaqus CAE procedure is listed below. The CAE file is available in [2, Ex_2.4.cae]. Poisson's ratio should not be needed for a 1D analysis of a bar, but it is required in Abaqus CAE for defining the isotropic material, so the value $\nu = 0.3$ is used.

   i. *Creating the geometry*

```
Module: Part
Menu:    Create, Name: [Part-1], 2D, Deformable, Wire,
   Approx size: [2000], Cont
Menu:    Add, Line, Connected Lines,
   # enter coordinates in the dialog box below the WS
   [0,0], Enter,
   [750,0], Enter, X, Done
```

  ii. *Entering the material and section properties*

```
Module: Property
Menu:    Material, Create, Name [Material-1],
   Mechanical, Elasticity, Elastic, Isotropic, [200E3, 0.3], OK
Menu:    Section, Create, Name: [Section-1], Beam, Truss, Cont
   Material-1, Cross-sectional area: [63.617], OK
Menu:    Assign, Section, # pick the truss, Done, OK
```

 iii. *Creating an assembly, which is trivial in this case*

```
Module: Assembly
Menu:    Instance, Create, [Part-1], Independent, OK
```

iv. Creating a step as part of the solution strategy

```
Module: Step
Menu:   Step, Create, Name [Step-1], Static, General, Cont, OK
```

v. Specifying loads and boundary conditions, as well as adjusting the display

```
Module: Load
Menu:   BC, Create, Name: [BC-1],
        Step: Initial, Category: Mechanical, Types: Symmetry, Cont
        # pick the origin, Done,
        Encastre, OK
Menu:   Load, Create, Name: [Load-1],
        Step: Step-1, Category: Mechanical, Types: Concentrated, Cont
        # pick the end @ x=750, Done,
        CF1 [100E3], OK
Menu:   View, Assembly display options,
        Attribute, Symbol, Size: [24], Arrows: [24], Apply, OK
        F6  # roll the mouse wheel forward to see the part smaller
```

vi. Meshing

```
Module: Mesh
Menu:   Seed, Instance, Apply, # to see the proposed mesh seeds
        # change it to get 3 elements
        Approx global size [250], Apply, OK
Menu:   Mesh, Element Type, Family: Truss, Linear, OK
Menu:   Mesh, Instance, Yes
Menu:   View, Assembly display options,
        Tab: Mesh, # checkmark Show node labels, Apply
        # checkmark Show element labels, Apply, OK
        # to save these options as default for the future
Menu: File, Save Options, Current, OK
```

vii. Defining a Job for the Abaqus processor to execute

```
Module: Job
Menu:   Job, Manager,
        Create, Name [Job-1], Cont, OK
        Submit,  # monitor the progress window below the WS
        Results, # switches to Module: Visualization
```

viii. Visualizing the results is easier by invoking the Results from the Job Manager as in the previous step. Once in the Visualization module

```
Module: Visualization
# to get arrows with magnitude proportional to the displacement
# max. value read on the color scale is 5.895 mm
    Menu:   Result, Field Output, Tab: Symbol Variable,
        Name: U, Vector: Resultant, Apply, OK
# to get the stress
# max. value read on the color scale is 1572 MPa
    Menu:   Result, Field Output, Tab: Symbol Variable,
```

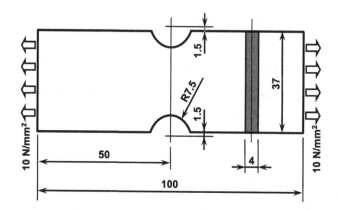

Fig. 2.28: Rectangular notched strap analyzed in Example 2.5.

```
    Name: S, Tensor quantity, Select direct components, S11, Apply, OK
# to get the strain
# max. value read on the color scale 7860E-3 (micro-strain)
   Menu:   Result, Field Output, Tab: Symbol Variable,
      Name: E, Tensor quantity, Select direct components, E11, Apply, OK
# further modify the way things are seen on the screen
# and save the settings
   Menu:   View, ODB Display Options, Tab: Entity Display,
         Symbol size [24]
 Menu:   File, Save Display Options, Current, OK
```

A convenient combination of units for this case is Newton, mm, and $MPa = N/mm^2$.
The analysis results can be easily verified by mechanics of material calculations, as follows

$$U_x = \frac{PL}{AE} = \frac{(750)(100000)}{(63.617)(200000)} = 5.894 \ mm$$

$$\sigma = \frac{P}{A} = \frac{100000}{63.617} = 1571.9 \ MPa$$

$$\epsilon = \frac{\sigma}{E} = 7.859 \cdot 10^{-3}$$

Example 2.5 *Use a commercial FE code to find the stress concentration factor of a rectangular notched strap. The dimensions and the load state are defined in Figure 2.28. Use eight node (quadratic) quadrilateral plane stress elements.*

Solution to Example 2.5 *Watch the video in [1].*
The Abaqus CAE procedure is listed below. The CAE file is available in [2, Ex_2.5.cae].

   i. Creating the geometry

```
   Module: Part,
   Menu: Part, Create, Name: [Part-1],
      2D, Deformable, Shell, Approx size [200], Cont
      # a quarter model of the Figure in the textbook
   Menu: Add, Point,
```

```
        [0,0]
        [50,18.5]
        X,
   Menu: Add, Line, Rectangle, #pick diagonal points, X
        Add, Point,
        [0,20]       # center
        [7.5,20]     # radius
        X,
   Menu: Add, Circle     # select center and edge point
        Edit, Auto trim # the unneeded lines/arches
        X,
        Done    # with the sketch. finalize the part.
   Menu: File, Save
```

ii. Entering the material and section properties

```
   Module: Property
   Menu: Material, Create, Name: [Material-1],
          Mechanical, Elasticity, Elastic, [190E3, 0.3], OK
   Menu: Section, Create, Name: [Section-1], Solid, Homogeneous, Cont
          # checkmark Plane stress/strain thickness: [4], OK
   Menu: Assign, Section, # pick the part, Done, OK
```

iii. Creating an assembly

```
   Module: Assembly
   Menu: Instance, Create, Name: [Part-1], Independent, OK
```

iv. Creating a step as part of the solution strategy

```
   Module: Step
   Menu: Step, Create, Name: [Step-1], Static, General, Cont, OK
```

v. Specifying loads and boundary conditions, as well as adjusting the display

```
   Module: Load
   Menu:   BC, Create, Name: [BC-1], Step: Initial, Mechanical, Symmetry,
   Cont
        # pick the horiz edge on bottom, Done, YSYMM, OK
   Menu:   BC, Create, Name: [BC-2], Step: Initial, Mechanical, Symmetry,
   Cont
        # pick the vert edge on left, Done, XSYMM, OK
   Menu:   Load, Create, Name: [Load-1],
        Step: Step-1, Mechanical, Types: Pressure, Cont
        # pick the vertical edge on the right, Done,
        Distribution: Uniform, Magnitude [-10], OK
        # mouse wheel fwd to zoom out on the WS
```

vi. Meshing

```
   Module: Mesh
   # Above the WS, select Object: Assembly
   Menu: Seed, Instance, Apply, #if you like it, OK,
        # otherwise Approx global size: [1.0], Apply, OK
```

```
Menu: Mesh, Controls, Quadratic, Structured, OK
Menu: Mesh, Element type, Standard, Quadratic,
        Family: Plane stress, Reduced Integration, OK
Menu: Mesh, Instance, Yes, # OK to mesh the part instance
```

vii. Defining a Job for the Abaqus processor to execute

```
Module: Job
Menu:    Job, Manager
    Create, Name: Job-1, Cont, OK
    Submit  # the job for execution by the Abaqus processor
    Results # switches to Module:Visualization
```

viii. Visualizing the results

```
Module: Visualization
# to get a report file
Menu:    Report, Field output,
    Tab: Variable, Position: Integ. point, expand S, # checkmark S11
    Tab: Setup, Name: [abaqus.rpt],
    OK, # immediately the .rpt file is written in the current WD
    # min S11 = 0 at intersection of hole and horz symmetry line
    # max S11 ~ 28.1 at intersection of hole and vert symmetry line
    # or
    # integration point location varies upon mesh refinement
    # the node with max stress is always the same
    # unique nodal
    # displays average stress from neighbor elements at node
    # element nodal also, but displays the element number as well
    Tab: Variable, Position: Unique nodal, expand S, check S11
    Tab: Setup, Name: [abaqus.rpt], OK
    # open abaqus.rpt with notepad++
    # min S11 = 0 @ node 5
    # max S11 ~ 27.9 @ node 7
# to see the node #'s
Module: Mesh, Object: Assembly
Menu:    View, Assembly display options,
    Tab: Mesh, # checkmark Show node labels, Apply
```

The stress in the net area without stress concentration is

$$\sigma_o = \frac{P}{A} = \frac{10 \cdot 37 \cdot 4}{25 \cdot 4} = 14.8 \; MPa$$

From the FE model, the maximum horizontal stress close to the notch is approximately 27.9 *MPa. Therefore, the stress concentration factor is*

$$k = \frac{\sigma_{max}}{\sigma_o} = 1.885$$

Example 2.6 *Using Abaqus CAE, generate a model for a dome with different types of elements (shell and beam elements), using different section properties and two materials. By solving this example the user will become familiar with the construction of 3D parts in Abaqus CAE.*

Solution to Example 2.6 *Watch the video in [1].*
To facilitate manipulating the part during the construction, activate the Views *toolbar, as follows*

```
Menu: View, Toolbars, Views.
# Integrate it next to the other toolbars above the WS area.
```

Creating the 3D dome

i. *Creating a solid semi-sphere:*

```
Module: Part
Menu: Part, Create, Name [Part-1], 3D, Deformable, Solid,
    Revolution, Approximate size [2000], Cont,
Menu: Add, Arc, Center/Endpoint,
    # in the dialog box below the WS enter the coordinates
    [0,0], [0,500], [500,0], X # to end the Arc command, F6 # zoom/fit
Menu: Add, Line, Connected Lines,
    # enter the coordinates
    [0,500], [0,0], [500,0], X # to end the Line command, Done,
    Angle [360], OK
```

A 3D semi-sphere should have been created.

ii. *Creating the side-walls:*

```
# Rotate the part so you can see the base of the semi-sphere,
Menu: View, Rotate, X # to end the Rotate command,
    # roll the mouse-wheel backward to zoom in,
    # save this view-point for use in future procedures,
    # click the 'Save View' icon from the 'Views toolbar',
    # in the pop-up window select [User 1], Save Current, OK,
Menu: Shape, Cut, Extrude, # pick the base of the semi-sphere,
    # pick the edge of the base of the semi-sphere,
    # Sketch mode is now active,
Menu: Add, Line, Rectangle, # enter the coordinates
    [300,300], [-300,-300],
    # create a new rectangle entering the coordinates
    [600,600], [-600,-600], X # to end the Line command, Done,
    Type: [Through All], OK
```

The semi-sphere should have been cut creating four identical side-walls as shown in Figure 2.29.

iii. *Reducing the height of the side-walls:*

```
Menu: Shape, Cut, Extrude, # pick one of the side-walls,
    # in the dialog box below the WS select [horizontal on bottom],
    # pick the longest straight edge of the side-wall selected,
    # Sketch mode is now active,
    # roll the mouse-wheel forward to see the figure smaller,
```

```
                   # Draw a rectangle in an area away from the figure,
        Menu: Add, Line, Rectangle, # pick any two point on the screen,
            X # to end Line command
```

Your construction should look similar to Figure 2.30.

```
        Menu: Add, Dimension, # pick a vertical line of the rectangle
            # may need to move pointer and right-click to create dimension
            # enter the height of the rectangle [200],
            # pick a horizontal line of the rectangle,
            # enter the width of the rectangle [600],
            X # to end the Dimension command
```

Your construction should look similar to Figure 2.31.

```
        Menu: Edit, Transform, Translate,
            # in the notification area below the WS select [Move],
            # pick the four lines forming the rectangle
            # hold the 'shift-key' to individually select the lines, Done,
            # pick the lower-left point of the rectangle,
            # pick the lower-left point of the figure,
            X # to end the Translate command, Done
```

The rectangle should have been moved as shown in Figure 2.32.

```
            Type: [Through All], OK
```

The height of the part should have been reduced as shown in Figure 2.33.

iv. *Converting the solid part to a shell part:*

```
        Menu: Shape, Shell, From Solid,
            # pick the part, Done, X # to end the shell command,
        Menu: Tools, Geometry Edit, Face, Remove,
            # pick the surface at the base of the part,
            Done, X # to end the command,  # close the pop-up window
```

The part should look like the one shown in Figure 2.34.

v. *Creating the supporting columns:*

```
        Menu: Shape, Wire, Sketch, # pick one of the side-walls,
            # in the dialog box below the WS select [horizontal on bottom],
            # pick the longest straight edge of the side-wall selected,
            # Sketch mode is now active,
            # roll the mouse-wheel forward to see the figure smaller,
        Menu: Add, Line, Connected Lines,
            # pick the lower-left corner of the part,
            # pick any point directly below (it must create a vertical line),
            X # to end the Line command,
        Menu: Add, Dimension, # pick the line,
```

```
        # enter the length of the line [200],
        X # to end the Dimension command,
        # Repeat the procedure to create a vertical line of length 200
        # starting at the lower-right point of the part,
        Done # to end the sketch
```

Create another sketch on the opposite side-wall, and repeat the procedure above to create the two additional vertical columns. The part should look like the one shown in Figure 2.35.

Materials, and Section Properties:

Module: Property

  i. Defining Materials

```
    Menu: Material, Manager,
        Create, Name [Dome], Mechanical, Elasticity, Elastic, Isotropic,
        Young's Modulus [190e3], Poisson's ratio [0.27], OK,
        Create, Name [Columns], Mechanical, Elasticity, Elastic, Isotropic,
        Young's Modulus [200e3], Poisson's ratio [0.29], OK,
        # close the Material Manager pop-up window
```

  ii. Defining Section Properties

```
    Menu: Section, Manager,
        Create, Name [Dome-Roof], Shell, Homogeneous, Cont,
        Shell thickness [6.0], Material [Dome], OK,
        Create, Name [Side-Walls], Shell, Homogeneous, Cont,
        Shell thickness [4.0], Material [Dome], OK,
        Create, Name [Columns], Beam, Truss, Cont, Material [Columns],
        Cross-sectional area [100], OK,
        # close the Material Manager pop-up window
```

  iii. Assign Sections

```
    Menu: Assign, Section,
        # pick the roof of the dome, Done, Section [Dome-Roof], OK,
        # pick the side-walls,
        # hold the 'shift-key' to individually select the surfaces,
        Done, Section [Side-Walls], OK,
        # pick the columns,
        # hold the 'shift-key' to individually select the lines,
        Done, Section [Columns], OK,
        X # to end the command
```

Assembly:

```
Module: Assembly
Menu: Instance, Create, Parts [Part-1], Independent, OK
```

Loads and BC:

   i. Creating a Load and BC Step

```
        Module: Step
        Menu: Step, Create, Name [Step-1], General, Static/Gen., Cont, OK,
            # Return to the saved view,
            # click the 'Apply User 1 View' icon from the 'Views toolbar'
```

   ii. Defining Boundary Conditions

```
        Module: Load
        Menu: BC, Manager,
            Create, Name [BC-1], Step: Initial, Mechanical, Sym/Ant/Enc, Cont,
            # pick the lower-end of the four columns,
            # hold the 'shift-key' to individually select the points, Done,
            Encastre, OK,
            Create, Name [BC-2], Step: Initial, Mechanical, Disp/Rota, Cont,
            # pick the vertical edges of the side-walls,
            # hold the 'shift-key' to individually select the sides, Done,
            # checkmark U1, and U3, OK,
            # Close the Boundary Condition Manager pop-up window,
        Menu: Load, Create, Name [Load-1], Step: Step-1,
            Mechanical, Pressure, Cont, # pick the dome-roof, Done,
            # in the notification area below the WS select [Brown],
            Distribution: Uniform, Magnitude [-100], OK,
            # click the 'Apply Iso View' icon from the 'Views toolbar',
            # roll the mouse-wheel backward to zoom in,
            # increase the size of the BC and Load markers,
        Menu: View, Assembly Display Options,
            Tab: Attribute, Symbol, Size [18], Arrows [18], Face Density [10],
            Apply, OK
```

The part should look similar to the one shown in Figure 2.36.

Meshing the Part:

```
# Return to the saved view,
# click the 'Apply User 1 View' icon from the 'Views toolbar',
Module: Mesh
Menu: Seed, Instance, Approximate Global Size [30], Apply, OK,
Menu: Mesh, Element Type,
    # pick the roof of the dome, Done, Standard, Linear, Family [Shell], OK,
    # pick the four side-walls, Done, Standard, Linear, Family [Shell], OK,
    # pick the four columns, Done, Standard, Linear, Family [Truss], OK,
    X # to end the command,
Menu: Mesh, Instance,
    # in the notification area below the WS select [Yes],
    # click the 'Apply Iso View' icon from the 'Views toolbar',
    # roll the mouse-wheel backward to zoom in
```

The part should look similar to the one shown in Figure 2.37.

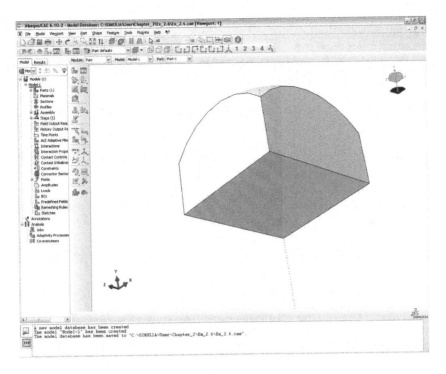

Fig. 2.29: Dome's side-walls.

Running the analysis and visualizing the results:

```
Module: Job
Menu: Job, Manager,
    Create, Name [Job-1], Cont, OK
    Submit, # when the Status message indicates 'Completed'
    Results
Module: Visualization
Menu: Plot, Contours, On Deformed Shape
```

The part should look similar to the one shown in Figure 2.38. The CAE file is available in [2, Ex_2.6.cae].

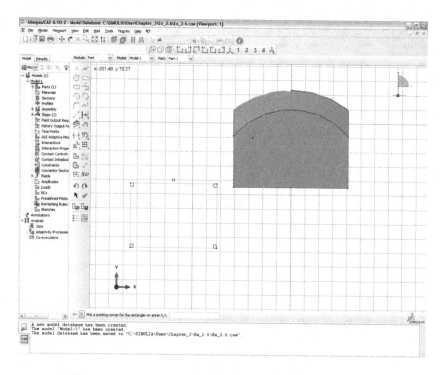

Fig. 2.30: Dome's side-walls height reduction step 1.

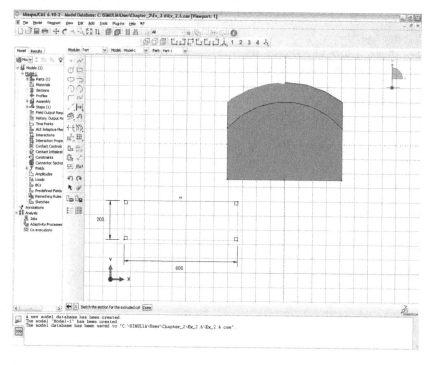

Fig. 2.31: Dome's side-walls height reduction step 2.

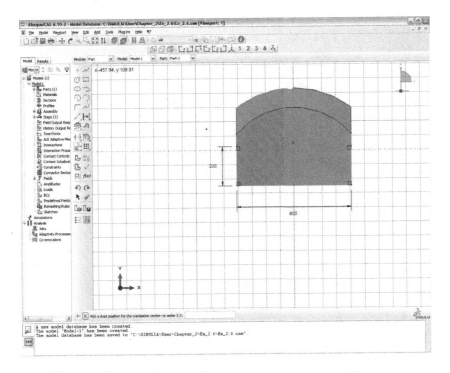

Fig. 2.32: Dome's side-walls height reduction step 3.

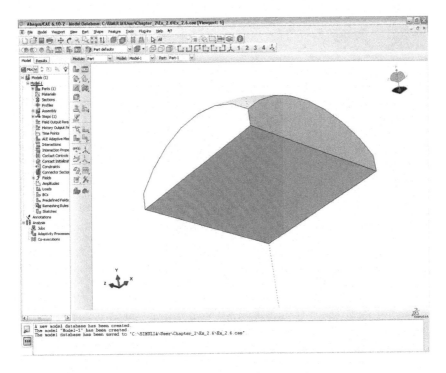

Fig. 2.33: Dome's side-walls height reduction step 4.

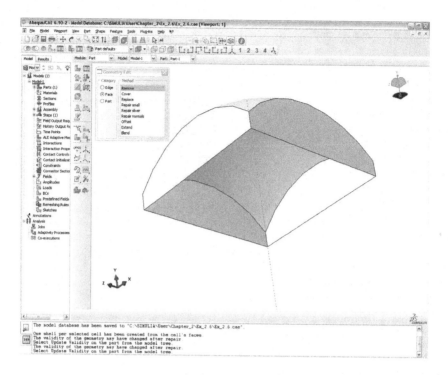

Fig. 2.34: Dome as shell structure.

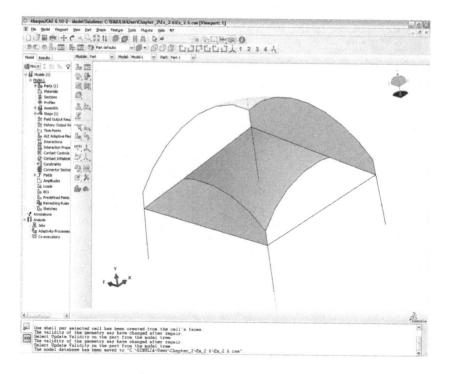

Fig. 2.35: Dome completed.

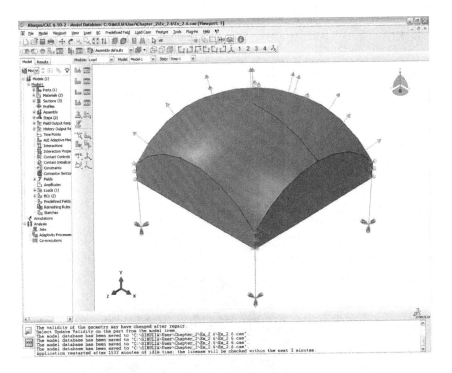

Fig. 2.36: Loaded Dome structure.

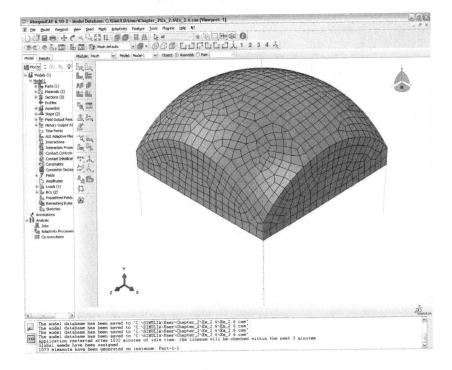

Fig. 2.37: Meshed Dome structure.

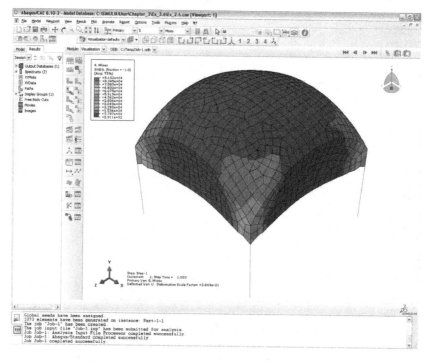

Fig. 2.38: Contour plot of von Mises stress. Maximum values are $91,020$ MPa in the columns and $66,922$ MPa in the shell.

2.4   Interactions and Constraints

Abaqus CAE does not recognize mechanical contact between part instances or regions of an assembly, even if they have features at the same location. For contact to be active the interaction must be explicitly defined in the Interaction module.

There are many types of interactions, including: General contact, Surface-to-surface, Fluid cavity, Fluid exchange, Fluid inflator, XFEM crack growth, Model change, Cyclic symmetry, Elastic foundations, Cavity radiation, Thermal film conditions, Radiation to and from the ambient environment, Abaqus/Standard to Abaqus/Explicit co-simulation, Incident waves, Acoustic impedance, and Actuator/sensor. In this textbook, we use Surface-to-surface interactions.

The Interaction module is used to define constraints on DOF between regions of a model. Since Interactions are step-dependent, they can be defined, activated and deactivated at different steps in the analysis.

Constraints is a subset of interactions. The following are some of the constraints available: Tie, Rigid body, Equations, Coupling, Display body, Adjust points, multi-point constraint (MPC), Shell-to-solid coupling, and Embedded region. In this textbook we use tie constraints, rigid body constraints, coupling constraints, and constraint equations (CE).

2.4.1   Tie Constraint

A tie constraint merges two regions through some of their surfaces. The meshes may be dissimilar. The surfaces may be at a distance from each other, as shown by *Position Tolerance, Specify distance* in Figure 2.39. The value of the specified distance should be equal or slightly larger than the actual distance between the surfaces. Nodes on the slave surface that are considered to be outside the position tolerance (specified distance) will not be tied.

While using *Discretization method: surface-to-surface* (Figure 2.39), the way the surfaces are selected has an influence on how rotational DOF are constrained and whether or not the meshes on the surfaces of the tied regions may be dissimilar. There are two types of *Region Selection: 'Surface' and 'Node Region'*, as it can be seen in Figure 2.40.

Surface regions and *Node regions* behave differently with respect to rotational DOF, and they behave differently in Abaqus Explicit and Abaqus Standard.

In response to *Choose the master/slave type : Surface* (Figure 2.40), the meshes on the surfaces of the tied regions may be dissimilar, in which case the slave and master nodes are mapped in an approximate way. Such approximation allows tie constraints between dissimilar meshes and may reduce noise in the solution, but it may lead to the deletion of some DOF. If deleted DOF are associated with boundary conditions, those boundary conditions may be lost. In Abaqus Standard R2017 and later, when the nodes/elements have no rotational DOF, the rotation of the surfaces are not tied. In Abaqus Explicit, the rotations are tied. See Appendix A for additional details. Due to these considerations, *master/slave type : Surface* are not used in this textbook.

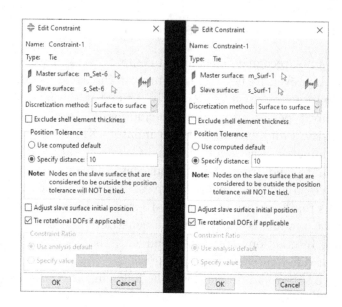

Fig. 2.39: Edit window, Tie constraint

Fig. 2.40: While defining a surface-to-surface tie constraint in Abaqus Standard Choose master/slave type: either 'Surface' or 'Node Region'.

In response to *Choose the master/slave type : Node Region* (Figure 2.40), the meshes on the tied surfaces of the regions must be identical because the master/slave nodes are tied one-on-one according to their positions on the tied surfaces. In both Standard and Explicit, the rotation of the two surfaces are tied whether the nodes have rotational DOF or not. Due to these considerations, *master/slave type : Node Region* are used in this textbook. See Example 6.3, Video [1, Example 6.3] at time 7:22, and Appendix A.

2.4.2   Rigid Body Constraint

A *rigid body constraint*, constrains the motion of regions of the assembly to the motion of a reference point (RP). The relative positions of the regions that are part of the rigid body remain constant throughout the analysis. See Example 3.11. There are four rigid body region types, as follows

Body elements All the elements of the geometric region selected behave as a rigid body.

Pin nodes Pins selected nodes to a RP. Only translational DOF are pinned to the RP. In Figure 2.41 the load is applied to the RP connected to the end section by a Pin Nodes constraint. The cantilever support is achieved by symmetry-z plus encastre of the center node at (0,0,0) to prevent rigid body motion

(RBM) and $y = 0$ on a side node (100,0,0) to prevent rotation. In Figure 2.42 the end of the beam is free to rotate to maintain a zero bending moment at the free end. This is achieved by pinning the nodes to the RP. All nodes translate as a rigid surface (the end section of the beam). If we were to apply the concentrated load to the center node we would get a stress concentration, as in Figure 2.43.

Tie nodes Ties selected nodes to an RP. Both translational and rotational DOF are pinned to the RP. Note that nodes associated with solid elements do not have rotational DOF.

Analytical surface Analytical rigid surfaces are 3D geometric surfaces. They are efficient for contact problems because they can be smoother than meshed surfaces, reducing contact noise caused by the roughness of meshed surfaces and they may reduce the computational cost incurred by the contact algorithm.

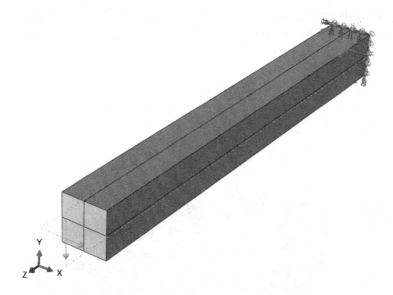

Fig. 2.41: A 200 × 200 mm square section, length = 2000 mm, loads $P_x = 300$ N and $P_y = -200$ N applied to an RP connected to the end section by a *Pin Nodes* constraint. The cantilever support is achieved by symmetry-z plus encastre of the center node to prevent RBM and $y = 0$ on a side node to prevent rotation.

2.4.3   Display Body

A *display body constraint* allows you to select a part instance that will be used for display only.

2.4.4   Coupling

A *coupling constraint,* constrains the motion of a surface to the motion of a single point. Unlike a rigid body constraint that makes the region (edge, surface, volume)

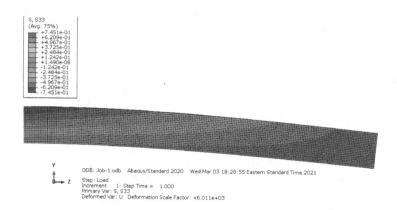

Fig. 2.42: Beam flexure with nodes at the loaded end rotating as a rigid surface via pin node constraint to an RP to which the loads are applied.

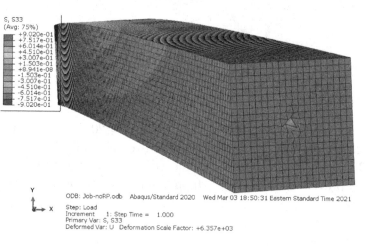

Fig. 2.43: A mesh-dependent stress concentration factor $K_t = 0.9/0.745 = 1.2$ appears when loads are applied directly to the center node at the free end of the beam. Compare the maximum stress S_{33} here to the maximum stress S_{33} in Figure 2.42.

into a rigid region, the coupling constraint allows the user to select individual DOF to be constrained. See Example 7.5 (step v).

2.4.5 Adjust Points

An *adjust points constraint* moves a point or points onto a specified surface.

2.4.6 MPC Constraint

An *multi-point constraint*, constrains the motion of the slave nodes of a region to the motion of a single point.

2.4.7 Shell-to-solid Coupling

A *shell-to-solid coupling constraint* couples the motion of a shell edge to the motion of an adjacent solid face.

2.4.8 Embedded Region

An *embedded region constraint* embeds a region of the model within a host region of the model or within the whole model.

2.4.9 Multi-point Constraint Equations

Constraint equations are linear, multi-point equation constraints to describe linear constraints between individual DOF. They are used in Examples 6.3, 6.4, 7.8 as well as in Sections 6.2 (Sixth Column), 6.3, 6.4, 12.4, and 12.4.2.

Suggested Problems

Problem 2.1 *Solve Example 2.4 (p. 76) explicitly as it is done in Section 2.1, using only two elements. Show all work.*

Problem 2.2 *From the solution of Problem 2.1, compute the axial displacement at (a) $x = 500$ mm, (b) $x = 700$ mm.*

Problem 2.3 *Using the same procedure in Example 2.1 (p. 45) calculate the element stiffness matrix and the equivalent force vector of a three-node element rod with quadratic interpolation functions. The interpolation functions are*

$$N_1^e = \frac{x - x_2}{x_1 - x_2}\frac{x - x_3}{x_1 - x_3} \qquad N_2^e = \frac{x - x_3}{x_2 - x_3}\frac{x - x_1}{x_2 - x_1} \qquad N_3^e = \frac{x - x_1}{x_3 - x_1}\frac{x - x_2}{x_3 - x_2}$$

where x_1, x_2, and x_3 are the coordinate positions of node 1, 2, and 3 respectively. Use $x_1 = 0$, $x_2 = h/2$, and $x_3 = h$, where h is the element length. Show all work.

Problem 2.4 *Program an FE code using the element formulation obtained in Example 2.1 (p. 45) and the assembly procedure shown in Section 2.1.6. With this code, solve Example 2.4, p. 76. Show all work in a report.*

Problem 2.5 *Program an FE code using the element formulation obtained in Problem 2.3 and the assembly procedure shown in Section 2.1.6. With this code, solve Example 2.4, p. 76. Show all work in a report.*

Chapter 3

Elasticity and Strength of Laminates

Most composite structures are built as assemblies of plates and shells. This is because the structure is more efficient when it carries membrane loads. Another important reason is that thick laminates are difficult to produce.

For example, consider a beam made of a homogeneous material with tensile and compressive strength σ_u subjected to bending moment M. Further, consider a solid beam of square cross-section (Figure 3.1), equal width and depth $2c$, with area A, moment of area I, and section modulus S given by

$$A = 4c^2$$
$$I = \frac{4}{3}c^4$$
$$S = \frac{I}{c} = \frac{4}{3}c^3 \tag{3.1}$$

When the stress on the surface of the beam reaches the failure stress σ_u, the bending moment per unit area is

$$m_u = \frac{M_u}{A} = \frac{S\sigma_u}{A} = \frac{1}{3}c\sigma_u \tag{3.2}$$

Now consider a square hollow tube (Figure 3.1) of dimensions $2c \times 2c$ and wall thickness t, with $2c \gg t$, so that the following approximations are valid

$$A = 4(2c)t = 8ct$$
$$I = 2\left[\frac{t(2c)^3}{12} + c^2(2ct)\right] = \frac{16}{3}tc^3$$
$$S = \frac{I}{c} = \frac{16}{3}tc^2 \tag{3.3}$$

Then

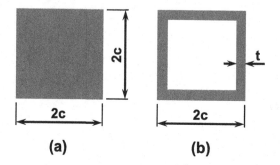

Fig. 3.1: Solid section (a) and hollow square tube (b).

$$m_u = \frac{M_u}{A} = \frac{S\sigma_u}{A} = \frac{\frac{16}{3}tc^2\sigma_u}{8ct} = \frac{2}{3}c\sigma_u \qquad (3.4)$$

The failure moment per unit area m_u is twice as large for the hollow square tube with thin walls than for the solid section.

Of course, the failure moment is limited by buckling of the thin walls (see Chapter 4). This is the reason buckling analysis is so important for composites. Most composite structures are designed under buckling considerations because the thicknesses are small and the material is very strong; so normally one does not encounter material failure as in metallic structures (e.g., yield stress) but structural failure such as buckling.

Plates are a particular case of shells, having no initial curvature. Therefore, only shells will be mentioned in the sequel. Shells are modeled as 2D structures because two dimensions (length and width) are much larger than thickness. The thickness coordinate is eliminated from the governing equations so that the 3D problem simplifies to 2D. In the process, the thickness becomes a parameter that is known and supplied to the model.

Modeling of laminated composites differs from modeling conventional materials in three aspects. First, the constitutive equations of each lamina are orthotropic (Section 1.12.3). Second, the constitutive equations of the element depend on the kinematic assumptions of the shell theory used and their implementation into the element. Finally, material symmetry is as important as geometric and load symmetry when trying to use symmetry conditions in the models.

3.1   Kinematic of Shells

Shell elements are based on various shell theories, which in turn are based on kinematic assumptions. That is, there are some underlying assumptions about the likely type of deformation of the material. These assumptions are needed to reduce the 3D governing equations to 2D. Such assumptions are more or less appropriate for various situations, as discussed next.

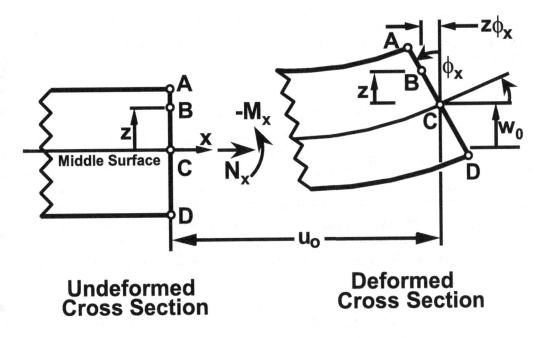

Fig. 3.2: Assumed deformation in FSDT.[1]

3.1.1 First-Order Shear Deformation Theory

The most popular composite shell theory is the first-order shear deformation theory (FSDT). It is based on the following assumptions:

i. A straight line drawn through the thickness of the shell in the undeformed configuration may rotate, but it will remain straight when the shell deforms. The angles it forms (if any) with the normal to the undeformed midsurface are denoted by ϕ_x and ϕ_y when measured in the $x - z$ and $y - z$ planes, respectively (Figures 2.19 and 3.2).

ii. The change of the shell thickness is negligible as the shell deforms.

These assumptions are verified by experimental observation in most laminated shells when the following are true:

– The aspect ratio $r = a/t$, defined as the ratio between the shortest surface dimension a and the thickness t, is larger than 10.

– The stiffness of the laminas in shell coordinates (x, y, z) does not differ by more than two orders of magnitude. This restriction effectively rules out sandwich shells, where the core is much softer than the faces.

---

[1] Reprinted from *Introduction to Composite Materials Design*, E. J. Barbero, Figure 6.2, copyright (1999), with permission from Taylor & Francis.

Based on the assumptions above, the displacement of a generic point B anywhere in the shell can be written in terms of the displacement and rotations at the midsurface C as

$$
\begin{aligned}
u(x, y, z) &= u_0(x, y) - z\phi_x(x, y) \\
v(x, y, z) &= v_0(x, y) - z\phi_y(x, y) \\
w(x, y, z) &= w_0(x, y)
\end{aligned}
\tag{3.5}
$$

The midsurface variables on the right-hand side of (3.5) are functions of only two coordinates (x and y); thus the shell theory is 2D. On the left-hand side, the displacements are functions of three coordinates, and thus correspond to the 3D representation of the material. At the 3D level, we use the 3D constitutive equations (1.68) and the 3D strain-displacement equations (1.5), which now can be written in terms of 2D quantities as

$$
\begin{aligned}
\epsilon_x(x, y, z) &= \frac{\partial u_0}{\partial x} - z\frac{\partial \phi_x}{\partial x} = \epsilon_x^0 + z\kappa_x \\
\epsilon_y(x, y, z) &= \frac{\partial v_0}{\partial y} - z\frac{\partial \phi_y}{\partial y} = \epsilon_y^0 + z\kappa_y \\
\gamma_{xy}(x, y, z) &= \frac{\partial u_0}{\partial y} + \frac{\partial v_0}{\partial x} - z\left(\frac{\partial \phi_x}{\partial y} + \frac{\partial \phi_y}{\partial x}\right) = \gamma_{xy}^0 + z\kappa_{xy} \\
\gamma_{yz}(x, y) &= -\phi_y + \frac{\partial w_0}{\partial y} \\
\gamma_{xz}(x, y) &= -\phi_x + \frac{\partial w_0}{\partial x} \\
\epsilon_z &= 0
\end{aligned}
\tag{3.6}
$$

where

- The midsurface strains ϵ_x^0, ϵ_y^0, γ_{xy}^0, also called membrane strains, represent stretching and in-plane shear of the midsurface.

- The change in curvature κ_x, κ_y, κ_{xy}, which are close but not exactly the same as the geometric curvatures of the midsurface. They are exactly that for the Kirchhoff theory discussed in Section 3.1.2.

- The intralaminar shear strains γ_{xz}, γ_{yz}, which are through-the-thickness shear deformations. These are small but not negligible for laminated composites because the intralaminar shear moduli G_{23}, G_{13} are small when compared with the in-plane modulus E_1. Metals are relatively stiff in shear ($G = E/2(1+\nu)$), and thus the intralaminar strains are negligible. In addition, the composite's intralaminar shear strength values F_4, F_5 are relatively small when compared to their in-plane strength values F_{1t}, F_{1c}, thus making evaluation of intralaminar strains (and possibly stresses) a necessity. On the other hand, the shear

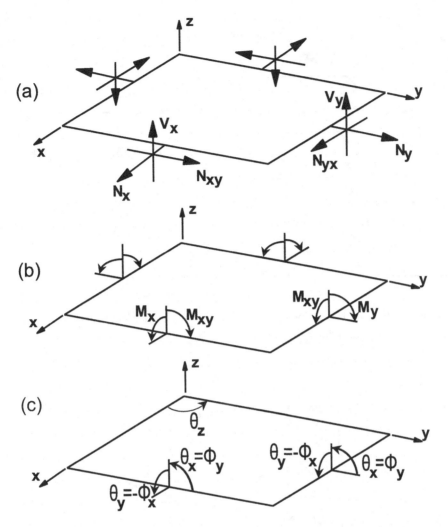

Fig. 3.3: Stress resultants acting on a plate or shell element: (a) forces per unit length, (b) moments per unit length, and (c) definition of shell theory rotations ϕ compared to mathematical angles θ.

strength of metals is comparable to their tensile strength, and since the intralaminar stress is always smaller than the in-plane stress, it is not necessary to check for intralaminar failure of metallic homogeneous shells. That is not the case for laminated metallic shells since the adhesive is not quite strong and it may fail by intralaminar shear.

While the 3D constitutive equations relate strains to stress, the laminate constitutive equations relate midsurface strains and curvatures. The laminate constitutive equations are obtained by using the definition of stress resultants. While in 3D elasticity every material point is under stress, a shell is loaded by stress resultants (Figure 3.3), which are simply integrals of the stress components through the thickness of the shell, as follows

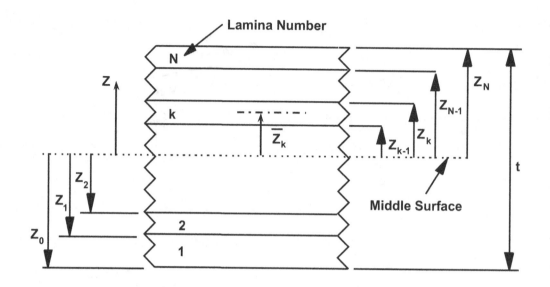

Fig. 3.4: Definition of the z-coordinate system to locate the interface between laminas (z_k) and the midsurface of the laminas ($\bar{z}_k$)

$$
\begin{Bmatrix} N_x \\ N_y \\ N_{xy} \end{Bmatrix} = \sum_{k=1}^{N} \int_{z_{k-1}}^{z_k} \begin{Bmatrix} \sigma_x \\ \sigma_y \\ \sigma_{xy} \end{Bmatrix}^k dz
$$

$$
\begin{Bmatrix} V_y \\ V_x \end{Bmatrix} = \sum_{k=1}^{N} \int_{z_{k-1}}^{z_k} \begin{Bmatrix} \sigma_{yz} \\ \sigma_{xz} \end{Bmatrix}^k dz
$$

$$
\begin{Bmatrix} M_x \\ M_y \\ M_{xy} \end{Bmatrix} = \sum_{k=1}^{N} \int_{z_{k-1}}^{z_k} \begin{Bmatrix} \sigma_x \\ \sigma_y \\ \sigma_{xy} \end{Bmatrix}^k z \, dz \tag{3.7}
$$

where N is the number of laminas, and z_{k-1} and z_k are the coordinates at the bottom and top surfaces of the k-th lamina, respectively (Figure 3.4). Replacing the plane stress version of the 3D constitutive equations in shell local coordinates (1.100-1.101) at each lamina and performing the integration we get

$$
\begin{Bmatrix} N_x \\ N_y \\ N_{xy} \\ M_x \\ M_y \\ M_{xy} \end{Bmatrix} = \begin{bmatrix} A_{11} & A_{12} & A_{16} & B_{11} & B_{12} & B_{16} \\ A_{12} & A_{22} & A_{26} & B_{12} & B_{22} & B_{26} \\ A_{16} & A_{26} & A_{66} & B_{16} & B_{26} & B_{66} \\ B_{11} & B_{12} & B_{16} & D_{11} & D_{12} & D_{16} \\ B_{12} & B_{22} & B_{26} & D_{12} & D_{22} & D_{26} \\ B_{16} & B_{26} & B_{66} & D_{16} & D_{26} & D_{66} \end{bmatrix} \begin{Bmatrix} \epsilon_x^0 \\ \epsilon_y^0 \\ \gamma_{xy}^0 \\ \kappa_x \\ \kappa_y \\ \kappa_{xy} \end{Bmatrix} \tag{3.8}
$$

$$
\begin{Bmatrix} V_y \\ V_x \end{Bmatrix} = \begin{bmatrix} H_{44} & H_{45} \\ H_{45} & H_{55} \end{bmatrix} \begin{Bmatrix} \gamma_{yz} \\ \gamma_{xz} \end{Bmatrix} = \begin{bmatrix} K_{22} & K_{12} \\ K_{11} & K_{11} \end{bmatrix} \begin{Bmatrix} \gamma_{yz} \\ \gamma_{xz} \end{Bmatrix}
$$

where K_{11}, K_{12}, K_{22} is the notation used by Abaqus, and

$$A_{ij} = \sum_{k=1}^{N} \left(\overline{Q}_{ij}\right)_k t_k; \quad i,j = 1,2,6$$

$$B_{ij} = \sum_{k=1}^{N} \left(\overline{Q}_{ij}\right)_k t_k \bar{z}_k; \quad i,j = 1,2,6$$

$$D_{ij} = \sum_{k=1}^{N} \left(\overline{Q}_{ij}\right)_k \left(t_k \bar{z}_k^2 + \frac{t_k^3}{12}\right); \quad i,j = 1,2,6$$

$$H_{ij} = \frac{5}{4} \sum_{k=1}^{N} \left(\overline{Q}_{ij}^*\right)_k \left[t_k - \frac{4}{t^2}\left(t_k \bar{z}_k^2 + \frac{t_k^3}{12}\right)\right]; \quad i,j = 4,5 \qquad (3.9)$$

where $\left(\overline{Q}_{ij}\right)_k$ are the coefficients in laminate coordinates of the plane-stress stiffness matrix for lamina number k, t_k is the thickness of lamina k, and $\bar{z}_k$ is the coordinate of the middle surface of the kth lamina. For an in-depth discussion of the meaning of various terms see [3, Section 6.3]. In summary, the A_{ij} coefficients represent in-plane stiffness of the laminate, the D_{ij} coefficients represent bending stiffness, the B_{ij} represent bending-extension coupling, and the H_{ij} represent intralaminar shear stiffness (see (3.9)).

In (3.9), $\bar{Q}_{ij}$ are the components of the in-plane Q matrix of particular lamina in [3, (5.25)], and $\bar{Q}*_{ij}$ are the components of the out-of-plane lamina stiffness matrix for the same lamina. In Abaqus CAE, H_{ij} are called $K_{11} = H_{55}$, $K_{22} = H_{44}$, and $K_{12} = H_{45}$. These values have to be provided to Abaqus as explained Example 3.1 and the corresponding video [1]. Referring to the last two lines of (3.8), where H_{44} is related to V_y and H_{55} to V_x and Table 1.1, we see that Abaqus uses the same $1,2,3$ and x,y,z notation as this textbook but Abaqus orders the contracted components differently, as it can be seen in the second column of Table 1.1. Note that in the column for Abaqus Standard, the fifth row is σ_{13}, which is the direction of V_y, and the sixth row corresponds to V_x. Then, using the last two lines in (3.8), we conclude that $K_{11} = H_{55}$ (it corresponds to V_x) and $K_{22} = H_{44}$ (it corresponds to V_y).

When membrane and bending deformations are uncoupled (e.g., symmetric laminates), the governing equations of FSDT involve three variables for solving the bending problem (w^0, ϕ_x, ϕ_y) and two to solve the membrane problem (u^0, v^0). Bending-extension coupling means that all five variables will have to be found simultaneously, which is what FEA software codes do for every case, whether the problem is coupled or not.

The equilibrium equations of plates can be derived by using the PVW (see (1.16)). Furthermore, the governing equations can be derived by substituting the constitutive equations (3.8) into the equilibrium equations.

3.1.2   Kirchhoff Theory

Historically, Kirchhoff theory was preferred because the governing equations can be written in terms of only one variable, the transverse deflection of the shell w_0. Since it is easier to obtain analytical solutions in terms of only one variable rather than in terms of the three variables needed in FSDT, a wealth of closed-form design equations and approximate solutions exist in engineering design manuals that are based on Kirchhoff theory [14]. Such simple design formulas can still be used for preliminary design of composite shells if we are careful and we understand their limitations. Metallic shells were and still are commonly modeled with Kirchhoff theory. The FSDT governing equations can be reduced to Kirchhoff governing equations, and closed form solutions can be found, as shown in [15].

In Kirchhoff theory the intralaminar shear strain is assumed to be zero. From the last two equations in (3.6) we get

$$\phi_x = \frac{\partial w_0}{\partial x}$$
$$\phi_y = \frac{\partial w_0}{\partial y} \tag{3.10}$$

and introducing them into the first three equations in (3.6) we get

$$\epsilon_x(x,y,z) = \frac{\partial u_0}{\partial x} - z\frac{\partial^2 w_0}{\partial x^2} = \epsilon_x^0 + z\kappa_x$$
$$\epsilon_y(x,y,z) = \frac{\partial v_0}{\partial y} - z\frac{\partial^2 w_0}{\partial y^2} = \epsilon_y^0 + z\kappa_y$$
$$\gamma_{xy}(x,y,z) = \frac{\partial u_0}{\partial y} + \frac{\partial v_0}{\partial x} - 2z\frac{\partial^2 w_0}{\partial x\partial y} = \gamma_{xy}^0 + z\kappa_{xy} \tag{3.11}$$

Notice that the variables ϕ_x, ϕ_y have been eliminated and Kirchhoff theory only uses three variables $u_0(x,y)$, $v_0(x,y)$, and $w_0(x,y)$. This makes analytical solutions easier to find, but numerically, Kirchhoff theory is more complex to implement. Since second derivatives of w_0 are needed to write the strains, the weak form (2.30) will have second derivatives of w_0. This will require that the interpolation functions (see Section 2.1.4) have C^1 continuity. That is, the interpolation functions must be such that not only the displacements but also the slopes be continuous across element boundaries. In other words, both the displacement w_0 and the slopes $\partial w_0/\partial x$, $\partial w_0/\partial y$ will have to be identical at the boundary between elements when calculated from either element sharing the boundary. This is difficult to implement.

Consider the case of beam bending. The ordinary differential equation (ODE) with an applied distributed load $\widehat{q}(x)$ is

$$EI\frac{d^4 w_0}{dx^4} = \widehat{q}(x) \tag{3.12}$$

The weak form is obtained as in (2.3).

$$0 = \int_{x_A}^{x_B} v \left[-EI \frac{d^4 w_0}{dx^4} + \widehat{q}(x) \right] dx \qquad (3.13)$$

Integrating by parts twice

$$0 = \int_{x_A}^{x_B} \frac{d^2 v}{dx^2} EI \frac{d^2 w_0}{dx^2} dx + v EI \frac{d^3 w_0}{dx^3} - \frac{dv}{dx} EI \frac{d^2 w_0}{dx^2} - \int_{x_A}^{x_B} v \widehat{q}(x) dx$$

$$0 = B(v, w_0) + [v Q_x]_{x_A}^{x_B} - \left[\frac{dv}{dx} M_x \right]_{x_A}^{x_B} - \int_{x_A}^{x_B} v \widehat{M}(x) dx$$

$$0 = B(v, w_0) + L(v) \qquad (3.14)$$

When the elements are assembled as in Section 2.1.6, it turns out that adjacent elements i and $i+1$ that share a node have identical deflection but opposite shear force Q_x and bending moment M_x at their common node, as follows

$$w^i = w^{i+1}$$
$$Q^i = -Q^{i+1}$$
$$M^i = -M^{i+1} \qquad (3.15)$$

For the shear forces to cancel as in (2.24), it is only required to have $v^i = v^{i+1}$, which is satisfied by C^0 continuity elements having $w^i = w^{i+1}$ at the common node. For the bending moments to cancel as in (2.24), it is required that $dw^i/dx = dw^{i+1}/dx$. This can only be done if the elements have C^1 continuity. That is, the slopes $dw^i/dx = dw^{i+1}/dx$ must be identical at the common node. Such elements are difficult to work with [16, p. 276].

In FSDT theory, only first derivatives are used in the strains (3.6). So, the weak form (2.30) has only first derivatives and, like (2.24), all the internal generalized forces cancel at common nodes with only C^0 element continuity.

3.1.3 Simply Supported Boundary Conditions

Composite plates with coupling effects may have bending, shear, and membrane deformations coupled even if loaded by pure bending, pure shear, or pure in-plane loads (see [3, Figure 6.7]). While the term *simply supported* always means to restrict the transverse deflection $w(x, y)$, it does not uniquely define the boundary conditions on the in-plane displacements u_n and u_s, normal and tangent to the boundary, respectively. In the context of analytical solutions, it is customary to restrict either u_n or u_s. Therefore, the following possibilities exist

- SS-1: $w = u_s = \phi_s = 0$; $N_n = \widehat{N}_n$; $M_n = \widehat{M}_n$

- SS-2: $w = u_n = \phi_s = 0$; $N_{ns} = \widehat{N}_{ns}$; $M_n = \widehat{M}_n$

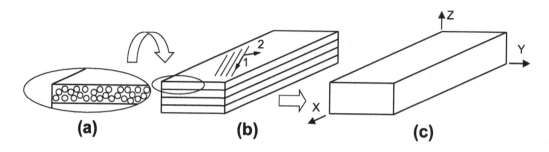

Fig. 3.5: (a) Micromechanics, (b) lamina level, and (c) laminate level approach.

In type SS-1, a normal force and a moment are specified. In SS-2, a shear force and a moment are specified. The naming convention for the rotations is the same as that used for moment resultants in Figure 3.3, where a subscript $()_n$ indicates the direction normal to the edge of the shell, and a subscript $()_s$ indicates the direction tangent to the edge (see also [15, Figure 6.2.1]). Note that the rotation vector ϕ_s is perpendicular to the direction s, resulting in $\phi_s \approx \frac{\partial w}{\partial s}$; therefore, both SS-1 and SS-2 set that rotation to zero. Finally, $\widehat{()}$ represents a fixed known value, which may or may not be zero.

Geometrically linear analysis of a symmetric laminate subjected to bending only will not develop noticeable u_n, u_s, displacements. Therefore, SS-1 and SS-2 should give virtually identical results. Differences may be important for other cases. Furthermore, the condition $\phi_s = 0$ should not be neglected. For example, the calculated center deflection was found to increase by 11.5% when the condition $\phi_s = 0$ was not included in the solution of a sandwich plate [17] using 36 S4R elements.

3.2   Finite Element Analysis of Laminates

Deformation and stress analysis of laminated composites can be done at different levels (Figure 3.5). The level of detail necessary for description of the material depends on the level of post-processing desired.

When a great level of detail is necessary (Figure 3.5.a), the strain and stress are computed at the constituent level, i.e. fiber and matrix. In this case, it is necessary to describe the microstructure, including the fiber shape and geometrical distribution, and the material properties of the constituents. More details are given in Chapter 6 where micromechanical modeling is used to generate properties for any combination of fibers and matrix. Also, when the composite material is a woven fabric, or the laminate is very thick, or when studying localized phenomena such as free edges effects, the composite should be analyzed as solid, as shown in Chapter 5. However, it must be noted that most of the laminated structures can be analyzed using the plates and shell simplifications explained in Section 3.1.

At the other end of the spectrum (Figure 3.5.c), the composite material can be considered as a homogeneous equivalent material. In this case, its structural behavior can be analyzed by using orthotropic properties shown in Chapter 1. If the whole laminate is analyzed as a homogeneous equivalent shell, using the macro-scale level approach (Figure 3.5.c), the stress distribution in the laminate cannot be obtained. However, this very simple description of the laminate is sufficient when only displacements, buckling loads and modes, or vibration frequencies and modes are required. In these cases, only the laminate stiffness (3.8) is needed (see Section 3.2.9). In certain cases, even a simpler material description will suffice. For example, when the laminate is only unidirectional, or if the laminate is balanced and symmetric (see [3, Section 6.3]), the laminate can be modeled as a single lamina of orthotropic material (Section 3.2.10).

In most cases, stress and strains need to be calculated for every lamina in the laminate. Then, the actual laminate stacking sequence (LSS) must be input to the program (Section 3.2.11). In this case, the elastic properties of each lamina, as well as thickness and fiber orientation of every lamina, must be given. This method is usually called the mesoscale level approach (Figure 3.5.b).

A unidirectional lamina can be satisfactorily approximated as being transversely isotropic. Then, it suffices to use $E_3 = E_2$, and $G_{23} = E_3/2(1 + \nu_{23})$ in the equations for an orthotropic material. The elastic properties of a unidirectional lamina can be computed using micromechanics (Chapter 6) or with experimental data of unidirectional composites. Material properties of some unidirectional composites are shown in Table 3.1.

In the analysis of most composite structures, it is usual to avoid the micromechanics approach and to obtain experimentally the properties of the unidirectional lamina, or even the whole laminate. However, the experimental approach is not ideal because a change of constituents or fiber volume fraction during the design process invalidates all the material data and requires a new experimental program for the new material. It is better to calculate the elastic properties of the lamina using micromechanics formulas [3] (see also Section 6.1). Unfortunately, micromechanics formulas are not accurate to predict strength, so experimental work cannot be ruled out completely.

In summary, laminate properties can be specified in two ways:

– by the constitutive matrices A, B, D, and H, or

– by specifying the LSS and properties for every lamina.

When the constitutive matrices A, B, D, H of the laminate are used to define the laminate, the shell element cannot distinguish between different laminas. It can only relate generalized forces and moments to generalized strains and curvatures. On the other hand, laminated shell elements have the capability to compute the laminate properties using the LSS and the properties of the laminas.

3.2.1 Element Types and Naming Convention

Shell elements allow one to model thin to moderately thick shells, down to a side-to-thickness ratio of 10. While some of them have three or four nodes, others have eight nodes, thus using interpolation functions of higher degree. Shell elements are

Table 3.1: Material properties of unidirectional carbon/epoxy composites.[a]

Property	Unit	AS4D/9310	T300/5208
E_1	[GPa]	133.86	136.00
$E_2 = E_3$	[GPa]	7.706	9.80
$G_{12} = G_{13}$	[GPa]	4.306	4.70
G_{23}	[GPa]	2.76	4.261
$\nu_{12} = \nu_{13}$		0.301	0.280
ν_{23}		0.396	0.150
V_f		0.55	
ρ	[g/cm^3]	1.52	1.54
α_1	[$10^{-6}/^\circ C$]	0.32	
α_2	[$10^{-6}/^\circ C$]	25.89	
F_{1t}	[MPa]	1830	1550
F_{1c}	[MPa]	1096	1090
$F_{2t} = F_{3t}$	[MPa]	57	59
$F_{2c} = F_{3c}$	[MPa]	228	207
F_4	[MPa]	141	128
F_6	[MPa]	71	75

[a] F_4 was calculated with $\alpha_0 = 54^\circ$ in [3, (4.84)]

defined in 3D space and have five or six DOF at each node (translations in the nodal x-, y-, and z-directions, and rotations about the nodal x-, y-, and z-axis). The 6th DOF (rotation about the z-axis) is included in the shell formulation to allow modeling of folded plates, but it would not be necessary if the shell surface is smooth.

The Abaqus documentation uses several terms that at first may seem confusing and/or overlapping. They can be summarized as follows

Conventional shell elements require the geometry (mesh) to represent a 2D flat/curved surface in 3D, as opposed to continuum shell elements. Conventional shell elements include thin shell elements, such as S8R5, and thick shell elements such as S8R.

Continuum shell elements require the geometry to model explicitly the thickness of the shell, as it would be done with 3D solid elements, but in contrast to the latter, continuum shell elements enforce the FSDT constraints (Section 3.1.1) through particular element interpolation functions [13]. Continuum shell elements include SC6R and SC8R.

Thin shell elements enforce the Kirchhoff constraints (Section 3.1.2), either theoretically (e.g., STRI3), or numerically when the shell is thin (e.g., S8R5). Therefore, the transverse shear deformations are assumed to be zero or negligible. They are not accurate for composites if the laminate is thick and/or the transverse shear moduli G_{23} of one or more laminas are small, in which case the shear deformation may be underestimated. All thin shell elements are conventional shell elements. Thin shell elements include STRI3, S4R5, STRI65, S8R5, S9R5, SAXA1N, and SAXA2N. They do not have a drilling rotation DOF.

Thick shell elements only enforce the FSDT constraints (Section 3.1.1). Therefore, the transverse shear deformations are not zero. Thick shell elements can be conventional (e.g., S3R, S4, S4R, S8R, SAX1, SAX2), or continuum, such as SC6R and SC8R.

General purpose shell elements are appropriate to model both thick and thin shells. They include thick, conventional shell elements (e.g., S3, S3R, S3RS, S4, S4R, S4RS, S8R, S4RSW, SAX1, SAX2, SAX2T), as well as thick, continuum shell elements such as SC6R and SC8R.

3D solid elements (also called *solid 3D continuum elements*) discretize the 3D body without using any assumptions of shell theory. Since they are not shell elements, the model may be computationally expensive. They can be used for detailed analysis in regions where a rapid variation of stress and strain is expected. They require a 3D mesh, explicitly modeling the 3 dimensions of the shell, including the thickness. Shell-to-solid coupling constraints (Section 2.4) can be used to transition from shell to solid elements.

Another source of confusion arises around the use of the terms 2D and 3D. In shell theory, a shell is a 2D surface because only 2 curvilinear coordinates are needed to locate any point on the reference surface of the shell. But in Abaqus documentation, a shell is said to be 3D because it occupies a portion of 3D space, as opposed to a 2D model that is always planar.

Furthermore, the Abaqus documentation uses the following naming convention for shell elements:

SnRsW Conventional shell

S shell

n number of nodes

R reduced integration

s small membrane strains

W warping included

SCnRT Continuum shell

S shell

C continuum

n number of nodes

R reduced integration

T thermomechanical coupling

STRInm Triangular shell

>   **S** shell
>
>   **TRI** triangular
>
>   **n** number of nodes
>
>   **m** number of DOF

Type SAXAxN axisymmetric shell

>   **S** shell
>
>   **AXA** Axisymmetric
>
>   **1N** meridional linear interpolation, N Fourier modes
>
>   **2N** meridional quadratic interpolation, N Fourier modes

The terms introduced in this section are further explained in the following sections.

3.2.2   Thin (Kirchhoff) Shell Elements

When a shell is thin and the material has a high shear modulus, lines normal to the middle surface remain normal during deformation. This is one of the assumptions of Kirchhoff shell theory, also called the Kirchhoff constraints (see Section 3.1.2).

The main Kirchhoff constraint states that a line normal to the reference surface in the undeformed configuration remains normal to the deformed reference surface. This implies that the transverse shear deformations γ_{xz}, γ_{yz} are zero. Additional assumptions that are common to both Kirchhoff theory and FSDT (Section 3.2.4) are that normals remain straight and inextensible ($\epsilon_{zz} = 0$).

Shell elements that enforce the Kirchhoff constraints are called *thin shell elements*, as opposed to *thick shell elements* that only enforce the FSDT constraints. All the thin shell elements are *conventional shell elements* in the sense that the geometry is represented by a 2D flat/curved surface in 3D, as opposed to *continuum shell elements* that require the geometry to model the thickness of the shell explicitly, as it would be done with 3D solid elements.

Thin shell elements such as STRI3, S4R5, STRI65, S8R5, S9R5, SAXA1N, and SAXA2N are specifically formulated for modeling thin shells. They discretize the reference surface of the shell, and the thickness is given as a section property. Element STRI3 enforces the Kirchhoff constraints analytically so it yields the thin-walled theory solution even for a thick shell. Elements S4R5, STRI65, S8R5, S9R5, SAXA1N, and SAXA2N impose the Kirchhoff constraints numerically. These elements should not be used for applications in which transverse shear deformation is important, either because the shell is thick, or the shear modulus is low, or both. See Example 3.4.a, p. 128.

All elements with a name ending with the number 5 have five DOF per node (three displacements and two in-plane rotations). They do not have drilling rotational DOF (see Section 3.2.8). Five degrees of freedom elements, such as S4R5,

STRI65, S8R5, S9R5, may be more economical than other elements, but they are available only for modeling thin shells and they should not be used to model thick shells.

3.2.3 Thick Shell Elements

Composites should be modeled as thick shells even if they are geometrically thin because they have relatively low shear modulus, thus necessitating inclusion of their shear deformation in the same way as thick shells do. Abaqus provides elements S3R, S3RS, S4, S4R, S4RS, S4RSW, S8R, SC6R, and SC8R to model laminated composite shells. Continuum shell elements, such as SC6R and SC8R, can be used when better resolution of the transverse shear deformation is needed. See Example 3.4.b, p. 128.

3.2.4 General-purpose (FSDT) Shell Elements

FSDT (Section 3.1.1) states that undeformed normals to the reference surface in the undeformed configuration remain straight and inextensible but not necessarily normal to the deformed reference surface. This allows for non-zero transverse shear deformation.

General-purpose shell elements, such as S3, S3R, S3RS, S4, S4R, S4RS, S8R, S4RSW, SAX1, SAX2, and SAX2T, as well as SC6R and SC8R, include transverse shear deformation but they can be used for thin shells as well. Elements S3, S3R, S3RS, S4, S4R, S4RS, S4RSW, S8R, SAX1, SAX2, and SAX2T discretize the reference surface of the shell, and the thickness is given as a section property. Elements S3, S3R, S3RS, S4, S4R, S4RS, S4RSW, and S8R have six DOF per node, with DOF number 6 being a drilling rotation that allows the element to be used to model folded shells. Elements SC6R and SC8R are continuum elements (see Section 3.2.5), but they qualify as general-purpose because they can be used for thin and thick shells.

3.2.5 Continuum Shell Elements

Continuum shell (CS) elements such as SC6R and SC8R are basically 3D solid elements where the FSDT constraints are enforced by special interpolation functions [13]. Compared to conventional shell elements, they have more nodes to depict the thickness in the model.

Continuum shell elements have only displacement degrees of freedom, no rotations. Therefore, they can be stacked and they can be connected to solid elements. Continuum shell elements discretize the 3D geometry of the shell in the same way as 3D solid elements do.

Continuum shell elements can be internally laminated to represent a laminate or sublaminate with just one element. Also, they can be stacked to provide a better representation of the shear deformation through the thickness. In this case, each of the stacked elements may represent a sublaminate. If only one internally laminated element is used to model the entire thickness of the laminate, the result is the same as that of a FSDT element. But CS elements can be stacked, and there lies their advantage. By stacking several CS elements, the actual deformation of the normal can be approximated to any precision. As the number of elements through the thickness approaches infinity, they yield the exact elasticity solution to the plate problem. But they are better than 3D solid elements because CS elements have no aspect ratio problems; i.e., the thickness direction can be very thin compared to the other two dimensions. This is because CS elements incorporate the incompressibility of the normals ($\epsilon_z = 0$,(3.6)), just like FSDT elements.

Meshing is more difficult with continuum shell elements because the thickness of the shell has to be meshed. Also, boundary conditions are more difficult to apply because one cannot apply the classical shell boundary conditions, but rather one has to simulate them by constraining the displacements at the top and bottom surface of the shell. For example, a simply supported boundary condition is rather difficult to apply if the continuum shell mesh does not have nodes located at the middle surface of the shell. Clamped (encastre) and symmetry boundary conditions are easy to apply. See Example 3.4.c, p. 132.

3.2.6   Sandwich Shells

For a sandwich shell, the core is much softer than the faces and the transverse shear deformations are significant regardless of the total thickness of the shell. Conventional shell elements may not have the shear flexibility required. Use continuum shell elements instead. For the most refined model possible, stack three continuum shell elements through the thickness, one for each face and the core. See Example 3.7, p. 145.

3.2.7   Nodes and Curvature

Three and four node elements such as STRI3 and S4R5 are flat. For curved shells, it is better to use six, eight, or nine node elements such as STRI65, S8R5, and S9R5. Element S8R5 has a hidden internal ninth-node, which may end up outside the reference surface for doubly curved shells. If this happens, buckling loads may be inaccurate and S9R5 would be better.

3.2.8   Drilling Rotation

Conventional shell elements are based on shell theory, which constrains the 3D continuum deformations according to some kinematic assumptions, such as Kirchhoff, FSDT, or one of many others. In continuum theory, the deformation of a point in 3D can be described in terms of the relative displacements of two points, thus requiring six degrees of freedom (three per point). Both Kirchhoff and FSDT theories formally reduce the requirement to three displacements and two in-plane rotations at a single point. The two rotations ϕ_x, ϕ_y, are rotations of the normal to the reference surface. These are called in-plane rotations because the rotation vectors lie on the surface of the shell. For a smoothly curved or flat shell, there is no need of tracking the rotation ϕ_z of the normal around itself. This last rotation is called

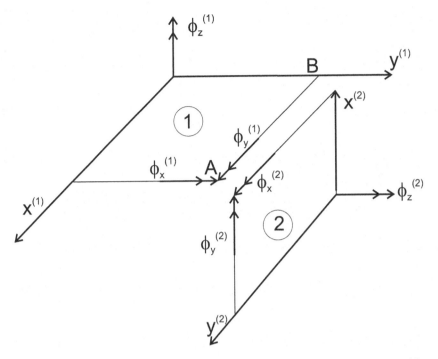

Fig. 3.6: Exploded view of a plate folded along AB. Drilling rotations $\phi_z^{(1)}$ and $\phi_z^{(2)}$ associated to elements 1 and 2, respectively.

drilling rotation. However, if the shell has a fold, an in-plane rotation on one side of the fold corresponds to a drilling rotation on the other side (Figure 3.6). Thus, to enforce compatibility of displacements, the drilling rotations become necessary. Since elements having five DOF do not have drilling rotation, they are suited to model smoothly curved shells but not folded shells.

Abaqus/Standard will automatically switch from five-DOF elements to six-DOF elements at any node when

- A kinematic boundary condition (displacement, velocity) is applied to rotational degrees of freedom at the node,
- The node is used in a multi-point constraint that involves rotational degrees of freedom,
- The node is shared with a beam element or a shell element that uses the three global rotation components at all nodes,
- The node is on a fold line in the shell, or
- The node is loaded with moments.

3.2.9   A, B, D, H Input Data for Laminate FEA

As previously mentioned, macro-scale level (laminate level) analysis is adequate if only deflections, modal analysis, or buckling analysis are to be performed, with no requirement for detailed stress analysis. Then, it is not necessary to specify the laminate stacking sequence (LSS), the thickness, and the elastic properties of each

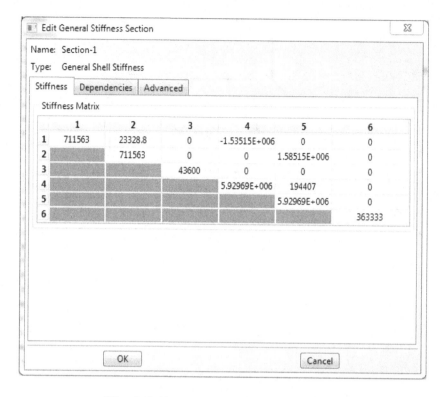

Fig. 3.7: Input A, B, and D matrices.

lamina of the laminate. Only the elastic laminate properties (A, B, D, H matrices) defined in (3.9) are required. This is convenient because it allows one to input the aggregate composite material behavior with few parameters. The reduction of the complexity of the input data allows modeling of laminates with an unlimited number of laminas, using only four matrices.

When the A, B, D, H matrices are used to define the FE analysis, the computer model knows the correct stiffness but it does not know the LSS. Therefore, the software can compute the deformation response (including buckling and vibrations) and even the strain distribution through the thickness of the shell, but it cannot compute the stress components because it does not know where the lamina material properties change from lamina to lamina.

The A, B, D, H input data can be found by using (3.9). Then, these are input into the FE software, as illustrated in Example 3.1.

In Abaqus, the A, B, D matrices can be input as **Edit General Stiffness Section** (Figure 3.7), which is invoked as follows

```
Module: Property
   Menu:   Section, Edit, [Section-name]
```

The A, B, D, matrices are entered on the **Stiffness** tab (Figure 3.7). The H matrix has to be converted into transverse shear stiffness values and entered using the **Advanced** tab (Figure 3.8). The transverse shear stiffness coefficients

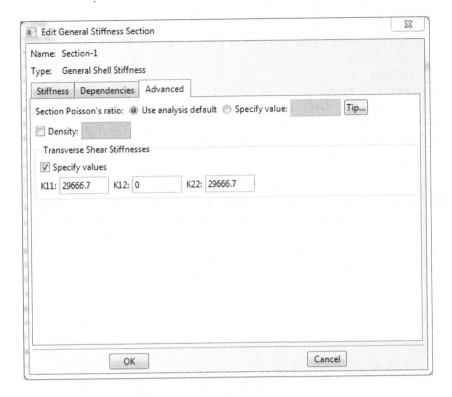

Fig. 3.8: Input H matrix.

H_{44}, H_{45}, H_{55} are defined in (3.9). Taking into account the way Abaqus handles contracted notation, as shown in Table 1.1 (p. 5), compared to the standard (Voigt) definition used in this textbook, it turns out that the notation for transverse shear coefficients in Figure 3.8 means that

$$K_{11} = H_{55}$$
$$K_{22} = H_{44}$$
$$K_{12} = H_{45} \tag{3.16}$$

Example 3.1 *Consider a simply supported square plate $a_x = a_y = 2000 \ mm$, laminated with AS4D/9310 (Table 3.1) in a $[0/90]_n$ configuration. The total laminate thickness is $t = 10 \ mm$ for all values of n. The plate is loaded in compression with an edge load $N_x = 1 \ N/mm$ and $(N_y = N_{xy} = M_x = M_y = M_{xy} = 0)$. Tabulate the center deflection perpendicular to the plate surface when the number of laminas is $n = 1, 5, 10, 15, 20$. Calculate the A, B, D, H matrices and enter them into Abaqus. Use symmetry to model one quarter of the plate.*

Solution to Example 3.1 *The procedure to solve the $[0/90]$ laminate ($n = 1$) is described in detail next. Watch the video in [1].*

i. *Ensure that your files will be stored in the selected folder*

```
Menu: File, Set Work Directory, [C:\SIMULIA\User\Ex_3.1]
Menu: File, Save as, [C:\SIMULIA\User\Ex_3.1\Ex_3.1.cae]
```

ii. Creating parts
   Due to the symmetry of the plate, only 1/4 of it will be modeled.

```
Module: Part
Menu: Part, Create,
      [Part-1], 3D, Deformable, Shell, Planar, Approx. size [4000], Cont
Menu: Add, Line, Rectangle,
      # Enter the points [0,0], [1000,1000], X # to end the command, Done
```

iii. Creating materials and sections, and assigning sections to parts
   In Abaqus the matrices A, B, D, H are entered as properties of the section. Notice
   that, since matrices A, B, D, H contain all the information relevant to the elastic
   behavior of the laminate, it is not necessary to define a material.

   The matrices A, B, D, H are calculated using (3.9). For the $[0/90]_1$ *laminate, the A,*
   B, D, H matrices are:

$$[A] = \begin{bmatrix} 711541 & 23317 & 0 \\ \cdot & 711541 & 0 \\ \cdot & \cdot & 43060 \end{bmatrix} [MPa * mm]$$

$$[B] = \begin{bmatrix} -1585193 & 0 & 0 \\ \cdot & 1585193 & 0 \\ \cdot & \cdot & 0 \end{bmatrix} [MPa * mm^2]$$

$$[D] = \begin{bmatrix} 5929510 & 194306 & 0 \\ \cdot & 5929510 & 0 \\ \cdot & \cdot & 358833 \end{bmatrix} [MPa * mm^3]$$

$$[H] = \begin{bmatrix} 29442 & 0 \\ \cdot & 29442 \end{bmatrix} [MPa * mm]$$

These are entered into Abaqus CAE as follows

```
Module: Property
Menu: Section, Create, [Section-1],
      Shell, General Shell Stiffness, Cont
      # In the pop-up window fill the stiffness matrix as follows
      Tab: Stiffness,
      # Upper part of matrix A in rows 1 to 3, columns 1 to 3
      # matrix B in rows 1 to 3, columns 4 to 6
      # Upper part of matrix D in rows 4 to 6, columns 4 to 6
      # The Stiffness matrix should be
      # 711541      23317        0   -1585193           0         0
      #             711541        0          0     1585193         0
      #                       43060          0           0         0
      #                               5929510      194306         0
      #                                           5929510         0
```

```
        #                                    358833
    Tab: Advanced,
    # The values corresponding to the H matrix are entered here
    Transverse Shear Stiffness, # checkmark: Specify values,
    K11: [29442], K12: [0], K22: [29442], OK
```

Menu: Assign, Section, # pick the part, Done, OK

iv. Creating the Assembly

```
    Module: Assembly
    Menu: Instance, Create, Independent, OK
```

v. Creating steps

```
    Module: Step
    Menu: Step, Create, Name [Step-1], Static/General, Cont, OK
```

vi. Adding BC and loads

```
    Module: Load
    Menu: BC, Manager,
        Create, Name [X-symm], Step: Initial,
        Mechanical, Symm/Anti/Enca, Cont
        # pick the left vertical line, Done, XSYMM, OK
        Create, Name [Y-symm], Step: Initial,
        Mechanical, Symm/Anti/Enca, Cont
        # pick the upper horizontal line, Done, YSYMM, OK
        Create, Name [Simple-Support],
        Step: Initial, Mechanical, Disp/Rota, Cont
        # pick right vertical and lower horizontal lines, Done,
        # checkmark: U3, OK, # close BC Manager
```

```
    Menu: Load, Create, Step: Step-1, Mechanical, Shell edge load, Cont
        # pick the right vertical line, Done, Magnitude [1], OK
```

vii. Creating mesh

```
    Module: Mesh
    Menu: Seed, Instance, Approximate global size [50], Apply, OK
    Menu: Mesh, Controls, Element Shape [Quad], Technique [Structured], OK
    Menu: Mesh, Instance, # At the bottom of the WS select [Yes]
```

viii. Solving and visualizing the solution

```
    Module: Job
    Menu: Job, Manager
        Create, Name [0-90_1], Cont, OK
        # Job name should be changed for each laminate so the user can
        # access results of different models without re-running
        Data Check # To check the model for errors,
        Submit # To run the model,
        Results # To visualize the solution
```

```
Toolbar:    Views, Apply Front View
Menu: Plot, Contours, On Deformed Shape,
    Results, Field Output,
    Output Variable [U], Component [U3], Apply, OK
    # Note that maximum deflection should be -2.196e-01
    # at the upper left corner of the plate
```

ix. Saving the model database

```
Menu: File, Save,
Menu: File, Save As [Ex_3.1(1).cae], OK
Menu: File, New Model Database, With Standard/Explicit Model
```

x. The "Save As" instruction allows the user to keep a copy of the model for the $[0/90]_1$ laminate, while the original model is modified to complete the analysis. Also notice that the results of the model can be accessed using

```
Menu: File, Open, # Browse to C:\SIMULIA\User\Ex_3.1
    File Filter, Output Database (*.odb*), # Select [0-90_1.odb], OK
```

The solution for the remaining laminates with $n > 1$ is similar to the procedure described above. Modifying the number of laminas used, without changing the thickness of the laminate, affects only the bending-extension coupling matrix, B. For the additional stacking sequences, the matrix B is found by using (3.9), which yields

For $[0/90]_5$:

$$[B] = \begin{bmatrix} -317039 & 0 & 0 \\ \cdot & 317039 & 0 \\ \cdot & \cdot & 0 \end{bmatrix} [MPa * mm^2]$$

For $[0/90]_{10}$:

$$[B] = \begin{bmatrix} -158519 & 0 & 0 \\ \cdot & 158519 & 0 \\ \cdot & \cdot & 0 \end{bmatrix} [MPa * mm^2]$$

For $[0/90]_{15}$:

$$[B] = \begin{bmatrix} -105468 & 0 & 0 \\ \cdot & 105468 & 0 \\ \cdot & \cdot & 0 \end{bmatrix} [MPa * mm^2]$$

For $[0/90]_{20}$:

$$[B] = \begin{bmatrix} -79260 & 0 & 0 \\ \cdot & 79260 & 0 \\ \cdot & \cdot & 0 \end{bmatrix} [MPa * mm^2]$$

Next, modify the model previously saved as Ex_3.1.cae to use each of the bending-extension coupling matrices for $n > 1$.

i. Edit the section property, submit, and visualize

```
Menu: File, Open, # Browse to ...\Ex_3.1, [Ex_3.1.cae], OK
```

```
Module: Property
Menu: Section, Edit, Section-1
    # In the pop-up window modify the cells of the new matrix B
    # rows 1 to 3, columns 4 to 6. The new B sub-matrix is
    # -317039          0    0
    #      0      317039    0
    #      0           0    0
OK
```

```
Module: Job
Menu: Job, Manager, Create, Name [0-90_5], Cont, OK
    Data Check # To check the model for errors,
    Submit # To run the model,
    Results # To visualize the solution
```

```
Toolbar:    Views, Apply Front View
Menu: Plot, Contours, On Deformed Shape,
    Results, Field Output,
    Output Variable [U], Component [U3], Apply, OK
```

The maximum deflection should be -2.109×10^{-2} mm at the upper-left corner of the plate.

ii. Save the model database

```
Menu: File, Save,
Menu: File, Save As [Ex_3.1(5).cae], OK
Menu: File, New Model Database, With Standard/Explicit Model
```

The solution is tabulated in Table 3.2. Bending extension coupling produces a lateral deflection, which diminishes for increasing number of laminas n. Note how the bending-extension coupling coefficient $-B_{11}$ decreases proportionally to the number of laminas.

Table 3.2: Lateral deflection vs. number of laminas in Example 3.1.

n	δ [mm]	$-B_{11}$	%
1	0.2196	1,585,193	100
5	0.0211	317,039	1/5
10	0.0104	158,518	1/10
15	0.0069	105,468	1/15
20	0.0052	79,260	1/20

3.2.10   Equivalent Orthotropic Input for Laminate FEA

Some FEA codes do not have laminated elements and do not accept the A, B, D, and H matrices as explained in Section 3.2.9. However, if they have orthotropic elements, it is still possible to perform deformation, vibration, and buckling analysis for laminated composites, as it is shown in this section.

Unidirectional Laminate FEA

To model a unidirectional laminate, standard shell elements can be used, even if they are not *laminated* shell elements, and still obtain correct results of displacements, strains, and stress. The geometry of shells is a surface that represents the mid-surface of the real shell. The mid-surface is located halfway through the thickness. The positive thickness coordinate points along a normal to the shell mid-surface, i.e., the local z-direction, which coincides with the 3-direction. This is the standard definition for shells and it is used in shell elements, as shown in Example 3.2.

Example 3.2 *Use Abaqus to model a simply supported rectangular plate with dimensions* $a_x = 4,000$ *mm,* $a_y = 2,000$ *mm, and thickness* $t = 10$ *mm. Apply a uniform transverse load* $q_0 = 0.12 \times 10^{-3}$ *MPa. The material is a unidirectional AS4D/9310 lamina (Table 3.1), with the fibers oriented in the x-direction. Determine the deflection of the center point of the plate. Use (1.102–1.104) to calculate the orthotropic stiffness matrix and enter the material as* **Type: Orthotropic**. *This example is continued in Example 3.8, p. 158.*

Solution to Example 3.2 *Watch the video in [1].*
The thickness coordinate is eliminated from the governing equations so that the 3D problem simplifies to 2D. In the process, the thickness becomes a parameter, which is known and supplied to the modeling software. The model can be constructed and solved as follows

```
Menu: File, Set Work Directory [C:\SIMULIA\User\Ex_3.2]
Menu: File, Save as [C:\SIMULIA\User\Ex_3.2\Ex_3.2.cae]
```

i. *Creating parts*
   Due to the symmetry of the plate, only 1/4 of it will be modeled.

```
Module: Part
Menu: Part, Create,
      [Part-1], 3D, Deformable, Shell, Planar, Approx. size [4000], Cont
Menu: Add, Line, Rectangle,
      # Enter the points [0,0], [2000,1000], X # ends the command, Done
```

ii. *Creating materials and sections, and assigning sections to parts*

```
Module: Property
Menu: Material, Create,
      Mechanical, Elasticity, Elastic, Type: Orthotropic,
      # Enter the values of the components of the stiffness matrix, OK
```

Notice that the stiffness matrix in Abaqus is built in a different order than the one shown in (1.68). In Abaqus

$$
\begin{Bmatrix} \sigma_1 \\ \sigma_2 \\ \sigma_3 \\ \sigma_6 \\ \sigma_5 \\ \sigma_4 \end{Bmatrix} = \begin{bmatrix} C_{11} & C_{12} & C_{13} & 0 & 0 & 0 \\ \cdot & C_{22} & C_{23} & 0 & 0 & 0 \\ \cdot & \cdot & C_{33} & 0 & 0 & 0 \\ \cdot & \cdot & \cdot & C_{66} & 0 & 0 \\ \cdot & \cdot & \cdot & \cdot & C_{55} & 0 \\ \cdot & \cdot & \cdot & \cdot & \cdot & C_{44} \end{bmatrix} \begin{Bmatrix} \epsilon_1 \\ \epsilon_2 \\ \epsilon_3 \\ \gamma_6 \\ \gamma_5 \\ \gamma_4 \end{Bmatrix}
$$

and $D_{1111} = C_{11}$, $D_{1122} = C_{12}$, $D_{2222} = C_{22}$, $D_{1133} = C_{13}$, $D_{2233} = C_{23}$, $D_{3333} = C_{33}$, $D_{1212} = C_{66}$, $D_{1313} = C_{55}$, $D_{2323} = C_{44}$. *Using the mechanical properties of AS4D/9310 (Table 3.1) to calculate the compliance matrix (1.104), and inverting it, the stiffness matrix is obtained*

$$\begin{bmatrix} 136222.40 & 3908.70 & 3908.70 & 0 & 0 & 0 \\ \cdot & 9251.35 & 3731.31 & 0 & 0 & 0 \\ \cdot & \cdot & 9251.35 & 0 & 0 & 0 \\ \cdot & \cdot & \cdot & 4306.00 & 0 & 0 \\ \cdot & \cdot & \cdot & \cdot & 4306.00 & 0 \\ \cdot & \cdot & \cdot & \cdot & \cdot & 2760.00 \end{bmatrix} [MPa]$$

```
Menu: Section, Create,
      Shell, Homogeneous, Cont
      Shell thickness [10], Material [Material-1], OK

Menu: Assign, Section, # pick the part, Done, OK
```

Another option would be to enter the material properties as **Engineering Constants** *(9 values) or* **Lamina** *(5 values[2]) in the* **Edit Materials** *window of Abaqus CAE.*

iii. *Creating the assembly*

```
Module: Assembly
Menu: Instance, Create, Independent, OK
```

iv. *Creating analysis steps*

```
Module: Step
Menu: Step, Create, Name [Step-1], Static/General, Cont, OK
```

v. *Adding BC and loads*

```
Module: Load
Menu: BC, Manager,
      Create, Name [X-symm], Step: Initial,
      Mechanical, Symm/Anti/Enca, Cont
      # pick the left vertical line, Done, XSYMM, OK
      Create, Name [Y-symm], Step: Initial,
      Mechanical, Symm/Anti/Enca, Cont
      # pick the upper horizontal line, Done, YSYMM, OK
      Create, Name [Simple-supp], Step: Initial,
      Mechanical, Disp/Rota, Cont
      # pick the right vertical and the lower horizontal lines, Done,
      # checkmark: U3, OK, # close BC Manager

Menu: Load, Create
```

---

[2]Considering that $G_{12} = G_{13}$ for a transversely isotropic material.

```
     Step: Step-1, Mechanical, Pressure, Cont
     # pick the part, Done,
     # Select side for the shell, Brown, Magnitude [0.12e-3], OK
```

vi. Creating the mesh

```
   Module: Mesh
   Menu: Seed, Instance, Approximate global size [50], Apply, OK
   Menu: Mesh, Controls, Element Shape [Quad], Technique [Structured], OK
   Menu: Mesh, Instance, # At the bottom of the WS select [Yes]
```

vii. Solving and visualizing the solution

```
   Module: Job
   Menu: Job, Manager
        Create, Cont, OK
        Data Check # To check the model for errors,
        Submit # To run the model
        Results # To visualize the solution

   Toolbar: Views, Apply Front View
   Menu: Plot, Contours, On Deformed Shape,
        Results, Field Output,
        Output Variable [U], Component [U3], Apply, OK
```

The maximum deflection is 17.43 mm at the center of the plate.

Symmetric Laminate FEA

If a multi-directional laminate is balanced and symmetric, the apparent laminate orthotropic properties can be found as explained in Section 1.15. The apparent laminate properties represent the stiffness of an equivalent (fictitious) orthotropic plate that behaves like the actual laminate under in-plane loads. These apparent properties should not be used to predict bending response. When the only important response is bending, e.g., a thick cantilever plate under bending, the formulation shown in [3, (6.43)] should be used to obtain the apparent laminate properties. However, in most of the structural design using composite shell, the laminate works under in-plane loads and the formulation in Section 1.15 should be used.

If the laminate is symmetric but not balanced, the axes of orthotropy are rotated with respect to the laminate coordinate system, but still the laminate is equivalent to an orthotropic material as per Section 1.15. For example, a unidirectional laminate oriented at an angle θ with respect to global axes should be modeled on a coordinate system oriented along the fiber direction (see Section 3.2.14).

Example 3.3 *Use Abaqus to model a simply supported rectangular plate with dimensions $a_x = 2000$ mm, $a_y = 2000$ mm, for a laminate $[\pm 45/0]_S$. Apply a tensile edge load $N_x = 200$ N/mm. Determine the maximum horizontal displacement. Each lamina is 1.0 mm thick with the following properties:*

$$E_1 = 37.88\ GPa \quad G_{12} = 3.405\ GPa \quad \nu_{12} = 0.299$$
$$E_2 = 9.407\ GPa \quad G_{23} = 3.308\ GPa \quad \nu_{23} = 0.422$$

Solution to Example 3.3 *Watch the video in [1].*
Since the laminate is balanced symmetric, compute the averaged laminate properties E_x, E_y and so on using Section 1.15 to yield

$$E_x = 20.104 \ GPa \quad G_{xy} = 8.237 \ GPa \quad \nu_{xy} = 0.532$$
$$E_y = 12.042 \ GPa \quad G_{yz} = 3.373 \ GPa \quad \nu_{yz} = 0.203$$
$$E_z = 10.165 \ GPa \quad G_{xz} = 3.340 \ GPa \quad \nu_{xz} = 0.307$$

The model is implemented in Abaqus following the steps:

```
Menu: File, Set Work Directory, [C:\SIMULIA\User\Ex_3.3]
Menu: File, Save as, [C:\SIMULIA\User\Ex_3.3\Ex_3.3.cae]
```

i. *Creating parts*
   Due to the assumed symmetry of the plate, only 1/4 of it will be modeled.

```
Module: Part
Menu: Part, Create,
      [Part-1], 3D, Deformable, Shell, Planar, Approx. size [4000], Cont
Menu: Add, Line, Rectangle,
      # Enter the points [0,0], [1000,1000], X # ends the command, Done
```

ii. *Creating materials and sections, and assigning sections to parts*

```
Module: Property
Menu: Material, Create,
      Mechanical, Elasticity, Elastic, Type [Orthotropic],
# Enter the values of the components of the stiffness matrix, OK
```

Notice that the stiffness matrix in Abaqus is built in a different order than the one shown in (1.68). In Abaqus

$$\begin{Bmatrix} \sigma_1 \\ \sigma_2 \\ \sigma_3 \\ \sigma_6 \\ \sigma_5 \\ \sigma_4 \end{Bmatrix} = \begin{bmatrix} C_{11} & C_{12} & C_{13} & 0 & 0 & 0 \\ \cdot & C_{22} & C_{23} & 0 & 0 & 0 \\ \cdot & \cdot & C_{33} & 0 & 0 & 0 \\ \cdot & \cdot & \cdot & C_{66} & 0 & 0 \\ \cdot & \cdot & \cdot & \cdot & C_{55} & 0 \\ \cdot & \cdot & \cdot & \cdot & \cdot & C_{44} \end{bmatrix} \begin{Bmatrix} \epsilon_1 \\ \epsilon_2 \\ \epsilon_3 \\ \gamma_6 \\ \gamma_5 \\ \gamma_4 \end{Bmatrix}$$

and $D_{1111} = C_{11}$, $D_{1122} = C_{12}$, $D_{2222} = C_{22}$, $D_{1133} = C_{13}$, $D_{2233} = C_{23}$, $D_{3333} = C_{33}$, $D_{1212} = C_{66}$, $D_{1313} = C_{55}$, $D_{2323} = C_{44}$. Using the mechanical properties given above to calculate the compliance matrix (1.104), and inverting it, the stiffness matrix C is (see Scilab code Ex-3-3.sce in [2])

$$C = \begin{bmatrix} 26564 & 10103 & 5350 & 0 & 0 & 0 \\ 10103 & 16927 & 5428 & 0 & 0 & 0 \\ 5350 & 5428 & 12122 & 0 & 0 & 0 \\ 0 & 0 & 0 & 3340 & 0 & 0 \\ 0 & 0 & 0 & 0 & 3373 & 0 \\ 0 & 0 & 0 & 0 & 0 & 8237 \end{bmatrix}$$

Note that row/columns 4 and 6 must be transposed for Abaqus (Section 1.5.1). Then, the stiffness matrix is entered into Abaqus CAE, as follows

```
Menu: Section, Create,
    Shell, Homogeneous, Cont
    Shell thickness [6], Material [Material-1], OK
Menu: Assign, Section, # pick the part, Done, OK
```

Alternatively, one could enter the material properties as **Engineering Constants** *(9 values) in the* **Edit Materials** *window of Abaqus CAE.*

iii. *Creating the assembly*

```
Module: Assembly
Menu: Instance, Create, Independent, OK
```

iv. *Creating analysis steps*

```
Module: Step
Menu: Step, Create, Name [Step-1], Static/General, Cont, OK
```

v. *Adding BC and loads*

```
Module: Load
Menu: BC, Manager,
    Create, Name [X-symm], Step: Initial, Mechanical,
    Symm/Anti/Enca, Cont
    # pick the left vertical line, Done, XSYMM, OK
    Create, Name [Y-symm], Step: Initial, Mechanical,
    Symm/Anti/Enca, Cont
    # pick the upper horizontal line, Done, YSYMM, OK
    Create, Name [Simple-supp], Step: Initial, Mechanical,
    Disp/Rota, Cont
    # pick the right vertical and the lower horizontal lines, Done,
    # checkmark: U3, OK, # close BC Manager

Menu: Load, Create
Step: Step-1, Mechanical, Shell edge load, Cont
    # pick the right vertical line, Done, Magnitude [-200], OK
```

vi. *Creating the mesh*

```
Module: Mesh
Menu: Seed, Instance, Approximate global size [50], Apply, OK
Menu: Mesh, Controls, Element Shape [Quad], Technique [Structured], OK
Menu: Mesh, Instance, # At the bottom of the WS select [Yes]
```

vii. *Solving and visualizing the solution*

```
Module: Job
Menu: Job, Manager
    Create, Cont, OK
    Data Check # To check the model for errors,
    Submit # To run the model
```

```
Results # To visualize the solution
```

```
# 'Apply Front View' from the 'Views toolbar',
Menu: Plot, Contours, On Deformed Shape,
      Results, Field Output,
      Output Variable [U], Component [U1], Apply, OK
```

The maximum displacement should be 1.667 at the right edge of the plate.

The resulting maximum horizontal displacement on a quarter-plate model is 1.667 mm. The planes $x = 0$ and $y = 0$ are not symmetry planes for a $[\pm 45/0]_S$ but once the laminate is represented by equivalent orthotropic properties, as is done in this example, the lack of symmetry at the lamina level is lost and it does not have any effect on the mid-surface displacements. Therefore, one quarter of the plate represents well the entire plate as long as no stress analysis is performed. Furthermore, displacement and mid-surface strain analysis can be done with the laminate replaced by an equivalent orthotropic material. However, even if the full plate were to be modeled, the stress values in the equivalent orthotropic material are not the actual stress values of the laminate. While the material analyzed in this example is not homogeneous, but laminated, the material in Example 3.2 is a homogeneous unidirectional material. Therefore, the stress values are not correct in this example but they are correct in Example 3.2, p. 120.

Asymmetric Laminate FEA

If the laminate is not symmetric, bending-extension coupling must be considered. Strictly speaking, such material is not orthotropic and should not be modeled with an equivalent laminate material. Even then, if only orthotropic shell elements are available and the bending-extension coupling effects are not severe, the material could be approximated by an orthotropic material by neglecting the matrices B and D. See [3, Section 6.5.1] for tools that can be used to assess the quality of the approximation obtained using apparent elastic properties. Care must be taken for unbalanced laminates that the A and H matrix are formulated in a coordinate system coinciding with the axes of orthotropy of the laminate.

3.2.11 LSS for Multi-directional Laminate FEA

For computation of strain and stress at the mesoscale (lamina level), it is necessary to know the description of the laminate and the properties of each lamina. The description of the multi-directional laminate includes the LSS, which specifies the angle of each lamina with respect to the x-axis of the laminate, the thickness, and the elastic material properties of each lamina. Then, the software computes the matrices A, B, D, and H internally. In this way, the software can compute the stress components in each lamina. This approach is illustrated in the Example 3.4, p. 128. In Abaqus CAE, the LSS can be specified in two ways, using Composite Sections or Composite Layups.

Composite Section

Composite Section is the basic tool that allows the user to define the LSS. It does not differ from the classical definition and assignment of all other types of sections such as *solid, shell, beam, truss,* and so on. First, one defines one or more materials, which in this case could be type *isotropic, engineering constants, lamina, orthotropic,* or *anisotropic.* In CAE, this is done as follows

```
Module: Property
Menu: Material, Create, Mechanical, Elasticity, Elastic, Type:Lamina
```

Second, one defines a laminated section, as follows

```
Module: Property
Menu: Section, Create, Category: Solid, Composite, Cont
      Material:[select], Thickness:[value], Orientation Angle:[value],
      Integration Points: [value], Ply name: [value]
```

Note that *category* can be *shell* or *solid* depending on what type of element you intend to use. *Continuum shell elements* (e.g., SC8R), *thin shell elements* (e.g., S8R5), and *general-purpose shell elements* (e.g., S8R) need a *shell, composite* section. Solid elements, such as C3D20, need a *solid, composite* section.

Solid elements can be laminated but the maximum order of variation of the displacements is quadratic. Therefore, the strain variation can be at most linear. This is often insufficient to model the variation of strain through the thickness of a laminate. To alleviate this problem, solid elements can be stacked, even at one element per lamina. However, due to aspect ratio limitations, a solid element with the thickness of a lamina should not have the other two dimensions larger than about 10 times the thickness. This would lead to a very refined mesh that would result in a computationally expensive solution.

For each section defined, the *Edit Section* window allows the user to define the LSS by giving the following data for each lamina in the laminate: *Material, Thickness, Orientation Angle, Integration Points,* and *Ply Name.* Ply Name is optional, but recommended because it helps understand the results. The material is chosen from the materials previously defined. Thickness and orientation are required input data. For integration points, it is best to use the default, which is three points using the Simpson rule. In this way, one point is on each surface and one in the middle of each lamina. This is optimal for results visualization provided one remembers to modify the *field output requests,* as follows

```
Module: Step
Menu:   Output, Field Output Requests, Edit, F-Output-1,
        Output at shell, Specify: 1,2,3
```

where $1, 2, 3, ..., 3N$, are the Simpson points through the thickness of the shell, where N is the number of laminas in the laminate. Then, visualization is done by selecting

```
Module: Visualization
Menu:   Results, Section Points, Selection method: Plies,
        Plies: [select],
        Ply result location: [Bottom-most, Middle/Single, Topmost,
          Top & Bottom]
```

If one selects Top & Bottom, the results on the faces of the laminate are shown. If a face is not in view, just rotate the model to see it.

Third, one must *assign* sections to *regions* in the classical way. If using *conventional shell elements* and *general-purpose shell elements*, the *regions* will be a flat or curved surfaces, i.e., 2D cells. If using *continuum shell elements*, the regions will be 3D cells.

If using *conventional shell elements* and *general-purpose shell elements*, the entire LSS is defined in a *composite section* and assigned to a *region*. *Continuum shell elements* may span the entire thickness of the shell, in which case the assignment is similar, with one composite section assigned to each flat/curved surface. However, continuum shell elements can be stacked. This is accomplished by partitioning the thickness of the 3D cell to delimit the regions for each continuum shell element. In this case, a different composite section can be assigned to each region (cell). Each continuum shell element will be laminated as mandated by its associated composite section definition.

Composite Layup

Composite Layup is an alternate way of defining the LSS. The concept of region is different than the usual concept of region in classical CAE. A *composite layup* must be associated to a single element through the thickness of a region. If you refine the mesh through the thickness of a region (cell) spanned by a composite layup, each element through the thickness will be laminated as specified by the composite layup, which would lead to erroneous results.

With composite layup, the user does not create a section nor assign a section to a region. Rather, the *composite layup* does the definition and assignment simultaneously. If by mistake, both *composite section* and *composite layup* are used for the same region, *composite layup* overwrites the former.

A composite layup window is invoked as follows

```
Module: Property
Menu:   Composite, Create
```

At this point one must decide what type of section to create: Conventional shell, Continuum shell, or Solid. *Continuum shell elements* (e.g., SC8R), *thin shell elements* (e.g., S8R5), and *general-purpose shell elements* (e.g., S8R) need a *shell, composite* section. Solid elements, such as C3D20, need a *solid, composite* section.

The Edit Composite Layup window (Figure 3.9) has four tabs: Plies, Offset, Shell Parameters, and Display. The LSS is specified in the Plies tab. The position of the reference surface is given in the Offset tab. The thickness modulus is specified in the Shell Parameters tab.

The Plies tab has seven columns: Ply Name, Region, Material, Relative Thickness, CSYS, Rotation Angle, and Integration Points. Ply Name is optional, but recommended because it helps understand the results. Region ties the ply to a region. The material is chosen from the materials previously defined. The ply relative thickness are proportionality factors used to determine the thickness of the ply as a function of the thickness of the element, which may be given by the mesh. They do not need to be actual thicknesses or add up to 1.0. CSYS is used to have additional flexibility in specifying the ply orientation with respect to the global coordinate system of the model. Basically, CSYS allows the user to rotate the laminate coordinate system as a whole. Rotation Angle is the ply angle θ_k with respect to the laminate coordinate system. For integration points, it is best to use the default, which is three points using the Simpson rule. In this way, one point is on each surface and one in the middle of each lamina. This is optimal for results visualization provided one remembers to modify the *field output requests* as explained in Section 3.2.11.

Example 3.4 *Consider a simply supported square plate $a_x = a_y = 2000 \ mm$, $t = 10 \ mm$ thick, laminated with AS4D/9310 (Table 3.1) in a $[0/90/\pm 45]_S$ symmetric laminate configuration. The plate is loaded with a tensile load $N_x = 100 \ N/mm$ ($N_y = N_{xy} = M_x = M_y = M_{xy} = 0$). Compute the in-plane shear stress in lamina coordinate system σ_6 using four modeling techniques:*

    a. with conventional shell elements,
    b. with general-purpose shell elements,
    c. with continuum shell elements in a composite section with a single element through the plate thickness, and
    d. with continuum shell elements in a composite layup with several elements through the plate thickness.

This example is continued in Example 3.12, p. 174.

Solution to Example 3.4 *Watch the video in [1].*
Parts (a) and (b) involve conventional shell elements *using a mesh without thickness. Therefore, most of the model for parts (a) and (b) is independent of the type of element. Note the LSS is given starting at lamina #1 at the bottom of the laminate. The full plate is modeled without taking advantage of symmetry, and nodes are constrained to illustrate how to constrain a model to prevent rigid body motion. The following pseudo-code illustrates the whole procedure. If in doubt, open* Ex_3.4.cae *available in [2].*

    i. *Make sure the .cae file is saved in the right place. You can also set the* Work Directory *to the same folder where you saved the* .cae *file*

```
Menu:   File, Set Work Directory [c:\simulia\user\Ex_3.4]
Menu:   File, Save as [c:\simulia\user\Ex_3.4\Ex_3.4.a.cae]
```

    ii. *Creating parts*

```
Module: Part,
Menu:   Part, Create,
```

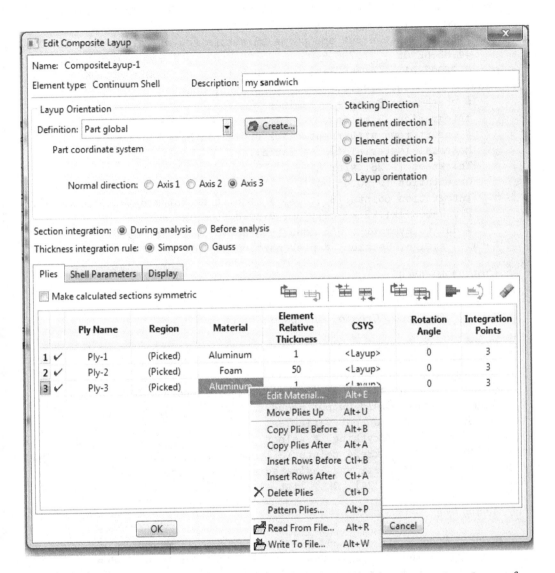

Fig. 3.9: Entering the laminate stacking sequence using the `Composite Layup` feature.

> Name [Part-1], 3D, Deformable, Shell, Planar, Approx size [2000], Cont
> Menu: Add, Line, Rectangle, [-1000,-1000], [1000,1000], X, Done

iii. Creating materials and sections, and assigning sections to parts

> Module: Property
> Menu: Material, Create, Name [Material-1], Mech., Elast., Elastic,
>   Type [Engineering Constants],
>   # Note: you can cut and paste an entire row of values from Excel
>   [1.3386E+05   7.7060E+03   7.7060E+03   3.0100E-01 3.0100E-01
>   3.9600E-01   4.3060E+03   4.3060E+03   2.7600E+03], OK
> Menu: Section, Create, Shell, Composite, Cont
>   # checkmark: Symmetric laminas
>   Integration [Simpson]
>   # insert rows with right click until you have 4 rows
>   Material [Material-1]   # for all
>   Thickness [1.25] #for all
>   Orientation [0,90,45,-45]
>   Integration points [3,3,3,3] # results at bot/mid/top of each ply
>   Ply Names [k1,k2,k3,k4], OK, # k1:bottom, k4:middle
>   # due to symmetry, CAE adds [-45,45,90,0] to the model,
> Menu: Assign, Section, # pick part, Done, OK

iv. Creating assembly

> Module: Assembly
> Menu: Instance, Create, Independent, OK

v. Creating analysis steps

> Module: Step
> Menu: Step, Create, Name [Step-1], Static/General, Cont, OK
> Menu: Output, Field Output Requests, Edit, F-Output-1
>   Step: Step-1, Output variables:
>   # unselect everything, # select only what you need: S, E, U, RF
>   # request at 8x3=24 points (8 laminas, 3 Simpson points per lamina)
>   # Output at shell and laminated section points:
>   Specify: [1,...24], OK # list all #'s from 1 to 24

vi. Adding loads

> Module: Load
>   # since the part is a shell, an edge load is: Shell edge load
>   Menu: Load, Create, Name [Load-1], Step: Step-1, Mechanical,
>   Shell edge load, Cont
>   # pick the edges x=-1000 and x=1000, Done, Magnitude [-100], OK

In order to identify sets of nodes to constrain, when setting the BC, it is required to mesh the model now.

vii. Creating mesh

> Module: Mesh
> Menu: Seed, Instance, Approx global size [100], Apply, OK

Menu: Mesh, Controls, Element Shape [Quad], Technique [Structured],
        OK
Menu: Mesh, Element Type, Std, Quad, DOF [5], OK # element S8R5
Menu: Mesh, Instance, Yes
Menu: Tools, Set, Manager
        Create, [center], Node, Cont, # center node, Done
        Create, [y-hold], Node, Cont, # nodes at (-1000,0) and (1000,0),
        Done
        Create, [edges], Node, Cont, # nodes at the edges of the plate,
        Done
        # close Set Manager

viii. Adding BC

Module: Load
Menu: BC, Manager
        Create, Name [edges], Step: Initial, Mechanical, Disp/Rota, Cont
        Sets, # in the Region Selection pop-up window, [edges], Cont
        # checkmark: U3, OK
        Create, Name [center], Step: Initial, Mechanical, Disp/Rota, Cont
        # in the Region Selection pop-up window, [center], Cont
        # checkmark: U1, U2, and U3, OK
        Create, Name [y-hold], Step: Initial, Mechanical, Disp/Rota, Cont
        # in the Region Selection pop-up window, [y-hold], Cont
        # checkmark: U2, and U3, OK
        # close BC Manager

ix. Solving and visualizing the solution

Module: Job
Menu: Job, Manager
        Create, Cont, OK,
        Submit # to run the model
        Results # to visualize the solution

Menu: Viewport, Viewport Annotation Options, Tab: State Block,
        Set font, Size [12], Apply to [# checkmark all], OK, OK
Menu: Plot, Undeformed Shape,
Menu: Plot, Allow Multiple Plot States,
Menu: Plot, Deformed Shape,
Menu: Plot, Material Orientations, On Deformed Shape
        # by default shows the 0-deg lamina on bottom
Menu: Result, Section Points, Plies, # select K1, Apply
        # repeat for all the plies to visualize their orientation, OK
Menu: Plot, Contour, On Deformed Shape,
        Field Output Toolbox, Primary, S, S12 # laminate stress_6
Menu: Result, Section Points, Plies, # select K1, Apply
        # repeat for all the plies to visualize S12, OK
        # select ply K1, Apply # sigma_6(K1) = 0.0
        # select ply K2, Apply # sigma_6(K2) = 0.0
        # select ply K3, Apply # sigma_6(K3) = -1.112
        # select ply K4, Apply # sigma_6(K4) = 1.122, OK

(a) Solution using thin shell elements

In the procedure given above, selection of a conventional shell element S8R5 is illustrated by:

```
Module: Mesh
Menu:   Mesh, Element Type, Standard, Quadratic, DOF [5], OK
```

which results in element S8R5 being chosen. It is convenient to save the solution of this problem, and modify the model to solve part (b).

```
Menu:   File, Save as [c:\simulia\user\Ex_3.4\Ex_3.4.a.cae], OK
```

Visualization of lamina stresses in the lamina coordinate system results in

$$\sigma_6 = 0, 0, -1.112, 1.112 \, MPa$$

in the $0, 90, 45°, -45°$ *plies, respectively. These results can be verified by comparison with classical lamination theory (CLT) [3].*

(b) Solution using general-purpose shell elements

To change the analysis from thin shell to general-purpose shell, one only needs to change the element type, by selecting DOF [6], *as follows*

```
Menu:   File, Save as [c:\simulia\user\Ex_3.4\Ex_3.4.b.cae], OK
Module: Mesh
Menu:   Mesh, Element Type, Standard, Quadratic, DOF [6], OK # S8R
```

This results in element S8R being chosen–a doubly-curved, general-purpose (thick/thin) shell with reduced integration. The job must be resubmitted for execution. Since this example does not involve bending, the results are identical to those of part (a).

(c) Solution using continuum shell elements in a composite shell

To use continuum shell elements [13], the geometric model must be that of a solid, in this case with dimensions $2,000 \times 2,000 \times 10 \, mm$.

i. Saving the model, and setting the Work Directory

```
Menu:   File, Save as [c:\simulia\user\Ex_3.4\Ex_3.4.c.cae]
Menu:   File, Set Work Directory [c:\simulia\user\Ex_3.4]
```

ii. Creating parts

```
Module: Part
Menu:   Part, Create,
        Name [Part-1], 3D, Deformable, Solid, Extrusion, size [2000], Cont
Menu:   Add, Line, Rectangle, [-1000,-1000], [1000,1000], X, Done,
        Depth [10], OK
```

iii. Creating materials and sections, and assigning sections to parts

```
Module: Property
Menu: Material, Create, Name [Material-1], Mech., Elast., Elastic,
      Type [Engineering Constants],
      # Note: you can cut and paste an entire row of values from Excel
```

        [1.3386E+05   7.7060E+03   7.7060E+03   3.0100E-01 3.0100E-01
        3.9600E-01   4.3060E+03   4.3060E+03   2.7600E+03], OK
    Menu: Section, Create, [Section-1], Shell, Composite, Cont
        # insert rows with right click until you have 8 rows
        Material [Material-1]   # for all, Thickness [1.25] # for all
        Orientation [0,90,45,-45,-45,45,90,0],
        Integration points [3] # for all
        Ply Names [PLY-1,PLY-2,PLY-3,PLY-4,PLY-5,PLY-6,PLY-7,PLY-8], OK
    Menu: Assign, Section, # pick the part, Done, OK

iv. Creating assembly

    Module: Assembly
    Menu: Instance, Create, Independent, OK

v. Creating analysis steps

    Module: Step
    Menu: Step, Create, Name [Step-1], Static/General, Cont, OK
    Menu: Output, Field Output Requests, Edit, F-Output-1
        Step: Step-1, Output variables:
        # unselect everything, # select only what you need: S, E, U, RF
        # request at 8x3=24 points (8 laminas, 3 Simpson pts per lamina)
        # Output at shell and laminated section points:
        Specify: [1,...24], OK # 1,2,3,4,...

vi. Adding loads

    Module: Load
    Menu: Load, Create
        Name [Load-1], Step: Step-1, Mechanical, Surface traction, Cont
        # an edge load 100 N/mm on a surface 10 mm-thick equals 10 N/mm2
        # Zoom into the upper-right corner of the plate
        # pick the surface x=1000, Done,
        Distribution [Uniform], Traction [General],
        Vector, Edit, [0,0,0], [1,0,0], Magnitude [10], OK
        # repeat the procedure to apply a surface traction of 10 N/mm2
        # on the x=-1000 surface, # Name [Load-2] and direction [-1,0,0]

In order to identify the nodes to constrain, when setting the BC, it is required to mesh the model now.

vii. Creating mesh

    Module: Mesh
    Menu: Seed, Instance, Approx global size [100], OK
    Menu: Mesh, Controls, # if required select the whole plate, Done,
        Element Shape [Hex], Technique [Sweep], OK
    Menu: Mesh, Element Type, # if required select whole plate, Done,
        Family [Continuum Shell], # element SC8R has been selected, OK
    Menu: Mesh, Instance, Yes
        # create sets of nodes to facilitate applying the BC to the model
        Views Toolbar, Front View # to visualize better
    Menu: Tools, Set, Manager

```
Create, Name [edges], Type [Node], Cont
# select all the nodes along the edges of the plate, Done
Create, Name [center], Type [Node], Cont
# select the nodes at the center of the plate (x=0,y=0), Done
Create, Name [y-hold], Type [Node], Cont
# select the nodes at (x=-1000,y=0) and (x=1000,y=0), Done
# close the Set Manager pop-up window
```

viii. *Applying BC*

```
Module: Load
Menu:   BC, Manager
    Create, Name [edges], Step: Initial, Mechanical, Disp/Rota, Cont
    Sets, # in the Region Selection pop-up window, [edges], Cont
    # checkmark: U3, OK
    Create, Name [center], Step: Initial, Mechanical, Disp/Rota, Cont
    # in the Region Selection pop-up window, [center], Cont
    # checkmark: U1, U2, and U3, OK
    Create, Name [y-hold], Step: Initial, Mechanical, Disp/Rota, Cont
    # in the Region Selection pop-up window, [y-hold], Cont
    # checkmark: U2, and U3, OK
    # close BC Manager
```

ix. *Solving and visualizing the solution*

```
Module: Job
Menu:   Job, Manager
    Create, Cont, OK,
    Submit  # to run the model
    Results # to visualize the solution

Menu:   Viewport, Viewport Annotation Options, Tab: State Block,
    Set font, Size [12], Apply to [# checkmark all], OK, OK
Menu:   Plot, Undeformed Shape,
Menu:   Plot, Allow Multiple Plot States,
Menu:   Plot, Deformed Shape,
Menu:   Plot, Material Orientations, On Deformed Shape
Menu:   Result, Section Points, Plies, # select PLY-1, Apply
    # repeat for all the plies to visualize their orientation, OK
Menu:   Plot, Contour, On Deformed Shape,
    Field Output Toolbox, Primary, S, S12 # laminate stress_6
Menu:   Result, Section Points, Plies, # select PLY-1, Apply
    # repeat for all the plies to visualize S12 stress, OK
```

(d) Solution using continuum shell elements in a composite layup

In order to use a 'composite layup' to model the laminate using CS elements [13], the solid part must be divided in regions, each one corresponding to a lamina of the laminate. The creation of the part must be modified, as follows

i. *Creating parts*

```
Module: Part
```

```
Menu:   Part, Create,
        Name [Part-1], 3D, Deformable, Solid, Extrusion, size [2000], Cont
Menu:   Add, Line, Rectangle, [-1000,-1000], [1000,1000], X, Done,
        Depth [10], OK
        # zoom into the upper-right corner of the plate
        # divide the thickness of the plate in 8 segments
Menu:   Tools, Partition, Type [Edge], Method [Enter parameter]
        # pick edge, Done, Normalized edge param [0.5], Create Partition
        # repeat procedure until the edge is divided in 8 segments, Done
        # in the Create Partition pop-up window select
        Type [Cell], Method [Define cutting plane], Point & Normal,
        # select one point defined previously (start at top of the plate),
        # select edge used as normal direction (along positive z-axis),
        Create Partition # create first region representing the top ply
        # pick region to partition, Done,
        Point & Normal, # repeat procedure,
        # once 8 regions have been created, Done, # close pop-up window
```

The definition of the material properties must be modified as follows.

ii. Creating materials and sections, and assigning sections to parts

```
Module: Property
Menu:   Material, Create, Name [Material-1], Mech., Elast., Elastic,
        Type [Engineering Constants],
        # Note: you can cut and paste an entire row of values from Excel
        [1.3386E+05   7.7060E+03   7.7060E+03   3.0100E-01   3.0100E-01
        3.9600E-01   4.3060E+03   4.3060E+03   2.7600E+03], OK
Menu:   Composite, Create
        Initial ply count [8], Element Type [Continuum Shell], Cont
        Thickness integration rule [Simpson]
        # in column named Region double-click cell corresponding Ply-1,
        # pick top region of laminate, Done, # repeat for other plies
        # in column named Material double-click cell corresponding Ply-1,
        Material [Material-1], OK
        # repeat to assign Material-1 to the other plies
        Element Relative Thickness [1] # for all plies
        Rotation Angle [0, 90, 45, -45, -45, 45, 90, 0] # one per ply
        Integration points [3] # for all plies, OK
```

The rest of the solution follows the same steps described in part (c).

3.2.12 FEA of Ply Drop-Off Laminates

Sometimes it is convenient to set the reference surface at the bottom (or top) of the shell. One such case is when the laminate has ply drop-offs, as shown in Figure 3.10. When the design calls for a reduction of laminate thickness, plies can be gradually terminated from the thick to the thin part of the shell. As a rule of thumb, ply drop-off should be limited to a 1:16 to 1:20 ratio ($Th : L$ ratio in Figure 3.10) unless detailed analysis and/or testing supports a steeper drop-off ratio. For this case, it is convenient to specify the geometry of the smooth surface, or tool surface.

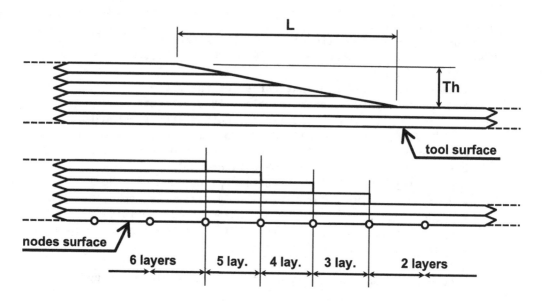

Fig. 3.10: Ply drop-off of length (L) and thickness (Th) and FE model simplifications.

Then, every time a ply or set of plies is dropped, the material and thickness for those elements is changed. This is illustrated in the next examples. Since not all software has this capability, it may be necessary to assume that the midsurface is smooth while in reality only the tool surface is smooth. As long as the thickness is small compared to the other two dimensions of the structure, such assumption is unlikely to have a dramatic effect in the results of a global analysis, such as deformation, buckling, and even membrane stress analysis. The exact description of the thickness geometry begins to play a role when detailed 3D stress analysis of the ply drop-off region is required, but at that point, a 3D local model is more adequate.

Example 3.5 *A composite strip* 120 *mm long and* 100 *mm wide has a ply drop-off between laminate A:* [90/0]$_S$*, and laminate B:* [90/0]*, with ply thickness* 0.75 *mm. A ply drop-off ratio 1:20 is used, meaning that the transition length is 20 times the dropped thickness. The strip is under tension N* = 10 *N/mm applied to the bottom edges on the strip. The material is AS4D/9310 (Table 3.1). Use symmetry to model 1/2 of the tape. Visualize and report the maximum deflection (in the Z-direction, normal to the plate). Also, tabulate the maximum values of* σ_{11} *at the top and bottom surface of each lamina (not the minimum, not the maximum absolute value).*

Solution to Example 3.5 *Watch the video in [1].*
Three different sections are defined: one for laminate A, one for laminate B, and one to model the ply drop-off between them (called laminate AB), where only the top 90° lamina has been dropped. The thickness of the drop-off is equivalent to two laminas, 0.75 × 2 = 1.5 *mm. With a ply drop-off ratio 1:20, the total length of ply drop-off is* 1.5 × 20 = 30 *mm. If the laminate A spans* 0 < x < 60 *mm, the transition goes from* x = 60 *mm to* x = 90 *mm, and laminate B from* x = 90 *mm to the loaded edge at* x = 120 *mm.*

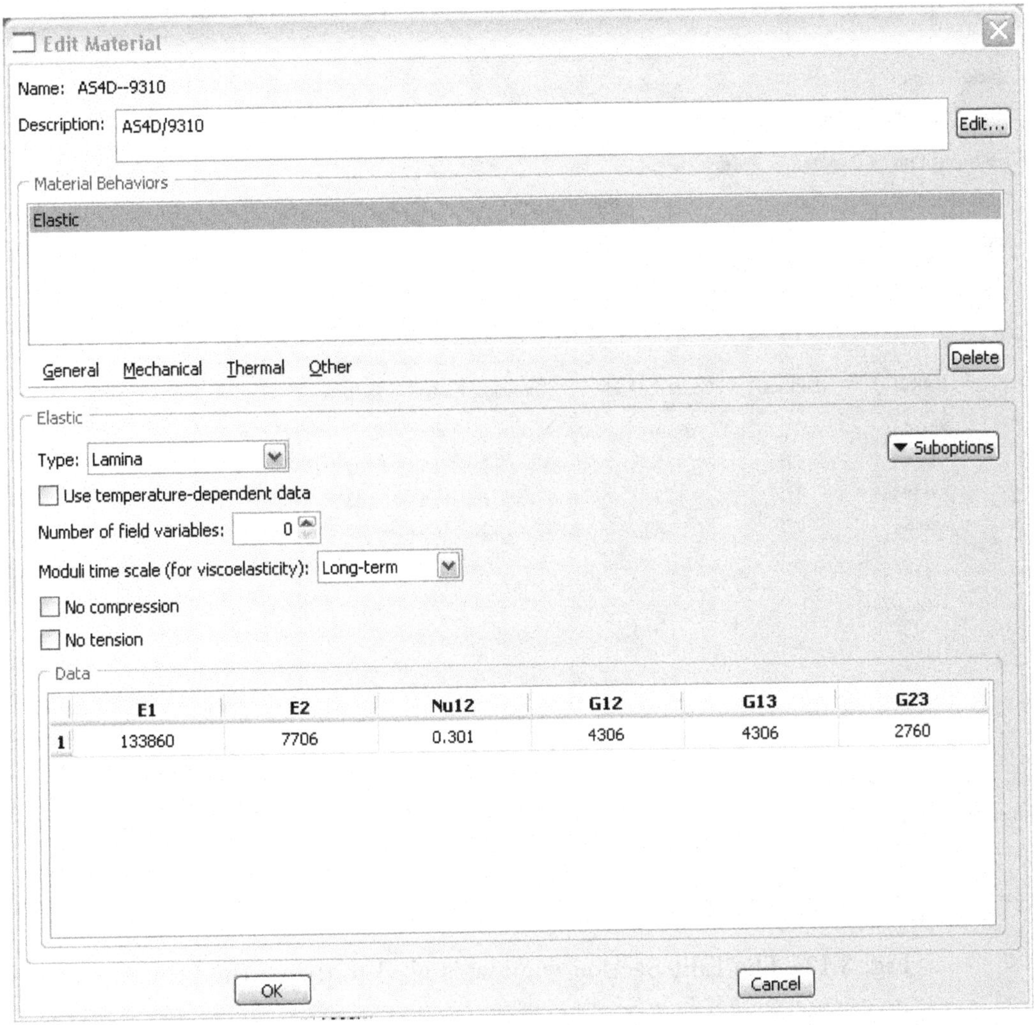

Fig. 3.11: The Edit Material window.

The midsurface of each laminate (A, B, and AB) is offset at different z-distances with respect to the flat bottom surface. Therefore, it is preferred to apply the load at the common bottom surface. This however introduces a bending moment because the traction load is offset with respect to the midsurface of the loading edge. This effect can be easily corrected by applying a bending moment equal to the product of the traction times one half the thickness of the laminate at the loading edge, but it has not been done in this example to emphasize the resulting deformation.

The bottom lamina is designated as #1. Additional laminas are stacked from bottom to top in the positive normal z-direction of the element coordinate system. With the surface of the plate resting on the x-y plane, the x-z plane is a plane of symmetry (YSYMM). Then, the thicker end of the strip (laminate A) on the left of the model can be properly supported by imposing an **Encastre-condition**; this will guarantee symmetry conditions with respect to the y-z plane.

The material properties are defined in the **Edit Material** window (Figure 3.11), but this is quite tedious. If the properties are listed in **Excel**, or in a Tab-separated file, one

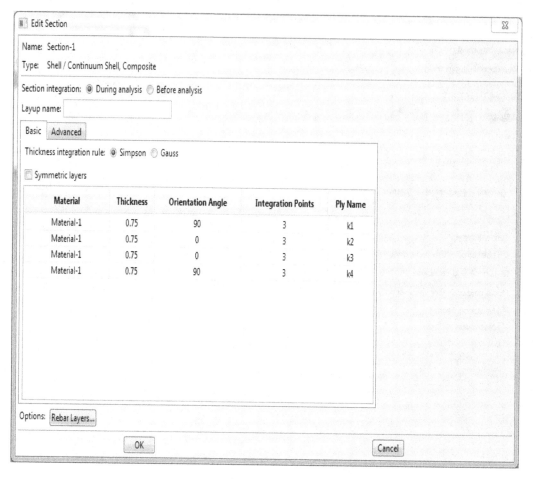

Fig. 3.12: The Edit Section window is used to define the LSS.

can cut and paste directly into the **Edit Material** *window.*

The laminate stacking sequence is defined in the **Edit Section** *window, which is reached as follows*

    Module: Property
    Menu: Section, Create, Shell, Composite, Cont

In the **Edit Section** *window (Figure 3.12), the first row corresponds to lamina #1, which is located at the bottom of the laminate. Subsequent rows are used to define LSS of a laminate with n laminas. The last row corresponds to the topmost lamina in the laminate. Note how we have assigned the names k1, k2, k3, k4, to the laminas, following the usual notation where lamina #1 is at the bottom of the laminate.*

When the results are later visualized, one has to select where to visualize them. That is, one has to select the lamina and the position through the thickness of the lamina. This is done by invoking the **Section Points** *window, as follows*

    Module: Visualization
    Menu: Results, Section Points, Selection method: [Plies]

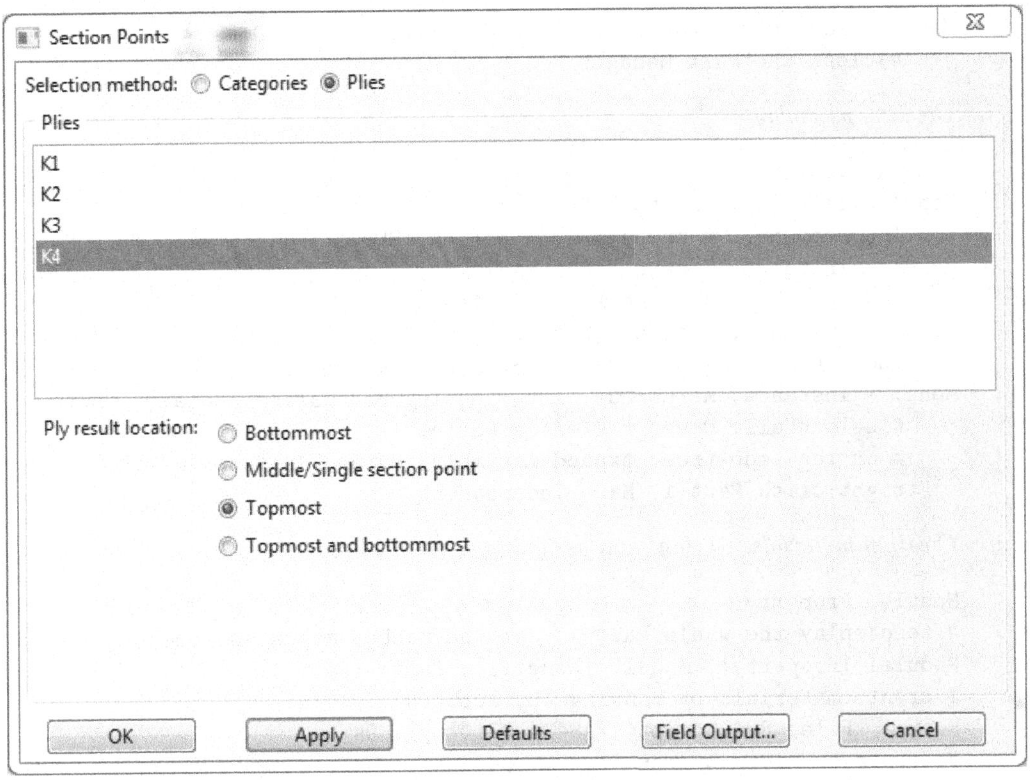

Fig. 3.13: Section Points are used to select visualization at top and/or bottom of each lamina.

In the Section Points *window (Figure 3.13),* K4, topmost *selects results at the top surface of the laminate. Similarly,* K1, bottom-most *selects the bottom surface of the laminate.*

The following pseudo-code illustrates the whole procedure. If in doubt, open Ex_3.5.cae *available in [2]. For the sake of space, the materials and sections are created by running .py scripts, but they could be easily defined by using the methods shown in previous examples.*

```
# Units: N, mm, MPa
Menu:   File, Save as [c:\simulia\user\Ex_3.5\Ex_3.5.cae]
Menu:   File, Set Work Directory [c:\simulia\user\Ex_3.5]
```

i. *Creating parts*

```
Module: Part
Menu:   Part, Manager
    Create, [Part-A], 3D, Deformable, Shell, Planar,
    Approx size [500], Cont
Menu:   Add, Line, Rectangle, [0,0], [60,50], X, Done
    Create, [Part-AB], 3D, Deformable, Shell, Planar,
    Approx size [500], Cont
Menu:   Add, Line, Rectangle, [60,0], [90,50], X, Done
    Create, [Part-B], 3D, Deformable, Shell, Planar,
    Approx size [500], Cont
```

```
Menu:    Add, Line, Rectangle, [90,0], [120,50], X, Done
    # close the Part Manager
```

ii. Creating assembly

```
Module: Assembly
Menu:    Instance, Create, [Part-A],
    Independent,  # uncheck Auto-offset, OK
Menu:    Instance, Create, [Part-AB],
    Independent,  # uncheck Auto-offset, OK
Menu:    Instance, Create, [Part-B],
    Independent,  # uncheck Auto-offset, OK
Menu:    Instance, Merge/Cut, Intersecting Boundaries: Retain, Cont
    # select all, Done, # creates Part-1
    # on the .mdb tree, expand Assembly, expand Instances,
    right-click Part-1, Make Independent
```

iii. Creating materials, section, and assigning sections to parts

```
Module: Property
# to display the whole Part-1, set the tabbed menus as follows
Module: [Property], Model: [Model-1], Part: [Part-1]
# create materials by running .py scripts
Menu:    File, Run Script, [AS4D--9310.py], OK
# create sections by running .py scripts
Menu:    File, Run Script, [laminate-A.py], OK
Menu:    File, Run Script, [laminate-AB.py], OK
Menu:    File, Run Script, [laminate-B.py], OK
# recall that from bottom to top, section A is [90/0/0/90],
# section B is [90/0], and section AB is [90/0/0]
Menu:    Section, Assignment Manager
    Create, # pick the region/part Section-A, Done
Section: [Section-A], Shell Offset Definition: [Bottom Surface], OK
    Create, # pick the region/part Section-AB, Done
Section: [Section-AB], Shell Offset Definition: [Bottom Surface], OK
    Create, # pick the region/part Section-B, Done
Section: [Section-B], Shell Offset Definition: [Bottom Surface],
    OK # close the Section Assignment Manager,
    Done # close current command
```

iv. Defining analysis steps

```
Module: Step
Menu:    Step, Create, Name [Step-1], Static/General, Cont, OK
Menu:    Output, Field Output Request, Edit, F-Output-1
    Output at shell, Specify [1,2,3,4,5,6,7,8,9,10,11,12], OK
```

v. Adding BC and loads

```
Module: Load
Menu:    BC, Manager
    Create, [BC-1], Step: Initial, Mechanical, Symm/Anti/Enca, Cont
    # pick edge @ x=0, Done, Encastre, OK
```

```
      Create, [BC-2], Step: Initial, Mechanical, Symm/Anti/Enca, Cont
      # pick the 3 segments @ y=0, Done, YSYMM, OK, # close BC Manager
      # since the part is a Shell, an edge load is: Shell edge load
      # load applies on the offset defined during Assign
Menu:   Load, Create
      [Load-1], Step: Step-1, Mechanical, Shell edge load, Cont
      # pick the edge at x=120, Done, Magnitude [-10], OK
```

vi. Creating mesh

```
Module: Mesh
Menu:   Seed, Instance, Approx global size: [5], Apply, OK
Menu:   Mesh, Controls, # select all the regions, Done
      Element Shape [Quad], Technique [Structured]
Menu:   Mesh, Element Type, # select all the regions, Done
      Standard, Quadratic, DOF [6], OK # element selected S8R
      Mesh, Instance, Yes
```

vii. Solving and visualizing the results

```
Module: Job
Menu:   Job, Manager
      Create, Cont, OK, Submit # when completed, Results


Module: Visualization
Menu:   Viewport, Viewport Annotation Options,
      Tab: State Block, Set Font, Size [10], Apply to [checkmark all],
      OK, OK
Menu:   File, Save Options, OK # to save options for future sessions
Menu:   Plot, Undeformed Shape
Menu:   Plot, Allow Multiple Plot States
Menu:   Plot, Deformed Shape # you will see both shapes
Menu:   Plot, Contours, On Deformed Shape
      Field Output Toolbar, [Primary], [U], [U3]
      # read the maximum value from the legend on top left, U3=2.578 mm
      Field Output Toolbar, [Primary], [S], [S11]
      # next to select the top or bottom surface of each lamina
Menu:   Result, Section Points
      Selection method: [Plies], K1, Bottommost, Apply
      # repeat for the other plies
      Selection method: [Plies], K1, topmost, Apply
      # repeat for the other plies
      # a gray region means the ply is not modeled in that region
```

The maximum deflection is $w = 2.578\ mm$. The maximum values of stress are reported in Table 3.3.

Example 3.6 *Model the laminate shown in Figure 3.14. The laminate in section A is a $[+45/-45/0/90/0]$. The thickness of each lamina is 1.2 mm. A ply drop-off ratio 1:10 is used. The strip is 120 mm long and 100 mm wide. It is loaded by tension $N_x = 10\ N/mm$ applied to the bottom edges on the strip. Use symmetry to model 1/2 of the tape. The*

Table 3.3: Stresses calculated in Example 3.5.

Ply #	z [mm]	$\theta[°]$	$\sigma_{11}[MPa]$
1 (bot)	0.00	90	1.233
1 (top)	0.75	90	0.6206
2 (bot)	0.75	0	73.17
2 (top)	1.50	0	6.596
3 (bot)	1.50	0	6.596
3 (top)	2.25	0	-19.15
4 (bot)	2.25	90	0.1907
4 (top)	3.00	90	0.4094

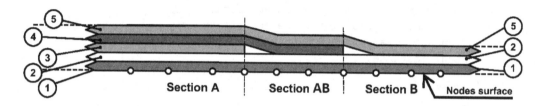

Fig. 3.14: Example 3.6. Laminate with dropped laminas.

material is AS4D/9310 (Table 3.1). Visualize and report the maximum value of transverse deflection U_3. Also, visualize and tabulate the maximum value of σ_{11} at the top surface of every lamina (k1, k2, k3, k4, k5). Note the absence of results for the laminas dropped in the region where those laminas have been dropped.

Solution to Example 3.6 *Watch the video in [1].*
In this example, we illustrate the use of partitions to separate the three regions. With a drop-off ratio 1:10, the transition length is $2 \times 1.2 \times 10 = 24$ mm, with drops at 12 mm on each side of the center line. With the center line at 60 mm from the left edge, the transitions are located at 48 mm and 72 mm. The bottom lamina is designated as lamina #1. Additional laminas are stacked from bottom to top in the positive normal direction of the element coordinate system. The dropped laminas are #3 and #4. The following pseudo-code illustrates the whole procedure. If in doubt, open `Ex_3.6.cae` *available in [2].*

```
# Units: N, mm, MPa
Menu:   File, Set Work Directory [c:\simulia\user\Ex_3.6]
Menu:   File, Save As [Ex_3.6.cae]
```

i. Creating parts

```
Module: Part
Menu:   Part, Create
     [Part-1], 3D, Deformable, Shell, Planar, Approx size [300], Cont
Menu:   Add, Line, Rectangle, [0,0], [120,50], X, Done
# creating 3 partitions for the different regions
Menu:   Tool, Datum, Plane, Offset from principal plane
     [YZ Plane], [48], Enter, [YZ Plane], [72], Enter, X
     # close the Create Datum pop-up window
Menu: Tools, Partition, Face, Use datum plane,
```

```
# pick a datum plane, Create Partition
# pick face to partition (area not partitioned yet), Done
# pick datum plane, Create Partition, Done
# close the Create Partition pop-up window
```

ii. Creating materials, sections, and assigning sections to parts

```
Module: Property
Menu:   Material, Create
    Name [AS4D--9310], Mechanical, Elasticity, Elastic, Type [Lamina]
    # you can cut and paste an entire row of values from Excel
    [1.3386E5, 7.706E3, 0.301, 4.306E3, 4.306E3, 2.76E3], OK
    # or use a .py file to create materials as in previous example
Menu:   Section, Manager
    Create, Name [Section-A], Shell, Composite, Cont
    # insert rows with right click until you have 5 rows
    Material [AS4D--9310] # for all, Thickness [1.2] # for all
    Orientation [45,-45,0,90,0], Integration points [3,3,3,3,3]
    Ply names [k1,k2,k3,k4,k5], OK
    Copy, [Section-AB], OK, Edit, # delete the center ply (k3), OK
    Copy, [Section-B],  OK, Edit, # delete the center ply (k4), OK
    # close Section Manager
Menu:   Section, Assignment Manager
    Create, # pick the region on the left, Done
    Section [Section-A], Shell Offset [Bottom surface], OK
    Create, # pick the region in the middle, Done
    Section [Section-AB], Shell Offset [Bottom surface], OK
    Create, # pick the region on the right, Done
    Section [Section-B], Shell Offset [Bottom surface], OK, Done
    # close Section Assignment Manager
```

iii. Creating assembly

```
Module: Assembly
Menu:   Instance, Create, Independent, OK
```

iv. Defining analysis steps

```
Module: Step
Menu:   Step, Create
    Name [Step-1], Procedure type: General, Static/General, Cont, OK
Menu:   Output, Field Output Request, Edit, F-Output-1
    Specify [1,2,3,4,5,6,7,8,9,10,11,12,13,14,15], OK
```

v. Adding BC and loads

```
Module: Load
Menu:   BC, Create
    Name [BC-1], Step: Initial, Mechanical, Symm/Anti/Enca, Cont
    # pick edge @ x=0, Done, Encastre, OK
Menu:   BC, Create
    Name [BC-2], Step: Initial, Mechanical, Symm/Anti/Enca, Cont
    # pick 3 segments @ y=0, Done, YSYMM, OK
```

```
# since the part is a Shell, an edge load is: Shell edge load
# load applied with an offset defined during Assign
Menu:   Load, Create
   Name [Load-1], Step: Step-1, Mechanical, Shell edge load, Cont
   # pick edge @ x=120, Done, Magnitude [-10], OK
```

vi. Creating Mesh

```
Module: Mesh
Menu:    Seed, Instance, Approx global size: [5], Apply, OK
Menu:    Mesh, Controls, # select all regions, Done,
   Element Shape [Quad], Technique [Structured], OK
Menu:    Mesh, Element Type, # select all regions, Done
   [Standard], [Quadratic], DOF [6], OK, # element used S8R
Menu:    Mesh, Instance, Yes
```

vii. Solving and visualizing the results

```
Module: Job
Menu:    Job, Manager
   Create, Cont, OK, Submit, # when completed, Results
Module: Visualization
Menu:    Viewport, Viewport Annotation Options
   Tab: State Block, Set Font, Size [10],
   Apply to [checkmark all], OK, OK
Menu:    File, Save Options, OK # to save options for future sessions
Menu:    Plot, Undeformed shape
Menu:    Plot, Allow Multiple Plot States
Menu:    Plot, Deformed Shape # you will see both shapes
Menu:    Plot, Material Orient., On Def. Shape # identify the 1-axis
Menu:    Result, Section Points, [Plies], K1, Bottommost
   # pick one ply at a time and comment on the results, OK
   Field Output Toolbar, [Primary], [S], [S11]
Menu:    Result, Section Points, [Plies], K1, Topmost
   # pick one ply at a time and tabulate the results, OK
```

The maximum transverse deflection is 0.6075 mm. The maximum values of stress are given in Table 3.4.

Table 3.4: Maximum values of stress at the top surface of each ply for Example 3.6.

Ply #	z [mm]	$\theta[°]$	$\sigma_{11}[MPa]$
1	1.2	45	10.1
2	2.4	−45	17.88
3	3.6	0	17.63
4	4.8	90	4.303
5	6.0	0	9.632

3.2.13   FEA of Sandwich Shells

Laminates should be modeled as *sandwich shells* when specifically designed for sandwich construction with thin face sheets and a thick, relatively weak, core (e.g., honeycomb, foam, balsa wood). The faces are intended to carry all, or almost all, of the bending and in-plane normal load. Conversely, the core is assumed to carry all of the transverse shear. Example 3.7 shows how to define and calculate a sandwich cantilever beam.

The following assumptions are customarily made for a sandwich shell:

- The terms H_{ij} in (3.9) depend only on the middle lamina (core) and they can be calculated as

$$H_{ij} = \left(\overline{Q}_{ij}^*\right)_{core} t_{core} \quad ; \quad i, j = 4, 5 \tag{3.17}$$

- The transverse shear moduli (G_{23} and G_{13}) are set to zero for the top and bottom laminas (non-core laminas) so they do not add artificial stiffness to the sandwich shell.

- The transverse shear strains and stresses in the face plate (non-core) laminas are neglected or assumed to be zero.

- The transverse shear strains and shear stresses in the core are assumed constant through the thickness.

Example 3.7 *Consider a sandwich plate clamped around the boundary and loaded with uniform pressure on the top surface.[3] The sandwich is made of two face plates and an inner core with properties, thicknesses, and pressure given in Table 3.5. Use symmetry conditions to model one quarter of the plate. Use a surface mesh of six by six elements. Compute the center deflection using four different methods:*

a. *Conventional shell elements S8R.*
b. *Continuum shell elements SC8R spanning the entire thickness, with the laminate defined inside the element.*
c. *Continuum shell elements SC8R stacked through the thickness, one for each face and one or more for the core.*
d. *Continuum shell elements SC8R for the core and conventional shell elements S8R for the faces.*

Solution to Example 3.7 *Here we solve the problem in four ways and the results are tabulated in Table 3.6. Watch the four videos in [1].*

The stacked continuum shell model (c) and stacked continuum shell plus skins (d) are more flexible than (a) and (b) models because in (c, d) the FSDT constraint (i.e., normals remain straight) is enforced separately in the core and two faces/skins. Thus, each lamina (core and each face) is able to rotate independently of the other laminas. This means that each lamina achieves its own value of transverse shear deformation. Since the core has lower transverse shear moduli (G_{13}, G_{23}), it can deform more. This additional flexibility impacts the transverse deflection and more notably the transverse shear strains and stresses. Increasing the number of elements through the thickness in the core does not change the results because the core is homogeneous (not laminated). Therefore, its normal deforms as

---

[3]This example is modified from [17, Test R0031/3] by making the edges clamped rather than simply supported. The change is made because it is difficult to impose a simple support with continuum shell elements [13]. Doing so requires to use constraint equations, which complicates the example too much at this stage.

Table 3.5: Data for Example 3.7.

Quantity	Units	Face	Core
E_1	MPa	68947.57	0.068948
E_2	MPa	27579.03	0.068948
ν_{12}		0.3000	0.0000
G_{12}	MPa	12927.67	0.068948
G_{13}	MPa	12927.67	209.6006
G_{23}	MPa	12927.67	82.73708
thickness	mm	0.7112	19.0500
half plate width	mm	127.00	
pressure	MPa	0.689476	
thickness modulus	MPa	689475.7	689475.7

a straight line regardless of how many elements are used to discretize its thickness. The conventional and single element continuum models yield similar results because the single, laminated SC8R element has the same kinematics as the conventional S4R (provided the transverse compressibility of SC8R is constrained by selecting a high value for the thickness modulus).

Continuum shell elements in Abaqus need a thickness modulus E_3 and they cannot read it from the given material properties, even if E_3 is given in Engineering Properties, *or it is implied for* Isotropic $(E_3 = E)$ *and* Lamina $(E_3 = E_2)$. *So, the user must remember to assign a thickness modulus in tab* Shell Parameters *of the Edit Composite Layup window (Figure 3.15). Since a shell is incompressible in the thickness direction, we chose a thickness modulus ten times the highest modulus of the face material,[4] i.e., $E_3 = 10 \times E_1 = 689475.7\ MPa$. No thickness modulus is needed for conventional shell elements because they are intrinsically incompressible through the thickness.*

(a) Solution with conventional shell elements

Here we show how to model the geometry of the sandwich plate as a 2D flat surface. Then, define the laminate inside the conventional shell element. The following pseudo-code illustrates the whole procedure. If in doubt, open Ex_3.7.a.cae *available in [2].*

```
# NAFEMS sandwich clamped metric conventional shell elements
Menu:   File, Set Work Directory [c:\simulia\user\Ex_3.7]
Menu:   File, Save As [Ex_3.7.a.cae]
```

i. Creating parts

```
Module: Part
Menu:   Part, Create, 3D, Deformable, Shell, Planar, Cont
Menu:   Add, Line, Rectangle, [0,0], [127,127], X, Done
     # name the part to make it easier to use Composite Modeler later
Menu:   Tools, Set, Create, Name [plate], Cont, # select all, Done
```

---

[4]This is in agreement with r313_std_sc8r_composite.inp in [18, 4.9.3 R0031(3)].

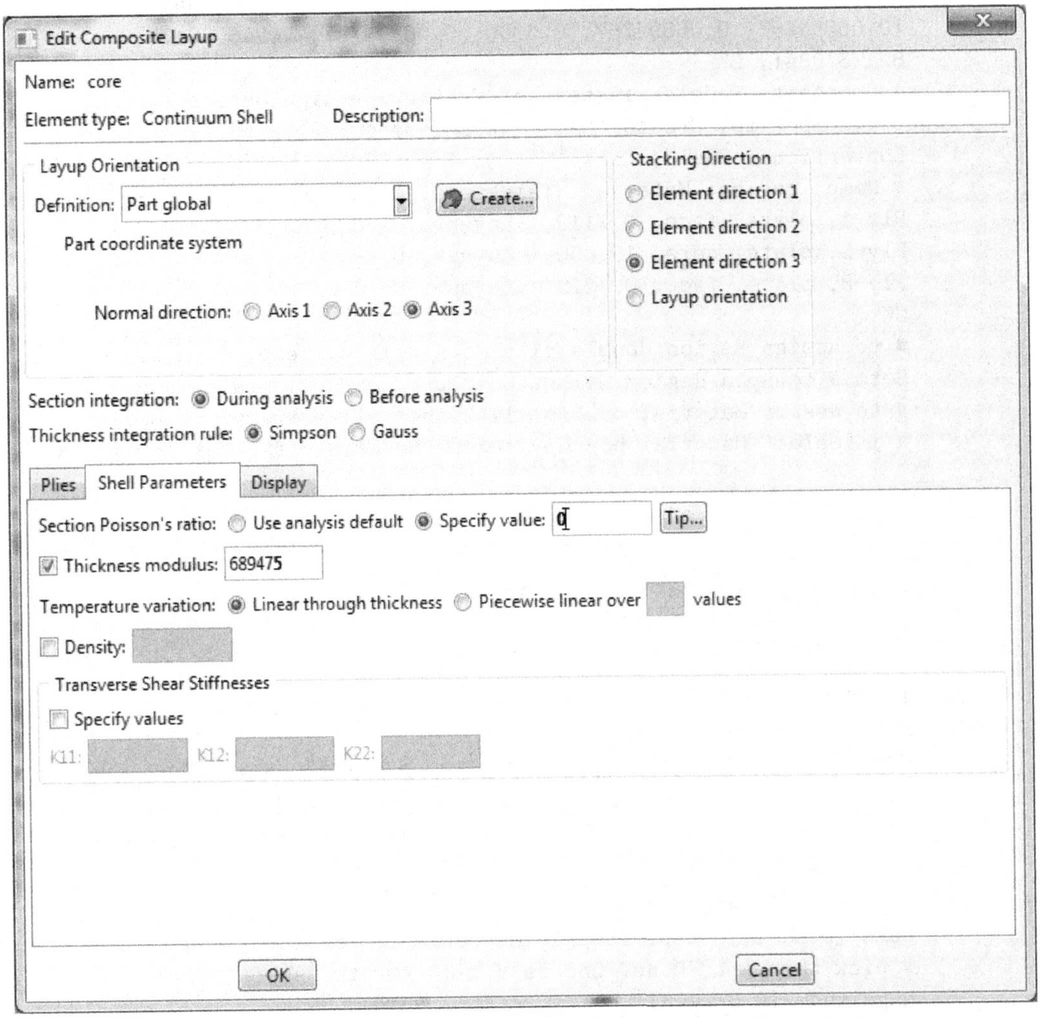

Fig. 3.15: Specify a thickness modulus for the shell section.

ii. Defining materials, sections, and assigning sections to parts

```
Module: Property
Menu:   Material, Create
Name [face], Mechanical, Elasticity, Elastic, Type [Lamina]
    [68947.57, 27579.028, 0.3000, 12927.66938, 12927.66938,
    12927.66938], OK
Menu:   Material, Create
Name [core], Mechanical, Elasticity, Elastic, Type [Lamina]
    [0.06894757, 0.06894757, 0.0000, 0.06894757, 209.6006128,
    82.737084], OK
# use Composite Modeler instead of Create + Assign Sections
Menu:   Composite, Create, Name [NAFEMS-sandwich],
    Conventional Shell, Cont
    # Name, region, Material, Thickness, CSYS, Angle, Integ Points
    Ply-1, plate, face, 0.7112, <Layup>, 0, 3
    Ply-2, plate, core, 19.050, <Layup>, 0, 3
    Ply-3, plate, face, 0.7112, <Layup>, 0, 3
    OK
    # to assign Region double click the cell corresp. to desired ply
    Sets # to open Region Selection pop-up window, plate, Cont
    # to assign Material double click the cell corresp. to desired ply
    # in Select Material pop-up window, Material [face or core], OK
```

iii. Creating assembly

```
Module: Assembly
Menu:   Instance, Create, Independent, OK
```

iv. Defining analysis steps

```
Module: Step
Menu:   Step, Create
    Procedure type: Linear perturbation, Static/General, Cont, OK
```

v. Adding BC and loads

```
Module: Load
Menu:   Load, Create
    Name [Load-1], Step: Step-1, Mechanical, Pressure, Cont,
    # pick the part, Done, Choose a side for the shell [Brown]
    Magnitude [0.689476], OK
Menu:   BC, Manager
    Create, Step: Initial, Mechanical, Symm/Anti/Enca, Cont
    # pick edges @ x=0 and y=0, Done, Encastre, OK
    Create, Step: Initial, Mechanical, Symm/Anti/Enca, Cont
    # pick edge @ x=127, Done, XSYMM, OK
    Create, Step: Initial, Mechanical, Symm/Anti/Enca, Cont
    # pick edge @ y=127, Done, YSYMM, OK
    # close BC Manager
```

vi. Meshing the model

```
Module: Mesh
Menu:   Seed, Instance, Approx global size [21], OK
Menu:   Mesh, Controls, Element Shape [Quad], Tech [Structured], OK
Menu:   Mesh, Element Type, # select all surfaces, Done
        Element Library [Standard], Geometric Order [Linear], OK # S4R
Menu:   Mesh, Instance, Yes
```

vii. Solving and visualizing the results

```
Module: Job
Menu:   Job, Manager
        Create, Cont, OK, Submit # when completed, Results
Module: Visualization
        Field Output Toolbar, [Primary], [U], [U3]
```

The displacement at the upper-right corner is −1.667 mm

(b) Solution with continuum shell elements

Here we show how to model the geometry of the sandwich plate as a 3D body. Then, define the laminate inside the continuum shell element [13]. Note that in the composite modeler, the relative thicknesses do not need to add to 1.0. The actual ply thicknesses are calculated in terms of the mesh thickness. To emphasize this, we have chosen to enter the thicknesses of the plies in inches (0.028",075",0.028"), even though the model is set up in SI units. The following pseudo-code illustrates the whole procedure. If in doubt, open Ex_3.7.b.cae *available in [2].*

```
# NAFEMS sandwich clamped metric laminated continuum shell elements
Menu:   File, Set Work Directory [c:\simulia\user\Ex_3.7]
Menu:   File, Save As [Ex_3.7.b.cae]
```

i. Creating parts

```
Module: Part
Menu:   Part, Create, 3D, Deformable, Solid, Extrusion, Cont
Menu:   Add, Line, Rectangle, [0,0], [127,127], X, Done
        Depth [20.4724], OK  # core + two faces
# name the part to make it easier to use Composite Modeler later
Menu:   Tools, Set, Create, Name [plate], Cont, # select all, Done
```

ii. Defining materials, sections, and assigning sections to parts

```
Module: Property
Menu:   Material, Create
        Name [face], Mechanical, Elasticity, Elastic, Type [Lamina]
        [68947.57, 27579.028, 0.3, 12927.66938, 12927.66938,
        12927.66938], OK
Menu:   Material, Create
        Name [core], Mechanical, Elasticity, Elastic, Type [Lamina]
        [0.06894757, 0.06894757, 0.0, 0.06894757, 209.6006128,
        82.737084], OK
```

```
# use Composite Modeler instead of Create + Assign Section
Menu: Composite, Create, Name [sandwich], Type: Continuum Shell, Cont
    # Name, region, Material, Relative Thickness, CSYS, Angle, Integ
    Points
    Ply-1, plate, face, 0.0347, <Layup>, 0, 3
    Ply-2, plate, core,  0.9305, <Layup>, 0, 3
    Ply-3, plate, face, 0.0347, <Layup>, 0, 3 # the sum is 1.0
    # to assign Region double click the cell corresp. to desired ply
    Sets # to open Region Selection pop-up window, plate, Cont
    # to assign Material double click the cell corresp. to desired ply
    # in Select Material pop-up window, Material [face or core], OK
    Tab: Shell Parameters
    Section Poisson's ratio: Specify value [0.0]
    # checkmark: Thickness modulus [689475.7], OK
    # the last two instructions are very important!
```

iii. *Creating assembly*

```
Module: Assembly
Menu:   Instance, Create, Independent, OK
```

iv. *Defining analysis steps*

```
Module: Step
Menu:   Step, Create
    Procedure type: Linear perturbation, Static/General, Cont, OK
Menu:   Output, Field Output Requests, Edit, F-Output-1
    Output variables: Edit variables [CF,CTSHR,LE,RF,S,SE,TSHR,U]
    Output at shell: Specify [1,2,3,4,5,6,7,8,9], OK
```

v. *Adding BC and loads*

```
Module: Load
Menu:   Load, Create
    Step: Step-1, Mechanical, Pressure, Cont
    # pick the top surface (z=20.4724), Done, Magnitude [0.689476], OK
Menu:   BC, Manager
    Create, Step: Initial, Mechanical, Symm/Anti/Enca, Cont
    # pick surfaces @ x=0 and y=0, Cont, Encastre, OK
    Create, Step: Initial, Mechanical, Symm/Anti/Enca, Cont
    # pick surface @ x=127, Cont, XSYMM, OK
    Create, Step: Initial, Mechanical, Symm/Anti/Enca, Cont
    # pick surface @ y=127, Cont, YSYMM, OK
    # close BC Manager
```

vi. *Meshing the model*

```
Module: Mesh
Menu:   Seed, Instance, Approx global size [21], OK
Menu:   Mesh, Controls, Element Shape [Hex], Technique [Sweep], OK
Menu:   Mesh, Element Type, # select all, Done
    Family [Continuum Shell], OK # notice element chosen is SC8R
Menu:   Mesh, Instance, Yes
```

Fig. 3.16: Seeding an edge **By number**, using 50 elements through the length of the edge.

vii. Solving and visualizing the results

```
Module: Job
Menu:   Job, Manager
      Create, Cont, OK, Submit # when completed, Results
Module: Visualization
      Field Output Toolbar, [Primary], [U], [U3]
```

The displacement at the upper-right corner is −1.667 mm.

(c) Solution with stacked continuum shell elements

Here we show how to model the geometry of the sandwich plate as a 3D body, partitioned into a core and two faces. Each region is then discretized with continuum shell elements [13]. Discretizing with 3D solid elements would require a very fine mesh of the plate dimensions (3000 × 600) to avoid locking. In fact, the element dimensions could not be smaller than 10 times the thickness of the aluminum face. Also in this example we show how to seed with a sweep and how to re-seed an edge, using the local seeds window (Figure 3.16). Remember to assign a thickness modulus in tab **Shell Parameters** *of the Edit Composite Layup window (Figure 3.15). The following pseudo-code illustrates the whole procedure. If in doubt, open* **Ex_3.7.c.cae** *available in [2]. Remeshing to accommodate four elements through the thickness of the core does not alter the results.*

```
# NAFEMS sandwich clamped metric stack 3 continuum shell elements
Menu:   File, Set Work Directory [c:\simulia\user\Ex_3.7]
Menu:   File, Save As [Ex_3.7.c.cae]
```

i. Creating parts

```
Module: Part
Menu:    Part, Create, 3D, Deformable, Solid, Extrusion, Cont
Menu:    Add, Line, Rectangle, [0,0], [127,127], X, Done
    Depth [20.4724], OK  # core + two faces
# partition the faces from the core
Menu:    Tool, Datum, Plane, Offset from principal plane
    [XY], [0.7112], Enter, [XY], [19.7612], Enter,
    X, # close Create Datum
Menu:    Tools, Partition, Cell, Use datum plane
    # the entire part is automatically selected
    Select datum plane # back datum plane, Create Partition
    # pick front cell to create the new partition, Done
    Select datum plane, Create Partition, Done
```

ii. Defining materials, sections, and assigning sections to parts

```
Module: Property
Menu:    Material, Manager
    Create, Name [face], Mechanical, Elasticity, Elastic,
    Type [Lamina]
    [68947.57, 27579.028, 0.3, 12927.66938, 12927.66938,
    12927.66938], OK
    Create, Name [core], Mechanical, Elasticity, Elastic,
    Type [Lamina]
    [0.06894757, 0.06894757, 0.0, 0.06894757, 209.6006128,
    82.737084], OK
    # close Material Manager, # zoom in to pick the regions easily
# use Composite Modeler instead of Create + Assign Section
Menu:    Composite, Create
    Name [front face], Initial ply count [1],
    Element Type [Continuum Shell], Cont
    # Name, region, Material, Rel. Thick., CSYS, Angle, Integ. Points
    Ply-1, <picked>, face, 1, <Layup>, 0, 3
    Tab: Shell Param., Section Poisson's ratio: Specify value [0.0]
    # checkmark: Thickness modulus: [689475.7], OK
    # the last two instructions are very important!
Menu:    Composite, Create
    Name [back face], Initial ply count [1],
    Element Type [Continuum Shell], Cont
    # Name, region, Material, Rel. Thick., CSYS, Angle, Integ. Points
    Ply-1, <picked>, face, 1, <Layup>, 0, 3
    Tab: Shell Param., Section Poisson's ratio: Specify value [0.0]
    # checkmark: Thickness modulus: [689475.7], OK
    # the last two instructions are very important!
Menu:    Composite, Create
    Name [core], Initial ply count [1], Type [Continuum Shell], Cont
    # Name, region, Material, Rel. Thick., CSYS, Angle, Integ. Points
    Ply-1, <picked>, core, 1, <Layup>, 0, 3
    Tab: Shell Param., Section Poisson's ratio: Specify value [0.0]
    # checkmark: Thickness modulus: [689475.7], OK
```

iii. Creating assembly

Module: Assembly
Menu: Instance, Create, Independent, OK

iv. Defining analysis steps

Module: Step
Menu: Step, Create
        Procedure type: Linear perturbation, Static/Linear pert., Cont, OK
Menu: Output, Field Output Requests, Edit, F-Opuput-1
        Output variables: Edit variables [CF,CTSHR,LE,RF,S,SE,TSHR,U]
        Output at shell: Specify [1,2,3], OK

v. Adding BC and loads

Module: Load
Menu: Load, Create
        Step: Step-1, Mechanical, Pressure, Cont
        # pick the top surface of the part (z=20.4724), Done
        Magnitude [0.689476], OK
Menu: BC, Manager
        Create, Step: Initial, Mechanical, Symm/Anti/Enca, Cont
        # pick 3 surfaces @ x=0 and the 3 @ y=0, Done, Encastre, OK
        Create, Step: Initial, Mechanical, Symm/Anti/Enca, Cont
        # pick the three surfaces @ x=127, Done, XSYMM, OK
        Create, Step: Initial, Mechanical, Symm/Anti/Enca, Cont
        # pick the three surfaces @ y=127, Done, YSYMM, OK
        # close the BC Manager pop-up window

vi. Meshing the model

Module: Mesh
Menu: Seed, Instance, Approx global size [21], OK
Menu: Seed, Edges
        # select the four vertical edges of the core part, Done
        Method [By number], Number of elements [1], OK, X
Menu: Mesh, Controls, # select all regions, Done, Sweep, OK
Menu: Mesh, Element Type, # select all regions, Done
        Family [Continuum Shell], OK # notice element chosen is SC8R
Menu: Mesh, Instance, Yes

vii. Solving and visualizing the results

Module: Job
Menu: Job, Manager
        Create, Cont, OK, Submit # when completed, Results
        Module: Visualization
        Field Output Toolbar, [Primary], [U], [U3]

*The displacement at the upper-right corner is −1.892 mm. For a more refined mesh
in the core part of the sandwich, modify the procedure above, as follows*

```
Module: Mesh
Menu:    Seed, Edges
    # select the four vertical edges of the core part, Done
    Method [By number], Number of elements [4], OK, X
Menu:    Mesh, Instance, Yes
```

(d) Solution with conventional and continuum shell elements

Here we show how to model the geometry of the core as a 3D body, then apply two skins to represent the faces. The core is discretized with continuum shell elements [13] and the faces with conventional shell elements. One must be careful to offset the location of the reference surface of the skins to coincide with the thickness of the core. Otherwise the results may be more or less wrong depending on how thick is the skin. To offset correctly, one has to place the reference at the top or bottom of the skin. This depends on the direction of the normal to the shell surface. Skins inherit the normal of the underlying shell. So, one has to visualize the normal to decide. If the normal is outward, then assigning the reference surface to the bottom of the skin will set the reference surface exactly where the surface nodes of the core are. The following pseudo-code illustrates the whole procedure. If in doubt, open Ex_3.7.d.cae *available in [2].*

```
# NAFEMS sandwich clamped metric using skins
Menu:    File, Set Work Directory [c:\simulia\user\Ex_3.7]
Menu:    File, Save As [Ex_3.7.d.cae]
```

i. Creating parts

```
Module: Part
Menu:    Part, Create, 3D, Deformable, Solid, Extrusion, Cont
Menu:    Add, Line, Rectangle, [0,0], [127,127], X, Done
    Depth [19.05], OK  # notice you created only the core part
# name the part to make it easier to use Composite Modeler later
Menu:    Tools, Set, Create, Name [plate], Cont, # select all, Done
```

ii. Defining materials, sections, and assigning sections to parts

```
Module: Property
Menu:    Special, Skin, Create, # pick top face, Done
Menu:    Special, Skin, Create, # pick back face, Done, X
Menu:    Material, Create
    Name [face], Mechanical, Elasticity, Elastic, Type [Lamina]
    [68947.57, 27579.028, 0.3, 12927.66938, 12927.66938,
    12927.66938], OK
Menu:    Material, Create
    Name [core], Mechanical, Elasticity, Elastic, Type [Lamina]
    [0.06894757, 0.06894757, 0.0, 0.06894757, 209.6006128,
    82.737084], OK
# Notice the type of element used for the core (Continuum Shell)
Menu:    Composite, Create
    Name [core], Initial ply count [1], Type [Continuum Shell], Cont
    # Name, Region, Mat., Relat. Thick., CSYS, Angle, Integ. Points
    Ply-1, plate, core, 1, <Layup>, 0, 3
    Tab: Shell Parameters,
```

        Section Poisson's ratio: Specify value [0.0]
        # checkmark: Thickness modulus: [689475.7], OK
    # Notice type of element used for skins/faces (Conventional Shell)
    Menu: Composite, Create, Name [back face],
                Initial ply count [1], Type [Conv. Shell], Cont
        # Name, Region, Material, Thick., CSYS, Angle, Integ. Points
        Ply-1, <picked>, face, 0.7112, <Layup>, 0, 3
        # use actual thickness of the laminate
        Tab: Offset,
        Shell Reference Surface and Offsets [Bottom surface], OK
    Menu: Composite, Create, Name [front face],
        Initial ply count [1], Type [Conventional Shell], Cont
        # Name, Region, Material, Thick., CSYS, Angle, Integ. Points
        Ply-1, <picked>, face, 0.7112, <Layup>, 0, 3
        # use the actual thickness of the laminate
        Tab: Offset,
        Shell Reference Surface and Offsets [Bottom surface], OK

iii. Creating assembly

    Module: Assembly
    Menu: Instance, Create, Independent, OK

iv. Defining analysis steps

    Module: Step
    Menu: Step, Create
        Procedure type: Linear perturbation, Static/Linear pert., Cont, OK
    Menu: Output, Field Output Requests, Edit, F-Opuput-1
        Output variables: Edit variables [CF,CTSHR,LE,RF,S,SE,TSHR,U]
        Output at: Specify [1,2,3,4,5,6,7,8,9], OK # important step!

v. Adding BC and loads

    Module: Load
    Menu: Load, Create
        Step: Step-1, Mechanical, Pressure, Cont
        # pick the top surface of the part (z=19.05), Done
        Magnitude [0.689476], OK
    Menu: BC, Manager
        Create, Step: Initial, Mechanical, Symm/Anti/Enca, Cont
        # pick the surfaces @ x=0 and @ y=0, Done, Encastre, OK
        Create, Step: Initial, Mechanical, Symm/Anti/Enca, Cont
        # pick the surface @ x=127, Done, XSYMM, OK
        Create, Step: Initial, Mechanical, Symm/Anti/Enca, Cont
        # pick the surface @ y=127, Done, YSYMM, OK
        # close BC Manager pop-up window

vi. Meshing the model

    Module: Mesh
    Menu: Seed, Instance, Approx global size [21], OK
    Menu: Mesh, Controls, Sweep, OK

Table 3.6: Center deflection for Example 3.7.

Method	Center deflection [mm]
Conventional	1.667
Single continuum	1.667
Stacked continuum	1.892
Continuum + skins	1.892

```
Menu:    Mesh, Element Type, Sets [plate],
     Family [Continuum Shell], OK # element SC8R was applied
     # the skins are meshed automatically
Menu:    Mesh, Instance, Yes
```

vii. Solving and visualizing the results

```
Module: Job
Menu:    Job, Manager
     Create, Cont, OK, Submit # when completed, Results
Module: Visualization
     Field Output Toolbar, [Primary], [U], [U3]
```

The displacement at the upper-right corner is -1.892 *mm.*

3.2.14   Element Coordinate System

In the preprocessor, during the definition of the laminate, it is very important to know the orientation of the laminate coordinate system. Material properties, the relative lamina orientation with respect to the laminate axis, and other parameters and properties are defined in the laminate coordinate system, unless specified otherwise. Also, it can be used to obtain the derived results (strains and stress) in these directions. In FEA, the laminate coordinate system is associated to the element coordinate system, with a unique right-handed orthogonal system associated to each element.

The element coordinate system orientation is associated with the element type. For bar or beam elements the orientation of the x-axis is generally along the line defined by the end nodes of the element. For solid elements in two and three dimensions, the orientation is typically defined parallel to the global coordinate system. For shell elements, axes x and y are defined on the element surface, with the z-axis always normal to the surface. The default orientation of x- and y- axes depends on the commercial code used and the element type.

There are various ways to define the default orientation of the *element coordinate system* in shell elements. In ANSYS (Figure 3.17.(a)), the x-axis is aligned with the edge defined by the first and second nodes of each element, the z-axis normal to the shell surface (with the outward direction determined by the right-hand rule), and the y-axis perpendicular to the x- and z-axis. In MSC-MARC, the x-axis is aligned with a line joining the middle points of two opposite edges (Figure 3.17.(b)).

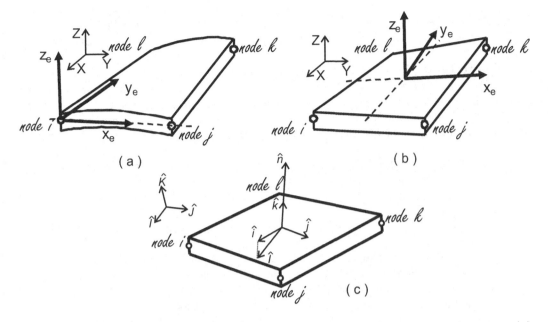

Fig. 3.17: Default orientations of element coordinate systems in shell elements: (a) ANSYS, (b) MSC-MARC, (c) Abaqus.

In Abaqus (Figure 3.17.(c)), the local x-direction is calculated projecting the global X-direction onto the surface of the element, i.e.,

$$\hat{i} = (\hat{I} - \hat{i}^*)/\|\hat{I} - \hat{i}^*\|$$
$$\hat{k} = \hat{n}$$
$$\hat{j} = \hat{k} \times \hat{i} \tag{3.18}$$

with $\hat{i}^* = (\hat{I} \cdot \hat{n})\hat{n}$ being the projection of the global X-direction, $\hat{I}$, along the normal of the element, $\hat{n}$. If the global X-direction, $\hat{I}$, is within 0.1° of the normal to the element, $\hat{n}$, i.e., they are almost parallel, the local x-direction is calculated using the global Z-direction, $\hat{K}$, instead of $\hat{I}$. The local coordinate system can be redefined using the procedures described in the Abaqus documentation [8, Section 2.2.5].

If no additional rotation is specified, the *laminate coordinate system* coincides with the element coordinate system. Additional rotations are specified by **ESYS** in ANSYS and **CSYS** in Abaqus, as shown in Examples 3.8 and 3.9.

In Examples 3.2–3.7, only rectangular plates with rectangular elements are analyzed. All of them have the first and the second node aligned with the global X-axis. Therefore, the material axes have been chosen parallel to the global axis. But this doesn't need to be the case. Most commercial codes have utilities to change the element coordinate system. Example 3.8 illustrates how to define a local coordinate system. Example 3.9 illustrates how it can be done in a shell with curvature. Example 3.10 illustrates how different orientations can be used in different locations of the structure.

Example 3.8 *Use a local coordinate system to model the plate of Example 3.2 (p. 120) if the orthotropic material is rotated +30° with respect to the x-direction. See also Section 12.2.*

Solution to Example 3.8 *Watch the video in [1].*
This example is a continuation of Example 3.2 (p. 120). However, the material is no longer symmetric so we cannot reuse Ex._3.2.cae *and we must construct a new full model. In the procedure presented below the mechanical properties of the material are entered using the definition of a* Lamina *instead of using the* [ABDH] *matrix.*

```
Menu: File, Set Work Directory, [C:\SIMULIA\User\Ex_3.8], OK
Menu: File, Save as, [Ex_3.8.cae], OK
```

   i. Creating parts

```
Module: Part
Menu:    Part, Create
    3D, Deformable, Shell, Planar, Approximate size [4000], Cont
Menu:    Add, Line, Rectangle
    # Enter the points [-2000,-1000], [2000,1000], X, Done
```

  ii. Defining materials

```
Module: Property
Menu:    Material, Create, Mechanical, Elast., Elastic, Type: Lamina
    [133860, 7706, 0.301, 4306, 4306, 2760], OK
```

 iii. Creating a local axis oriented at 30° and assigning material orientation to regions

```
Menu:    Tools, Datum
    # enter: origin, point along new x-axis, point on xy-plane
    # cos(30)=0.866, sin(30)=0.5
    CSYS, 3 Points, Name [csys-1], Rectangular, Cont
    [0,0,0], [0.866,0.5,0], [1,1,0], # close pop-up windows
Menu:    Assign, Material Orientation, # pick the region, Done
    Datum CSYS List, Names: csys-1, OK, # look at the figure, OK
```

  iv. Creating sections and assigning sections to parts

```
Menu:    Section, Create
    Shell, Homogeneous, Cont
    Section integration: Before analysis, Thickness: Value [10], OK
Menu:    Assign, Section, # pick part, Done, OK
```

  v. Creating assembly

```
Module: Assembly
Menu:    Instance, Create, Independent, OK
    # the global and user-defined coordinate systems are displayed
```

  vi. Defining analysis steps

```
Module: Step
Menu:    Step, Create, Name [Step-1], Static/General, Cont, OK
```

vii. Adding BC and loads

```
Module: Load
Menu:   BC, Manager
        Create, Name [SS1-xx], Step: Initial, Mech., Disp/Rota, Cont
        # pick the lines @ x=-2000 and x=2000, Done, U2, U3, UR1, OK
        Create, Name [SS1-yy], Step: Initial, Mech., Disp/Rota, Cont
        # pick the lines @ y=-1000 and y=1000, Done, U1, U3, UR2, OK
        # close BC Manager
Menu:   Load, Create
        Step [Step-1], Mech., Pressure, Cont, # pick the part, Done
        # Choose a side for the shell: Brown, Magnitude [0.12e-3], OK
```

viii. Meshing the model

```
Module: Mesh
Menu:   Seed, Instance, Approximate global size [500], Apply, OK
Menu:   Mesh, Controls, Elem. Shape: Quad, Technique: Structured, OK
Menu:   Mesh, Element Type, Geometric Order: Quadratic, DOF: 5, OK
Menu:   Mesh, Instance, Yes
```

ix. Solving and visualizing the solution

```
Module: Job
Menu:   Job, Manager
        Create, Cont, OK, Submit # when completed, Results
Module: Visualization
Menu:   Plot, Contours, On Deformed Shape
        Field Output Toolbox, Primary, U, U3
```

The maximum deflection of the model is 8.704 mm.

Example 3.9 *Align the element coordinate system (laminate coordinate system) of a 3D curved shell with the projection of the global Y-axis onto the surface of the shell.*

Solution to Example 3.9 *Watch the video in [1].*
The material used is AS4D/9310 in a [90/45/-45] laminate configuration. This example is solved in two steps. First, the laminate coordinate system will be aligned with the element coordinate system by selecting CSYS: <Layup> *in* Composite Layup, *The orientation of the element coordinate system is explained in Section 3.2.14. Second, an user-defined orientation will be created by a rotation of 90° around the z-axis of the element coordinate system, thus aligning the laminate coordinate system with the projection of the global Y-axis on the surface of the shell.*

```
Menu: File, Set Work Directory, [C:\SIMULIA\User\Ex_3.9], OK
Menu: File, Save as, [Ex_3.9.cae], OK
```

i. Creating parts

```
Module: Part
Menu:   Part, Create, 3D, Deformable, Point, Approx. size [500], Cont
        [0,0,0] # creates a reference point denoted by RF
Menu:   Tools, Datum, Point, Enter coordinates
```

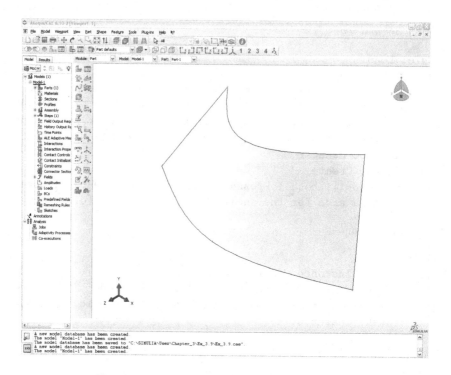

Fig. 3.18: Part geometry in Example 3.9.

```
[300,0,135], [0,0,235], [-300,0,135], [200,200,0],
[0,200,135], [-200,200,0], X, # close pop-up, Views: Iso View
Menu:    Shape, Wire, Point to Point, Geometry Type [spline]
    # click on green + sign,
    # leftmost and topmost points, Done, X #to close the popup
    # close the popup after each spline
    # repeat 3 times for a total of 4 splines
Menu:    Shape, Shell, Loft
    Insert Before, # pick the leftmost straight line, Done
    Insert After, # pick the rightmost straight line, Done
    Tab: Transition, Method: Select path
    Add, # pick lower curve, Done, Add, # pick upper curve, Done, OK
```

The part should look like the one displayed in Figure 3.18.

ii. Defining materials, section, and assigning sections to parts

```
Module: Property
Menu:    Material, Create, Mechanical, Elast., Elastic, Type: Lamina
    [133860, 7706, 0.301, 4306, 4306, 2760], OK
Menu:    Tools, Datum, # create a CSYS to be used later
    CSYS, 3 Points, Name [user-csys], Rectangular, Cont
    [0,0,0], [0,1,0], [-1,1,0], # close pop-up windows
Menu:    Composite, Create, Element Type: Conventional Shell, Cont
    Region # double click the cell corresponding to each ply
    # pick the part, Done # the cell must read (Picked)
    Material # double click the cell corresponding to each ply, OK
```

```
Thickness [1.05] # all plies, CSYS # <Layup> all plies
Rotation Angle [90,45,-45], Integ. Points [3] # all plies, OK
```

Note: Although the displayed triad changes with each input from the user (see video on [1, Example 3.9]), you must provide all input with respect to the global coordinate system. That is, the origin {0,0,0}, a point on the CSYS x-axis {0,1,0}, and a point on the CSYS xy-plane {-1,+1,0}, are all values with respect to the fixed, immutable, global coordinate system.

iii. Creating assembly

```
Module: Assembly
Menu:   Instance, Create, Independent, OK
        # the global and user-defined coordinate systems are displayed
```

iv. Defining analysis steps

```
Module: Step
Menu:   Step, Create, Name [Step-1], Static/General, Cont, OK
Menu:   Output, Field Output Requests, Edit, F-Output-1
        Output at shell: Specify [1,2,3,4,5,6,7,8,9], OK
```

v. Meshing the model

```
Module: Mesh
Menu:   Seed, Instance, Approximate global size [100], Apply, OK
Menu:   Mesh, Controls, Element Shape: Quad, Technique: Structure, OK
Menu:   Mesh, Instance, Yes
```

Since the goal of this example is to show the user how the element and laminate coordinate systems are oriented, it is not necessary to apply loads, BC, or to solve the problem. However, doing a Data Check of the model allows the user to visualize the orientation of the element coordinate systems.

vi. Visualizing laminate coordinate systems

```
Module: Job
Menu:   Job, Manager
        Create, Cont, OK, Data Check # when completed, Results
        # to visualize the coordinate system of the elements
Module: Visualization
Toolbar: Render Style, Render Model: Wireframe
Menu:   Plot, Material Orientations, On Undeformed Shape
        # select different plies and visualize their orientations
Menu:   Results, Section Points, Selection method: Plies
```

Note: To install the rendering tool, watch the video, at 6',55" on [1, Example 3.9], or use the following pseudo-code

```
Menu:   View, Toolbars [RenderView]
```

By selecting <Layup> as the option for the CSYS of the Composite Layup, the user is choosing to define the laminate coordinate system aligned with the element coordinate system. To use an user-defined orientation, the CSYS in the

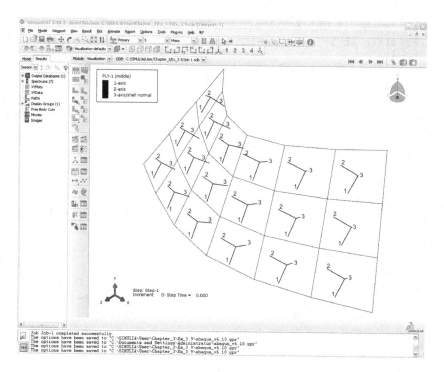

Fig. 3.19: Element coordinate system defined using Material Layup for the 90° lamina of a $[90/45/-45]$ laminate.

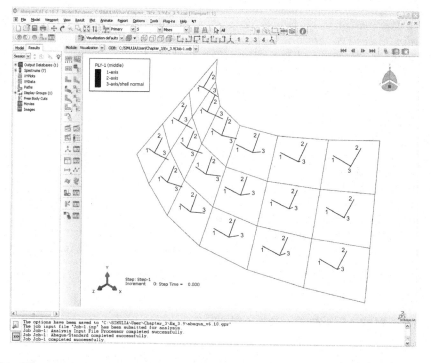

Fig. 3.20: Element coordinate system defined using user-defined CSYS for the 90° lamina of a $[90/45/-45]$ laminate.

Composite Layup must be changed. In the pseudo-code for Example 3.9, an user-defined orientation was defined and named user-csys. The pseudo-code below shows how to modify the problem to use an user-defined orientation.

```
# Defining materials, sections, and assigning sections to parts
Module: Property
Menu:   Composite, Edit, CompositeLayup-1
    Region # must read (Picked) for all the plies
    Material # must read Material-1 for all the plies
    Thickness [1.05] # for all plies
    CSYS # double the cell corresponding to each ply
    Base Orient.: CSYS, Select, Datum CSYS List, user-csys, OK, OK
    # must read user-csys.3 for all the plies
    Rotation Angle [90,45,-45], Integ.Points [3] # all plies, OK
```

The lamina (ply) coordinate system for the 90° lamina, before and after rotation, is shown in Figures 3.19 and 3.20. By visualizing the lamina coordinate system for various laminas, one can see how the coordinate systems are affected by the user-defined orientation.

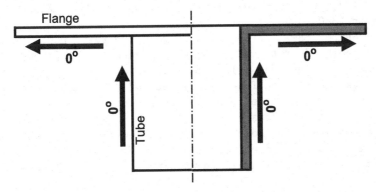

Fig. 3.21: Example 3.10. Reference axis in a flange tube.

Example 3.10 *Model in Abaqus a flanged tube with axial and radial laminate orientation. In the cylindrical part, the reference axis will be in the longitudinal direction. In the flange, the reference axis will be radial (see Figure 3.21).*

Solution to Example 3.10 *Watch the video in [1].*
The material used is AS4D/9310 in a [0/45/-45] laminate configuration. To satisfy the required orientations, a cylindrical coordinate system will be used. The pseudo-code below shows how to align the coordinate system of the elements of the cylinder in the axial direction (z-axis of the cylindrical coordinate system), and the coordinate system of the elements of the flange in the radial direction of the cylindrical coordinate system.

```
Menu: File, Set Work Directory, [C:\SIMULIA\User\Ex_3.10], OK
Menu: File, Save as, [Ex_3.10.cae], OK
```

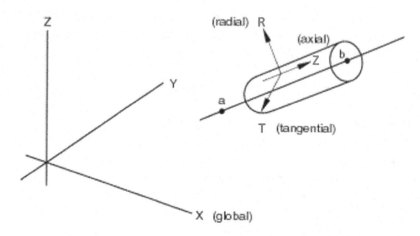

Fig. 3.22: Example 3.10. Datum CSYS: cylindrical.

i. Creating parts

```
Module: Part
Menu:   Part, Create, 3D, Solid, Extrusion, Approx size [600], Cont
Menu:   Add, Circle, [0,0], [0,175], [0,0], [0,275], X, Done
        Depth [350], OK
Menu:   Shape, Shell, From Solid, # pick the part, Done, X
Menu:   Tools, Geometry Edit, Category: Face, Method: Remove
        # pick the external cylindrical surface, Done
        # pick the flat surface at the back of the model, Done
        X, # close the Geometry Edit pop-up window
Menu:   Tools, Set, Manager
        Create, Name [cylinder], Cont, # pick cylindrical surface, Done
        Create, Name [flange], Cont, # pick the flat surface, Done
        # close the Set Manager pop-up window, # sets will be used later
```

ii. Defining materials, section, and assigning sections to parts. See Figure 3.22 where R=1, T=2, Z=3.

```
Module: Property
Menu: Material, Create, Mechanical, Elasticity, Elastic, Type: Lamina
      [133860, 7706, 0.301, 4306, 4306, 2760], OK
Menu: Tools, Datum, CSYS, 3 Points, Name [cyl-csys], Cylindrical, Cont
      [0,0,0], [1,0,0], [1,1,0], # close pop-up windows
Menu:   Composite, Manager
        Create, Name [cyl-zone], Element Type: Conventional Shell, Cont
        Region # double click the cell corresponding to each ply
        Sets, cylinder, Cont, # the cell must read cylinder for all plies
        Material # double click the cell corresponding to each ply, OK
        # must read Material-1 for all plies, Thickness [1.05] # all plies
        CSYS # double click the cell corresponding to each ply
        Base Orientation: CSYS, Select, Datum CSYS List, cyl-csys, OK
        Normal Direction: Axis 1, OK, # must read cyl-csys.1 for all plies
        Rotation Angle [0,45,-45], Integration Points [3] # all plies, OK
```

```
Create, Name [flange-zone], Element Type: Conventional Shell, Cont
Region # double click the cell corresponding to each ply
Sets, flange, Cont, # the cell must read flange for all plies
Material # double click the cell corresponding to each ply, OK
# must read Material-1 for all cells, Thickness [1.05] # all plies
CSYS # double click the cell corresponding to each ply
Base Orientation: CSYS, Select, Datum CSYS List, cyl-csys, OK
Normal Direction: Axis 3, OK, # must read cyl-csys.3 for all plies
Rotation Angle [0,45,-45], Integration Points [3] # all plies, OK
# close Composite Layup Manager
```

iii. Creating assembly

```
Module: Assembly
Menu:   Instance, Create, Independent, OK
        # the global and user-defined coordinate systems are displayed
```

iv. Defining analysis steps

```
Module: Step
Menu:   Step, Create, Name [Step-1], Static/General, Cont, OK
Menu:   Output, Field Output Requests, Edit, F-Output-1
        Output at shell: Specify [1,2,3,4,5,6,7,8,9], OK
```

v. Meshing the model

```
Module: Mesh
Menu:   Seed, Instance, Approximate global size [75], Apply, OK
Menu:   Mesh, Controls, # pick all the surfaces, Done
        Element Shape: Quad, Technique: Free, OK
Menu:   Mesh, Instance, Yes
```

The goal of this example is to show the user how to align the element's orientation using a cylindrical coordinate system. Therefore, it is not necessary to apply loads, BC, or to solve the problem. However, doing a Data Check of the model allows the user to visualize the orientation of the element coordinate systems.

vi. Visualizing coordinate element systems

```
Module: Job
Menu:   Job, Manager
        Create, Cont, OK, Data Check # when completed, Results
Module: Visualization
        Toolbar: Render Style, Render Model: Shaded
        # to easily visualize the coordinate system of all the elements
Menu:   Plot, Material Orientations, On Undeformed Shape
Menu:   Results, Section Points, Selection method: Plies
        # select the different plies and visualize their orientations
```

Example 3.11 *Create a FE model for a pultruded composite column under axial compression load $P = 11,452$ N [19] and calculate the end axial displacement $u(L/2)$, where $x = 0$ is located at the half-length of the column. The column is simply supported (pinned) at both ends $x = (-L/2, L2)$ but only one half needs to be discretized because the model is symmetric at half the length of the column. Its length is $L = 1.816$ m. The cross-section of the column*

is that of a wide-flange I-beam (also called H-beam) with equal outside height and width,
$H = W = 304.8$ mm. The thickness of both the flange and the web is $t_f = t_w = 12.7$ mm.
The material properties are given by the A, B, D, H matrices, with units [mm MPa],
[mm^2 MPa], [mm^3 MPa], and [mm MPa], respectively.
   For the flange:

$$[A] = \begin{bmatrix} 335{,}053 & 47{,}658 & 0 \\ 47{,}658 & 146{,}155 & 0 \\ 0 & 0 & 49{,}984 \end{bmatrix} \quad ; \quad [B] = \begin{bmatrix} -29{,}251 & -1{,}154 & 0 \\ -1{,}154 & -5{,}262 & 0 \\ 0 & 0 & -2{,}274 \end{bmatrix}$$

$$[D] = \begin{bmatrix} 4{,}261{,}183 & 686{,}071 & 0 \\ 686{,}071 & 2{,}023{,}742 & 0 \\ 0 & 0 & 677{,}544 \end{bmatrix} \quad ; \quad [H] = \begin{bmatrix} 34{,}216 & 0 \\ 0 & 31{,}190 \end{bmatrix}$$

For the web:

$$[A] = \begin{bmatrix} 338{,}016 & 44{,}127 & 0 \\ 44{,}127 & 143{,}646 & 0 \\ 0 & 0 & 49{,}997 \end{bmatrix} \quad ; \quad [B] = \begin{bmatrix} -6{,}088 & -14{,}698 & 0 \\ -14{,}698 & -6{,}088 & 0 \\ 0 & 0 & 0 \end{bmatrix}$$

$$[D] = \begin{bmatrix} 4{,}769{,}538 & 650{,}127 & 0 \\ 650{,}127 & 2{,}155{,}470 & 0 \\ 0 & 0 & 739{,}467 \end{bmatrix} \quad ; \quad [H] = \begin{bmatrix} 34{,}654 & 0 \\ 0 & 31{,}623 \end{bmatrix}$$

Solution to Example 3.11 *Watch the video in [1].*
The geometry of the I-beam is defined by the mid-surfaces of the web and the flanges. Notice
that the geometry defined in the procedure below corresponds to an I section 292.1 mm in
height and 304.8 in width. The column is placed along the global Z-axis, with one-half of
its length on each side of the global XY plane. Therefore, only one half of the length of
the column is modeled. The load is applied as a concentrated load. This requires the use of
a rigid body constraint of the type `tie` *nodes on the loaded end to guarantee that all the*
points at the loading edge follow the displacements and rotations of the point where the load
is applied.

```
Menu: File, Set Work Directory, [C:\SIMULIA\User\Ex_3.11], OK
Menu: File, Save as, [Ex_3.11.cae], OK
```

i. Creating parts

```
Module: Part
Menu:   Part, Create
     3D, Deformable, Shell, Extrusion, Approx size [500], Cont
Menu:   Add, Line, Connected Lines, [-152.4,-146.05],
             [152.4,-146.05], X
Menu:   Add, Line, Connected Lines, [-152.4,146.05],
             [152.4,146.05], X
Menu:   Add, Line, Connected Lines, [0,-146.05],
             [0,146.05], X, Done
     Depth [908], OK
```

ii. Defining materials, section, and assigning sections to parts

```
Module: Property
Menu:   Section, Create, Name [flange],
            Shell, General Shell Stiffness, Cont
    # In the pop-up window fill the stiffness matrix as follows
    Tab: Stiffness
    # Upper part of matrix A in rows 1 to 3, columns 1 to 3
    # matrix B in rows 1 to 3, columns 4 to 6
    # Upper part of matrix D in rows 4 to 6, columns 4 to 6
    # The Stiffness matrix should be
    # 335053    47658        0    -29251     -1154          0
    #          146155        0     -1154     -5262          0
    #                    49984         0         0      -2274
    #                            4261183    686071          0
    #                                       2023742         0
    #                                                  677544
    Tab: Advanced
    # The values corresponding to the H matrix are entered here
    Transverse Shear Stiffness: # checkmark: Specify values
    K11 [34216], K12 [0], K22 [31190], OK
Menu:   Section, Create, Name [web],
            Shell, General Shell Stiffness, Cont
    Tab: Stiffness
    # The Stiffness matrix should be
    # 338016    44127        0     -6088    -14698          0
    #          143646        0    -14698     -6088          0
    #                    49997         0         0          0
    #                            4769538    650127          0
    #                                       2155470         0
    #                                                  739467
    Tab: Advanced
    Transverse Shear Stiffness: # checkmark: Specify values
    K11 [34654], K12 [0], K22 [31623], OK
Menu:   Assign, Section
    # pick the four regions forming the flanges of the beam, Done
    Section: flange, OK
    # pick the region forming the web of the beam, Done
    Section: web, OK, X # to end the command
# Create orientations to align the material accordingly
Menu:   Tools, Datum, Type: CSYS, Method: 2 lines
    Name [web-csys], Rectangular, Cont
    # pick a line along Z-global, pick a line along Y-global
    Name [flange-csys], Rectangular, Cont
    # pick a line along Z-global, pick a line along X-global
    # close the pop-up windows
Menu:   Assign, Material Orientation
    # pick the web, Done, Datum CSYS List, web-csys, OK,
        Normal: Axis 3, OK
    # pick the four regions forming the flange, Done
    Datum CSYS List, # pick flange-csys, OK, Normal: Axis 3, OK
```

iii. Creating the assembly

```
Module: Assembly
Menu:   Instance, Create, Independent, OK
```

iv. Defining analysis steps

```
Module: Step
Menu:   Step, Create, Cont, OK
```

v. Create constraints (see Section 2.4)

```
Module: Interaction
# create a reference point (RP) at the loaded end of the I-beam
Menu:   Tools, Reference Point, [0,0,0], Enter, X
# define rigid body constraints for the loaded end
# Tie nodes makes all disp and rota equal to those of the RP
Menu:   Constraint, Create, Rigid Body, Cont
   Tie (nodes), Edit, # pick the 5 lines at the loaded end, Done
   Reference Point: Edit, # pick the RP created, OK
```

vi. Adding loads and BC

```
Module: Load
Menu:   Load, Create
   Step: Step-1, Mechanical, Concentrated force, Cont
   # select the RP, Done, CF3 [11452], OK
Menu:   BC, Create
   Step: Initial, Mechanical, Symm/Anti/Enca, Cont
   # pick lines opposite to loaded end, Done, ZSYMM, OK
```

vii. Meshing the model

```
Module: Mesh
Menu:   Seed, Edges
   # pick the six longitudinal lines, Done
   Method: By Number, Bias: None, Number of elements [10], Apply, OK
   # pick edges of the flanges at both ends of the beam, Done
   Method: By Number, Bias: None, Number of elements [3], Apply, OK
   # pick the edges of the web at both ends of the beam, Done
   Method:By Number, Bias:None, Number of elements [4], Apply, OK, X
Menu:   Mesh, Controls, # select all, Done, Technique: Structured, OK
Menu:   Mesh, Element Type, # select all, Done
   Geometric Order: Quadratic, DOF: 6, Family: Shell, OK # S8R in use
Menu:   Mesh, Instance, Yes
```

Create a set containing the central point of the end opposite to the loaded end. This point will be used to apply an extra BC to prevent the movement on the symmetry plane.

```
Module: Mesh
Menu:   Tools, Set, Create, Name [Cpoint], Type: Node, Cont
   # pick the point at the center of unloaded end, Done
Module: Load
```

```
Menu:    BC, Create
    Step: Initial, Mechanical, Disp/Rota, Cont
    Sets, Cpoint, Cont, # checkmark: U1,U2,UR3, OK
```

viii. Solving and visualizing the results

```
Module: Job
Menu:    Job, Manager
    Create, Cont, OK, Submit, # when completed, Results
Module: Visualization
Menu:    Plot, Allow Multiple Plot States
Menu:    Plot, Deformed Shape
    Toolbar: Field Output, Primary, U, U3
```

Notice that the displacement at the loaded end is 0.03584 mm.

3.3 Failure Criteria

Failure criteria are curve fits of experimental data that attempt to predict failure under multiaxial stress based on experimental data obtained under uniaxial stress. All failure criteria described in this section predict the first occurrence of failure in one of the laminas but are unable to track failure propagation until complete laminate failure. Damage mechanics is used in Chapters 8 and 9 to track damage evolution up to laminate failure. The truncated-maximum-strain criterion estimates laminate failure without tracking damage evolution by making certain approximations and assumptions about the behavior of the laminate [3, Section 7.3.2].

In this section, failure criteria are presented using the notion of failure index, which is used for several FEA packages, and it is defined as

$$I_F = \frac{stress}{strength} \tag{3.19}$$

Failure is predicted when $I_F \geq 1$. The strength ratio [3, Section 7.1.1] is the inverse of the failure index

$$R = \frac{1}{I_F} = \frac{strength}{stress} \tag{3.20}$$

Failure is predicted when $R \leq 1$.

3.3.1 2D Failure Criteria

Strength-based failure criteria are commonly used in FEA to predict failure events in composite structures. Numerous criteria exist for unidirectional (UD) laminas subjected to a state of plane stress ($\sigma_3 = 0$). The most commonly used are described in [3, Section 7.1, Section 7.3.2]. They are

- Maximum stress criterion
- Maximum strain criterion
- Truncated maximum strain criterion, and

– Interacting failure criterion

A few additional criteria are presented in this section.

Hashin Failure Criterion

The Hashin failure criterion (HFC) proposes four separate modes of failure:

- Fiber tension
- Fiber compression
- Matrix tension
- Matrix compression

that are predicted by four separate equations, as follows[5]

$$I_{Fft}^2 = \left(\frac{\sigma_1}{F_{1t}}\right)^2 + \alpha\left(\frac{\sigma_6}{F_6}\right)^2 \qquad \text{if } \sigma_1 \geq 0 \tag{3.21}$$

$$I_{Ffc}^2 = \left(\frac{\sigma_1}{F_{1c}}\right)^2 \qquad \text{if } \sigma_1 < 0 \tag{3.22}$$

$$I_{Fmt}^2 = \left(\frac{\sigma_2}{F_{2t}}\right)^2 + \left(\frac{\sigma_6}{F_6}\right)^2 \qquad \text{if } \sigma_2 \geq 0 \tag{3.23}$$

$$I_{Fmc}^2 = \left(\frac{\sigma_2}{2F_4}\right)^2 + \left[\left(\frac{F_{2c}}{2F_4}\right)^2 - 1\right]\frac{\sigma_2}{F_{2c}} + \left(\frac{\sigma_6}{F_6}\right)^2 \qquad \text{if } \sigma_2 < 0 \tag{3.24}$$

where α is a weight factor to give more or less emphasis to the influence of shear on fiber failure. With $\alpha = 0$, Hashin and maximum stress failure criteria would predict longitudinal tensile failure of the unidirectional lamina at the same stress σ_1. Note (3.22) predicts longitudinal compressive failure without influence of the shear stress, although it is well known that the state of shear has a strong influence on the longitudinal compression failure [20].

Equations (3.21–3.24) define the *square* of failure indexes according to HFC. Abaqus documentation refers to the square of failure indexes for HFC, Tsai-Hill, Azzi-Tsai-Hill, and Tsai-Wu[6], in contrast to the (*not squared*) failure indexes for maximum stress (MSTRS) and maximum strain (MSTRN) failure criteria.

The values of F_{1c} and F_{2c} are considered positive throughout this textbook and most of the literature, but in Abaqus they must be given with different signs in different areas of Abaqus. Specifically, when entering strength values in Edit Material, Material Behavior: Elastic, Sub-option: Fail Stress, F_{1c} and F_{2c} must be entered as negative values. However, when entering the same values to define the threshold of damage initiation in Edit Material, Material Behavior: Hashin Damage, they must be given as positive quantities.

---

[5]The Hashin failure equations on the right-hand side of (3.21–3.24) yield squares of failure indexes I_F. Compare to the maximum stress failure criterion (i.e., $I_{Fft} = \sigma_1/F_{1T}$) and so on.

[6]The Tsai-Hill, Azzi-Tsai-Hill, and Tsai-Wu failure criteria are not recommended because they overemphasize the interaction between fiber (σ_1) and transverse matrix (σ_2) damage modes.

Although the threshold of damage initiation is calculated in Abaqus by the Hashin failure criterion, the subsequent analysis of damage accumulation is done by a decohesion methodology that is independent of the damage initiation criterion used. Despite this, Abaqus documentation and Abaqus CAE refer to the whole damage methodology as *Hashin Damage*.

Puck Failure Criterion

The Puck failure criterion [21] distinguishes between fiber failure (FF) and matrix failure (MF). In the case of plane stress, the MF criteria discriminates three different modes. Mode A is when transverse cracks appear in the lamina under transverse tensile stress with or without in-plane shear stress. Mode B also denotes transverse cracks, but in this case they appear under in-plane shear stress with small transverse compression stress. Mode C indicates the onset of oblique cracks (typically with an angle of 53° in carbon epoxy laminates) when the material is under significant transverse compression.

The FF and the three MF modes yield separate failure indexes. The Puck criterion assumes that FF only depends on longitudinal tension. Therefore, the failure index for FF is defined as

$$I_{FF} = \begin{cases} \sigma_1/F_{1t} & \text{if } \sigma_1 > 0 \\ -\sigma_1/F_{1c} & \text{if } \sigma_1 < 0 \end{cases} \tag{3.25}$$

The MF failure indexes have different expressions depending on the mode that becomes active. With positive transverse stress, mode A is active. In this case, the failure index for matrix-dominated tensile failure (mode A) is

$$I_{MF,A} = \sqrt{\left(\frac{\sigma_6}{F_6}\right)^2 + \left(1 - p_{6t}\frac{F_{2t}}{F_6}\right)^2 \left(\frac{\sigma_2}{F_{2t}}\right)^2} + p_{6t}\frac{\sigma_2}{F_6} \quad \text{if } \sigma_2 \geq 0 \tag{3.26}$$

where p_{6t} is a fitting parameter. Lacking experimental values, it is assumed that $p_{6t} = 0.3$ [21].

Under negative transverse stress, either mode B or mode C is active, depending on the relationship between in-plane shear stress and transverse shear stress. The limit between mode B and C is defined by the relation F_{2A}/F_{6A}, where

$$F_{2A} = \frac{F_6}{2p_{6c}}\left[\sqrt{1 + 2p_{6c}\frac{F_{2c}}{F_6}} - 1\right] \tag{3.27}$$

$$F_{6A} = F_6\sqrt{1 + 2p_{2c}} \tag{3.28}$$

and p_{2c} is defined as

$$p_{2c} = p_{6c}\frac{F_{2A}}{F_6} \tag{3.29}$$

and p_{6c} is another fitting parameter. Lacking experimental values, it is assumed that $p_{6c} = 0.2$ [21].

Finally, the failure index for matrix-dominated shear (mode B) is

$$I_{MF,B} = \frac{1}{F_6}\left[\sqrt{\sigma_6^2 + (p_{6c}\sigma_2)^2} + p_{6c}\sigma_2\right] \quad \text{if} \quad \left\{ \begin{array}{l} \sigma_2 < 0 \\ \left|\frac{\sigma_2}{\sigma_6}\right| \le \frac{F_{2A}}{F_{6A}} \end{array}\right. \tag{3.30}$$

and for matrix-dominated compression (mode C) is

$$I_{MF,C} = -\frac{F_{2c}}{\sigma_2}\left[\left(\frac{\sigma_6}{2(1+p_{2c})F_6}\right)^2 + \left(\frac{\sigma_2}{F_{2c}}\right)^2\right] \quad \text{if} \quad \left\{ \begin{array}{l} \sigma_2 < 0 \\ \left|\frac{\sigma_2}{\sigma_6}\right| \ge \frac{F_{2A}}{F_{6A}} \end{array}\right. \tag{3.31}$$

3.3.2   3D Failure Criteria

Failure criteria presented here are 3D generalizations of the ones presented in [3, Section 7.1]. The 2D version of the criteria presented in this section can be recovered by setting $\sigma_4 = \sigma_5 = 0$. The user of FEA packages should be careful because some packages use only the in-plane stress components for the computation of the failure index (e.g., Abaqus), even though all six stress components may be available from the analysis. In those cases, the intralaminar and thickness components of stress should be evaluated separately to see if they lead to failure.

In this section, the numerical subscript denotes the directions of 1) fiber, 2) in-plane transverse to the fibers, and 3) through the thickness of the lamina. The letter subscript denotes t) tensile and c) compressive. Contracted notation is used for the shear components as described in Section 1.5.

Maximum Strain Criterion

The failure index is defined as

$$I_F = \max \left\{ \begin{array}{ll} \epsilon_1/\epsilon_{1t} & \text{if } \epsilon_1 > 0 \quad \text{or} \quad -\epsilon_1/\epsilon_{1c} \text{ if } \epsilon_1 < 0 \\ \epsilon_2/\epsilon_{2t} & \text{if } \epsilon_2 > 0 \quad \text{or} \quad -\epsilon_2/\epsilon_{2c} \text{ if } \epsilon_2 < 0 \\ \epsilon_3/\epsilon_{3t} & \text{if } \epsilon_3 > 0 \quad \text{or} \quad -\epsilon_3/\epsilon_{3c} \text{ if } \epsilon_3 < 0 \\ abs(\gamma_4)/\gamma_{4u} \\ abs(\gamma_5)/\gamma_{5u} \\ abs(\gamma_6)/\gamma_{6u} \end{array}\right. \tag{3.32}$$

The quantities in the denominator are the ultimate strains of the unidirectional lamina. Note that compression ultimate strains in (3.32) are positive numbers.

Maximum Stress Criterion

The failure index is defined as

$$I_F = \max \left\{ \begin{array}{ll} \sigma_1/F_{1t} & \text{if } \sigma_1 > 0 \quad \text{or} \quad -\sigma_1/F_{1c} \text{ if } \sigma_1 < 0 \\ \sigma_2/F_{2t} & \text{if } \sigma_2 > 0 \quad \text{or} \quad -\sigma_2/F_{2c} \text{ if } \sigma_2 < 0 \\ \sigma_3/F_{3t} & \text{if } \sigma_3 > 0 \quad \text{or} \quad -\sigma_3/F_{3c} \text{ if } \sigma_3 < 0 \\ abs(\sigma_4)/F_4 \\ abs(\sigma_5)/F_5 \\ abs(\sigma_6)/F_6 \end{array}\right. \tag{3.33}$$

The letter F is used here to denote a strength value for a unidirectional lamina as in [22]. Note that compression strength in (3.33) are positive numbers.

Tsai-Wu Criterion

Using the Tsai-Wu criterion the failure index is defined as

$$I_F = \frac{1}{R} = \left[-\frac{B}{2A} + \sqrt{\left(\frac{B}{2A}\right)^2 + \frac{1}{A}} \right]^{-1} \tag{3.34}$$

with

$$A = \frac{\sigma_1^2}{F_{1t}F_{1c}} + \frac{\sigma_2^2}{F_{2t}F_{2c}} + \frac{\sigma_3^2}{F_{3t}F_{3c}} + \frac{\sigma_4^2}{F_4^2} + \frac{\sigma_5^2}{F_5^2} + \frac{\sigma_6^2}{F_6^2}$$
$$+ c_4 \frac{\sigma_2\sigma_3}{\sqrt{F_{2t}F_{2c}F_{3t}F_{3c}}} + c_5 \frac{\sigma_1\sigma_3}{\sqrt{F_{1t}F_{1c}F_{3t}F_{3c}}} + c_6 \frac{\sigma_1\sigma_2}{\sqrt{F_{1t}F_{1c}F_{2t}F_{2c}}} \tag{3.35}$$

and

$$B = \left(F_{1t}^{-1} - F_{1c}^{-1}\right)\sigma_1 + \left(F_{2t}^{-1} - F_{2c}^{-1}\right)\sigma_2 + \left(F_{3t}^{-1} - F_{3c}^{-1}\right)\sigma_3 \tag{3.36}$$

where c_i, with $i = 4..6$, are the Tsai-Wu coupling coefficients that by default are taken to be -1. Note that compression strength in (3.35) and (3.36) are here positive numbers.

The through-the-thickness strength values F_{3t} and F_{3c} are seldom available in the open literature, so it is common practice to use the corresponding in-plane transverse values of strength. Also, the intralaminar strength F_5 is commonly assumed to be equal to the in-plane shear strength. The intralaminar strength F_4 can be estimated [3, (4.84)] in terms of transverse compression strength F_{2c} and the angle of the fracture plane α_0. The latter is consistently reported to be close to 53°.

The maximum strain (MSTRN), maximum stress (MSTRS), and Hashin failure criteria are adequate, but Tsai-Hill (TSAIH) and Tsai-Wu (TSAIW) failure criteria are not recommended because they overestimate the interaction between σ_1 (S11) and σ_2 (S22).

Unlike the definition used in this textbook (3.34), Abaqus defines the Tsai-Wu failure index as:

$$TSAIW = f_1 S_{11} + f_2 S_{22} + f_{11} S_{11}^2 + f_{22} S_{22}^2 + f_{66} S_{12}^2 + 2 f_{12} S_{11} S_{22} \tag{3.37}$$

which is a *square* of a failure index. Therefore, the values of TSAIW do not compare with the failure indices for MSTRN (3.32) and MSTRS (3.33) unless a square root of TSAIW is taken. The same happens for the TSAIH failure index and for the four Hashin failure indexes in Abaqus (compare the Abaqus documentation to (3.21–3.24)).

The strength values F_{1t}, F_{1c}, F_{2t}, F_{2c}, F_6, f_{12}, and S_{lim} are supplied to CAE using

```
Module: Property
Menu:    Material, Edit,
         Material Behavior: Elastic, Sub-option: Fail Stress
```

Note that F_{1c} and F_{2c} must be given as negative numbers for TSAIW but as positive numbers for Hashin Damage (see Section 3.3.1). Also for TSAIW only, f_{12} and the bi-axial stress S_{lim} default to zero if not given, which is good because interaction f_{12} overestimates the interaction between σ_1 (S11) and σ_2 (S22).

Example 3.12 *Compute the failure index I_F and strength ratio R in each lamina of Example 3.4 (p. 128) using the maximum stress failure criterion (MSTRS). The elastic moduli and strength values for the lamina are given in Table 3.1, p. 108.*

Solution to Example 3.12 *Watch the video in [1].*
To solve this example, open Ex_3.4.a.cae *and Save As* Ex_3.12.cae*. Then, do the following:*

   i. Set Work Directory

```
   Menu:   File, Set Work Directory, [C:\SIMULIA\User\Ex_3.12], OK
```

  ii. Provide strength values for the material.

```
   Module: Property
   Menu:    Material, Manager
      Edit, Suboptions, Fail Stress, # you can copy/paste from excel
      [1830    -1096    57    -228    71    -1    0], OK, OK
      # close Material Manager pop-up window
```

 iii. Defining analysis step to create a field output for the failure index

```
   Module: Step
   Step:    Step-1
   Menu:    Output, Field Output Requests, Edit, F-Output-1
      # expand: Failure/Fracture, # checkmark: CFAILURE
      # uncheck: everything else we don't need, OK
```

  iv. Solving and visualizing the results

```
   Module: Jobs
   Menu:    Job, Manager
      Submit, # when completed, Results
   Module: Visualization
   Menu:    Plot, Contours, On Deformed Shape
   Menu:    Result, Field Output, MSTRS, OK
   Menu:    Results, Section Points, Method: Plies
      # pick a ply, Apply, # read the value, OK
```

The results are tabulated in Table 3.7. Compare the results to the classical lamination theory (CLT) solution [3].

Table 3.7: MSTRS (maximum stress failure criterion) calculated in Example 3.12.

Ply #	MSTRS	
	I_F	R
1	0.01442	69.34813
2	0.02430	41.15226
3	0.01566	63.85696
4	0.01566	63.85696

3.4 Predefined Fields

The following example illustrates the analysis of thermal expansion and how to introduce a **Predefined Field** in general.

Example 3.13 *Display the deformed shape of a* $[0/90]_T$ *laminate illustrating thermal expansion coupling (discussed in [3, Section 6.1.4, Figure 6.7]). Each lamina is 1.2 mm thick, made of AS4D/9310 (Table 3.1). The geometry of the plate is square with* $a_x = a_y = 1000$ *mm. The thermal loading is* $\Delta T = -125°\ C$*. Use element S8R5 in Abaqus.*

Solution to Example 3.13 *Watch the video in [1].*
A $[0/90]_T$ *flat laminate under uniform thermal loading* ΔT*, with square geometry* $(a_x = a_y)$ *on the x-y plane, is symmetric with respect to the x-z and y-z planes, so only one quarter of the plate needs to be modeled. Note that the fact that the LLS is not symmetric does not affect the symmetry of the discretization, which is determined by three factors: (a) symmetry of geometry, (b) symmetry of load, and (c) symmetry of material, the latter on the x-y plane only. The following pseudo-code illustrates the procedure. The saddle shape of the deformed laminate is shown in Figure 3.23. The CAE file is available in [2, Ex_3_13.cae].*

i. Geometry

```
Module: Part
Menu:   Part, Create, [Part-1], 3D, Deformable,
        Shell, Planar, Approximate size [4000], Cont
Menu:   Add, Line, Rectangle,
        # Enter the points [0,0], [1000,1000], X, Done
```

ii. Material and Section Properties

```
Module: Property
Menu:   Material, Create,
        Name [Material-1], Mechanical, Elasticity, Elastic, Lamina,
        [13386000 7706 0.301 4306 4306 2760]
        Mechanical, Expansion, Type, Orthotropic,
        [13.23672E-007 2.58908E-005 2.58908E-0051], OK
Menu:   Section, Create, Name [Section-1], Shell, Composite, Cont
        Material: Material-1 # for all
        Thickness [1.2] # for all
        Orientation [0,90]
Menu:   Assign, Section
        # pick part, Done, OK
```

iii. Assembly and Step

```
Module: Assembly
Menu:   Instance, Independent, OK

Module: Step
Menu:   Step, Create, Name [Step-1], Static/General, Cont, OK
```

iv. Load and Boundary Conditions

```
Module: Load
Menu:   BC, Create, Name [BC-1], Step: Initial,
   Mechanical, Symmetry, Cont
   # pick left vertical line, Done, XSIMM, OK
Menu:   BC, Create, Name [BC-2], Step: Initial,
   Mechanical, Symmetry, Cont
   # pick lower horizontal line, Done, YSIMM, OK
Menu: BC, Create, Displacement/Mechanical, Cont,
   # pick bottom left point, Done
   # checkmark all (U1 to UR3), OK
Menu:   Predefined Field, Create, Other, Temperature, Cont
   # Select rectangle, Done, Magnitude [0], OK
Menu:   Predefined Field, Manager
   # Double click cell Predefined Field-2, Step-1 (Propagated)
   Status, Modified, Magnitude [-150], OK
```

v. Mesh

```
Module: Mesh
Menu:   Seed, Instance, Apply, OK
Menu:   Mesh, Element type, Std, Quad, DOF [5], OK # S8R5
Menu:   Mesh, Instance, Yes
```

vi. Job and Visualization

```
Module: Jobs
Menu:   Job manager
   Create, Cont, OK
   Submit,
   Results

Module: Visualization
Menu:   Plot,
   Allow multiple plot states, Undeformed shape, Deformed shape
Menu:   Options, Common,
   Tab: Basic, Render style: Shaded, Visible Edges: Free edges,
   Tab: Other, Scale coordinates, Z: [3.0], OK
Menu:   Options, Superimpose plot options,
   Visible Edges: Feature edges, OK
Menu:   View, Odb display options,
   Tab: Mirror/Pattern, Mirror planes: XZ and YZ
```

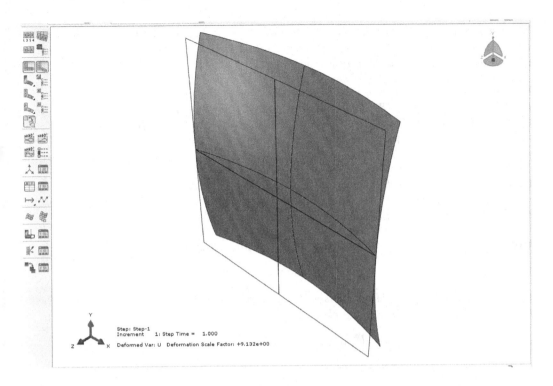

Fig. 3.23: Deformed shape in Example 3.13 under uniform thermal load.

Suggested Problems

Problem 3.1 *Compute the maximum bending moment per unit cross-sectional area m_u that can be applied to a beam of circular hollow cross-section of outside radius r_o and inner radius r_i. The loading is pure bending, no shear. The material is homogeneous and failure occurs when the maximum stress reaches the strength σ_u of the material. The hollow section is filled with foam to prevent buckling. Derive an expression for the efficiency of the cross-section as the ratio of m_u of the hollow beam by m_u of a solid rod of same outside radius. Faced with the problem of using a strong and relatively expensive material, would you recommend a small or large radius?*

Problem 3.2 *Compute the maximum outside radius for a cantilever beam of length L, loaded by a tip load P, otherwise similar to the beam in Problem 3.1 but subjected to pure shear loading. The shear strength is $\tau_u = \sigma_u/2$. Consider only shear. Buckling of the thin wall is likely to limit further the practical thickness of the wall.*

Problem 3.3 *Compute the maximum deflection per unit volume δ_V that can be applied to a beam of circular hollow cross-section of outside radius r_o and inner radius r_i. This is a cantilever beam of length L, loaded by a tip load P. The hollow section is made of an homogeneous material with moduli E and $G = E/2.5$, filled with foam to prevent buckling. Derive an expression for the efficiency of the cross-section as the ratio of δ_V between the hollow cross-section and a solid rod of the same outside radius. Faced with the problem of using a relatively expensive and not quite stiff material, would you recommend a small or large radius?*

Problem 3.4 *Write a computer program to evaluate (3.9). The program data input is the LSS, the thickness of the laminas, and the material elastic properties. The output should be written in a file. Show all work in a report.*

Problem 3.5 *Using the program of Problem 3.4, compute the $A, B, D,$ and H matrices for the following laminates. The material is AS4D/9310 and all laminas are 0.85 mm thick. Comment on the coupling of the constitutive equations for each case: (a) one lamina $[0]$, (b) one lamina $[30]$, (c) $[0/90]_2$, (d) $[0/90]_s$, (e) $[0/90]_8$, (f) $[\pm45]_2 = [+45/-45/+45/-45]$, (g) $[\pm45]_s = [+45/-45/-45/+45]$, (h) $[\pm45/0/90/\pm30]$. Show all work in a report.*

Problem 3.6 *Compute the value and location of the absolute maximum transverse shear strain γ_{23} in Example 3.2, p. 120. At that location, plot the distribution of γ_{23} through the thickness of the plate. Is that distribution a reasonable answer?*

Problem 3.7 *Recompute Example 3.2, p. 120, with a doubly sinusoidal load*

$$q(x, y) = q_0 \sin(\pi x/2a) \sin(\pi x/2b)$$

where $2a, 2b$ are the plate dimensions in x and y, respectively. Compare the result with the exact solution at the center of the plate, that is

$$w_0 = 16q_0 b^4 / [\pi^4 (D_{11} s^4 + 2(D_{12} + 2D_{66}) s^2 + D_{22})]$$

where $s = b/a$ ([15, (5.2.8–5.2.10)]).

Problem 3.8 *Calculate the first vibration frequency ϖ_{11} of the plate with the analytical solution $\varpi_{mn}^2 = \pi^4 [D_{11} m^4 s^4 + 2(D_{12} + 2D_{66}) m^2 n^2 s^2 + D_{22} n^4]/(16\rho h b^4)$, where ρ, h are the density and thickness of the plate, respectively [15, (5.7.8)].*

Problem 3.9 *Consider a rectangular plate with $a_x = 1000$ mm and $b_y = 100$ mm. Each lamina is 1.2 mm thick, made of AS4D/9310 (Table 3.1).*

Propose three different LSS, each having one of the following couplings (shown in [3, Figure 6.7]); that is: (a) bending-extension coupling, (b) torsion-extension coupling, and (c) shear-extension coupling.

Next, consider reduced FEA models applying symmetry conditions; either (i) one-half plate, 500×100 mm, or (ii) one-quarter plate, 500×50 mm. Use the smallest (quarter, half, or full plate) FEA model that is valid for each of the cases (a) through (c). Apply membrane loading $N_x = 1000$ N/mm (all other zero). Show all work in a report including pseudo-code for each case (a) through (c). Use element S8R5 in Abaqus. See Example 3.13, p. 175.

Problem 3.10 *A failure envelope is a 3D surface for which $I_f = 1$. The intersection of this surface with each coordinate plane is a 2D curve. Using a program such as MATLAB, plot the failure envelope in the plane $\sigma_1 - \sigma_2$ for maximum stress, Tsai-Wu, and Puck failure criteria in the same plot. Do a similar plot in the plane $\sigma_2 - \sigma_6$.*

Problem 3.11 *(a) Compute the failure index I_F on each lamina of Example 3.12 (p. 174) using the maximum stress failure criterion and the Hashin failure criterion. The lamina strength values are given in Table 3.1.*

Problem 3.12 *Using an UMAT subroutine, compute the failure index I_F on each lamina of Example 3.12 (p. 174) using the Puck failure criterion. The elastic moduli and strength values for the lamina are given in Table 3.1, p. 108. Values for the Puck parameters are suggested in Section 3.3.1, p. 171. Show all work in a report.*

Problem 3.13 *Write on a text file the nodal stress results at the top and bottom of each lamina of Problem 3.11. Then, using an external program such as MATLAB, compute the same failure indexes as in Problem 3.11 in terms of the nodal stress values at the center of the plate, and compare them to the results of Problem 3.11. Show all work in a report.*

Chapter 4

Buckling

Most composite structures are thin-walled. This is a natural consequence of the following facts:

- Composites are stronger than conventional materials. Then, it is possible to carry very high loads with a small area, and thus small thickness in most components.

- Composites are expensive when compared to conventional materials. Therefore, there is strong motivation to reduce the volume, and thus the thickness as much as possible.

- The cost of polymer matrix composites increases with their stiffness. The stiffness in the fiber direction can be estimated by using the fiber dominated rule of mixtures, $E_1 = E_f V_f$. For example, when glass fibers are combined with a polymer matrix, the resulting composite stiffness is lower than that of aluminum. Using Aramid yields a stiffness comparable to aluminum. Carbon fibers yield composite stiffness lower than steel. Therefore, there is strong motivation to increase the moment of area I of beams and stiffeners without increasing the cross-sectional area. The best option is to increase I by enlarging the cross-section dimensions and reducing the thickness.

All the above factors often lead to design of composite structures with larger, thin-walled cross-sections, with modes of failure likely to be controlled by buckling.

4.1 Eigenvalue Buckling Analysis

Buckling is loss of stability due to geometric effects rather than material failure. But it can lead to material failure and collapse if the ensuing deformations are not restrained. Most structures can operate in a linear elastic range. That is, they return to the undeformed configuration upon removal of the load. Permanent deformations result if the elastic range is exceeded, as when matrix cracking occurs in a composite.

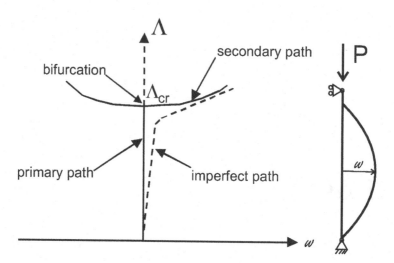

Fig. 4.1: Equilibrium paths for the perfect column.

Consider a simply supported column of area A, length L, and moment of inertia I, made of homogeneous material with modulus E and strength F along the length of the column. The column is loaded by a compressive load P acting on the centroid of the cross-section [5]. If the column geometry, loading, and material have no imperfections, the axial deformation is

$$u = PL/EA \tag{4.1}$$

with no lateral deformation $w = 0$. The deformation of the structure (u, v, w) before buckling occurs is called the *primary path* (Figure 4.1). The slightest imperfection will make the column buckle when

$$P_{cr} = \pi^2(EI)/L^2 \tag{4.2}$$

The load capacity for long slender columns will be controlled by buckling, as opposed to the crushing strength of the material. What happens after the column reaches its critical load depends largely on the support conditions. For the simply supported column, the lateral deflection[1]

$$w = A \sin(\pi x/L) \tag{4.3}$$

will grow indefinitely $(A \to \infty)$ when the load just barely exceeds P_{CR}. Such large lateral deflections will cause the material to fail, and the column will collapse. The behavior of the structure after buckling has occurred is called *post-buckling*.

The simply supported column in Figure 4.1 experiences no deformations in the shape of the buckling mode (4.3) before buckling actually happens. In this case, it is said that the structure has a *trivial* primary path. This is a consequence of having a perfect structure with perfectly aligned loading. For these types of structures,

---

[1]With x measured from one end of the column.

Table 4.1: Lamina elastic properties for AS4/9310, $V_f = 0.6$.

Young's Moduli	Shear Moduli	Poisson's Ratio
$E_1 = 145880$ MPa	$G_{12} = G_{13} = 4386$ MPa	$\nu_{12} = \nu_{13} = 0.263$
$E_2 = E_3 = 13312$ MPa	$G_{23} = 4529$ MPa	$\nu_{23} = 0.470$

buckling occurs at a bifurcation point. A bifurcation point is the intersection of the primary path with the secondary path, i.e., the post-buckling path [23].

The bifurcation loads, one for every possible mode of buckling, are fairly easy to obtain using commercial software. The geometry of the structure is that of the perfect undeformed configuration, loaded with the nominal loads, and the material is elastic. Such analysis requires a minimum of effort on the part of the analyst. Commercial programs refer to this analysis as an *eigenvalue buckling analysis*, because the critical loads are the eigenvalues λ_i of the discretized system of equations

$$([K] - \lambda[K_s])\{v\} = 0 \qquad (4.4)$$

where K and K_s are the stiffness and stress stiffness matrix, respectively, and v is the column of eigenvectors (buckling modes) [23].

Example 4.1 *Consider a simple supported plate, with side dimensions $a_x = 1000$ mm, $a_y = 500$ mm, edge-wise loaded in compression with $N_x = N_y = 1$ N/mm. The plate is made of $[(0/90)_3]_S$, AS4/9310 (Table 4.1) with fiber volume fraction 0.6 and total thickness $t_T = 10.2$ mm. Compute the critical load of the lowest four modes using eigenvalue analysis. Visualize the lower four modes.*

Solution to Example 4.1 *Here we model the whole plate to capture all buckling modes, not just the symmetric ones. Element S8R5 is used because the we are modeling a thin plate. Since the laminate is symmetric and stress computation lamina by lamina is not required, the critical loads can be obtained using three different approaches. Note that for element S8R5, you are not required to specify* Transverse Shear Stiffness *when the shell section is* Homogeneous *or* Composite *(see cases (a) and (c)), but they have to be entered for a* General Stiffness *section (see case (b)). Watch the two videos in [1].*

(a) Using Equivalent Laminate Moduli

In this first approach, the equivalent laminate moduli are calculated and used along with an orthotropic shell element. In this case, laminate moduli represent the stiffness of an equivalent orthotropic plate that behaves like the actual laminate under in-plane loads (see Section 3.2.10). Laminate moduli can be found as explained in Section 1.15. Introduce the lamina properties (Table 4.1) into (1.91), rotate each lamina (1.53), add them (1.102) to get the laminate moduli (1.105) listed in Table 4.2. Bending moduli [3, (6.43)] may give better results for this example.

The following pseudo code illustrates the modeling, solution, and visualization procedure using the GUI in Abaqus CAE. The corresponding Ex_4.1.a.cae *file is available on the website [2]. The results are tabulated in Table 4.3.*

```
# Ex.4.1.a using equivalent laminate moduli
Menu:   File, Set Work Directory [C:\SIMULIA\user\Ex_4.1], OK
Menu:   File, Save As [C:\SIMULIA\user\Ex_4.1\Ex_4.1.a.cae], OK
```

Table 4.2: Equivalent laminate moduli for $[(0/90)_3]_S$.

Young's Moduli	Shear Moduli	Poisson's Ratio
$E_x = 79985$ MPa	$G_{xy} = 4386$ MPa	$\nu_{xy} = 0.044$
$E_y = 79985$ MPa	$G_{yz} = 4458$ MPa	$\nu_{yz} = 0.415$
$E_z = 16128$ MPa	$G_{xz} = 4458$ MPa	$\nu_{xz} = 0.415$

i. Creating the part

```
Module: Part
Menu:   Part, Create
    3D, Deformable, Shell, Planar, Approx size [2000], Cont
Menu:   Add, Line, Rectangle, [-500,-250], [500,250]
    X # to close command, Done
```

ii. Defining materials, sections, and assigning sections to parts

```
Module: Property
Menu:   Material, Create
    Name [Material-1], Mechanical, Elasticity, Elastic
    Type: Engineering Constants
    [79985, 79985, 16128, 0.044, 0.415, 0.415, 4386, 4458, 4458], OK
Menu:   Section, Create
    Name [Section-1], Shell, Homogeneous, Cont
    Shell thickness: Value [10.2], Material: Material-1, OK
Menu:   Assign, Section
    # pick the part, Done, OK
```

iii. Creating assembly

```
Module: Assembly
Menu:   Instance, Create, Independent, OK
```

iv. Defining the analysis steps

```
Module: Step
Menu:   Step, Create
    Procedure type: Liner Perturbation, Buckle, Cont
    Number of eigenvalues requested [10], Vectors/iteration [10], OK
```

v. Adding BC and loads

```
Module: Load
Menu:   Load, Create
    Step: Step-1, Mechanical, Shell Edge Load, Cont
    # pick: edges @ x=500 & y=250, Done, Magnitude [1.0], OK
Menu: BC, Manager
    Create, Step: Initial, Mechanical, Disp/Rota, Cont
    # pick: the 4 edges, Done, # checkmark: U3, OK
    Create, Step: Initial, Mechanical, Disp/Rota, Cont
    # pick: edge @ x=-500, Done, # checkmark: U1, OK
    Create, Step: Initial, Mechanical, Disp/Rota, Cont
    # pick: edge @ y=-250, Done, # checkmark: U2, OK, # close Manager
```

vi. *Meshing the model*

```
Module: Mesh
Menu:   Seed, Instance, Size: [100], OK
Menu:   Mesh, Controls, Technique: Structured, OK
Menu:   Mesh, Element Type, Geometric Order: Quadratic, DOF: 5, OK
Menu:   Mesh, Instance, Yes
```

vii. *Solving and visualizing the results*

```
Module: Job
Menu:   Job, Manager
        Create, Name [Ex-4-1-a], Cont, OK, Submit, # when completed,
        Results
Module: Visualization
Menu:   Plot, Deformed Shape # buckling mode 1 will be displayed
Menu:   Result, Step/Frame, # pick a mode, Apply # to visualize
        # eigenvalues are shown in the WS and recorded in Ex-4-1-a.dat
```

(b) Using $A - B - D - H$ Matrices

In this second approach, $A - B - D - H$ matrices are used. To get the laminate properties $(A, B, D,$ and H matrices), introduce the lamina properties (Table 4.1) into (3.9), p. 103. The resulting laminate matrices are

$$\begin{bmatrix} A & B \\ B & D \end{bmatrix} = \begin{bmatrix} 817036 & 35937.6 & 0 & 0 & 0 & 0 \\ 35937.6 & 817036 & 0 & 0 & 0 & 0 \\ 0 & 0 & 44737.2 & 0 & 0 & 0 \\ 0 & 0 & 0 & 8.55845\ 10^6 & 311579 & 0 \\ 0 & 0 & 0 & 311579 & 5.60896\ 10^6 & 0 \\ 0 & 0 & 0 & 0 & 0 & 387872 \end{bmatrix}$$

$$[H] = \begin{bmatrix} 37812.8 & 0 \\ 0 & 37964.7 \end{bmatrix}$$

The following pseudo code illustrates how to modify Ex_4.1.a.cae to use the A–B–D–H matrices. The corresponding file is named Ex_4.1.b.cae and it is available on the website [2]. The results are tabulated in Table 4.3.

```
# Ex.4.1.b using ABDH matrices
Menu:   File, Set Work Directory [C:\SIMULIA\user\Ex_4.1], OK
Menu:   File, Open [C:\SIMULIA\user\Ex_4.1\Ex_4.1.a.cae], OK
Menu:   File, Save As [C:\SIMULIA\user\Ex_4.1\Ex_4.1.b.cae], OK
```

i. *Defining materials, sections, and assigning sections to parts*

```
Module: Property
Menu:   Material, Delete, Material-1, Yes
Menu:   Section, Delete, Section-1, Yes
Menu:   Section, Create, Shell, General Shell Stiffness, Cont
        Tab: Stiffness
        [817036  35937.6      0       0       0       0
                 817036       0       0       0       0
```

```
                    44737.2        0        0        0
                              8.56e6   311579        0
                                       5.61e6        0
                                              387872]
```

Tab: Advanced
Transverse Shear Stiffnesses: # checkmark: Specify values
    K11 [37812.8], K12 [0], K22 [37964.7], OK

ii. Defining analysis step

Module: Step
Menu: Step, Edit, Step-1
    Maximum number of interactions [70], OK
    # solve the problem using a smaller number for this quantity
    # the following error message may be displayed
    # TOO MANY ITERATIONS NEEDED TO SOLVE THE EIGENVALUE PROBLEM

iii. Solving and visualizing the results

Module: Job
Menu: Job, Manager
    Create, Name [Ex-4-1-b], Cont, OK, Submit, # when completed,
    Results
    # visualize the results as it is indicated in part (a)

(c) Using the Laminate Stacking Sequence

*In this third approach, the laminate stacking sequence (LSS) and the lamina properties
(Table 4.1) are entered. The following pseudo code illustrates how to change* Ex_4.1.a.cae
to solve this part of the problem. The corresponding Ex_4.1.c.cae *file is available on the
website [2]. The results are tabulated in Table 4.3.*

```
    # Ex_4.1.c using LSS
Menu:    File, Set Work Directory [C:\SIMULIA\user\Ex_4.1], OK
Menu:    File, Open [C:\SIMULIA\user\Ex_4.1\Ex_4.1.a.cae], OK
Menu:    File, Save As [C:\SIMULIA\user\Ex_4.1\Ex_4.1.c.cae], OK
```

i. Defining materials, sections, and assigning sections to parts

Module: Property
Menu: Material, Delete, Material-1, Yes
Menu: Section, Delete, Section-1, Yes
Menu: Material, Create
    Name [Material-1], Mechanical, Elasticity, Elastic, Type: Lamina
    [1.4588e5, 1.3312e4, 2.63e-1, 4.386e3, 4.386e3, 4.529e3], OK
Menu: Section, Create
    Shell, Composite, Cont
    # checkmark: Symmetric laminas
    # right-click: Insert Row After, # until there are six rows
    Material: Material-1 # for all laminas, Thickness [0.85]
    Orientation Angle: [0,90,0,90,0,90], Integration Points [3]
    Ply Name [k1, k2, k3, k4, k5, k6], OK

Table 4.3: Bifurcation loads [N/mm].

Mode #	(a)	(b)	(c)
1	253.27	210.1	210.04
2	320.05	319.87	319.75
3	573.36	643.18	642.81
4	998.64	831.65	831.11

ii. Defining analysis step

```
Module: Step
Menu:   Step, Edit, Step-1
    Maximum number of iteractions [150]
    # solve the problem using a smaller number for this quantity
    # the following error message may be displayed
    # TOO MANY ITERATIONS NEEDED TO SOLVE THE EIGENVALUE PROBLEM
```

iii. Solving and visualizing the results

```
Module: Job
Menu:   Job, Manager
    Create, Name [Ex-4-1-c], Cont, OK, Submit,
    # when Completed, Results
    # visualize the results as it is indicated in part (a)
```

4.1.1 Imperfection Sensitivity

To illustrate the influence of imperfections in buckling, let us consider the solid lines in Figure 4.1. The lateral deflection is zero for any load below the bifurcation load P_{CR} that is on the primary path of the perfect structure. The primary path intersects the secondary path at the bifurcation point, for which the load is P_{CR}. The post-critical behavior of the column is indifferent and slightly stable. Indifferent means that the column can deflect right or left. Stable post-critical path means that the column can take a slightly higher load once it has buckled. For a column, this stiffening behavior is so small that one cannot rely upon it to carry any load beyond P_{CR}. In fact, the column will deform laterally so much that the material will fail and the system will collapse. Unlike columns, simply supported plates experience significant stiffening on the secondary path.

4.1.2 Asymmetric Bifurcation

Consider the frame illustrated in Figure 4.2. An eigenvalue analysis using one finite element per bar [23, Sections 5.9 and 7.8] reveals the bifurcation load

$$P_{CR} = 8.932(10^{-6})AE \qquad (4.5)$$

but gives no indication about the nature of the critical state: whether it is stable or not, whether the post-critical path is symmetric or not, and so on. We shall see

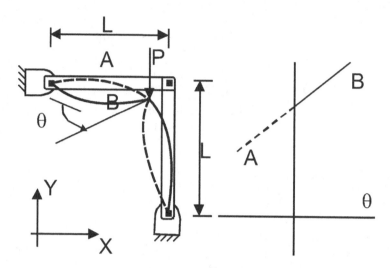

Fig. 4.2: Two bar frame.

later on, that the frame has an asymmetric, and thus unstable, post-critical path, as represented in Figure 4.2. That is, the post-critical path has a slope

$$P^{(1)} = 18.73(10^{-9})AE \ (1/\text{rad}) \tag{4.6}$$

in the force-rotation diagram in Figure 4.2, where θ is the rotation of the joint at the load point.

In general, the problem with eigenvalue analysis is that it provides no indication as to the nature of the post-critical path. If the post-critical path is stiffening and symmetric as in Figure 4.1, the real structure may have a load capacity close to the bifurcation load. But if the post-critical path is unstable and/or asymmetric, as in Figure 4.2, or if there is mode interaction [19, 24–28], the real structure may have a load capacity much smaller than the bifurcation load. In order to use the information provided by eigenvalue analysis, it is necessary to understand and quantify the post-buckling behavior.

4.1.3   Post-critical Path

One way to investigate the post-buckling behavior is to perform a continuation analysis of the imperfect structure, as presented in Section 4.2. This is perfectly possible, but complicated and time consuming, as it will be seen later in this chapter. A more expedient solution can be obtained using software capable of predicting the nature of the post-critical path, including symmetry, curvature, and mode interaction. If the secondary path is stable and symmetric, the bifurcation load can be used as a good estimate of the load capacity of the structure. The curvature of the post-critical path gives a good indication of the post-buckling stiffening, and it can be used to a certain extent to predict post-buckling deformations.

The bifurcation load, slope, and curvature of the post-critical path emerging from the bifurcation (4.1) can be computed with BMI3 [25–27] available in [2]. The

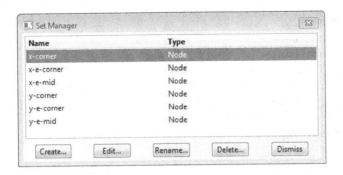

Fig. 4.3: Node sets defined after meshing to be used for applying concentrated loads at nodes.

post-buckling behavior is represented by the following formula

$$\Lambda = \Lambda^{(cr)} + \Lambda^{(1)}s + \frac{1}{2}\,\Lambda^{(2)}s^2 + \ldots \tag{4.7}$$

where s is the perturbation parameter, which is chosen as one component of the displacement of one node, $\Lambda^{(cr)}$ is the bifurcation multiplier, $\Lambda^{(1)}$ is the slope, and $\Lambda^{(2)}$ is the curvature of the post-critical path [24, (43)], see also [19,25–28]. When the slope is zero, the post-critical path is symmetric. Therefore, buckling is indifferent, and the real structure will buckle to either side. There is no way to predict which way it is going to buckle, unless of course one knows the shape of the imperfections on the real plate, which is seldom the case. A positive curvature denotes stiffening during post-buckling, and a negative one indicates that the stiffness decreases.

Example 4.2 *Consider the simple supported plate of Example 4.1, p. 183. Compute the bifurcation multiplier $\Lambda^{(cr)}$, the critical load N_{cr}, the slope $\Lambda^{(1)}$, and the curvature $\Lambda^{(2)}$ of the post-critical path. Estimate the load when the maximum lateral deflection is equal to the thickness of the plate. As perturbation parameter, use the largest displacement component of the buckling mode with lowest buckling load.*

Solution to Example 4.2 *Watch the two videos in [1].*
The program BMI3 [2] is used in this case to compute the bifurcation multiplier $\Lambda^{(cr)}$, the slope $\Lambda^{(1)}$, and the curvature $\Lambda^{(2)}$ of the post-critical path. Refer to Appendix E for a description of the software interface and operation procedure. Since BMI3 requires the A-B-D-H matrices, we can use the `.cae` *file generated during the solution of Example 4.1(b) (available as* `Ex_4.1.b.cae` *in [2]), but a few changes are needed to satisfy the current limitations of BMI3, as follows*

    i. Open the `Ex_4.1.b.cae` *and save it as* `Ex_4.2.cae`

```
Menu:   File, Set Work Directory [C:\SIMULIA\user\Ex_4.2], OK
Menu:   File, Open [C:\SIMULIA\user\Ex_4.1\Ex_4.1.b.cae], OK
Menu:   File, Save As [C:\SIMULIA\user\Ex_4.2\Ex_4.2.cae], OK
```

    ii. Delete the mesh. Then re-seed and re-mesh with a 2×4 mesh of S8R5 elements.

```
Module: Mesh
```

```
Object: Assembly
Menu:    Mesh, Delete Instance Mesh, Yes
Menu:    Seed, Edges
    # pick: horizontal lines, Done
    Method: By Number, Number of Elements [4], OK
    # pick: vertical lines, Done
    Method: By Number, Number of Elements [2], OK
Menu:    Mesh, Instance, Yes
```

iii. Step

```
Module: Step
Menu:    Step, Edit, Step-1
Maximum number of iteractions [300]
# If you use less iterations
# the following error message may be displayed
# TOO MANY ITERATIONS NEEDED TO SOLVE THE EIGENVALUE PROBLEM
```

iv. BMI3 only accepts concentrated loads. Therefore, the distributed edge loads must be replaced by a set of concentrated loads. Since S8R5 are quadratic elements, a uniform (constant) distributed load on the edge is equivalent to three loads concentrated at the nodes. For a unit edge load ($N_n = 1.0 \ N/mm$), the element corner nodes get 1/6 each and the mid-node gets 4/6. At a node where two elements contribute, the contributions from both elements add up to 1/3. To do this change, first create node sets to later apply the concentrated loads. Six sets are needed as illustrated in the **Set Manager** *window shown Figure in 4.3. A set is created as follows*

```
Module: Load
Menu:    Tools, Set, Manager
    Create, Name [x-corner], Type: Node, Cont
    # pick: the two corners of the plate on the edge at x=500, Done
```

v. Using the same procedure, create the additional sets. The **x-e-corner** *contains the nodes where pairs of elements join (at the edge at $x = 500$ mm). The* **x-e-mid** *contains the mid–side nodes of elements on the edge at $x = 500$ mm. The remaining sets, for the edge at $y = 250$ mm, are created similarly.*

vi. Delete all the edge loads and replace them by concentrated loads, as follows

```
Module: Load
Menu:    Load, Manager
    # delete the existing load
    Create, Step: Step-1, Mechanical, Concentrated force, Cont
    Sets, x-corner, Cont, CF1 [-41.67], OK
    Create, Step: Step-1, Mechanical, Concentrated force, Cont
    Sets, x-e-corner, Cont, CF1 [-83.33], OK
    Create, Step: Step-1, Mechanical, Concentrated force, Cont
    Sets, x-e-mid, Cont, CF1 [-166.67], OK
    # Analogously add compressive loads on the edge y=250
    # close the Load Manager
```

vii. Run the model to verify it produces the right results. Notice the eigenvalues are slightly different than those obtained in Example 4.1; this is due to the coarseness of the mesh used.

Module: Job
Menu: Job, Manager
    Create, Name [Job-1], Cont, OK, Submit, # when Completed, Results
    # visualize the mode shapes as it was indicated in Example 4.1

viii. Copy the filtering program inp2bmi3.exe *(available in [2]) to the* Work Directory. *Execute this file by double-clicking on it. It will create the files* ABQ.inp, BMI3.inp, *and* BMI3.dat, *based on the file* Job-1.inp. *The file* ABQ.inp *is a filtered copy that Abaqus can execute.*

ix. Create a new Job to run ABQ.inp *to verify everything went well with the filtering. Note that unlike previous examples where a Job was created to run a model, in this case a Job is created to run an input file. This is done as follows*

Module: Job
Menu: Job, Manager
    Create, Name [ABQ], Source: Input file # not Model
    Select, ABQ.inp, OK, Cont, OK
    Submit, # when Completed, Results

x. Copy the program bmi3.exe *(available in [2]) to the* Work Directory *and execute it by double-clicking on it. Take note of the results, either from the display output or from the* BMI3.out *file.*

By default, BMI3 chooses as perturbation parameter the largest displacement component of the buckling mode with lowest buckling load, but that can be changed by editing the file BMI3.dat. *For this example, the default perturbation corresponds to the first buckling mode, the node in the middle of the plate, and the deflection direction U3 perpendicular to the surface of the plate. The results, specially* $\Lambda^{(2)}$, *are sensitive to mesh refinement.*
Since BMI3 solves the problem using reversed loads (see Appendix E), then (4.7) becomes

$$-N = \quad \Lambda^{(cr)} + \Lambda^{(1)}s + \tfrac{1}{2}\,\Lambda^{(2)}s^2$$
$$N = \quad -\Lambda^{(cr)} - \Lambda^{(1)}s - \tfrac{1}{2}\,\Lambda^{(2)}s^2$$

and, in this case, the perturbation direction is $s = -\delta$, *so*

$$N = \quad -\Lambda^{(cr)} - \Lambda^{(1)}(-\delta) - \tfrac{1}{2}\,\Lambda^{(2)}(-\delta)^2$$
$$N = \quad -\Lambda^{(cr)} + \Lambda^{(1)}\delta - \tfrac{1}{2}\,\Lambda^{(2)}\delta^2$$

Therefore, using the results from a more refined mesh, the secondary path is

$$N = -(-209.0) + (0)\,\delta - (-0.1154)\,\delta^2 = 209.0 + 0.1154\,\delta^2$$

Since the slope $\Lambda(1)$ *is zero, the post-critical path is symmetric. The post-buckling load when the lateral deflection* (w) *is equal to the thickness* $(s = Th = 10.2\ mm)$ *is equal to 221 N/mm, as shown in Figure 4.4.*
Since BMI3 reverses the loads that it receives from the user, or from Abaqus, it creates some confusion for interpretation of the results. In Module: Load *we applied a compression load. BMI3 reverses that load, transforming it into a tensile load, before performing any calculations. Then, the solution is* $\Lambda^{(cr)} = -209$, *which means compression critical load and* $\Lambda^2 = -0.1154$ *means a stable post-critical path, as in Figure 4.4.*

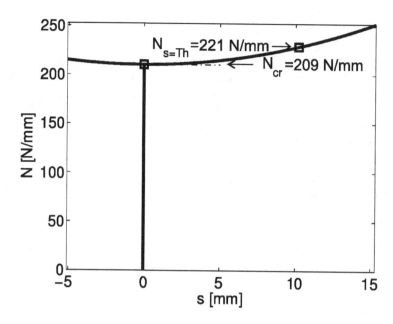

Fig. 4.4: Equilibrium paths for a perfect plate.

4.2   Continuation Methods

The strain to failure of polymer matrix composites (PMC) is high. Compare 1.29% for AS4/3501 and 2.9% for S-glass/epoxy with only 0.2% for steel and 0.4% for aluminum. That means that buckling deformations can go into a post-buckling regime while the material remains elastic. However, great care must be taken that no matrix-dominated degradation mode takes place, in which case the material will not remain elastic (see Chapters 8 and 9). Eigenvalue buckling analysis is relatively simple as long as the material remains elastic, because classical theory of elastic stability can be used, as was done in Section 4.1. Material nonlinearity is one reason that motivates an incremental analysis. Another reason is to evaluate the magnitude of the buckling load for an *imperfection sensitive* structure.

In an incremental analysis, also called continuation analysis, the load is increased gradually step by step. At each step, the deformation, and possibly the changing material properties, is evaluated. Incremental analysis must include some type of imperfection in the geometry, material, or alignment of loads. Lacking any imperfection, incremental analysis will track the linear solution, revealing no bifurcations or limit points.

Continuation methods are a form of geometrically nonlinear analysis. The system must have a nontrivial fundamental path, such as a flat plate with asymmetric laminate stacking sequence (LSS) under edge loads.

If the system has a trivial fundamental path, such as a flat plate with symmetric LSS under edge load, the nontrivial fundamental path can be forced by introducing an imperfection. Several types of imperfections are possible, including material imperfections (e.g., unsymmetrical LSS), geometric imperfections, or load eccentricity.

Since the real geometric imperfections are seldom known, the preferred artificial geometric imperfection is in the form of the bifurcation mode having the lowest bifurcation load. This is true in most cases; however, in some cases, a second mode associated with imperfections that are more damaging to the structure should be used [29]. Also, if the structure has an asymmetric post-buckling path, as the two-bar example in Figure 4.2, care must be taken not to force the structure along the stiffening path.

FEA codes allow the user to modify a mesh by superposing an imperfection in the shape of any mode from a previous bifurcation analysis onto the perfect geometry (see Example 4.3).

Example 4.3 *Apply a geometric imperfection* $w_p(x,y) = \delta_0\ \phi(x,y)$ *to Example 4.1 (p. 183) and plot the load-multiplier vs. maximum lateral deflection for an imperfection magnitude* $\delta_0 = Th/10$ *and* $\delta_0 = Th/100$, *where* Th *is the total laminate thickness, and* $\phi(x,y)$ *is the buckling mode corresponding to the lowest bifurcation load found in Example 4.1, p. 183.*

Solution to Example 4.3 *Watch the two videos in [1].*
The buckling modes were found in Example 4.1, p. 183. Here we will be using the model created as `Ex_4.1.b.cae`. *The solution procedure is summarized below:*

i. *Get* `Ex_4.1.b.cae`, *save as* `Ex_4.3.cae`

```
Menu:   File, Set Work Directory, [C:\SIMULIA\user\Ex_4.3], OK
Menu:   File, Open, [C:\SIMULIA\user\Ex_4.1\Ex_4.1.b.cae], OK
Menu:   File, Save As, [C:\SIMULIA\user\Ex_4.3\Ex_4.3.cae], OK
```

ii. *Delete the old mesh, then re-mesh. Using the [Module Model Object Part] toolbar (2nd from top Menu)*

```
# select
Module: Mesh, Model: Model-1, Object: Assembly (shows current mesh)
#
Menu:   Mesh, Delete Instance Native Mesh, Yes
Menu:   Seed, Instance, Approx. Global Size [50]
Menu:   Mesh, Instance, Yes
```

iii. *Delete the old jobs, create a new one, and write the input file*

```
Module: Job
Menu:   Job, Manager
          # Select Ex-4-1-a, Delete, Yes
          # Select Ex-4-1-b, Delete, Yes
          Create, Name [Job-1], Cont, OK, Write Input
          # input file Job-1.inp has been created in the Work Directory
```

iv. *Compute and save the mode shapes (displacements) of the first few modes. For this, open* `Job-1.inp` *with Notepad++ (or your preferred text editor). Save the file as* `Job-pert.inp`.

```
**Data lines to specify the reference load,
*NODE FILE, GLOBAL=YES, LAST MODE=4
U
**
```

Then, add the following lines to the *Buckle *step before the text*
**BOUNDARY CONDITIONS
and save the file. This will save the mode shapes of the first four modes to the .fil
results file. The .fil *file is binary, so you cannot open it with Notepad; it is to be
read by Abaqus only.*

v. *In Abaqus CAE, create a new job and submit it, as follows*

```
Module: Job
Menu:   Job, Manager
    Create, Name: [Job-pert], Source: Input file
    Input file: Select, [Job-pert.inp], OK, Cont, OK
    Submit
```

If the execution is completed without errors, the file Job-pert.fil *should have been
created in the* Work Directory. *Since the mode shapes are normalized, the maximum
displacement in the results file is 1.0.*

vi. *Set up a Riks nonlinear analysis.*

*Using the following procedure, the displacements from a previous analysis can be used
to update the geometry of the finite element model. In this example, the multiplier
factor for the first mode is chosen to be* $\delta_1 = 0.05$. *The contributions of modes 2 and
3 are commented out in the* .inp *file (as denoted by lines starting with double star
**). Therefore, an initial deflection equal to the first mode of buckling with a central
deflection* δ_1 *is forced on the structure. To do this, open* Job-1.inp *with Notepad++
and save it as* Job-riks.inp. *Then, replace the following three lines*

```
*Step, name=Step-1, perturbation
*Buckle
10, , 10, 70
```

with the following:

```
**IMPERFECTION, FILE=Job-pert, STEP=n
** n : step number where the modes were calculated in the previous run
**      if there was only an Initial Step followed by a Buckle Step,n=1
*IMPERFECTION, FILE=Job-pert, STEP=1
** Multiple data lines to specify the contributions of each mode
** m, delta_m
** m: mode number
** delta_m : scale factor for mode m
1,0.05
**2,0.02
**3,0.01
*STEP, NLGEOM
** next 2 lines (2nd one is empty) define increment/stopping criteria
*STATIC, RIKS
```

followed by the same multiple data lines to specify reference loading used in previously
Job-pert.inp,

```
**Data lines to specify the reference load,
*NODE FILE, GLOBAL=YES, LAST MODE=4
U
**
```

Be sure to delete all the lines related to load case=2 *and save the file. Otherwise, you will get an error message. The lines to be eliminated are similar to:*

```
*Boundary, op=NEW, load case=2
_PickedSet5, 3, 3
*Boundary, op=NEW, load case=2
_PickedSet6, 1, 1
*Boundary, op=NEW, load case=2
_PickedSet7, 2, 2
```

vii. *In Abaqus CAE, create a new job and submit it, as follows*

```
Module: Job
Menu:   Job, Manager
     Create, name: [Job-riks], Source: Input file
     Input file: Select, [Job-riks.inp], OK, Cont, OK
  Submit, #  when Completed, Results
     # this finalizes the Riks analysis
Module: Visualization
Menu:   Plot, Contours, On Deformed Shape
     Toolbar: Field Output: Primary, U, U3
Menu:   Result, Step/Frame, Increment 25, Apply
Menu:   Result, Step/Frame, Increment 30, Apply
Menu:   Result, Step/Frame, Increment 35, Apply
Menu:   Result, Step/Frame, Increment 40, Apply
```

Using a continuation method with this non-perfect geometry, the continuation equilibrium paths shown in Figure 4.5 are obtained. It can be seen that eventually the continuation solution approaches the secondary path of the perfect structure, shown by dashed lines in Figure 4.5. For smaller imperfections, the continuation solution follows more closely the primary path, then the secondary path. A structure with large imperfections deviates more from the behavior of the perfect structure, as shown by the solution corresponding to an imperfection $\delta_0 = Th/10$.

Example 4.4 *Find the buckling load multiplier and the first mode shape for the column analyzed in Example 3.11, p. 165.*

Solution to Example 4.4 *Watch the video in [1].*
This example is solved simply by modifying the file Ex_3.11.cae. *The dead-load step must be redefined as a buckling step file. The pseudo code below describes the necessary changes.*

i. *Open the* Ex_3.11.cae *and save it as* Ex_4.4.cae

```
Menu:   File, Set Work Directory [C:\SIMULIA\user\Ex_4.4], OK
Menu:   File, Open [C:\SIMULIA\user\Ex_3.11\Ex_3.11.cae], OK
Menu:   File, Save As [C:\SIMULIA\user\Ex_4.4\Ex_4.4.cae], OK
```

ii. *Create a buckling step to replace the dead-weight step used in Example 3.11*

```
Module: Step
Menu:   Step, Manager
     # select: Step-1, Delete, Yes
     Create, Procedure type: Linear perturbation, Buckle, Cont
     Number of eigenvalues requested [1], OK, # close Step Manager
```

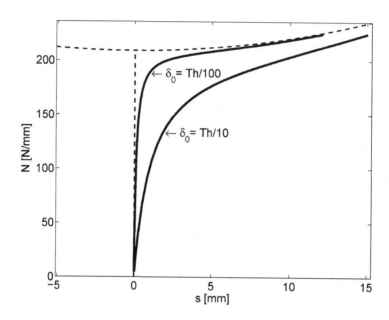

Fig. 4.5: Equilibrium paths for a $[(0/90)_3]_S$ plate, with $\delta_0 = Th/10$ and $\delta_0 = Th/100$.

iii. Add a unitary load. The load created to be applied the Step-1 *of Example 3.11 was deleted when the analysis step was deleted in the previous instruction.*

```
Module: Load
Menu:   Load, Create
    Step: Step-1, Category: Mechanical, Type: Concentrated force, Cont
    # pick: RP-1 in the WS, Done,
        CF3 [+1.0], OK #+Z-direction, compression
```

iv. Solve the model and visualize the results

```
Module: Job
Menu:   Job, Manager
    # select: existing job (Job-1), Delete, Yes
    Create, Cont, OK, Submit, # when Completed, Results
Module: Visualization
Menu:   Plot, Deformed Shape
```

In the WS, notice that the eigenvalue is displayed as 6.227E+05. The deformed shape should look similar to the one in Figure 4.6.

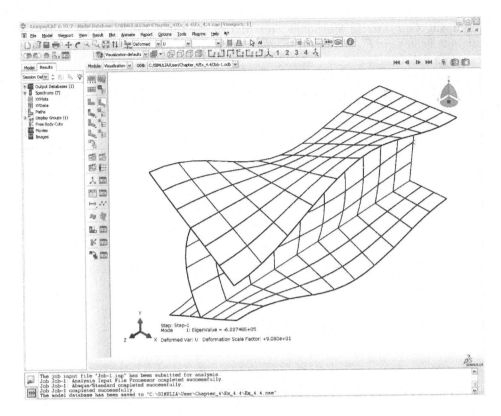

Fig. 4.6: Buckling Mode 1 in Example 4.4.

Suggested Problems

Problem 4.1 *Compute the bifurcation load P^c of the two-bar frame in Figure 4.2 using one quadratic beam element per bar. Each bar has length $L = 580$ mm, area $A = 41$ mm^2, moment of area $I = 8.5$ mm^4, height $H = 10$ mm, and modulus $E = 200$ GPa. The connection between the two bars is rigid.*

Problem 4.2 *Perform a convergence study on the bifurcation load P^c of the two-bar frame in Problem 4.1 by increasing the number of elements per bar N until the bifurcation load converges within 2%. Plot P^c vs. N.*

Problem 4.3 *Recalculate **Example** 4.2 when the LSS changes to $[(0/90)_6]_T$, thus becoming asymmetric. Do not introduce any imperfection but rather analyze the perfect system, which in this case is asymmetric.*

Problem 4.4 *Recalculate **Example** 4.2 with $[(0/90)_6]_T$, and $N_x = 1, N_y = N_{xy} = 0$. Do not introduce any imperfection but rather analyze the perfect system, which in this case is asymmetric.*

Problem 4.5 *Using an FEA code, plot the continuation solution for $\delta_0 = Th/100$, as in Figure 4.5, for a cylindrical shell with distributed axial compression on the edges. The cylinder has a length of $L = 965$ mm and a mid-surface radius of $a = 242$ mm. The LSS is $[(0/90)_6]_S$, with lamina thickness $t = 0.127$ mm. The laminas are of E-glass/epoxy with $E_1 = 54$ GPa, $E_2 = 18$ GPa, $G_{12} = 9$ GPa, $\nu_{12} = 0.25$, and $\nu_{23} = 0.38$.*

Problem 4.6 *Compute the maximum stress failure index I_f of Problem 4.5 at $P = \Lambda^{(cr)}$. The strength values are $F_{1t} = 1034$ MPa, $F_{1c} = 1034$ MPa, $F_{2t} = 31$ MPa, $F_{2c} = 138$ MPa, and $F_{6t} = 41$ MPa.*

Problem 4.7 *Plot the imperfection sensitivity of the cylindrical shell of Problem 4.5, for imperfections in the range $(Th/200) < s < Th$.*

Chapter 5

Free Edge Stresses

In-plane loading N_x, N_y, N_{xy} of balanced-symmetric laminates induces only in-plane stress σ_x, σ_y, σ_{xy} in the interior of the laminate. But near the free edges, interlaminar stresses σ_z, σ_{yz}, σ_{xz}, are induced due to the imbalance of the in-plane stress components at the free edge.

For illustration, consider a long laminated strip of length $2L$, width $2b << 2L$, and thickness $2H < 2b$ (Figure 5.1). The strip is loaded by an axial force N_x only. For a balanced, symmetric laminate, the mid-plane strains and curvatures (3.6) are uniform over the entire cross-section and given by

$$\epsilon_x^0 = \alpha_{11} N_x$$
$$\epsilon_y^0 = \alpha_{12} N_x$$
$$\gamma_{xy}^0 = 0$$
$$k_x = k_y = k_{xy} = 0 \tag{5.1}$$

where α_{11}, α_{12} are in-plane laminate compliances, which are obtained by inverting (3.8); see also [3, (6.25)]. From the constitutive equation [3, (6.28)] for lamina k, we get

$$\sigma_x^k = \left(\overline{Q}_{11}^k \alpha_{11} + \overline{Q}_{12}^k \alpha_{12}\right) N_x$$
$$\sigma_y^k = \left(\overline{Q}_{12}^k \alpha_{11} + \overline{Q}_{22}^k \alpha_{12}\right) N_x$$
$$\sigma_{xy}^k = \left(\overline{Q}_{16}^k \alpha_{11} + \overline{Q}_{26}^k \alpha_{12}\right) N_x$$
$$\sigma_z^k = \sigma_{xz}^k = \sigma_{yz}^k = 0 \tag{5.2}$$

A piece of laminate taken out of the interior of the laminate will have balanced σ_y and σ_{xy} on opposite faces; the free body diagram (FBD) is in equilibrium without the need for any additional forces. In this case we say the stress components are self-equilibrating. At the free edge in Figure 5.1, $\sigma_y = \sigma_{xy} = \sigma_{yz} = 0$. If σ_y and

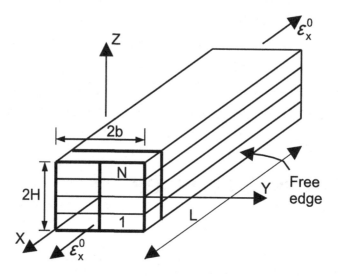

Fig. 5.1: Tensile coupon.[1]

σ_{xy} are not zero in the interior of the laminate, but are zero at the free edge, then some other stresses must equilibrate them.

5.1 Poisson's Mismatch

A lamina subjected to tensile loading in one direction will contract in the direction perpendicular to the load. If two or more laminas with different Poisson's ratios are bonded together, interlaminar stress will be induced to force all laminas to deform equally at the interfaces (Figure 5.2). Over the entire laminate thickness, these stresses add up to zero since there is no transverse loading N_y applied. In other words, they are self-equilibrating in such a way that

$$\int_{z_0}^{z_N} \sigma_y dz = 0 \tag{5.3}$$

where z_0 and z_N are the coordinates of the bottom and top surfaces, respectively.

5.1.1 Interlaminar Force

As noted in (5.3), the in-plane stress σ_y calculated with classical lamination theory (CLT) [3, Chapter 6] is self-equilibrating when added through the whole thickness of the laminate. But on a portion of the laminate (above z_k in Figure 5.3), the stresses σ_y may not be self-equilibrating. Therefore, the contraction or expansion of one or more laminas must be equilibrated by interlaminar shear stress σ_{yz}. Since there is no shear loading on the laminate, the integral of σ_{yz} over the entire width of the sample must vanish. Over half the width of the laminate, however, an interlaminar

---

[1]Reprinted from *Mechanics of Fibrous Composites*, C. T. Herakovich, Figure 8.1, copyright (1998), with permission from John Wiley & Sons, Inc.

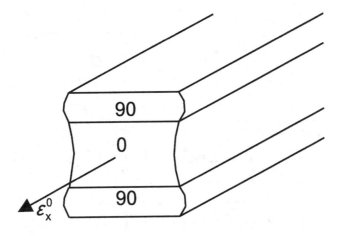

Fig. 5.2: Poisson's effect.[2]

shear force exists if the stress σ_y above or below the surface is not self-equilibrating. The magnitude of these per-unit-length forces can be estimated by integrating the interlaminar shear stress σ_{yz} over half the width of the laminate $(0 < y < b)$. By equilibrium

$$F_{yz}(z_k) = \int_0^b \sigma_{yz_{(z=z_k)}} dy = -\int_{z_k}^{z_N} \sigma_y dz \qquad (5.4)$$

The interlaminar shear stress σ_{yz} is not available from classical lamination theory but the transverse stress σ_y is. Therefore, the magnitude of the interlaminar shear force can be computed anywhere through the thickness of a laminate in terms of the known transverse stress distribution σ_y.

The in-plane stress σ_y in a balanced, symmetric laminate under tensile load is constant in each lamina. Therefore, when the interlaminar force is evaluated at an interface (located at $z=z_k$), the integration above reduces to

$$F_{yz}(z_k) = -\sum_{i=k}^{N} \sigma_y^i t_i \qquad (5.5)$$

where t_i are the thicknesses of the laminas.

The magnitude of the interlaminar shear force F_{yz} can be used to compare different stacking sequences in an effort to minimize the free-edge interlaminar shear stress σ_{yz}. However, the force does not indicate how large the actual stress is. Therefore, it can be used to compare different LSS but not as a failure criterion.

5.1.2 Interlaminar Moment

The interlaminar shear stress σ_{yz} produces shear strain γ_{yz}, which must vanish at the center line of the sample because of symmetry. Therefore, $\sigma_{yz} = 0$ at the

[2]Reprinted from *Mechanics of Fibrous Composites*, C. T. Herakovich, Figure 8.14, copyright (1998), with permission from John Wiley & Sons, Inc.

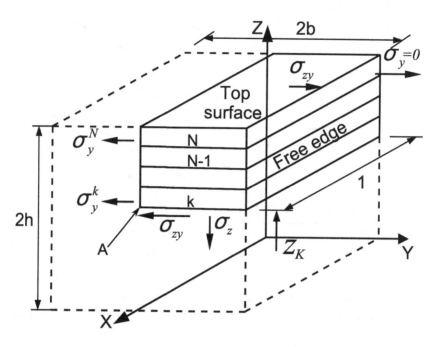

Fig. 5.3: Free body diagram of sublaminate used for computation of Poisson-induced forces F_{yz} and moments M_z.

center line. Also, at the free edge, σ_{yz} must vanish because σ_{zy} vanishes there. But for any position z_k above which σ_y is not self-equilibrating, σ_{yz} must be non-zero somewhere between the edge and the center line. A numerical solution of σ_{yz} is plotted in Figure 5.5 in terms of the distance y/b from the free edge. It reveals that σ_{yz} grows rapidly near the free edge and then tapers out at the interior of the laminate.

A not self-equilibrating distribution of stress yields both a force F_{yz} (5.5) and a moment. To compute the moment M_z, take moments of the stress σ_y with respect to point A in Figure 5.3. A non-vanishing moment produced by σ_y can only be equilibrated by a moment produced by transverse stress σ_z. Therefore, the moment M_z is defined as

$$M_z(z_k) = \int_0^b \sigma_{z(z=z_k)} y\,dy = \int_{z_k}^{z_N} (z - z_k)\sigma_y dz \qquad (5.6)$$

where z_k is the coordinate of the top surface of lamina k, and z_N is the coordinate of the top surface of the laminate (see [3, Figure 6.6] for the definition of the coordinate system through the thickness of the laminate).

The existence of σ_z is corroborated by free-edge delamination during a tensile test, at a much lower load than the failure load of the laminate. The magnitude of the moment can be used to compare different stacking sequences in an effort to minimize the stress σ_z. However, the moment does not indicate how large the actual stress is, and thus it cannot be used as a failure criterion.

The in-plane stress σ_y in a balanced laminate under tensile load is constant in each lamina. Therefore, when the interlaminar moment is evaluated at an interface (located at $z = z_k$), the integration above reduces to

$$M_z(z_k) = \sum_{i=k}^{N} \sigma_y^i \left(z_i t_i + \frac{t_i^2}{2} - z_k t_i \right) \tag{5.7}$$

Since σ_z is a byproduct of σ_{yz}, which vanishes at $y = 0$ due to symmetry, then σ_z must vanish at the center line of the specimen ($y = 0$) but it is large near the edge. Since no vertical load is applied, the integral of σ_z must be zero. Therefore, it must be tensile (positive) on some regions and compressive (negative) at others. A numerical solution reveals that σ_z grows rapidly near the free edge, dips to negative values, and then tapers out at the interior of the laminate. A numerical solution of σ_z is plotted in Figure 5.5 in terms of the distance y/b from the free edge. However, $\sigma_z \to \infty$ as $y \to b$. This is a singularity that is not handled well by FEA. Therefore the results, even for $y < b$, will be very dependent on mesh refinement. Furthermore, since $\sigma_z \to \infty$, the results cannot be used in a failure criterion without further consideration.

Example 5.1 *Compute F_{yz} and M_z at all interfaces of a balanced $[0_2/90_2]_s$ symmetric laminate (Figure 5.1) loaded with $N_x = 175$ KN/m. Use unidirectional lamina carbon/epoxy properties $E_1 = 139$ GPa, $E_2 = 14.5$ GPa, $G_{12} = G_{13} = 5.86$ GPa, $G_{23} = 5.25$ GPa, $\nu_{12} = \nu_{13} = 0.21$, $\nu_{23} = 0.38$. The lamina thickness is $t_k = 0.127$ mm.*

Solution to Example 5.1 *The in-plane stress distribution σ_y through the thickness can be obtained by the procedure described in [3, § 6.2]. The stress values are shown in Table 5.1.*

To calculate F_{yz}, compute the contribution of all laminas above a given interface using (5.5). The in-plane stress σ_y in a balanced laminate under in-plane load is constant in each lamina, so (5.5) applies. For other cases, (5.4) can be integrated exactly since σ_y is linear in z, or F_{yz} can be approximated by (5.5) using the average σ_y in each lamina.

Since the laminate is balanced and loaded with in-plane loads only, M_z can be computed using (5.7). Otherwise, use (5.6) or approximate M_z by using the average σ_z in each lamina into (5.7). The results are shown in Table 5.1 and Figure 5.4.

Example 5.2 *Plot the Poisson-induced stresses σ_z and σ_{yz} as a function of y for $0 < y < b$ (Figure 5.3) at the 90/0 interface above the middle surface of a $[0/90]_s$ laminate with properties $E_1 = 139$ GPa, $E_2 = 14.5$ GPa, $\nu_{12} = \nu_{13} = 0.21$, $\nu_{23} = 0.38$, $G_{12} = G_{13} = 5.86$ GPa, $G_{23} = 5.25$ GPa. Take $2b = 20$ mm, length of the sample $2L = 80$ mm, thickness of each lamina $t_k = 1.25$ mm. Load the sample with a uniform strain $\epsilon_x = 0.01$ by applying a uniform displacement at $x = L$. Use orthotropic solid elements on each lamina. Refine the mesh towards the free edge. Use at least two quadratic elements through the thickness of each lamina and an element aspect ratio of approximately one near the free edge.*

Solution to Example 5.2 *Watch the video in [1].*
Note that it is not necessary to model the whole geometry. Symmetry can be used to model only the quadrant with $x > 0, y > 0, z > 0$ (i.e., one-eighth of the plate, Figure 5.3). Since any cross-section $y - z$ has the same behavior, only a short segment between $x = 0$ and

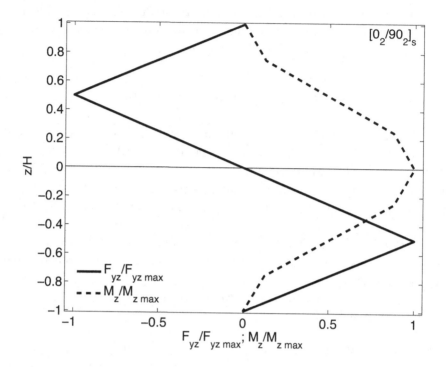

Fig. 5.4: Interlaminar force F_{yz} and moment M_z due to Poisson's effect.

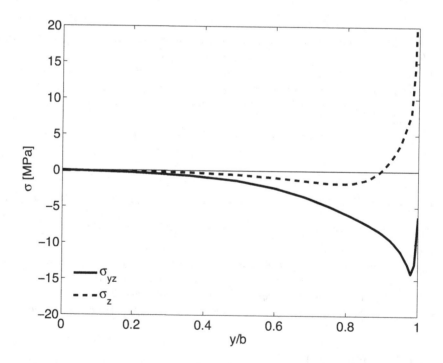

Fig. 5.5: Interlaminar stress σ_{yz} and σ_z at the 90/0 interface of a carbon/epoxy $[0/90]_S$ laminate (FEA).

Table 5.1: Poisson's interlaminar force F_{yz}.

k	Pos	σ_y [MPa]	t_k [mm]	z [mm]	F_{yz} [kN/m]	M_z [N m/m]
8	TOP	5.55 10^{-3}		0.508	0.000	
8	BOT	5.55 10^{-3}	0.127	0.381	-0.705	0.045
7	TOP	5.55 10^{-3}		0.381	-0.705	
7	BOT	5.55 10^{-3}	0.127	0.254	-1.410	0.179
6	TOP	-5.55 10^{-3}		0.254	-1.410	
6	BOT	-5.55 10^{-3}	0.127	0.127	-0.705	0.313
5	TOP	-5.55 10^{-3}		0.127	-0.705	
5	BOT	-5.55 10^{-3}	0.127	0.000	0.000	0.358
4	TOP	-5.55 10^{-3}		0.000	0.000	
4	BOT	-5.55 10^{-3}	0.127	-0.127	0.705	0.313
3	TOP	-5.55 10^{-3}		-0.127	0.705	
3	BOT	-5.55 10^{-3}	0.127	-0.254	1.410	0.179
2	TOP	5.55 10^{-3}		-0.254	1.410	
2	BOT	5.55 10^{-3}	0.127	-0.381	0.705	0.045
1	TOP	5.55 10^{-3}		-0.381	0.705	
1	BOT	5.55 10^{-3}	0.127	-0.508	0.000	0.000

$x = L\star$ *needs to be modeled. Since free edge effects also occur at $x = 0$ and $x = L\star$, take $L\star = 8h = 32t_k$, $t_k = 2.5$ mm, and plot the results at $x = L \star /2$ to avoid free edge effects at the two loaded ends of the model. The solution is shown in Figure 5.5. See the pseudo code below and Figure 5.5.*

i. *Setting* Work Directory

```
Menu:    File, Set Work Directory, [C:\SIMULIA\user\Ex_5.2]
Menu:    File, Save As, [C:\SIMULIA\user\Ex_5.2\Ex_5.2.cae]
```

ii. *Creating the part*

```
Module: Part
        # Look at Fig. 5.3, lamina thickness is tk=1.25 mm along Z-global
        # b=10 mm, length along x is L=40 mm (shown as 1 in Fig. 5.3)
        # Draw laminas 3 and 4 above the middle surface
Menu:    Part, Create
        3D, Deformable, Solid, Extrusion, Approx size [100], Cont
Menu:    Add, Line, Rectangle, [0,0], [40,10], X, Done, Depth [2.5], OK
        # We need to cut the volume in 2 laminas
Menu:    Tools, Datum, Plane, Offset from ppal plane, XY Plane, [1.25]
        X, # close the Create Datum pop-up window
Menu:    Tools, Partition, Cell, Use datum plane
        # pick the plane parallel to the XY plane, Create partition, Done
        # close the Create Partition pop-up window
```

iii. *Creating assembly*

```
Module: Assembly
Menu:    Instance, Create, Independent, OK
```

iv. Defining materials, sections, and assigning sections to parts

```
Module: Property
Menu:   Material, Create
    Mechanical, Elasticity, Elastic, Type: Engineering Constants
    [1.39E5    1.45E4    1.45E4    0.21    0.21
     0.38      5.86E3    5.86E3    5.2536E3], OK
Menu:   Section, Create, Solid, Homogeneous, Cont, OK
Menu:   Assign, Section, # pick: lamina 3, Done, Section-1, OK
    # pick: lamina 4, Done, Section-1, OK, Done
    # assign material orientation for 90-deg lamina (near mid surface)
Menu:   Assign, Material Orientation, # pick: lamina 3, Done
    Use Default Orientation, Definition: Coordinate system
    Additional Rotation Direction: Axis 3
    Additional Rotation: Angle [90], OK
    # assign material orientation for 0-degree lamina (top surface)
Menu:   Assign, Material Orientation, #pick lamina 4, Done,
    Use Default Orientation, Definition: Coordinate system
    Additional Rotation Direction: Axis 3
    Additional Rotation: None, OK
    # to see the orientations do this:
    # on the left menu tree,
    # Models, Model-1, Parts, Part-1, Orientations
    # you should have only 2 SYSTEM: <Global> entries, one per lamina
```

v. Defining analysis steps

```
Module: Step
Menu:   Step, Create
    Procedure type: General, Static/General, Cont, OK
```

vi. Adding loads and BC

```
Module: Load
Menu:   BC, Manager
    Create, Name [DISP], Step: Step-1, Disp/Rota, Cont
    # pick: surface x=40, Done, # checkmark: U1 [0.4], OK
    # you may need to rotate the model to pick the surfaces
    Create, Name [XSYMM], Step: Initial, Symm/Anti/Enca, Cont
    # pick: surface x=0, Done, XSYMM, OK
    Create, Name [YSYMM], Step: Initial, Symm/Anti/Enca, Cont
    # pick: surface y=0, Done, YSYMM, OK
    Create, Name [ZSYMM], Step: Initial, Symm/Anti/Enca, Cont
    # pick surface z=0, Done, ZSYMM, OK
    # close BC Manager pop-up window
```

vii. Meshing the model

```
Module: Mesh
    # Bias towards the edge y=10 is needed
Menu:   Seed, Edges
    # pick: 6 edges parallel to Y-global, Done
    Method: By number, Bias: Single
```

```
        Number of elements [10], Bias ratio [20], Apply, OK
        Flip bias, # if required adjust bias direction, point toward y=10
        # pick 6 edges parallel to X-global, Done
        Method: By number, Bias: Double
        Flip bias, # pick: lines, # direct the bias toward the center
        Number of elements: [20], Bias ratio: [10], Apply, OK
        # pick 8 edges parallel to Z-global, Done
        Method: By number, Bias: None, Number of elements [6], OK
    Menu:    Mesh, Element Type, # select all regions, Done
        Geometric Order: Quadratic # element C3D20R is assigned, OK
    Menu:    Mesh, Instance, Yes
```

viii. Solving

```
    Module: Job
    Menu:    Job, Manager
        Create, Cont, OK, Submit, # when Completed, Results
```

*ix. Visualize stress and strain in **lamina** coordinate system. By default, Abaqus displays stress and strain in the lamina coordinate system.*

```
    Module: Visualization
    Menu:    Plot, Contours, On Deformed Shape
        Toolbox: Field Output, Primary, U, U1  # global coord system
        Toolbox: Field Output, Primary, E, E11 # lamina coord system
        Toolbox: Field Output, Primary, S, S33 # lamina coord system
```

*x. Visualize stress and strain in **global** coordinate system. This method can be used to display stress and strain along any user-defined coordinate system. In this example, we use the fact that one of the laminas is oriented in the same direction as the global coordinate system, so the required coordinate system is already available.*

```
    Menu:    Result, Options
        Tab: Transformation, Transformation Type: User-specified
        # pick: ASSEMBLY_PART-1-1_ORI-2, Apply, # displays c.s., OK
        Toolbox: Field Output, Primary, E, E11 # global coord system
        Toolbox: Field Output, Primary, S, S33 # global coord system
    Menu:    Results, Options
        Tab: Transformation, Transformation Type: Default, OK
```

xi. Create a path for plotting S33 slightly below the 0/90 interface.

```
    Menu:    Tools, Path, Create
        Name [Path-1], Type: Point List, Cont
        [20,0.0,1.24
         20,1.0,1.24
         20,2.0,1.24
         20,3.0,1.24
         20,4.0,1.24
         20,5.0,1.24
         20,6.0,1.24
         20,7.0,1.24
         20,8.0,1.24
```

```
        20,9.0,1.24
        20,9.5,1.24
        20,9.8,1.24
        20,9.9,1.24
        20,10.,1.24]
        # visualize the path created on the WS, OK
```

xii. *Create an XY data set to visualize it and/or save it to a file later.*

```
    Menu:   Tools, XY Data, Create
            Source: Path, Cont, Path: Path-1, Model shape: Undeformed
            X Values: True distance, Y Values: Field output: S, S33, OK
            Save As [path1-S33], OK, Plot, # visualize on screen
            # close XY Data pop-up window
```

xiii. *Save the results on the path to a report file in the Work Directory.*

```
    Menu:   Report, XY
            # pick: path1-S33, Tab: Setup, File name [S33path1.rpt], OK
```

xiv. *Repeat the visualization and report for a path slightly above the interface (defined below).*

```
        [20,0.0,1.26
        20,1.0,1.26
        20,2.0,1.26
        20,3.0,1.26
        20,4.0,1.26
        20,5.0,1.26
        20,6.0,1.26
        20,7.0,1.26
        20,8.0,1.26
        20,9.0,1.26
        20,9.5,1.26
        20,9.8,1.26
        20,9.9,1.26
        20,10.,1.26]
```

Then, save the results in the file [S33path2.rpt]

The stress values along the paths defined above are plotted in Figure 5.6. Notice the results of stress are virtually identical above and below the interface.

5.2   Coefficient of Mutual Influence

In classical lamination theory, it is assumed that the portion of the laminate being analyzed is far from the edges of the laminate. Stress resultants N and M are then applied to a portion of the laminate, and these induce in-plane stress σ_x, σ_y, σ_{xy} on each lamina. In the interior of the laminate, interlaminar stresses σ_{xz}, σ_{yz} are induced only if shear forces are applied.

For uniaxial loading N_x, the transverse stresses generated in each lamina as a result of Poisson's effect must cancel out to yield a null laminate force N_y. Also, the

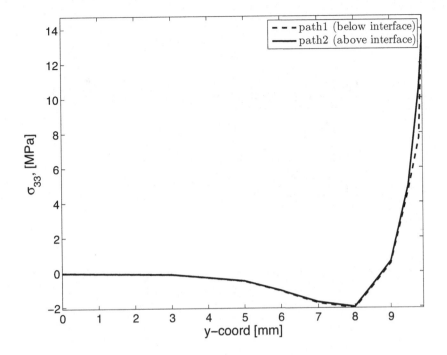

Fig. 5.6: Stress σ_{33} below and above the 90/0 interface.

in-plane shear stress on off-axis laminas must cancel out with those of other laminas to yield zero shear force N_{xy} for the laminate. The situation is more complex near the edges as the various components of in-plane stress do not cancel each other across the lamina interfaces. For the time being, let us revisit the concept of laminate engineering properties. In material axes, the plane-stress compliance equations are

$$
\left\{ \begin{array}{c} \epsilon_1 \\ \epsilon_2 \\ \gamma_6 \end{array} \right\} = \left[\begin{array}{ccc} S_{11} & S_{12} & 0 \\ S_{12} & S_{22} & 0 \\ 0 & 0 & S_{66} \end{array} \right] \left\{ \begin{array}{c} \sigma_1 \\ \sigma_2 \\ \sigma_6 \end{array} \right\} \tag{5.8}
$$

It is also known that the compliance coefficients can be written in terms of engineering properties as

$$
[S] = \left[\begin{array}{ccc} 1/E_1 & -\nu_{12}/E_1 & 0 \\ -\nu_{12}/E_1 & 1/E_2 & 0 \\ 0 & 0 & 1/G_{12} \end{array} \right] \tag{5.9}
$$

For an off-axis lamina (oriented arbitrarily with respect to the global axes), we have

$$
\left\{ \begin{array}{c} \epsilon_x \\ \epsilon_y \\ \gamma_{xy} \end{array} \right\} = \left[\begin{array}{ccc} \overline{S}_{11} & \overline{S}_{12} & \overline{S}_{16} \\ \overline{S}_{12} & \overline{S}_{22} & \overline{S}_{26} \\ \overline{S}_{16} & \overline{S}_{26} & \overline{S}_{66} \end{array} \right] \left\{ \begin{array}{c} \sigma_x \\ \sigma_y \\ \sigma_{xy} \end{array} \right\} \tag{5.10}
$$

Here it can be seen that uniaxial load ($\sigma_y = \sigma_{xy} = 0$) yields shear strain as a result of the shear-extension coupling

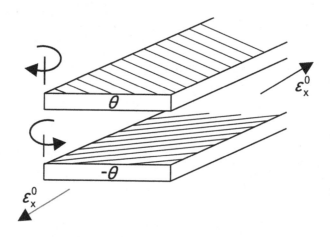

Fig. 5.7: Deformation caused by mutual influence.[3]

$$\gamma_{xy} = \overline{S}_{16}\sigma_x \tag{5.11}$$

where

$$\overline{S}_{16} = (2S_{11} - 2S_{12} - S_{66})\sin\theta\cos^3\theta \tag{5.12}$$
$$- (2S_{22} - 2S_{12} - S_{66})\sin^3\theta\cos\theta$$

Now, the coefficients of $[\overline{S}]$ can be defined in term of the engineering properties for the off-axis lamina as

$$\overline{S}_{11} = 1/E_x \; ; \; \overline{S}_{12} = -\nu_{xy}/E_x = -\nu_{yx}/E_y \tag{5.13}$$
$$\overline{S}_{22} = 1/E_y \; ; \; \overline{S}_{66} = 1/G_{xy}$$

To complete the definition of $[\overline{S}]$ in (5.10), two new engineering properties describing shear-extension coupling, $\eta_{xy,x}$ and $\eta_{xy,y}$, are defined as

$$\overline{S}_{16} = \frac{\eta_{xy,x}}{E_x} \quad ; \quad \overline{S}_{26} = \frac{\eta_{xy,y}}{E_x} \tag{5.14}$$

The engineering properties $\eta_{xy,x}$ and $\eta_{xy,y}$ are called coefficients of mutual influence and they represent the shear caused by stretching. Their formal definition is obtained by imposing an axial stress and measuring the resulting shear strain

$$\eta_{ij,i} = \frac{\gamma_{ij}}{\epsilon_i} \tag{5.15}$$

Alternatively, two other coefficients of mutual influence could be defined to represent the stretching caused by shear

---

[3]Reprinted from *Mechanics of Fibrous Composites*, C. T. Herakovich, Figure 8.14, copyright (1998), with permission from John Wiley & Sons, Inc.

$$\overline{S}_{16} = \frac{\eta_{x,xy}}{G_{xy}} \; ; \; \overline{S}_{26} = \frac{\eta_{y,xy}}{G_{xy}} \qquad (5.16)$$

These are defined by imposing a shear stress and measuring the axial strain

$$\eta_{i,ij} = \frac{\epsilon_i}{\gamma_{ij}} \qquad (5.17)$$

5.2.1 Interlaminar Stress due to Mutual Influence

Off-axis laminas induce in-plane shear stress when subject to axial loading because the natural shear deformations that would occur on an isolated lamina (Figure 5.7) are constrained by the other laminas. Through the whole thickness of the laminate, these stresses cancel out, but over unbalanced sublaminates (e.g., the top lamina in Figure 5.7), they amount to a net shear.

That shear can only be balanced by interlaminar stress σ_{zx} at the bottom of the sublaminate (Figure 5.8). Then, summation of forces along x leads to a net force

$$F_{xz}(z_k) = \int_0^b \sigma_{zx(z=z_k)} dy = - \int_{z_k}^{z_N} \sigma_{xy} dz \qquad (5.18)$$

Once again, the in-plane shear stress calculated with classical lamination theory [3, Chapter 6] can be used to compute the interlaminar force per unit length F_{xz}. For in-plane loading, CLT yields constant shear stress in each lamina. When the interlaminar force is evaluated at an interface (located at $z = z_k$), the integration above reduces to

$$F_{xz}(z_k) = - \sum_{i=k}^{N} \sigma_{xy}^i t_i \qquad (5.19)$$

The force F_{xz}, as well as the values of the coefficients of mutual influence, can be used to qualitatively select the LSS with the least interlaminar stress. Actual values of interlaminar stresses can be found by numerical analysis. However, $\sigma_z \to \infty$ as $y \to b$. This is a singularity that is not handled well by FEA. Therefore the results, even for $y < b$, are very dependent on mesh refinement. Furthermore, since $\sigma_z \to \infty$, the results cannot be used in a failure criterion without further consideration. A numerical approximation of σ_{xz} for a $[\pm 45]_S$ laminate is plotted in Figure 5.10 in terms of the distance y' from the free edge.

Example 5.3 *Compute F_{xz} at all interfaces of a $[30_2/-30_2]_s$ balanced symmetric laminate (Figure 5.1) loaded with $N_x = 175$ KN/m. The material properties are given in Example 5.1, p. 203. The lamina thickness is $t_k = 0.127$ mm.*

Solution to Example 5.3 *In-plane shear stress σ_{xy} through the thickness of the laminate can be obtained following the same procedure used to obtain σ_y in Example 5.1, p. 203.*

For a symmetric balanced laminate under in-plane loads, use (5.19). For a general laminate under general load, use (5.18) or approximate F_{xz} by (5.19) taking the average of σ_{xy} in each lamina.

The results are obtained with a spreadsheet and shown in Table 5.2 and Figure 5.9.

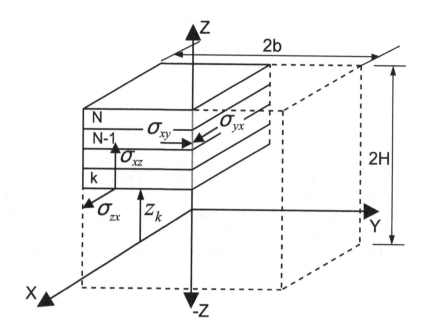

Fig. 5.8: Free body diagram of sublaminate to compute the interlaminar force F_{xz} due to mutual influence.

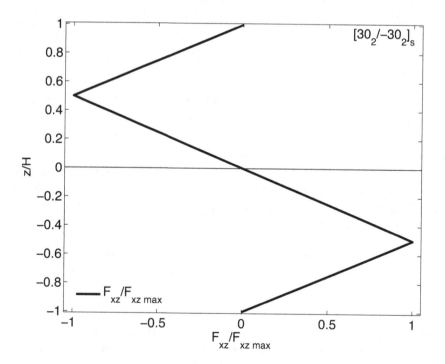

Fig. 5.9: Interlaminar shear force due to mutual influence F_{xz}.

Table 5.2: Interlaminar force F_{xz} due to mutual influence.

k	Pos	σ_{xy} [MPa]	t_k [mm]	z [mm]	F_{xz} [kN/m]
8	TOP	78.6 10^{-3}		0.508	0.000
8	BOT	78.6 10^{-3}	0.127	0.381	-9.982
7	TOP	78.6 10^{-3}		0.381	-9.982
7	BOT	78.6 10^{-3}	0.127	0.254	-19.964
6	TOP	-78.6 10^{-3}		0.254	-19.964
6	BOT	-78.6 10^{-3}	0.127	0.127	-9.982
5	TOP	-78.6 10^{-3}		0.127	-9.982
5	BOT	-78.6 10^{-3}	0.127	0.000	0.000
4	TOP	-78.6 10^{-3}		0.000	0.000
4	BOT	-78.6 10^{-3}	0.127	-0.127	9.982
3	TOP	-78.6 10^{-3}		-0.127	9.982
3	BOT	-78.6 10^{-3}	0.127	-0.254	19.964
2	TOP	78.6 10^{-3}		-0.254	19.964
2	BOT	78.6 10^{-3}	0.127	-0.381	9.982
1	TOP	78.6 10^{-3}		-0.381	9.982
1	BOT	78.6 10^{-3}	0.127	-0.508	0.000

Example 5.4 *Plot σ_{13} at the interface above the middle surface of a $[\pm45]_S$ laminate using the material properties, geometry, and loading of Example 5.2, p. 203.*

Solution to Example 5.4 *Watch the video in [1].*
Since the LSS is symmetric, it is possible to model half of the laminate ($z > 0$). However, it is not possible to use the symmetry conditions used in Example 5.2 (p. 203) because the material (say, a $30°$ lamina) is not symmetric with respect to the $x - z$ and $y - z$ planes (see Figure 5.8). Instead, the plane $y - z$ at $x = 0$ is not a symmetry plane but rather a plane with $\epsilon_x = 0$. Also, the edge effects at the ends of the model at $x = 0$ and in $x = L\star$ are important, so the results must be plotted at $x = L \star /2$ to avoid free edge effects at the loaded ends. The solution is shown in Figure 5.10.

i. *Setting the* Work Directory

```
Menu: File, Set Work Directory, [C:\SIMULIA\user\Ex_5.4]
Menu: File, Save As, [C:\SIMULIA\user\Ex_5.4\Ex_5.4.cae]
```

ii. *Creating the part*

```
Module: Part
Menu:   Part, Create
    3D, Deformable, Solid, Extrusion, Approx size [100], Cont
Menu:   Add, Line, Rectangle, [0,0], [80,20], X, Done, Depth [2.5], OK
    # We need to cut the volume in 2 laminas
Menu:   Tools, Datum, Plane,
    Offset from principal plane, XY Plane, [1.25]
    X, # close the pop-up window
Menu:   Tools, Partition, Cell, Use datum plane
    # pick the plane parallel to the XY plane, Create partition, Done
    # close pop-up window
```

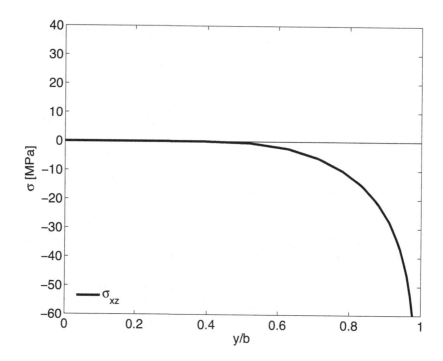

Fig. 5.10: Interlaminar shear stress σ_{xz} at the interface above the middle-surface of a carbon/epoxy $[\pm45]_S$ laminate (FEA).

iii. Creating assembly

```
Module: Assembly
Menu:   Instance, Create, Independent, OK
```

iv. Defining materials, sections, and assigning sections to parts

```
Module: Property
Menu:   Material, Create
    Mechanical, Elasticity, Elastic, Type: Engineering Constants
    [1.39E5    1.45E4    1.45E4    0.21    0.21
     0.38      5.86E3    5.86E3    5.2536E3], OK
Menu:   Section, Create, Solid, Homogeneous, Cont, OK
Menu:   Assign, Section, # pick: lamina 3, Done, Section-1, OK
    # pick: lamina 4, Done, Section-1, OK, Done
    # assign material orientation -45 lamina (near middle surface)
Menu:   Assign, Material Orientation, # pick: lamina 3, Done
    Use Default Orientation, Definition: Coordinate system
    Additional Rotation Direction: Axis 3
    Additional Rotation: Angle [-45], OK
    # assign material orientation for +45-degree lamina (top surface)
Menu:   Assign, Material Orientation, #pick lamina 4, Done,
    Use Default Orientation, Definition: Coordinate system
    Additional Rotation Direction: Axis 3
    Additional Rotation: Angle [45], OK
    # to see the orientations do this:
```

```
# on the left menu tree,
# Models, Model-1, Parts, Part-1, Orientations
# you should have only 2 SYSTEM: <Global> entries, one per lamina
```

v. Defining analysis steps

```
Module: Step
Menu:   Step, Create
    Procedure type: General, Static/General, Cont, OK
```

vi. Adding loads and BC

```
Module: Load
Menu:   BC, Manager
    Create, Name [DISP], Step: Step-1, Disp/Rota, Cont
    # pick: surface x=80, Done, # checkmark: U1 [0.8], OK
    # you may need to rotate the model to pick the surfaces
    Create, Name [XEND], Step: Initial, Rota/Disp, Cont
    # pick: surface x=0, Done, # checkmark: U1, OK
    Create, Name [ZSYMM], Step: Initial, Symm/Anti/Enca, Cont
    # pick: surface z=0, Done, # checkmark: U3, OK
    # close BC Manager pop-up window
```

vii. Meshing the model

```
Module: Mesh
Menu:   Seed, Edges
    # pick: 6 edges parallel to Y-global, Done
    Method: By number, Bias: Double
    # make sure the bias is directed toward the end of the lines
    Number of elements [20], Bias ratio [10], Apply, OK
    # pick 6 edges parallel to X-global, Done
    Method: By number, Bias: Double
    Flip bias, # all the lines toward the center
    Number of elements: [20], Bias ratio: [10], Apply, OK
    # pick 8 edges parallel to Z-global, Done
    Method: By number, Bias: None, Number of elements [6], Apply, OK
Menu:   Mesh, Element Type, # select all regions, Done
    Geometric Order: Quadratic, # C3D20R is assigned, OK
Menu:   Mesh, Instance, Yes
```

viii. Solving

```
Module: Job
Menu:   Job, Manager
    Create, Cont, OK, Submit,
    # when Completed, Results
```

ix. Visualizing the results. Unlike in Example 5.2 (step x), here we do not transform local tensors to global c.s.

```
Module: Visualization
Menu:   Plot, Contours, On Deformed Shape
# create a path for plotting S13 slightly above the -45/+45 interface
```

```
Menu:    Tools, Path, Create
   Name [Path-1], Type: Point List, Cont
   [40, 0.0,   1.26
    40, 0.5,   1.26
    40, 1.0,   1.26
    40, 1.5,   1.26
    40, 2.0,   1.26
    40, 2.5,   1.26
    40, 3.0,   1.26
    40, 3.5,   1.26
    40, 4.0,   1.26
    40, 4.5,   1.26
    40, 5.0,   1.26
    40, 6.0,   1.26
    40, 7.0,   1.26
    40, 8.0,   1.26
    40, 9.0,   1.26
    40, 10.0,  1.26
    40, 11.0,  1.26
    40, 12.0,  1.26
    40, 13.0,  1.26
    40, 14.0,  1.26
    40, 15.0,  1.26
    40, 15.5,  1.26
    40, 16.0,  1.26
    40, 16.5,  1.26
    40, 17.0,  1.26
    40, 17.5,  1.26
    40, 18.0,  1.26
    40, 18.5,  1.26
    40, 19.0,  1.26
    40, 19.5,  1.26
    40, 20.0,  1.26]
   # visualize the path created on the WS, OK
Menu:    Tools, XY Data, Create
   Source: Path, Cont, Path: Path-1, Model shape: Undeformed
   X Values: True distance, Y Values: Field output: S, S13
   Save As [S13plot], OK, Plot, # visualize result on screen
   # close XY Data pop-up window
Menu:    Report, XY
   # pick: S13plot, Tab: Setup, File name [S13plot.rpt], OK
   # the file S13plot.rpt must have been saved in the Work Directory
```

The stress values along the path are plotted in Figure 5.11.

Suggested Problems

Problem 5.1 *Write a computer program to use tabulated data of σ_y and σ_{xy} (at the top and bottom of every lamina) to compute F_{yz}, F_{xz}, and M_z, for all locations through the thickness of a laminate with any number of laminas. Using the program, plot F_{yz}, F_{xz}, and M_z, through the thickness $-4t < z < 4t$ of a $[\pm 45/0/90]_s$ laminate with lamina thickness*

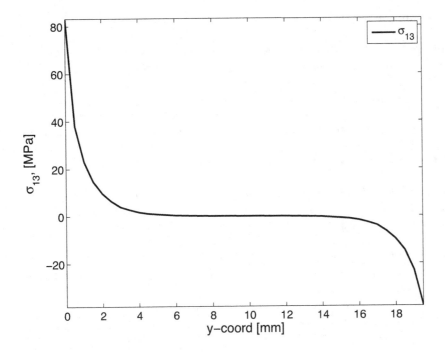

Fig. 5.11: Stress σ_{13} below and above the $-45/+45$ interface.

$t = 0.125$ mm, loaded with $N_x = 100$ kN/m. Use carbon/epoxy properties $E_1 = 139$ GPa, $E_2 = 14.5$ GPa, $G_{12} = G_{13} = 5.86$ GPa, $G_{23} = 5.25$ GPa, $\nu_{12} = \nu_{13} = 0.21$, $\nu_{23} = 0.38$. Submit a report including the source code of the program.

Problem 5.2 *Repeat Problem 5.1 for $M_x = 1$ Nm/m. Submit a report including the source code of the program.*

Problem 5.3 *Plot σ_z/σ_{x0} and σ_{yz}/σ_{x0} vs. y/b $(0 < y/b < 1)$ at $x = L/2$, at the first interface above the mid-surface of a $[0/90/90/0]$ laminate with lamina thickness $t = 0.512$ mm, loaded with $\epsilon_x = 0.01$. Compute the far-field uniform stress σ_{x0} in terms of the applied strain. Use quadratic solid elements and a mesh biased toward the free edge (bias 0.1) to model 1/8 of a tensile specimen (see Example 5.2, p. 203), of width $2b = 25.4$ mm and length $2L = 20$ mm. Use carbon/epoxy properties $E_1 = 139$ GPa, $E_2 = 14.5$ GPa, $G_{12} = G_{13} = 5.86$ GPa, $G_{23} = 5.25$ GPa, $\nu_{12} = \nu_{13} = 0.21$, $\nu_{23} = 0.38$. Attempt to keep the aspect ratio of the elements near the free edge close to one. Submit the input command file to obtain the solution and the plot. In addition, submit the plot.*

Problem 5.4 *For the laminate and loading described in Problem 5.3, plot σ_z/σ_{x0} and σ_{yz}/σ_{x0} versus z/t_k $(0 < z/t_k < 2)$ above the mid-surface, at a distance $0.1t_k$ from the free edge and $x = L/2$. Study the effect of mesh refinement by providing four curves with different number of divisions along the z-direction. Attempt to keep the aspect ratio of the elements near the free edge close to one. Submit the input command file to obtain the solution and the plot. In addition, submit the plot.*

Problem 5.5 *Plot σ_{xz}/σ_{x0} as in Problem 5.3 for all the interfaces above the middle surface of a $[\pm 10_2]_S$ laminate.*

Problem 5.6 *Plot* σ_{xz}/σ_{x0} *as in Problem 5.4 for a* $[\pm 10_2]_S$ *laminate.*

Problem 5.7 *Use solid elements and a biased mesh to model 1/8 of a tensile specimen (see Example 5.2, p. 203), of width* $2b = 24$ *mm and length* $2L = 20$ *mm. The laminate is* $[\pm 45/0/90]_s$ *with lamina thickness* $t = 0.125$ *mm, loaded with* $N_x = 175$ *KN/m. Use carbon/epoxy properties* $E_1 = 139$ *GPa,* $E_2 = 14.5$ *GPa,* $G_{12} = G_{13} = 5.86$ *GPa,* $G_{23} = 5.25$ *GPa,* $\nu_{12} = \nu_{13} = 0.21$, $\nu_{23} = 0.38$. *Plot the three interlaminar stress components, from the edge to the center line of the specimen, at the mid-surface of each lamina. Lump all four plots of the same stress into a single plot. Submit the input command file to obtain the solution and the three plots. In addition, submit the three plots.*

Problem 5.8 *Plot* E_x/E_2, G_{xy}/G_{12}, $10\nu_{xy}$, $-\eta_{xy,x}$, *and* $-\eta_{x,xy}$ *in the same plot vs* θ *in the range* $-\pi/2 < \theta < \pi/2$ *for a unidirectional single lamina oriented at an angle* θ. *The material is S-glass/epoxy [3, Tables 1.1–1.4].*

Problem 5.9 *Using the plot from Problem 5.8 and considering a* $[\theta_1/\theta_2]_S$ *laminate, what is the worst combination of values* θ_1, θ_2 *for two cases: (a) Poisson's mismatch and (b) shear mismatch.*

Problem 5.10 *In a single plot, compare* $-\eta_{xy,x}$ *of E-glass/epoxy, Kevlar-49/epoxy, and T800/3900-2 in the range* $-\pi/2 < \theta < \pi/2$ *([3, Tables 1.1–1.4]).*

Problem 5.11 *Obtain contour plots of the three deformations* u_x, u_y, u_z *(independently) on the top surface of a* $[\pm 45]_s$ *laminate. Use dimensions, load, and material properties of Problem 5.7. Explain your findings.*

Problem 5.12 *Repeat Problem 5.11 for a* $[0/90]_s$ *laminate. Explain your findings.*

Problem 5.13 *Use solid elements and a biased mesh to model 1/4 of a tensile specimen (Figure 5.1) of a total width* $2b = 12$ *mm and length* $2L = 24$ *mm. Compare in the same plot* σ_z *vs* z/H *for* $[\pm 15/\pm 45]_s$ *and* $[\pm(15/45)]$ *of SCS-6/aluminum with 50% fiber volume. Use micromechanics (6.8) to predict the unidirectional composite properties. The lamina thickness* $t_k = 0.25$ *mm. The laminate is loaded with* $\epsilon_x = 0.01$.

	Al-2014-T6 [30, App. B]	SCS-6 [3, Tables 2.1–2.4]
E [GPa]	75.0	427.0
G [GPa]	27.0	177.9

Problem 5.14 *Use an FEA model similar to Problem 5.13 to plot* $\sigma_{xz}/\sigma_{zx_{\max}}$ *vs* θ ($0 < \theta < \pi/2$) *for a* $[\pm \theta]_s$ *SCS-6/ Al laminate with* $\epsilon_x = 0.01$.

Problem 5.15 *Use the FEA model of Problem 5.13 to plot* σ_z *vs* y/b ($0 < y < 0.95b$) *at the mid-surface of the* $[\pm 15/\pm 45]_s$ *laminate. Note* $\sigma_z \to \infty$ *near* $y = b$, *so the actual value from FEA at* $y = b$ *is mesh dependent. Investigate mesh dependency at* $y = 0.95b$ *by tabulating the result using different mesh densities.*

Problem 5.16 *Use an FEA model similar to Problem 5.13 to plot* σ_x, σ_{xy}, *and* σ_{xz} *vs* y/b ($0 < y < b$) *when a* $[\pm \theta]_s$ *SCS-6/Al laminate is subjected to 1% axial strain* ($\epsilon_x = 0.01$).

Problem 5.17 *A* $[0/90]_s$ *laminate with properties* $E_1 = 139$ *GPa,* $E_2 = 14.5$ *GPa,* $G_{12} = G_{13} = 5.86$ *GPa,* $G_{23} = 5.25$ *GPa,* $\nu_{12} = \nu_{13} = 0.21$, $\nu_{23} = 0.38$ *is shown in Figure 5.1. The strength properties of the lamina are* $F_{1t} = 1550$ *MPa,* $F_{1c} = 1090$ *MPa,* $F_{2t} = F_{2c} = 59$ *MPa, and* $F_6 = 75$ *MPa. Take* $2b = 20$ *mm, length of the sample* $2L = 200$ *mm, thickness of each lamina* $t_k = 1.25$ *mm. Load the sample with a uniform strain* $\epsilon_x = 0.01$ *by applying a uniform displacement. Use symmetry to model only the quadrant with* $x > 0$, $y > 0$, $z > 0$. *Use orthotropic solid elements on each lamina, with at least two quadratic elements through the thickness of each lamina. Compute the 3D Tsai-Wu failure index* I_F *using a UMAT subroutine for solid elements. Obtain the contour plot of* I_F *in each lamina (do not use results averaging). Show all work in a report.*

Chapter 6

Computational Micromechanics

In Chapter 1, the elastic properties of composite materials were assumed to be available in the form of elastic modulus E, shear modulus G, Poisson's ratio ν, and so on. For heterogeneous materials such as composites, a large number of material properties are needed, and experimental determination of these many properties is a tedious and expensive process. Furthermore, the values of these properties change as a function of the volume fraction of reinforcement and so on. An alternative, or at least a complement to experimentation, is to use homogenization techniques to predict the elastic properties of the composite in terms of the elastic properties of the constituents (matrix and reinforcements). Since homogenization models are based on more or less accurate modeling of the microstructure, these models are also called micromechanics models, and the techniques used to obtain approximate values of the composite's properties are called micromechanics methods or techniques [3, Chapter 4]. Micromechanics models can be classified into empirical, semi empirical, analytical, and numerical.

This book deals only with strictly analytical or numerical models that do not require empirical adjusting factors, so that no experimentation is required. Since most of this book deals with 3D analysis, emphasis is placed on micromechanics models that can estimate the whole set of elastic properties using a single model, rather than using a disjointed collection of models based on different assumptions to assemble the set of properties needed. Many analytical techniques of homogenization are based on the equivalent eigenstrain method [31, 32], which considers the problem of a single ellipsoidal inclusion embedded in an infinite elastic medium. The Eshelby solution is used in [33] to develop a method that takes into account, approximately, the interactions among the inclusions. One of the more used homogenization techniques is the self-consistent method [34], which considers a random distribution of inclusions in an infinite medium. The infinite medium is assumed to have properties equal to the unknown properties sought. Therefore, an iterative procedure is used to obtain the overall moduli. Homogenization of composites with periodic microstructure has been accomplished by using various techniques including an extension of the Eshelby inclusion problem [31, 32], the Fourier series technique (see Section 6.1.3 and [35, 36]), and variational principles. The periodic eigenstrain method was further developed to determine the overall relaxation moduli of linear viscoelastic composite materials (see Section 7.6 and [37,38]). A particular case, the cell method for periodic media, considers a unit cell with a square inclusion [39].

The analytical procedures mentioned so far yield approximate estimates of the exact solution of the micromechanics problem. These estimates must lie between lower and upper bounds for the solution. Several variational principles were developed to evaluate bounds on the homogenized elastic properties of macroscopically isotropic heterogeneous materials [40]. Those bounds depend only on the volume fractions and the physical properties of the constituents.

In order to study the nonlinear material behavior of composites with periodic microstructure, numerical methods, mainly the finite element method, are employed. Nonlinear finite element analysis of metal matrix composite has been studied by looking at the behavior of the microstructure subjected to an assigned load history [41]. Bounds on overall instantaneous elastoplastic properties of composites have been derived by using the finite element method [42].

6.1   Analytical Homogenization

As discussed in the introduction, estimates of the average properties of heterogeneous media can be obtained by various analytical methods. Detailed derivations of the equations fall outside the scope of this book.

Available analytical models vary greatly in complexity and accuracy. Simple analytical models yield formulas for the stiffness $\mathbf{C}$ and compliance $\mathbf{S}$ tensors of the composite [39, (2.9) and (2.12)], such as

$$\mathbf{C} = \sum V_i \, \mathbf{C}^i \mathbf{A}^i \quad ; \quad \sum V_i \mathbf{A}^i = \mathbf{I}$$
$$\mathbf{S} = \sum V_i \, \mathbf{S}^i \mathbf{B}^i \quad ; \quad \sum V_i \mathbf{B}^i = \mathbf{I} \qquad (6.1)$$

where $V_i, \mathbf{C}^i, \mathbf{S}^i$ are the volume fraction, stiffness, and compliance tensors (in contracted notation)[1] of the i-th phase in the composite, respectively, and $\mathbf{I}$ is the 6×6 identity matrix. Furthermore, $\mathbf{A}^i, \mathbf{B}^i$ are the strain and stress concentration tensors (in contracted notation) of the i-th phase [39]. For fiber-reinforced composites, $i = f, m$ represent the fiber and matrix phases, respectively.

6.1.1   Reuss Model

The *Reuss model* (also called rule of mixtures) assumes that the strain tensors[2] in the fiber, matrix, and composite are the same $\varepsilon = \varepsilon^f = \varepsilon^m$; so, the strain concentration tensors are all equal to the 6×6 identity matrix $\mathbf{A}^i = \mathbf{I}$. The rule of mixtures (ROM) formulas for E_1 and ν_{12} are derived and computed in this way.

---

[1]Fourth-order tensors with minor symmetry are represented by a 6×6 matrix taking advantage of contracted notation.

[2]Tensors are indicated by boldface type, or by their components using index notation.

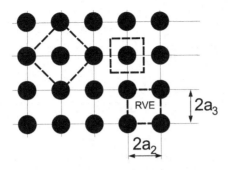

Fig. 6.1: Three possible representative volume elements (RVE) for a composite material with a periodic, square fiber array.

6.1.2 Voigt Model

The *Voigt model* (also called inverse rule of mixtures) assumes that the stress tensors in the fiber, matrix, and composite are the same $\boldsymbol{\sigma} = \boldsymbol{\sigma}^f = \boldsymbol{\sigma}^m$; so, the stress concentration tensors are all equal to the 6×6 identity matrix $\mathbf{B}^i = \mathbf{I}$. The inverse rule of mixtures (IROM) formulas for E_2 and G_{12} are derived and computed in this way. More realistic concentration tensors are given in [9, Appendix B].

6.1.3 Periodic Microstructure Micromechanics

If the composite has periodic microstructure, or if it can be approximated as having such a microstructure (see Section 6.1.4), then Fourier series can be used to estimate all the components of the stiffness tensor of a composite. Explicit formulas for a composite reinforced by isotropic, circular-cylindrical fibers, which are periodically arranged in a square array (Figure 6.1), were developed by [36] and they are presented here. The fibers are aligned with the x_1-axis, and they are equally spaced ($2a_2 = 2a_3$). If the fibers are randomly distributed in the cross-section, the resulting composite has transversely isotropic properties, as explained in Section 6.1.4. The case of a composite reinforced with transversely isotropic fibers is presented in [38] and the resulting equations are implemented in Appendix D.1.

Because the microstructure has square symmetry, the stiffness tensor has six unique coefficients given by

$$C_{11}^* = \lambda_m + 2\,\mu_m - \frac{V_f}{D}\left[\frac{S_3^2}{\mu_m^2} - \frac{2S_6 S_3}{\mu_m^2 g} - \frac{aS_3}{\mu_m\,c} + \frac{S_6^2 - S_7^2}{\mu_m^2 g^2} + \frac{aS_6 + bS_7}{\mu_m\,gc} + \frac{a^2 - b^2}{4\,c^2}\right]$$

$$C_{12}^* = \lambda_m + \frac{V_f}{D}b\left[\frac{S_3}{2c\mu_m} - \frac{S_6 - S_7}{2c\mu_m\,g} - \frac{a+b}{4\,c^2}\right]$$

$$C_{23}^* = \lambda_m + \frac{V_f}{D}\left[\frac{aS_7}{2\,\mu_m\,gc} - \frac{ba + b^2}{4\,c^2}\right]$$

$$C_{22}^* = \lambda_m + 2\,\mu_m - \frac{V_f}{D}\left[-\frac{aS_3}{2\,\mu_m\,c} + \frac{aS_6}{2\,\mu_m\,gc} + \frac{a^2 - b^2}{4\,c^2}\right]$$

$$C_{44}^* = \mu_m - V_f \left[-\frac{2\,S_3}{\mu_m} + (\mu_m - \mu_f)^{-1} + \frac{4\,S_7}{\mu_m(2 - 2\nu_m)} \right]^{-1}$$

$$C_{66}^* = \mu_m - V_f \left[-\frac{S_3}{\mu_m} + (\mu_m - \mu_f)^{-1} \right]^{-1} \tag{6.2}$$

where

$$D = \frac{a S_3^2}{2\,\mu_m^2 c} - \frac{a S_6 S_3}{\mu_m^2 g c} + \frac{a(S_6^2 - S_7^2)}{2\mu_m^2 g^2 c} + \tag{6.3}$$
$$+ \frac{S_3(b^2 - a^2)}{2\,\mu_m c^2} + \frac{S_6(a^2 - b^2) + S_7(ab + b^2)}{2\,\mu_m g c^2} + \frac{(a^3 - 2b^3 - 3\,ab^2)}{8\,c^3}$$

and

$$a = \mu_f - \mu_m - 2\,\mu_f\,\nu_m + 2\,\mu_m\,\nu_f$$
$$b = -\mu_m\,\nu_m + \mu_f\,\nu_f + 2\,\mu_m\,\nu_m\,\nu_f - 2\,\mu_f\,\nu_m\,\nu_f$$
$$c = (\mu_m - \mu_f)(\mu_f - \mu_m + \mu_f\,\nu_f - \mu_m\,\nu_m + 2\,\mu_m\,\nu_f - 2\,\mu_f\,\nu_m +$$
$$+ 2\,\mu_m\,\nu_m\,\nu_f - 2\,\mu_f\,\nu_m\,\nu_f)$$
$$g = (2 - 2\nu_m) \tag{6.4}$$

The subscripts $()_m$, $()_f$ refer to matrix and fiber, respectively. Assuming the fiber and matrix are both isotropic (Section 1.12.5), Lamé constants of both materials are obtained by using (1.75) in terms of the Young's modulus E, the Poisson's ratio ν, and the shear modulus $G = \mu$.

For a composite reinforced by long, circular-cylindrical fibers, periodically arranged in a square array (Figure 6.1), aligned with x_1-axis, with $a_2 = a_3$, the constants S_3, S_6, S_7 are given in [36], as follows

$$S_3 = 0.49247 - 0.47603V_f - 0.02748V_f^2$$
$$S_6 = 0.36844 - 0.14944V_f - 0.27152V_f^2 \tag{6.5}$$
$$S_7 = 0.12346 - 0.32035V_f + 0.23517V_f^2$$

The resulting tensor $\mathbf{C}^*$ has square symmetry due to the microstructural periodic arrangement in the form of a square array. The tensor $\mathbf{C}^*$ is therefore described by six constants. However, most composites have random arrangement of the fibers (see Figure 1.12), resulting in a transversely isotropic stiffness tensor. A generalization for transversely isotropic materials is presented in Section 6.1.4, next.

6.1.4   Transversely Isotropic Averaging

In order to obtain a transversely isotropic stiffness tensor (Section 1.12.4), equivalent in the average sense to the stiffness tensor with square symmetry, the following averaging procedure is used. A rotation θ of the tensor $\mathbf{C}^*$ about the x_1-axis produces

$$\mathbf{B}(\theta) = \overline{T}^T(\theta)\mathbf{C}^*\overline{T}(\theta) \tag{6.6}$$

where $\overline{T}(\theta)$ is the coordinate transformation matrix (see (1.50)). Then, the equivalent transversely isotropic tensor is obtained by averaging, as follows

$$\overline{\mathbf{B}} = \frac{1}{\pi}\int_0^\pi \mathbf{B}(\theta)d\theta \tag{6.7}$$

Then, using the relations between the engineering constants and the components of the $\overline{\mathbf{B}}$ tensor, the following expressions are obtained explicitly in terms of the coefficients (6.2-6.5) of the tensor $\mathbf{C}^*$

$$E_1 = C_{11}^* - \frac{2\,C_{12}^{*2}}{C_{22}^* + C_{23}^*}$$

$$E_2 = \frac{\left(2\,C_{11}^*\,C_{22}^* + 2\,C_{11}^*\,C_{23}^* - 4\,C_{12}^{*2}\right)\left(C_{22}^* - C_{23}^* + 2\,C_{44}^*\right)}{3\,C_{11}^*\,C_{22}^* + C_{11}^*\,C_{23}^* + 2\,C_{11}^*\,C_{44}^* - 4\,C_{12}^{*2}}$$

$$G_{12} = G_{13} = C_{66}^*$$

$$\nu_{12} = \nu_{13} = \frac{C_{12}^*}{C_{22}^* + C_{23}^*}$$

$$\nu_{23} = \frac{C_{11}^*\,C_{22}^* + 3\,C_{11}^*\,C_{23}^* - 2\,C_{11}^*\,C_{44}^* - 4\,C_{12}^{*2}}{3\,C_{11}^*\,C_{22}^* + C_{11}^*\,C_{23}^* + 2\,C_{11}^*\,C_{44}^* - 4\,C_{12}^{*2}} \tag{6.8}$$

Note that the transverse shear modulus G_{23} can be written in terms of the other engineering constants as

$$G_{23} = \frac{C_{22}^*}{4} - \frac{C_{23}^*}{4} + \frac{C_{44}^*}{2} = \frac{E_2}{2(1 + \nu_{23})}$$

or directly in terms of μ_m, μ_f as

$$G_{23} = \mu_m - \frac{f}{4D}\,[(-\frac{aS_3}{2\,\mu_m\,c} + \frac{a(S_7 + S_6)}{2\,\mu_m\,gc} - \frac{ba + 2b^2 - a^2}{4\,c^2})$$

$$+2(-\frac{2\,S_3}{\mu_m} + (\mu_m - \mu_f)^{-1} + \frac{4\,S_7}{\mu_m(2 - 2\nu_m)})^{-1}] \tag{6.9}$$

where D is given by (6.3), $a, b, c,$ and g are given by (6.4), and S_3, S_6, and S_7 can be evaluated by (6.5). These equations are implemented in PMMIE.m and PMMIE.xls [2]. For the case of transversely isotropic fibers, they are implemented in Appendix D.1.

Example 6.1 *Compute the elastic properties of a composite material reinforced with parallel cylindrical fibers randomly distributed in the cross-section. The constituent properties are $E_f = 241\,GPa$, $\nu_f = 0.2$, $E_m = 3.12\,GPa$, $\nu_m = 0.38$, fiber volume fraction $V_f = 0.4$.*

Solution to Example 6.1 *The results shown in Table 6.1 are obtained using Appendix D.1, which implements the periodic microstructure micromechanics (PMM) equations for the case of transversely isotropic fibers.*

Table 6.1: Lamina elastic properties for $V_f = 0.4$.

Young's Moduli	Poisson's Ratio	Shear Moduli
$E_1 = 98,306\ MPa$	$\nu_{12} = \nu_{13} = 0.298$	$G_{12} = G_{13} = 2,594\ MPa$
$E_2 = E_3 = 6,552\ MPa$	$\nu_{23} = 0.600$	

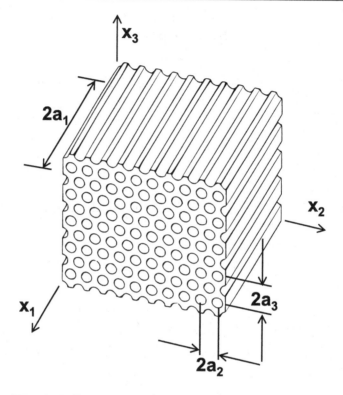

Fig. 6.2: Composite material with hexagonal array.

6.2 Numerical Homogenization

The composite material considered in this section has cylindrical fibers of infinite length, embedded in an elastic matrix, as shown in Figure 6.2. The cross-section of the composite obtained by intersecting with a plane orthogonal to the fiber axis is shown in Figure 6.3, which clearly shows a periodic microstructure. Because of the periodicity, the 3D representative volume element (RVE) shown in Figure 6.4 can be used for FE analysis.

In general, composites reinforced with parallel fibers display orthotropic material properties (Section 1.12.3) at the mesoscale (lamina level). In special cases, such as the hexagonal array shown in Figures 6.2 and 6.3, the properties become transversely isotropic (Section 1.12.4). In most commercially fabricated composites, it is impossible to control the placement of the fibers so precisely and most of the time the resulting microstructure is random, as shown in Figure 1.12. A random microstructure results in transversely isotropic properties at the mesoscale. The analysis of composites with random microstructure still can be done using a

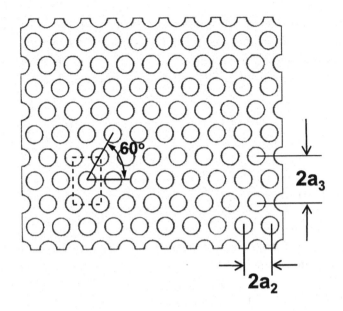

Fig. 6.3: Cross-section of the composite material.

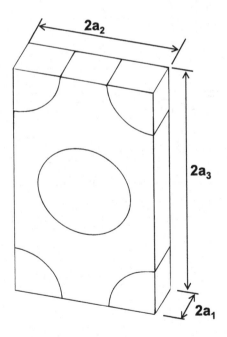

Fig. 6.4: Representative volume element (RVE).

fictitious periodic microstructure, such as that shown in Figure 6.1, then averaging the stiffness tensor $\mathbf{C}$ as in Section 6.1.4 to obtain the stiffness tensor of a transversely isotropic material. A simpler alternative is to assume that the random microstructure is well approximated by the hexagonal microstructure displayed in Figure 6.3. Analysis of such microstructure directly yields a transversely isotropic stiffness tensor, represented by (1.70), which is reproduced here for convenience

$$
\left\{
\begin{array}{c}
\bar{\sigma}_1 \\
\bar{\sigma}_2 \\
\bar{\sigma}_3 \\
\bar{\sigma}_4 \\
\bar{\sigma}_5 \\
\bar{\sigma}_6
\end{array}
\right\}
=
\left[
\begin{array}{cccccc}
C_{11} & C_{12} & C_{12} & 0 & 0 & 0 \\
C_{12} & C_{22} & C_{23} & 0 & 0 & 0 \\
C_{12} & C_{23} & C_{22} & 0 & 0 & 0 \\
0 & 0 & 0 & \frac{1}{2}(C_{22}-C_{23}) & 0 & 0 \\
0 & 0 & 0 & 0 & C_{66} & 0 \\
0 & 0 & 0 & 0 & 0 & C_{66}
\end{array}
\right]
\left\{
\begin{array}{c}
\bar{\epsilon}_1 \\
\bar{\epsilon}_2 \\
\bar{\epsilon}_3 \\
\bar{\gamma}_4 \\
\bar{\gamma}_5 \\
\bar{\gamma}_6
\end{array}
\right\}
\tag{6.10}
$$

where the 1-axis aligned with the fiber direction and an over-bar indicates the average computed over the volume of the RVE. Once the components of the transversely isotropic tensor $\mathbf{C}$ are known, the five elastic properties of the homogenized material can be computed by (6.11), i.e., the longitudinal and transversal Young's moduli E_1 and E_2, the longitudinal and transversal Poisson's ratios ν_{12} and ν_{23}, and the longitudinal shear modulus G_{12}, as follows

$$
\begin{aligned}
E_1 &= C_{11} - 2C_{12}^2/(C_{22}+C_{23}) \\
\nu_{12} &= C_{12}/(C_{22}+C_{23}) \\
E_2 &= \left[C_{11}(C_{22}+C_{23}) - 2C_{12}^2\right](C_{22}-C_{23})/(C_{11}C_{22}-C_{12}^2) \\
\nu_{23} &= \left[C_{11}C_{23} - C_{12}^2\right]/(C_{11}C_{22}-C_{12}^2) \\
G_{12} &= C_{66}
\end{aligned}
\tag{6.11}
$$

The shear modulus G_{23} in the transversal plane can be obtained by the classical relation (1.74), or directly as follows

$$
G_{23} = C_{44} = \frac{1}{2}(C_{22}-C_{23}) = \frac{E_2}{2(1+\nu_{23})}
\tag{6.12}
$$

In order to evaluate the overall elastic matrix $\mathbf{C}$ of the composite, the RVE is subjected to an average strain $\bar{\epsilon}_\beta$ [43]. The six components of strain ε_{ij}^0 are applied by enforcing the following boundary conditions (BC) on the displacement components

$$
u_i(a_1, x_2, x_3) - u_i(-a_1, x_2, x_3) = 2a_1\varepsilon_{i1}^0 \qquad
\begin{array}{c}
-a_2 \le x_2 \le a_2 \\
-a_3 \le x_3 \le a_3
\end{array}
\tag{6.13}
$$

$$
u_i(x_1, a_2, x_3) - u_i(x_1, -a_2, x_3) = 2a_2\varepsilon_{i2}^0 \qquad
\begin{array}{c}
-a_1 \le x_1 \le a_1 \\
-a_3 \le x_3 \le a_3
\end{array}
\tag{6.14}
$$

$$
u_i(x_1, x_2, a_3) - u_i(x_1, x_2, -a_3) = 2a_3\varepsilon_{i3}^0 \qquad
\begin{array}{c}
-a_1 \le x_1 \le a_1 \\
-a_2 \le x_2 \le a_2
\end{array}
\tag{6.15}
$$

Note that tensor components of strain, defined in (1.5), are used in (6.13-6.15). Also, note that a superscript $()^0$ indicates an *applied* strain, while an over-line indicates a volume average. Furthermore, $2a_j \, \varepsilon_{ij}^0$ is the displacement necessary to enforce a strain ε_{ij}^0 over a distance $2a_j$ (Figure 6.4).

The strain ε_{ij}^0 applied on the boundary by using (6.13-6.15) results in a complex state of strain inside the RVE. However, the volume average of the strain in the RVE equals the applied strain, i.e.,

$$\bar{\varepsilon}_{ij} = \frac{1}{V} \int_V \varepsilon_{ij} dV = \varepsilon_{ij}^0 \tag{6.16}$$

For the homogeneous composite material, the relationship between average stress and strain is

$$\bar{\sigma}_\alpha = C_{\alpha\beta} \, \bar{\varepsilon}_\beta \tag{6.17}$$

where the relationship between $i, j = 1..3$ and $\beta = 1..6$ is given by the definition of contracted notation in (1.9). Thus, the components of the tensor $\mathbf{C}$ are determined solving six elastic models of the RVE subjected to the boundary conditions (6.13-6.15), where only one component of the strain ϵ_β^0 is different from zero for each of the six problems.

By choosing a unit value of applied strain, and once the problem defined by the boundary conditions (6.13-6.15) is solved, it is possible to compute the stress field σ_α, whose average gives the required components of the elastic matrix, one column at a time, as

$$C_{\alpha\beta} = \bar{\sigma}_\alpha = \frac{1}{V} \int_V \sigma_\alpha \, (x_1, x_2, x_3) \, dV \quad \text{with } \epsilon_\beta^0 = 1 \tag{6.18}$$

where $\alpha, \beta = 1 \dots 6$ (see Section 1.5). The integrals (6.18) are evaluated within each finite element using the Gauss-Legendre quadrature. Commercial programs, such as Abaqus, have the capability to compute the average stress and volume, element by element. Therefore, computation of the integral (6.18) is a trivial matter. For more details see Example 6.2, p. 232.

The coefficients in $\mathbf{C}$ are found by setting a different problem for each column in (6.10), as follows

First Column of C

In order to determine the components C_{i1}, with $i = 1, 2, 3$, the following strain is applied to stretch the RVE in the fiber direction (x_1-direction)

$$\epsilon_1^0 = 1 \qquad \epsilon_2^0 = \epsilon_3^0 = \gamma_4^0 = \gamma_5^0 = \gamma_6^0 = 0 \tag{6.19}$$

Thus, the displacement boundary conditions (6.13-6.15) for the RVE in Figure 6.4 become

$$\begin{aligned} u_1\left(+a_1, x_2, x_3\right) - u_1\left(-a_1, x_2, x_3\right) &= 2a_1 \\ u_2\left(+a_1, x_2, x_3\right) - u_2\left(-a_1, x_2, x_3\right) &= 0 \\ u_3\left(+a_1, x_2, x_3\right) - u_3\left(-a_1, x_2, x_3\right) &= 0 \end{aligned} \qquad \begin{aligned} -a_2 \le x_2 \le a_2 \\ -a_3 \le x_3 \le a_3 \end{aligned}$$

$$u_i\left(x_1, +a_2, x_3\right) - u_i\left(x_1, -a_2, x_3\right) = 0 \qquad \begin{aligned} -a_1 \le x_1 \le a_1 \\ -a_3 \le x_3 \le a_3 \end{aligned} \qquad (6.20)$$

$$u_i\left(x_1, x_2, +a_3\right) - u_i\left(x_1, x_2, -a_3\right) = 0 \qquad \begin{aligned} -a_1 \le x_1 \le a_1 \\ -a_2 \le x_2 \le a_2 \end{aligned}$$

The conditions (6.20) are constraints on the relative displacements between opposite faces of the RVE. Because of the symmetries of the RVE and symmetry of the constraints (6.20), only one-eighth of the RVE needs to be modeled in FEA. Assuming the top-right-front portion is modeled (Figure 6.5), the following equivalent external boundary conditions, i.e., boundary conditions on components of displacements and stresses, can be used

$$\begin{aligned} u_1\left(a_1, x_2, x_3\right) &= a_1 \\ u_1\left(0, x_2, x_3\right) &= 0 \\ \sigma_{12}\left(a_1, x_2, x_3\right) &= 0 \\ \sigma_{12}\left(0, x_2, x_3\right) &= 0 \\ \sigma_{13}\left(a_1, x_2, x_3\right) &= 0 \\ \sigma_{13}\left(0, x_2, x_3\right) &= 0 \end{aligned} \qquad \begin{aligned} 0 \le x_2 \le a_2 \\ 0 \le x_3 \le a_3 \end{aligned}$$

$$\begin{aligned} u_2\left(x_1, a_2, x_3\right) &= 0 \\ u_2\left(x_1, 0, x_3\right) &= 0 \\ \sigma_{21}\left(x_1, a_2, x_3\right) &= 0 \\ \sigma_{21}\left(x_1, 0, x_3\right) &= 0 \\ \sigma_{23}\left(x_1, a_2, x_3\right) &= 0 \\ \sigma_{23}\left(x_1, 0, x_3\right) &= 0 \end{aligned} \qquad \begin{aligned} 0 \le x_1 \le a_1 \\ 0 \le x_3 \le a_3 \end{aligned} \qquad (6.21)$$

$$\begin{aligned} u_3\left(x_1, x_2, a_3\right) &= 0 \\ u_3\left(x_1, x_2, 0\right) &= 0 \\ \sigma_{31}\left(x_1, x_2, a_3\right) &= 0 \\ \sigma_{31}\left(x_1, x_2, 0\right) &= 0 \\ \sigma_{32}\left(x_1, x_2, a_3\right) &= 0 \\ \sigma_{32}\left(x_1, x_2, 0\right) &= 0 \end{aligned} \qquad \begin{aligned} 0 \le x_1 \le a_1 \\ 0 \le x_2 \le a_2 \end{aligned}$$

These boundary conditions are very easy to apply. Symmetry boundary conditions are applied on the planes $x_1 = 0$, $x_2 = 0$, $x_3 = 0$. Then, a uniform displacement is applied on the plane $x_1 = a_1$. The null stress boundary conditions do not need to be applied explicitly in a displacement-based formulation. The displacement components in (6.21) represent strains that are non-zero along the x_1-direction and zero along the other two directions. The stress boundary conditions listed in (6.21) reflect the fact that, in the coordinate system used, the composite material

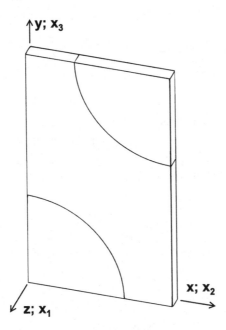

Fig. 6.5: One-eighth model of the RVE. Note that the model is set up with the fiber along the z-axis, which corresponds to the x_1-direction in the equations.

is macroscopically orthotropic and that the constituent materials are orthotropic, too. Therefore, there is no coupling between extension and shear strains. This is evidenced by the zero coefficients above the diagonal in columns 4 to 6 in (6.10).

The coefficients in the first column of (6.10) are found by using (6.18), as follows

$$C_{\alpha 1} = \overline{\sigma}_\alpha = \frac{1}{V} \int_V \sigma_\alpha (x_1, x_2, x_3) \ dV \qquad (6.22)$$

Second Column of C

The components $C_{\alpha 2}$, with $\alpha = 1, 2, 3$, are determined by setting

$$\epsilon_2^o = 1 \qquad \epsilon_1^o = \epsilon_3^o = \gamma_4^o = \gamma_5^o = \gamma_6^o = 0 \qquad (6.23)$$

Thus, the following boundary conditions on displacements can be used

$$
\begin{aligned}
u_1 (a_1, x_2, x_3) &= 0 \\
u_1 (0, x_2, x_3) &= 0 \\
u_2 (x_1, a_2, x_3) &= a_2 \\
u_2 (x_1, 0, x_3) &= 0 \\
u_3 (x_1, x_2, a_3) &= 0 \\
u_3 (x_1, x_2, 0) &= 0
\end{aligned}
\qquad (6.24)
$$

The trivial stress boundary conditions have not been listed because they are automatically enforced by the displacement-based FEA formulation. Using (6.18), the stiffness terms in the second column of **C** are computed as

$$C_{\alpha 2} = \overline{\sigma}_\alpha = \frac{1}{V} \int_V \sigma_{\alpha 2}\, (x_1, x_2, x_3) \ dV \qquad (6.25)$$

Third Column of C

Because the material is transversely isotropic (6.10), the components of the third column of the matrix $\mathbf{C}$ can be determined from the first and the second column, so no further computation is required. However, if desired, the components $C_{\alpha 3}$, with $\alpha = 1, 2, 3$, can be found by applying the following strain

$$\epsilon_3^o = 1 \qquad \epsilon_1^o = \epsilon_2^o = \gamma_4^o = \gamma_4^o = \gamma_5^o = 0 \qquad (6.26)$$

Thus, the following boundary conditions on displacement can be used

$$
\begin{aligned}
u_1\,(a_1, x_2, x_3) &= 0 \\
u_1\,(0, x_2, x_3) &= 0 \\
u_2\,(x_1, a_2, x_3) &= 0 \\
u_2\,(x_1, 0, x_3) &= 0 \\
u_3\,(x_1, x_2, a_3) &= a_3 \\
u_3\,(x_1, x_2, 0) &= 0
\end{aligned}
\qquad (6.27)
$$

The required components of $\mathbf{C}$ are determined by averaging the stress field as in (6.18).

Example 6.2 *Compute E_1, E_2, ν_{12}, and ν_{23} for a unidirectional composite with isotropic fibers $E_f = 241$ GPa, $\nu_f = 0.2$, and isotropic matrix $E_m = 3.12$ GPa, $\nu_m = 0.38$ with fiber volume fraction $V_f = 0.4$. The fiber diameter is $d_f = 7$ μm, placed in an hexagonal array as shown in Figure 6.3.*

Solution to Example 6.2 *Watch the video in [1].*
The dimensions a_2 and a_3 of the RVE, as shown in Figure 6.4, are chosen to obtain $V_f = 0.4$ with an hexagonal array microstructure. The fiber volume and the total volume of the RVE are

$$v_f = 4 a_1 \pi \left(\frac{d_f}{2} \right)^2 \quad ; \quad v_t = 2 a_1\, 2 a_2\, 2 a_3$$

The ratio between both is the volume fraction. Therefore,

$$V_f = \pi \frac{(d_f/2)^2}{2\, a_2\, a_3} = 0.4$$

Additionally, the relation between a_2 and a_3 is established by the hexagonal array pattern

$$a_3 = a_2 \tan(60°)$$

These two relations yield a_2 and a_3, while the a_1 dimension can be chosen arbitrarily. In this case, the RVE dimensions are

$$a_1 = a_2/4 = 1.3175\ \mu m \quad ; \quad a_2 = 5.270\ \mu m \quad ; \quad a_3 = 9.128\ \mu m$$

Equations (6.18) will be used to obtain the coefficients $C_{11}, C_{12}, C_{13}, C_{22}, C_{23},$ and C_{33} by applying unit strains $\epsilon_1^0, \epsilon_2^0, \epsilon_3^0$, one at a time. In the absence of applied shear deformations, the average deformations are symmetric with respect to the three coordinate planes. The RVE in Figure 6.4 is also geometrically symmetric with respect to the three coordinate planes. Therefore, it is possible to model one-eighth of the RVE, as shown in Figure 6.5. The procedure used to model one-eighth of the RVE in Abaqus is given below.

i. Setting the Work Directory

```
Menu: File, Set Work Directory, [C:\SIMULIA\user\Ex_6.2]
Menu: File, Save As, [C:\SIMULIA\user\Ex_6.2\Ex_6.2.cae]
```

ii. Creating parts

```
# Values used below, replace the variables accordingly.
# values in microns
# a_1 = 1.3175        # half RVE along x_1 (extrusion direction)
# a_2 = 5.270         # half RVE along x_2  (micrometers)
# a_3 = 9.128         # half RVE along x_3  (micrometers)
# r_f = 3.5           # fiber radius (micrometers)
Module: Part
Menu:   Part, Create
    3D, Deformable, Solid, Extrusion, Cont
Menu:   Add, Line, Rectangle, [0,0], [a_2, a_3], X, Done
    Depth [a_1], OK
Menu:   Tools, Partition
    Cell, Sketch planar partition, Sketch Origin: Specify
    # pick: front face (z=a_1), Sketch origin X,Y,Z: [0,0,a_1]
    Select and edge: vertical and on the right, # pick: right vertical
Menu:   Add, Circle
    [0,0], [0,r_f]
    [a_2,a_3], [a_2+r_f,a_3], X, Done #a_2+r_f=8.77
    Extrude/Sweep, # pick: lines forming lower quarter-circle, Done
    Extrude Along Direction, # pick edge parallel to Z-global, OK
    Create Partition
    # pick: top cell (larger cell), Done
    # pick: the lines forming the upper quarter-circle, Done
    Extrude Along Direction, # pick edge parallel to Z-global, OK
    Create Partition, Done, # close Create Partition pop-up window
```

iii. Defining materials, sections, and assigning sections to parts

```
Module: Property
Menu:   Material, Create
    Name [Fiber], Mechanical, Elasticity, Elastic, Type: Isotropic
    [0.241, 0.2], OK #TPa
Menu:   Material, Create
    Name [Matrix], Mechanical, Elasticity, Elastic, Type: Isotropic
    [0.00312, 0.38], OK #TPa
Menu:   Section, Create
    Name [Fiber], Solid, Homogeneous, Cont, Material: Fiber, OK
Menu:   Section, Create
    Name [Matrix], Solid, Homogeneous, Cont, Material: Matrix, OK
```

```
Menu:    Assign, Section
     # pick: the matrix, Done, Section: Matrix, OK
     # pick: both fibers, Done, Section: Fiber, OK, Done
```

iv. *Creating assembly*

```
Module: Assembly
Menu:    Instance, Create, Independent, OK
```

v. *Defining analysis steps. We need three steps, one for each column in Section 6.2. To solve this example, it will be necessary to include output variables* S *and* IVOL *on the Output Requests.*

```
Module: Step
Menu:    Step, Manager
     Create, Name [Column-1], Insert new step after: Initial
     Procedure type: Linear perturbation, Static, Cont, OK
     Create, Name [Column-2], Insert new step after: Column-1
     Procedure type: Linear perturbation, Static, Cont, OK
     Create, Name [Column-3], Insert new step after: Column-2
     Procedure type: Linear perturbation, Static, Cont, OK
     # close Step Manager pop-up window
Step:    Column-1
Menu:    Output, Field Output Requests, Edit, F-Output-1
         Edit variables, [S,E,U,IVOL], OK
Step:    Column-2
Menu:    Output, Field Output Requests, Edit, F-Output-1
         Edit variables, [S,E,U,IVOL], OK
Step:    Column-3
Menu:    Output, Field Output Requests, Edit, F-Output-1
         Edit variables, [S,E,U,IVOL], OK
```

vi. *Adding loads and boundary conditions.*
The boundary conditions are defined in three load steps, which are then used to obtain the coefficients $C_{\alpha\beta}$ in columns one, two, and three. A unit strain is applied along each direction, each time. Equations (6.13–6.15,6.18) are then used to obtain the stiffness coefficients. The procedure is shown below.

```
Module: Load
     # each step has its own set of BC
     # for step: Column-1
Menu:    BC, Manager
     Create, Name [xsymm-C1], Step: Column-1, Mechanical,
         Symm/Anti/Enca, Cont
     # pick: faces @x=0 and a_2, Done, XSYMM, OK
     #
     # for step: Column-1
     Create, Name [ysymm-C1], Step: Column-1, Mechanical,
         Symm/Anti/Enca, Cont
     # pick: faces @y=0 and a_3, Done, YSYMM, OK
     Create, Name [zsymm-C1], Step: Column-1, Mechanical,
         Symm/Anti/Enca, Cont
     # pick: face @z=0, Done, ZSYMM, OK, # note only one face is picked
```

```
# imposed displacement(unit strain)
Create, Name [disp-C1], Step: Column-1, Mech., Disp/Rota, Cont
# pick: faces @z=a_1, Done, # Checkmark: U3 [a_1], OK
#
# for step: Column-2
Create, Name [zsymm-C2], Step: Column-2, Mechanical,
    Symm/Anti/Enca, Cont
# pick: faces @z=0, and a_1, Done, ZSYMM, OK
Create, Name [ysymm-C2], Step: Column-2, Mechanical,
    Symm/Anti/Enca, Cont
# pick: faces @y=0 and a_3, Done, YSYMM, OK
Create, Name [xsymm-C2], Step: Column-2, Mechanical,
    Symm/Anti/Enca, Cont
# pick: face @x=0, Done, XSYMM, OK, # note only one face is picked
# imposed displacement(unit strain)
Create, Name [disp-C2], Step: Column-2, Mech., Disp/Rota, Cont
# pick: face @x=a_2, Done, # Checkmark: U1 [a_2], OK
#
# for step: Column-3
Create, Name [xsymm-C3], Step: Column-3, Mechanical,
    Symm/Anti/Enca, Cont
# pick: faces @x=0, and a_2, Done, XSYMM, OK
Create, Name [zsymm-C3], Step: Column-3, Mechanical,
    Symm/Anti/Enca, Cont
# pick: faces @z=0, and a_1, Done, ZSYMM, OK
Create, Name [ysymm-C3], Step: Column-3, Mechanical,
    Symm/Anti/Enca, Cont
# pick: face @y=0, Done, YSYMM, OK, # note only one face is picked
# imposed displacement(unit strain)
Create, Name [disp-C3], Step: Column-3, Mech., Disp/Rota, Cont
# pick: face @y=a_3, Done, # Checkmark: U2 [a_3], OK
# close the BC Manager pop-up window
```

vii. Meshing the model

```
Module: Mesh
Menu:   Seed, Instance, Approximate global size [0.5], OK
Menu:   Mesh, Controls, # select all cells, Done
    Element Shape: Hex, Technique: Structured, OK
Menu:   Mesh, Element Type, # select all cells, Done
    Geometric Order: Linear, # checkmark: Reduced integration, OK
Menu:   Mesh, Instance, Yes
```

viii. Solving and visualizing the results

```
Module: Job
Menu:   Job, Manager
    Create, Cont, OK, Submit, # when Completed, Results
```

The coefficients $C_{\alpha\beta}$ and the laminate elastic constants are computed using the Python script given below, which can be executed by typing or pasting in the Python shell. Alternatively, one can execute any Python script by doing **Menu: File, Run Script**. The script *srecover.py is available in [2] and it is used in this example. Now, click >>> to switch from the Progress Window to the Python shell. Then, type or copy the lines below to the Python shell.*

```
# srecover.py
from visualization import *
# Open the Output Data Base for the current Job
odb = openOdb(path='Job-1.odb');
myAssembly = odb.rootAssembly;
#
# Creating a temporary variable to hold the frame repository
# provides the same functionality and speeds up the process
# Column-1
frameRepository = odb.steps['Column-1'].frames;
frameS=[];
frameIVOL=[];
# Get only the last frame [-1]
frameS.insert(0,frameRepository[-1].fieldOutputs['S']
    .getSubset(position=INTEGRATION_POINT));
frameIVOL.insert(0,frameRepository[-1].fieldOutputs['IVOL']
    .getSubset(position=INTEGRATION_POINT));
# Total Volume
Tot_Vol=0;
# Stress Sum
Tot_Stress=0;
#
for II in range(0,len(frameS[-1].values)):
    Tot_Vol=Tot_Vol+frameIVOL[0].values[II].data;
    Tot_Stress=Tot_Stress+frameS[0].values[II].data * frameIVOL[0]\
        .values[II].data;

# Calculate Average
Avg_Stress = Tot_Stress/Tot_Vol;
print 'Abaqus/Standard Stress Tensor Order:'
# from Abaqus Analysis User's Manual - 1.2.2 Conventions -
# Convention used for stress and strain components
print 'Average stresses Global CSYS: 11-22-33-12-13-23';
print Avg_Stress;
C11 = Avg_Stress[2]#z-component,1-direction
C21 = Avg_Stress[0]#x-component,2-direction
C31 = Avg_Stress[1]#y-component,3-direction in Fig. 6.5

# Column-2
frameRepository = odb.steps['Column-2'].frames;
frameS=[];
frameIVOL=[];
# Get only the last frame [-1]
frameS.insert(0,frameRepository[-1].fieldOutputs['S']
    .getSubset(position=INTEGRATION_POINT));
frameIVOL.insert(0,frameRepository[-1].fieldOutputs['IVOL']
    .getSubset(position=INTEGRATION_POINT));
```

```
# Total Volume
Tot_Vol=0;
# Stress Sum
Tot_Stress=0;
#
for II in range(0,len(frameS[-1].values)):
     Tot_Vol=Tot_Vol+frameIVOL[0].values[II].data;
     Tot_Stress=Tot_Stress+frameS[0].values[II].data * frameIVOL[0]\
         .values[II].data;

# Calculate Average
Avg_Stress = Tot_Stress/Tot_Vol;
print 'Abaqus/Standard Stress Tensor Order:'
print 'Average stresses Global CSYS: 11-22-33-12-13-23';
print Avg_Stress;
C12 = Avg_Stress[2]#z-component,1-direction
C22 = Avg_Stress[0]#x-component,2-direction
C32 = Avg_Stress[1]#y-component,3-direction in Fig. 6.5

# Column-3
frameRepository = odb.steps['Column-3'].frames;
frameS=[];
frameIVOL=[];
# Get only the last frame [-1]
frameS.insert(0,frameRepository[-1].fieldOutputs['S']
     .getSubset(position=INTEGRATION_POINT));
frameIVOL.insert(0,frameRepository[-1].fieldOutputs['IVOL']
     .getSubset(position=INTEGRATION_POINT));
# Total Volume
Tot_Vol=0;
# Stress Sum
Tot_Stress=0;
#
for II in range(0,len(frameS[-1].values)):
     Tot_Vol=Tot_Vol+frameIVOL[0].values[II].data;
     Tot_Stress=Tot_Stress+frameS[0].values[II].data * frameIVOL[0]\
         .values[II].data;

# Calculate Average
Avg_Stress = Tot_Stress/Tot_Vol;
print 'Abaqus/Standard Stress Tensor Order:'
print 'Average stresses in Global CSYS: 11-22-33-12-13-23';
print Avg_Stress;
C13 = Avg_Stress[2]#z-component,1-direction
C23 = Avg_Stress[0]#x-component,2-direction
C33 = Avg_Stress[1]#y-component,3-direction in Fig. 6.5
#
EL=C11-2*C12*C21/(C22+C23)              # Longitudinal E1 modulus
nuL=C12/(C22+C23)                       # 12 Poisson coefficient
ET=(C11*(C22+C23)-2*C12*C12)*(C22-C23)/(C11*C22-C12*C21)      #E2
nuT=(C11*C23-C12*C21)/(C11*C22-C12*C21) # 23 Poisson coefficient
```

Table 6.2: Calculated elastic properties of the unidirectional lamina.

Property	PMM [44]	Abaqus
E_1 [TPa]	0.098306	0.098197
E_2 [TPa]	0.006552	0.007472
ν_{12}	0.298	0.299
ν_{23}	0.6	0.540
G_{12} [TPa]	0.002594	(*)
G_{23} [TPa]	0.002906	0.002406

(*) Not possible with the boundary conditions used in this example. See Example 6.3.

```
GT=(C22-C23)/2 # or GT=ET/2/(1+nuT)      # 23 Shear stiffness
#
print "If Moduli are in TPa and dimensions in microns, results are in TPa"
print "E1=",EL
print "E2=",ET
print "PR12=",nuL
print "PR23=",nuT
# end srecover.py
```

The results are shown in Table 6.2. If you get any errors, check that the blank lines following each of the **for** loops are completely empty. These blank lines are needed as terminators for the **for** loops in *Python*. Otherwise, try running the script as follows: Menu: File, Run Script: [srecover.py], OK.

The modeling procedure for Example 6.2 is quite involved. If one wishes to perform a parametric study, repeating this process on an interactive session using the CAE graphical user interface (GUI) would be very time consuming and prone to error. Instead, it is possible to capture the Python script generated by CAE during an interactive session and use it to automate the process.

Fourth Column of C

For a transversely isotropic material, according to (6.10), only the term C_{44} is expected to be different from zero and it can be determined as a function of the other components, so no further computation is needed. Therefore, it can be determined as

$$C_{44} = \frac{1}{2}(C_{22} - C_{23}) \tag{6.28}$$

If the material is orthotropic, a procedure similar to that used for the sixth column must be used.

Fifth Column of C

For a transversely isotropic material, according to (6.10), only the term C_{55} is different from zero, and it is equal to C_{66}, which can be found from the sixth column. If the material is orthotropic, a procedure similar to that used for the sixth column must be used.

Sixth Column of C

Because of the lack of symmetry of the loads, in this case it is not possible to use boundary conditions as was done for the first three columns. Thus, the boundary conditions must be enforced by using coupling constraint equations (called CE in most FEA commercial packages).

According to (6.10), only the term C_{66} is different from zero. The components $C_{\alpha 6}$ are determined by setting

$$\gamma_6^0 = \varepsilon_{12}^0 + \varepsilon_{21}^0 = 1.0 \qquad \epsilon_1^0 = \epsilon_2^0 = \epsilon_3^0 = \gamma_4^0 = \gamma_5^0 = 0 \qquad (6.29)$$

Note that $\varepsilon_{12}^0 = 1/2$ is applied between $x_1 = \pm a_1$ and another one-half is applied between $x_2 = \pm a_2$. In this case, the CE applied between two periodic faces (except points in the edges and vertices) is given as a particular case of (6.13-6.15), as follows

$$
\begin{aligned}
u_1\left(a_1, x_2, x_3\right) - u_1\left(-a_1, x_2, x_3\right) &= 0 & &-a_2 < x_2 < a_2 \\
u_2\left(a_1, x_2, x_3\right) - u_2\left(-a_1, x_2, x_3\right) &= a_1 & &-a_3 < x_3 < a_3 \\
u_3\left(a_1, x_2, x_3\right) - u_3\left(-a_1, x_2, x_3\right) &= 0 & & \\
\\
u_1\left(x_1, a_2, x_3\right) - u_1\left(x_1, -a_2, x_3\right) &= a_2 & &-a_1 < x_1 < a_1 \\
u_2\left(x_1, a_2, x_3\right) - u_2\left(x_1, -a_2, x_3\right) &= 0 & &-a_3 < x_3 < a_3 \\
u_3\left(x_1, a_2, x_3\right) - u_3\left(x_1, -a_2, x_3\right) &= 0 & & \\
\\
u_1\left(x_1, x_2, a_3\right) - u_1\left(x_1, x_2, -a_3\right) &= 0 & &-a_1 < x_1 < a_1 \\
u_2\left(x_1, x_2, a_3\right) - u_2\left(x_1, x_2, -a_3\right) &= 0 & &-a_2 < x_2 < a_2 \\
u_3\left(x_1, x_2, a_3\right) - u_3\left(x_1, x_2, -a_3\right) &= 0 & &
\end{aligned}
\qquad (6.30)
$$

Note that (6.30) are applied between opposite points on the faces of the RVE but not on edges and vertices. In FEA, CE are applied between degrees of freedom (DOF). Once a DOF has been used in a CE, it cannot be used in another CE. For example, the first of (6.30) for $x_2 = a_2$ becomes

$$u_1(a_1, a_2, x_3) - u_1(-a_1, a_2, x_3) = 0 \qquad (6.31)$$

The DOF associated to $u_1(a_1, a_2, x_3)$ (for all $-a_3 < x_3 < a_3$) are eliminated because they are identical to $u_1(-a_1, a_2, x_3)$, as required by (6.31) and enforced by a CE based on the same. Once the DOF are eliminated, they cannot be used in another CE. For example, the fourth of (6.30) at $x_1 = a_1$ is

$$u_1(a_1, a_2, x_3) - u_1(a_1, -a_2, x_3) = a_2 \qquad (6.32)$$

but this CE cannot be enforced because the DOF associated to $u_1(a_1, a_2, x_3)$ have been eliminated by the CE associated to (6.31). As a corollary, constraint equations on the edges and vertices of the RVE must be written separately from (6.30).

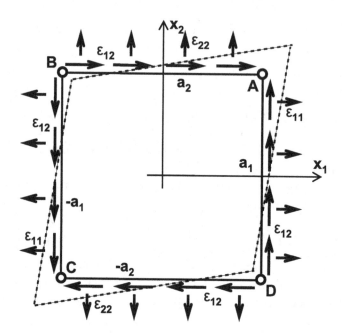

Fig. 6.6: Top view of the RVE showing that two displacements (vertical and horizontal) must be applied at edges to impose shear strain (shown as points A, B, C, and D in the figure).

Furthermore, only three equations, one for each component of displacement u_i, can be written between a pair of edges or pair of vertices. Simply put, there are only three displacements that can be used to enforce periodicity conditions.

For pairs of edges, the task at hand is to reduce the first six equations of (6.30) to three equations that can be applied between pairs of edges for the interval $-a_3 < x_3 < a_3$. Note that the new equations will not be applied at $x_3 = \pm a_3$ because those are vertices, which will be dealt with separately. Therefore, the last three equations of (6.30) are inconsequential at this point.

The only way to reduce six equations to three, in terms of six unique DOF, is to add the equations for diagonally opposite edges. Figure 6.6 is a top view of the RVE looking from the positive x_3-axis. Point A in Figure 6.6 represents the edge formed by the planes $x_1 = a_1$ and $x_2 = a_2$. This location is constrained by the first of (6.30) at that location, which is precisely (6.31). Point C in Figure 6.6 represents the edge formed by the planes $x_1 = -a_1$ and $x_2 = -a_2$. This location is constrained by the fourth of (6.30), which at that location reduces to

$$u_1(-a_1, a_2, x_3) - u_1(-a_1, -a_2, x_3) = a_2 \qquad (6.33)$$

Adding (6.31) and (6.33) yields a single equation, as follows

$$u_1(a_1, a_2, x_3) - u_1(-a_1, -a_2, x_3) = a_2 \qquad (6.34)$$

Repeating the procedure for the components u_2 and u_3, and grouping the resulting equations with (6.34), results in

$$
\begin{aligned}
u_1\left(a_1, a_2, x_3\right) - u_1\left(-a_1, -a_2, x_3\right) &= a_2 \\
u_2\left(a_1, a_2, x_3\right) - u_2\left(-a_1, -a_2, x_3\right) &= a_1 \qquad -a_3 < x_3 < a_3 \qquad (6.35)\\
u_3\left(a_1, a_2, x_3\right) - u_3\left(-a_1, -a_2, x_3\right) &= 0
\end{aligned}
$$

Considering (6.30) between edges B and D in Figure 6.6 results in

$$
\begin{aligned}
u_1\left(a_1, -a_2, x_3\right) - u_1\left(-a_1, a_2, x_3\right) &= -a_2 \\
u_2\left(a_1, -a_2, x_3\right) - u_2\left(-a_1, a_2, x_3\right) &= a_1 \qquad -a_3 < x_3 < a_3 \qquad (6.36)\\
u_3\left(a_1, -a_2, x_3\right) - u_3\left(-a_1, a_2, x_3\right) &= 0
\end{aligned}
$$

The planes $x_1 = \pm a_1$ and $x_3 = \pm a_3$ define two pairs of edges restrained by the following six CE

$$
\begin{aligned}
u_1\left(+a_1, x_2, +a_3\right) - u_1\left(-a_1, x_2, -a_3\right) &= 0 \\
u_2\left(+a_1, x_2, +a_3\right) - u_2\left(-a_1, x_2, -a_3\right) &= a_1 \qquad -a_2 < x_2 < a_2 \\
u_3\left(+a_1, x_2, +a_3\right) - u_3\left(-a_1, x_2, -a_3\right) &= 0 \\
&\qquad\qquad\qquad\qquad\qquad\qquad (6.37)\\
u_1\left(+a_1, x_2, -a_3\right) - u_1\left(-a_1, x_2, +a_3\right) &= 0 \\
u_2\left(+a_1, x_2, -a_3\right) - u_2\left(-a_1, x_2, +a_3\right) &= a_1 \qquad -a_2 < x_2 < a_2 \\
u_3\left(+a_1, x_2, -a_3\right) - u_3\left(-a_1, x_2, +a_3\right) &= 0
\end{aligned}
$$

The six CE for the two pairs of edges defined by the planes $x_2 = \pm a_2$ and $x_3 = \pm a_3$ are

$$
\begin{aligned}
u_1\left(x_1, +a_2, +a_3\right) - u_1\left(x_1, -a_2, -a_3\right) &= a_2 \\
u_2\left(x_1, +a_2, +a_3\right) - u_2\left(x_1, -a_2, -a_3\right) &= 0 \qquad -a_1 < x_1 < a_1 \\
u_3\left(x_1, +a_2, +a_3\right) - u_3\left(x_1, -a_2, -a_3\right) &= 0 \\
&\qquad\qquad\qquad\qquad\qquad\qquad (6.38)\\
u_1\left(x_1, +a_2, -a_3\right) - u_1\left(x_1, -a_2, +a_3\right) &= a_2 \\
u_2\left(x_1, +a_2, -a_3\right) - u_2\left(x_1, -a_2, +a_3\right) &= 0 \qquad -a_1 < x_1 < a_1 \\
u_3\left(x_1, +a_2, -a_3\right) - u_3\left(x_1, -a_2, +a_3\right) &= 0
\end{aligned}
$$

Note that (6.35-6.38) are not applied at the vertices because redundant CE would appear among pairs of vertices that are located symmetrically with respect to the center of the RVE's volume. Therefore, each of the four pairs of vertices need to be constrained one at a time. The resulting CE are

$$u_1(+a_1, +a_2, +a_3) - u_1(-a_1, -a_2, -a_3) = a_2$$
$$u_2(+a_1, +a_2, +a_3) - u_2(-a_1, -a_2, -a_3) = a_1$$
$$u_3(+a_1, +a_2, +a_3) - u_3(-a_1, -a_2, -a_3) = 0$$

$$u_1(+a_1, +a_2, -a_3) - u_1(-a_1, -a_2, +a_3) = a_2$$
$$u_2(+a_1, +a_2, -a_3) - u_2(-a_1, -a_2, +a_3) = a_1$$
$$u_3(+a_1, +a_2, -a_3) - u_3(-a_1, -a_2, +a_3) = 0$$

$$\quad (6.39)$$

$$u_1(-a_1, +a_2, +a_3) - u_1(+a_1, -a_2, -a_3) = a_2$$
$$u_2(-a_1, +a_2, +a_3) - u_2(+a_1, -a_2, -a_3) = -a_1$$
$$u_3(-a_1, +a_2, +a_3) - u_3(+a_1, -a_2, -a_3) = 0$$

$$u_1(+a_1, -a_2, +a_3) - u_1(-a_1, +a_2, -a_3) = -a_2$$
$$u_2(+a_1, -a_2, +a_3) - u_2(-a_1, +a_2, -a_3) = a_1$$
$$u_3(+a_1, -a_2, +a_3) - u_3(-a_1, +a_2, -a_3) = 0$$

Equations (6.30-6.39) constrain the volume of the RVE with a unit strain given by (6.29). The FEA of this model yields all the components of stress. As discussed previously, element by element averages of these components of stress are available from the FEA (see macro `srecover` in Example 6.2, p. 232) or they can be easily computed by post-processing. Therefore, the coefficient C_{66}, for this case is found using (6.18) written as

$$C_{66} = \overline{\sigma}_6 = \frac{1}{V} \int_V \sigma_6 (x_1, x_2, x_3) \ dV \quad \text{with } \gamma_6^0 = 1 \quad (6.40)$$

Finally, the elastic properties of the composite are determined using (6.11).

Example 6.3 *Compute G_{12} for the composite in Example 6.2 (p. 232), with properties $E_f = 0.241$ TPa, $\nu_f = 0.2$, $G_f = 0.100417$ TPa, $E_m = 3.12 \times 10^{-3}$ TPa, $\nu_f = 0.38$, $G_m = 1.13 \times 10^{-3}$ TPa. The fiber diameter is $d_f = 7$ μm. Note that μm$=10^{-3}$mm, N/mm^2=MPa, thus, N/μm^2=TPa.*

Solution to Example 6.3 *Watch the video in [1].*
First note that for the composite reinforced with a hexagonal array of fibers, $G_{12} = G_{13}$. Therefore, to compute $G_{13} = C_{55}$ we can apply a deformation $\gamma_5^0 = 2\epsilon_{13}^0 = 1.0$. Such state of average deformation is antisymmetric with respect to the (x_1, x_2) coordinate plane $(X, Z$ plane in the Abaqus model shown in Figure 6.7) and symmetric with respect to the (x_1, x_3) plane $(Y, Z$ plane in the model). Therefore, we choose to model the full RVE. The deformed shape and contour plot of strain γ_{13} (E23 in the model) is shown in Figure 6.7. Such plot corresponds to an average deformation $\epsilon_{13} = 1.0$ and average stress $\sigma_{13} = 2,577$ MPa, thus yielding $G_{13} = 2,577$ MPa according to

$$C_{55} = \overline{\sigma}_5 = \frac{1}{V} \int_V \sigma_5 (x_1, x_2, x_3) \ dV \quad \text{with } \gamma_5^0 = 1 \quad (6.41)$$

The fiber diameter is $d_f = 7$ μm and the RVE dimensions are:

$$a_1 = a_2/4 = 1.3175 \ \mu m \quad ; \quad a_2 = 5.270 \ \mu m \quad ; \quad a_3 = 9.128 \ \mu m$$

The geometry (Module: Part), material and sections (Module: Property), assembly (Module: Assembly), step (Module: Step), and mesh (Module: Mesh) can be created

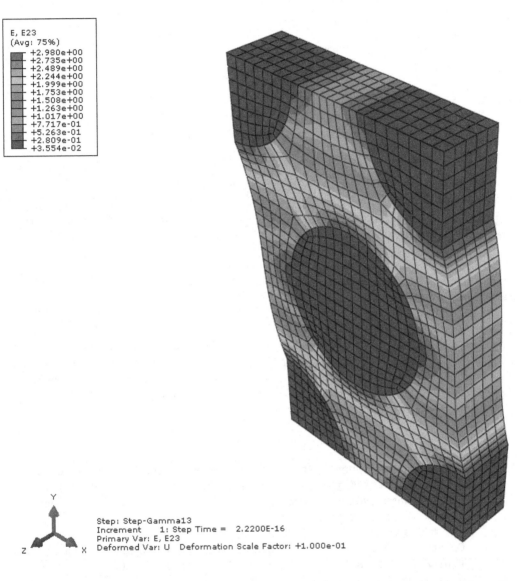

E, E23
(Avg: 75%)
+2.980e+00
+2.735e+00
+2.489e+00
+2.244e+00
+1.999e+00
+1.753e+00
+1.508e+00
+1.263e+00
+1.017e+00
+7.717e-01
+5.263e-01
+2.809e-01
+3.554e-02

Step: Step-Gamma13
Increment   1: Step Time =   2.2200E-16
Primary Var: E, E23
Deformed Var: U   Deformation Scale Factor: +1.000e-01

Fig. 6.7: Deformed shape and contour plot of strain E23 for Example 6.3. Note the coordinate system X, Y, Z.

Fig. 6.8: Enforce U2=0 on the top surface so that it moves parallel to the bottom surface in Figure 6.7.

in the same fashion as it was done in Example 6.2, p. 232, but meshing must be done with **Hex, Sweep** *controls. See also App. A.*

Model coordinates X, Y, Z (see Figure 6.7) are used in this section to describe the boundary conditions. The bottom of the RVE, at $Y = -a_3$ in the model, is fixed. The top is subjected to a displacement $U3 = 2 \times a_3$ to represent the applied shear strain $\epsilon_{13} = 1.0$. The sides of the RVE, at $X = \pm a_2$, are subject to symmetry boundary conditions, which emulate the periodicity conditions in this case because the deformation is symmetric with respect to the plane Y, Z in the model. The pseudo-code below describes how to apply the boundary conditions (see also Figure 6.8).

```
Module: Load
Menu:    BC, Manager
    Create, Name [ysupp], Step: Step-1, Mechanical, Disp/Rota, Cont
    # pick the face @ y=-a_3, Done, # Checkmark: U2 [0] and U3 [0], OK
    Create, Name [ydisp], Step: Step-1, Mechanical, Disp/Rota, Cont
    # pick the face @ y=a_3, Done,
        # Checkmark: U2[0], U3 [2*a_3] #18.256, OK
    Create, Name [xsymm], Step:Step-1, Mechanical, Symm/Anti/Enca, Cont
    # pick the faces @ x=-a_2 and x=a_2, Done, XSYMM, OK
    # close BC Manager pop-up window
```

Enforcing periodicity conditions on the faces at $Z = \pm a_1$ is more difficult. Theses faces cannot be let free because normal strains will develop and the applied boundary condition $U3 = 2a_3$ would not lead to a unit average strain. Although the average deformation at the mesoscale (composite) is pure shear γ_{13}, the deformation at the microscale (fiber and matrix) would include components $E11, E22$, and $E33$. Therefore, periodic conditions on the faces at $Z = \pm a_1$ must be enforced. Since there is no applied strain $E33$, we can use tie constraints.

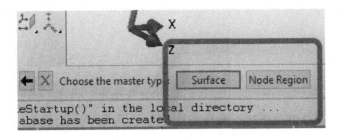

Fig. 6.9: The tie constraint must be defined in terms of `Node region`, not surface.

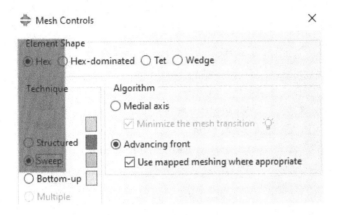

Fig. 6.10: The mesh must be done with `Hex, Sweep` controls.

Tie constraints are applied in the interaction module, as follows. For Abaqus 2016 and newer, be sure to select Master type: Node region, not Surface (Figure 6.9.) See Appendix A.

```
Module: Interaction
Menu:   Constraint, Create
    Type: Tie, Cont
    Choose the master type: Node region #not Surface
    # pick: front face (surfaces z=2*a_1), Done
    Choose the slave type: Node region #not Surface
    # pick: back face (surfaces z=0), Done
    Position Tolerance: Specify distance: [2*a_1] #2.635
    # uncheck: Adjust slave surface initial position
    # uncheck: Tie rotational DOF #solid elements do not have UR's
```

Also, the mesh must be done with Hex, Sweep controls so that nodes are paired at the same in-plane position on both node regions, see Figure 6.10.

It is very important to specify the distance between the surfaces (position tolerance) so that CAE can find the nodes on both surfaces. Fig 6.11.

You can extract the Python script corresponding to the pseudo-code shown above by looking at the .rec file. The complete script to solve this example is available in [2, Ex_6.3-TC.py]. The script can be executed inside Abaqus CAE by doing Menu: File, Run script: Ex_6.3-TC.py. It includes [2, srecover_6.3.py] to calculate the average strains and average stresses, from which the result displayed in Table 6.3 is obtained.

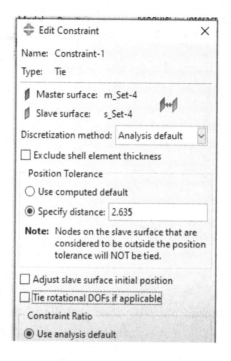

Fig. 6.11: "Specify distance" is the distance between the master and slave nodes.

Table 6.3: In-plane shear modulus of the unidirectional lamina.

Property	PMM	FEA
$G_{12}\ [MPa]$	2,594	2,577

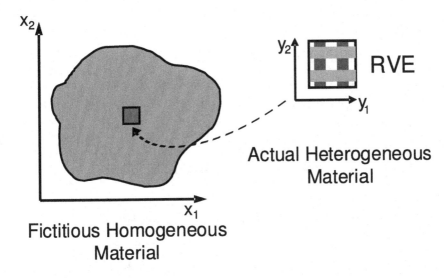

Fig. 6.12: Local-global analysis using RVE.

6.3   Local-Global Analysis

In local-global analysis (Figure 6.12), an RVE is used to compute the stress for a given strain at each Gauss integration point in the *global model*. The global model is used to compute the displacements and resulting strains, assuming that the material is homogeneous. The *local model* takes the inhomogeneities into account by modeling them with an RVE.

Equations (6.13-6.15) are used in Section 6.2 to enforce one component of strain at a time, with the objective of finding the equivalent elastic properties of the material. Equations (6.13-6.15) are still valid for a general state of strain applied to the RVE but care must be taken with the specification of periodic boundary conditions at the edges and vertices, as discussed in page 239. Equations (6.13-6.15) are nine constraint equations that can be imposed between all the pairs of periodic points on the faces of the RVE except on the edges and vertices.

On the faces $x_1 = \pm a_1$, u_1 is used to impose ε_{11}^0, u_2 is used to impose $\varepsilon_{21}^0 = \gamma_6/2$, and u_3 is used to impose $\varepsilon_{31}^0 = \gamma_5/2$. To achieve this, (6.13) is expanded into its three components, using tensor notation for strains, as follows

$$\begin{aligned}
u_1(a_1, x_2, x_3) - u_1(-a_1, x_2, x_3) &= 2a_1\varepsilon_{11}^0 \\
u_2(a_1, x_2, x_3) - u_2(-a_1, x_2, x_3) &= 2a_1\varepsilon_{21}^0 \\
u_3(a_1, x_2, x_3) - u_3(-a_1, x_2, x_3) &= 2a_1\varepsilon_{31}^0
\end{aligned} \tag{6.42}$$

On the faces $x_2 = \pm a_2$, u_1 is used to impose $\varepsilon_{12}^0 = \gamma_6/2$, u_2 is used to impose ε_{22}^0, and u_3 is used to impose $\varepsilon_{32}^0 = \gamma_4/2$. Therefore, (6.14) is expanded into its three components, using tensor notation for strains, as follows

$$u_1(x_1, a_2, x_3) - u_1(x_1, -a_2, x_3) = 2a_2\varepsilon_{12}$$
$$u_2(x_1, a_2, x_3) - u_2(x_1, -a_2, x_3) = 2a_2\varepsilon_{22}$$
$$u_3(x_1, a_2, x_3) - u_3(x_1, -a_2, x_3) = 2a_2\varepsilon_{32} \qquad (6.43)$$

On the faces $x_3 = \pm a_3$, u_1 is used to impose $\varepsilon_{13}^0 = \gamma_5/2$, u_2 is used to impose $\varepsilon_{23}^0 = \gamma_4/2$, and u_3 is used to impose ε_{33}^0. Therefore, (6.15) is expanded into its three components, using tensor notation for strains, as follows

$$u_1(x_1, x_2, a_3) - u_1(x_1, x_2, -a_3) = 2a_3\varepsilon_{13}$$
$$u_2(x_1, x_2, a_3) - u_2(x_1, x_2, -a_3) = 2a_3\varepsilon_{23}$$
$$u_3(x_1, x_2, a_3) - u_3(x_1, x_2, -a_3) = 2a_3\varepsilon_{33} \qquad (6.44)$$

Since each edge belongs to two faces on every edge, it would seem that each component of displacement would be used to impose two CE, one from each face, as given by (6.42-6.44). However, as discussed in page 239, only one CE can be written for each component of displacement. Therefore, edges must be dealt with separately. Similarly, since three faces converge at a vertex, three periodic CE, one from each face, need to be imposed using a single component of displacement. Following a derivation similar to that presented in page 239, the following is obtained.

The planes $x_1 = \pm a_1$ and $x_2 = \pm a_2$ define two pairs of edges, for which (6.42-6.44) reduce to the following six equations (with $i = 1, 2, 3$), as follows

$$u_i(+a_1, +a_2, x_3) - u_i(-a_1, -a_2, x_3) - 2a_1\varepsilon_{i1} - 2a_2\varepsilon_{i2} = 0$$
$$u_i(+a_1, -a_2, x_3) - u_i(-a_1, +a_2, x_3) - 2a_1\varepsilon_{i1} + 2a_2\varepsilon_{i2} = 0 \qquad (6.45)$$

The planes $x_1 = \pm a_1$ and $x_3 = \pm a_3$ define two pairs of edges, for which (6.42-6.44) reduce to the following six equations (with $i = 1, 2, 3$), as follows

$$u_i(+a_1, x_2, +a_3) - u_i(-a_1, x_2, -a_3) - 2a_1\varepsilon_{i1} - 2a_3\varepsilon_{i3} = 0$$
$$u_i(+a_1, x_2, -a_3) - u_i(-a_1, x_2, +a_3) - 2a_1\varepsilon_{i1} + 2a_3\varepsilon_{i3} = 0 \qquad (6.46)$$

The planes $x_2 = \pm a_2$ and $x_3 = \pm a_3$ define two pairs of edges, for which (6.42-6.44) reduce to the following six equations (with $i = 1, 2, 3$), as follows

$$u_i(x_1, +a_2, +a_3) - u_i(x_1, -a_2, -a_3) - 2a_2\varepsilon_{i2} - 2a_3\varepsilon_{i3} = 0$$
$$u_i(x_1, +a_2, -a_3) - u_i(x_1, -a_2, +a_3) - 2a_2\varepsilon_{i2} + 2a_3\varepsilon_{i3} = 0 \qquad (6.47)$$

Four pairs of corners need to be analyzed one at a time. For each pair, the corners are located symmetrically with respect to the center of the RVE located at coordinates $(0, 0, 0)$. The resulting CE are

$$u_i(+a_1, +a_2, +a_3) - u_i(-a_1, -a_2, -a_3) - 2a_1\varepsilon_{i1} - 2a_2\varepsilon_{i2} - 2a_3\varepsilon_{i3} = 0$$

$$u_i(+a_1, +a_2, -a_3) - u_i(-a_1, -a_2, +a_3) - 2a_1\varepsilon_{i1} - 2a_2\varepsilon_{i2} + 2a_3\varepsilon_{i3} = 0$$

$$u_i(-a_1, +a_2, +a_3) - u_i(+a_1, -a_2, -a_3) + 2a_1\varepsilon_{i1} - 2a_2\varepsilon_{i2} - 2a_3\varepsilon_{i3} = 0$$

$$u_i(+a_1, -a_2, +a_3) - u_i(-a_1, +a_2, -a_3) - 2a_1\varepsilon_{i1} + 2a_2\varepsilon_{i2} - 2a_3\varepsilon_{i3} = 0$$

$$(6.48)$$

Example 6.4 *Apply $\epsilon_2^0 = 0.2\%$ and $\gamma_4^0 = 0.1\%$ simultaneously to the composite in Example 6.2 (p. 232). Compute the average stresses in the RVE and the maximum stresses in the matrix.*

Solution to Example 6.4 *An FEA solution for this example is implemented using three Python scripts: `Ex_6.4.py`, `PBC_2D.py`, and `srecover2D.py`. The details of these scripts, available in [2, Ex_6.4.py], are explained in Section 12.4. Here we focus on the modeling issues.*

The geometry of the representative volume element (RVE) is shown in Figure 6.4.

This example requires us to apply $\epsilon_2^0 = 0.2\%$ and $\gamma_4^0 = 0.1\%$ simultaneously. Unfortunately, `Tie` constraints cannot be used for periodicity conditions when normal strains are applied, because the faces will separate as a result of the strain. In this example $\epsilon_2^0 \neq 0$.

Symmetry boundary conditions cannot be used, either, because the shear deformation γ_{23} is not symmetric with respect to the coordinate planes (x_1, x_2) and (x_1, x_3) $(Z, X$ and $Z, Y)$. Instead, the RVE shown in Figure 6.4 must be used along with the constraint equations described by (6.42)–(6.48).

However, a 2D model (Figure 6.13) can be used because the geometry is prismatic, i.e., all cross-sections normal to the x_1-axis are identical and deform identically. Since $\gamma_{12} = \gamma_{13} = 0$, the periodic conditions on the faces at $\pm a_1$ can be represented by a 2D model with generalized plane strain. Furthermore, since ϵ_1 is also zero, we can use regular plane strain elements.

The fiber diameter is $d_f = 7 \ \mu m$ and the RVE dimensions are:

$$a_1 = a_2/4 = 1.3175 \ \mu m \quad ; \quad a_2 = 5.270 \ \mu m \quad ; \quad a_3 = 9.128 \ \mu m$$

The geometry (`Module: Part`), material and sections (`Module: Property`), assembly (`Module: Assembly`), step (`Module: Step`), and mesh (`Module: Mesh`) can be defined in the same fashion as it was done in Example 6.2 but for a 2D model representing the RVE with one full fiber and four quarter fibers as shown in Figure 6.4. Instead of listing the pseudo-code for model creation, we captured the underlining script into `Ex_6.4.py`.

There are no loads per se, but rather the constant terms on the right-hand side (RHS) of (6.42)–(6.48) yield the loading for the model. These terms are generated by the applied strains $\epsilon_2^0, \epsilon_3^0, \epsilon_{23}^0$.

To write the constraint equations, one needs to define a reference point, as well as master and slave nodes. Then, the master and slave nodes are connected by constraint equations representing exactly (6.42)-(6.48). As noted in Section 6.3, the equations for the vertices are dealt with separately from the equations for the faces. Since this is too tedious to do it manually, the task is automated in a Python script [2, PBC_2D.py]. The script `PBC_2D.py` needs all faces, edges, and vertices to be declared into sets. Since this too complicated to be done manually with CAE, it is coded into `Ex_6.4.py`.

The script `Ex_6.4.py` can be executed inside Abaqus CAE by doing

```
Menu: File, Run script, [Ex\_6.4.py], OK
```

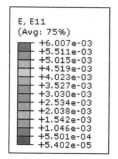

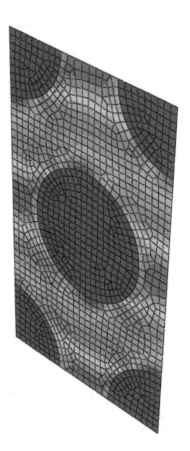

Fig. 6.13: Contour plot of ϵ_{22} (E11 in model coordinates X, Y, Z) for Example 6.4.

Table 6.4: Average stress [MPa] in the RVE and Maximum stress on the matrix for Example 6.4 (values from visualization).

Stress Component	Model Labels	Direction	Average	Maximum in Matrix
σ_{11}	S33	Fiber	10.013	16.630
σ_{22}	S11	Transverse	21.587	28.770
σ_{33}	S22	Transverse	11.904	16.050
σ_{23}	S12	Transverse	2.423	5.525

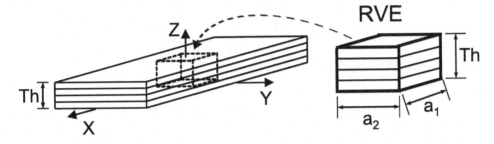

Fig. 6.14: Representative volume element (RVE) for a laminate.

Furthermore, *Ex_6.4.py* calls *PBC_2D.py* to write the CE and *srecover2D.py* to calculate the average stresses. The maximum stress components in the matrix can be obtained from the contour plots in the visualization. The results are shown in Table 6.4.

The output from *srecover2D.py* is:

```
2D Abaqus/Standard Stress Tensor Order: 11-22-33-12
Average stresses Global CSYS: 11-22-33-12
[2.1587221e-05 1.1903510e-05 1.0013077e-05 2.4229214e-06]   TPa
Average strain Global CSYS: 11-22-33-12
[ 1.9999992e-03 -1.9936441e-11  0.0000000e+00  9.9999970e-04]
```

6.4 Laminated RVE

A similar procedure to that used to obtain the RVE at the microscale can be used to analyze laminates on the mesoscale. In this case the RVE represents a laminate. Therefore, the through-thickness direction should remain free to expand along the thickness. For example, with laminas parallel to the x-y plane, then $\sigma_z = 0$ and (6.15) is not enforced, so that the thickness coordinate is free to contract (see Figure 6.14). In general, the RVE must include the whole thickness. For symmetrical laminates subjected to in-plane loads, the RVE can be defined with half the thickness using symmetry boundary conditions (see Example 6.5, p. 252).

The CE for a laminated RVE are simpler. Only (6.13) and (6.14) must be enforced. In an hexahedral RVE, such as shown in Figure 6.14, only four faces ($x_1 = \pm a_1$ and $x_2 = \pm a_2$) and the four edges defined by these faces need to be considered.

Therefore, in a laminated RVE the constraint equations (6.13) and (6.14) become the following. On the periodic pair of faces $x_1 = \pm a_1$, the CE are derived from (6.13) as

$$
\begin{aligned}
u_1(a_1, x_2, x_3) - u_1(-a_1, x_2, x_3) - 2a_1\varepsilon_{11} &= 0 \\
u_2(a_1, x_2, x_3) - u_2(-a_1, x_2, x_3) - 2a_1\varepsilon_{21} &= 0 \\
u_3(a_1, x_2, x_3) - u_3(-a_1, x_2, x_3) - 2a_1\varepsilon_{31} &= 0
\end{aligned}
\tag{6.49}
$$

and on the pair of faces $x_2 = \pm a_2$, the CE are derived from (6.14)

$$
\begin{aligned}
u_1(x_1, a_2, x_3) - u_1(x_1, -a_2, x_3) - 2a_2\varepsilon_{12} &= 0 \\
u_2(x_1, a_2, x_3) - u_2(x_1, -a_2, x_3) - 2a_2\varepsilon_{22} &= 0 \\
u_3(x_1, a_2, x_3) - u_3(x_1, -a_2, x_3) - 2a_2\varepsilon_{32} &= 0
\end{aligned}
\tag{6.50}
$$

The planes $x_1 = \pm a_1$ and $x_2 = \pm a_2$ define two pairs of periodic edges, for which (6.13-6.14) reduce to the following equations

$$
\begin{aligned}
u_1(+a_1, +a_2, x_3) - u_1(-a_1, -a_2, x_3) - 2a_1\varepsilon_{11} - 2a_2\varepsilon_{12} &= 0 \\
u_2(+a_1, +a_2, x_3) - u_2(-a_1, -a_2, x_3) - 2a_1\varepsilon_{21} - 2a_2\varepsilon_{22} &= 0 \\
u_3(+a_1, +a_2, x_3) - u_3(-a_1, -a_2, x_3) - 2a_3\varepsilon_{31} &= 0
\end{aligned}
\tag{6.51}
$$

and

$$
\begin{aligned}
u_1(+a_1, -a_2, x_3) - u_1(-a_1, +a_2, x_3) - 2a_1\varepsilon_{11} + 2a_2\varepsilon_{12} &= 0 \\
u_2(+a_1, -a_2, x_3) - u_2(-a_1, +a_2, x_3) - 2a_1\varepsilon_{21} + 2a_2\varepsilon_{22} &= 0 \\
u_3(+a_1, -a_2, x_3) - u_3(-a_1, +a_2, x_3) - 2a_3\varepsilon_{32} &= 0
\end{aligned}
\tag{6.52}
$$

For in-plane analysis, $\varepsilon_{31} = \varepsilon_{32} = 0$ and the third equation in (6.49)-(6.52) are automatically satisfied.

Example 6.5 *Compute G_{xy} for a $[0/90/-45/45]_S$ laminate with properties $E_1 = 139$ GPa, $E_2 = 14.5$ GPa, $G_{12} = G_{13} = 5.86$ GPa, $G_{23} = 2.93$ GPa, $\nu_{12} = \nu_{13} = 0.21$, $\nu_{23} = 0.38$, and lamina thickness $t_k = 1.25mm$.*

Solution to Example 6.5 *Watch the video in [1].*
As a result of laminate symmetry, an RVE of one-half the laminate thickness and symmetry boundary conditions at $z = 0$ can be used. The model of the RVE is discretized with 3D solid elements. The pseudo-code below can be followed to build the model using the Abaqus CAE GUI.

    i. Setting the `Work Directory`

```
Menu:   File, Set Work Directory, [C:\Simulia\User\Ex_6.5], OK
Menu:   File, Save As, [C:\Simulia\User\Ex_6.5\Ex_6.5.cae], OK
```

    ii. Creating parts

```
Module: Part
Menu:   Part, Create
        3D, Deformable, Solid, Extrusion, Approx. size [10], Cont
Menu:   Add, Line, Rectangle, [0,0], [1,1], X, Done, Depth [5], OK
Menu:   Tools, Datum
```

Type: Plane, Method: Offset from principal plane
XY Plane, Offset [1.25]
XY Plane, Offset [2.5]
XY Plane, Offset [3.75], X, # close Create Datum pop-up window
Menu: Tools, Partition
Cell, Use datum plane, # pick: Datum plane z=1.25,
Create Partition
pick: larger cell, Done, # pick: Datum plane z=2.5,
Create Partition
pick: larger cell, Done, # pick: Datum plane z=3.75,
Create Partition
Done, # close Create partition pop-up window

iii. Defining materials, sections, and assigning sections to parts

Module: Property
Menu: Material, Create
Mechanical, Elasticity, Elastic, Type: Engineering Constants
[139000, 14500, 14500, 0.21, 0.21, 0.38, 5860, 5860, 2930], OK
Menu: Assign, Material Orientation
pick: lamina 1 on Z=4*1.25 (free surface), Done
Use Default Orientation or Other Method
Definition: Coordinate system,
Additional Rotation Direction: Axis 3
Additional Rotation: Angle [0], OK
pick: lamina 2 on Z=3*1.25, Done
Use Default Orientation or Other Method
Definition: Coordinate system,
Additional Rotation Direction: Axis 3
Additional Rotation: Angle [90], OK
pick: lamina 3 on Z=2*1.25, Done
Use Default Orientation or Other Method
Definition: Coordinate system,
Additional Rotation Direction: Axis 3
Additional Rotation: Angle [-45], OK
pick: lamina 4 on Z=1*1.25, Done
Use Default Orientation or Other Method
Definition: Coordinate system,
Additional Rotation Direction: Axis 3
Additional Rotation: Angle [45], OK, Done
Menu: Section, Create
Solid, Homogeneous, Cont, Material: Material-1, OK
Menu: Assign, Section, # select all, Done, OK, Done

iv. Creating assembly

Module: Assembly
Menu: Instance, Create, Independent, OK

v. Defining analysis steps

Module: Step
Menu: Step, Create

```
            Procedure type: Linear perturbation, Static, Cont, OK
     Menu:   Output, Field Output Requests, Edit, F-Output-1
            Output Variables: [S,E,U,IVOL], OK
```

vi. *Adding boundary conditions and loads*

An in-plane shear strain $\gamma_{xy}^0 = 1$ must be applied to the RVE, and the model solved. From the results, the laminate shear stiffness G_{xy} can be obtained directly by computing the average stress in the RVE. The boundary conditions (6.49-6.52) are introduced, as follows

```
# Apply simple shear
Module: Load
Menu:   BC, Manager
        Create, Name [yfix], Step: Initial, Disp/Rota, Cont
        # pick: surface @ y=0, Done, # Checkmark: U1 and U2, OK
        Create, Name [disp], Step: Step-1, Disp/Rota, Cont
        # pick: surface @ Y=1, Done
        # checkmark: U1 [1], # checkmark: U2 [0], OK
        Create, Name [zsymm], Step: Initial, Symm/Anti/Enca, Cont
        # pick: surfaces @ Z=0, Done, ZSYMM, OK
        # close BC Manager pop-up window
```

vii. *Defining constraints*

Apply *Tie constraint* $x = constant$ surfaces to get $\epsilon_{xx} = 0$. For Abaqus 2016 and newer, be sure to select **Node region**, not **Surface** (Figure 6.9) and furthermore, the mesh must be done with **Hex, Sweep** controls. See App. A and Figure 6.10.

```
Module: Interaction
Menu:   Constraint, Create, Tie, Cont
        Choose the master type: Node region, # pick: region @ x=0, Done
        Choose the slave type: Node region, # pick: region @ x=1, Done
        Position Tolerance: Specify distance [1]
        # uncheck: Adjust slave surface initial position # !!!
        # uncheck: Tie rotational DOFs if applicable, OK
```

viii. *Meshing the model*

```
Module: Mesh
Menu:   Seed, Instance
        Approx global size [1.25], OK
        # one element/lamina-thickness enough for pure shear
Menu:   Mesh, Instance, Yes
```

ix. *Solving and visualizing the results*

```
Module: Job
Menu:   Job, Manager
        Create, Cont, OK, Submit, # when Completed, Results
Menu:   File, Save
Menu:   File, Run Script, [Ex_6.5-post.py], OK
```

The script [2, Ex_6.5-post.py] can be used to calculate the average shear stress, and thus the shear modulus. On account of the applied strain being equal to unity, the computed average stress σ_{12} is equal to C_{66}. Therefore, the last line in the script yields $G_{xy} = 21,441\ MPa$.

```
""" Post-processing Ex. 6.5 """

# make sure the Work Directory is OK
import os
os.chdir(r'C:\Simulia\User\Chapter_6\Ex_6.5')

# Open the Output Database for the current Job
from visualization import *
odb = openOdb(path='Job-1.odb');
myAssembly = odb.rootAssembly;

# Create temporary variable to hold frame repository speeds up the process
frameRepository = odb.steps['Step-1'].frames;
frameS=[];
frameIVOL=[];

# Create a Coordinate System in the Laminate direction (Global)
coordSys = odb.rootAssembly.DatumCsysByThreePoints(name='CSYSLAMINATE',
    coordSysType=CARTESIAN, origin=(0,0,0),
    point1=(1.0, 0.0, 0), point2=(0.0, 1.0, 0.0) )

# Transform stresses from Lamina Coordinates
# to Laminate Coordinate System defined in CSYSLAMINATE
# stressTrans=odb.steps['Step-1'].frames[-1].fieldOutputs['S']\
#   .getTransformedField(datumCsys=coordSys)
stressTrans=frameRepository[-1].fieldOutputs['S']\
    .getTransformedField(datumCsys=coordSys)

# Insert transformed stresses into frameS
frameS.insert(0,stressTrans.getSubset(position=INTEGRATION_POINT));
frameIVOL.insert(0,frameRepository[-1].fieldOutputs['IVOL']\
    .getSubset(position=INTEGRATION_POINT));

Tot_Vol=0;       # Total Volume
Tot_Stress=0;    # Stress Sum

for II in range(0,len(frameS[-1].values)):
    Tot_Vol=Tot_Vol+frameIVOL[0].values[II].data;
    Tot_Stress=Tot_Stress+frameS[0].values[II].data * frameIVOL[0]\
        .values[II].data;

#Calculate Average
Avg_Stress=Tot_Stress/Tot_Vol;
print 'Abaqus/Standard Stress Tensor Order:'
# From Abaqus Analysis User's Manual - 1.2.2 Conventions
print ' 11-22-33-12-13-23';
print Avg_Stress;

print 'Gxy=',Avg_Stress[3]
```

Suggested Problems

Problem 6.1 *Consider a unidirectional composite with isotropic fibers $E_f = 241$ GPa, $\nu_f = 0.2$, and isotropic matrix $E_m = 3.12$ GPa, $\nu_m = 0.38$ with fiber volume fraction $V_f = 0.4$. The fiber diameter is $d_f = 7$ μm, placed in a square array as shown in Figure 6.1. Choose an RVE including one full fiber in the center, with vertical faces spaced $2a_2$ and horizontal faces spaced $2a_3$. Part a: compute the first 3 columns of the stiffness matrix in (6.10). Part b: compute C_{66}.*

Problem 6.2 *Consider the same material and fiber distribution of Problem 6.1, but choose an RVE with faces rotated $45°$ with respect to the horizontal and vertical direction in Figure 6.1. Therefore, the RVE size will be $2\sqrt{2}a_2$ and horizontal faces $2\sqrt{2}a_3$ and it will include two fibers (one full and four quarters). Be careful to select a correct RVE that is periodic.*
  Part a: compute the first 3 columns of the stiffness matrix in (6.10).
  Part b: compute C_{66}.

Problem 6.3 *Compute $E_1, E_2, \nu_{12}, \nu_{23}, G_{12},$ and G_{23} using the stiffness matrices calculated in Problem 6.1.*

Problem 6.4 *Compute $E_1, E_2, \nu_{12}, \nu_{23}, G_{12},$ and G_{23} using the stiffness matrices calculated in Problem 6.2.*

Problem 6.5 *Show that removing the tie constraints from Example 6.3 (p. 242) leads to non-zero normal strains and to a wrong result for G_{13}.*

Problem 6.6 *Compute G_{12} as in Example 6.3 (p. 242) but using symmetry boundary conditions to discretize only one quarter of the RVE.*

Problem 6.7 *Compute G_{23} by modifying the boundary conditions and tie constraints of Example 6.3, p. 242.*

Problem 6.8 *Compute G_{23} by using Example 6.4 (p. 249) as a guide.*

Chapter 7

Viscoelasticity

Our interest in viscoelasticity is motivated by observed creep behavior of polymer matrix composites (PMC), which is a manifestation of viscoelasticity. The time-dependent response of materials can be classified as elastic, viscous, and viscoelastic. On application of a sudden load, which is then held constant, an elastic material undergoes instantaneous deformation. In a *1D state of stress*, the elastic strain is $\varepsilon = D\sigma$, where $D = 1/E$ is the compliance or inverse of the modulus E. The deformation then remains constant. Upon unloading, the elastic strain reverses to its original value; thus all elastic deformation is recovered.

A viscous material flows at a constant rate $\dot{\varepsilon} = \sigma/\eta$ where $\eta = \tau E_0$ is the Newton viscosity, E_0 is the initial modulus, and τ is the time constant of the material. The accumulated strain $\varepsilon = \int \dot{\varepsilon}dt$ cannot be recovered by unloading.

A viscoelastic material combines the behavior of the elastic and viscous material in one, but the response is more complex than just adding the viscous strain to the elastic strain. Let H be the Heaviside function defined as

$$H(t - t_0) = 0 \text{ when } t < t_0$$
$$H(t - t_0) = 1 \text{ when } t \geq t_0 \tag{7.1}$$

Upon step loading $\sigma = H(t - t_0)\,\sigma_0$, with a constant load σ_0, the viscoelastic material experiences a sudden elastic deformation, just like the elastic material. After that, the deformation grows by a combination of recoverable and unrecoverable viscous flow.

A simple series addition of viscous flow and elastic strain (*Maxwell model*, Figure 7.1(a), with $\eta = \tau E_0$) yields totally unrecoverable viscous flow plus recoverable elastic deformation as

$$\dot{\varepsilon}(t) = \frac{\sigma(t)}{\tau E_0} + \frac{\dot{\sigma}(t)}{E_0} \tag{7.2}$$

A simple parallel combination of elastic and viscous flow (*Kelvin model*, Figure 7.1(b), with $\eta = \tau E$) yields totally recoverable deformation with no viscous flow as

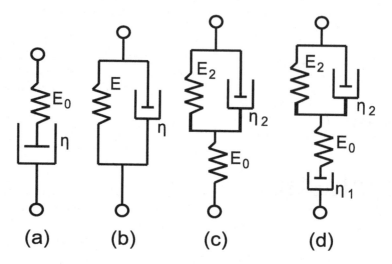

Fig. 7.1: Viscoelastic modes: (a) Maxwell, (b) Kelvin, (c) standard solid, (d) Maxwell-Kelvin.

$$\sigma(t) = \tau E\dot{\varepsilon}(t) + E\varepsilon(t) \tag{7.3}$$

but the deformation does not recover instantaneously.

Materials with unrecoverable viscous flow, such as (7.2), are called *liquids* even though the flow may occur very slowly. Glass is a liquid material over the time span of centuries; the thickness of window panes in medieval cathedrals is thicker at the bottom and thinner at the top, thus revealing the flow that took place over the centuries under the load imposed by gravity. Materials with fully recoverable viscous deformations, such as (7.3), are called *solids*. We shall see that structural design is much easier with solid materials than with liquid materials.

Please take heed of the common misconception introduced in early mechanics of materials courses that most structural materials are elastic. Only perfectly crystalline materials are elastic. Most materials are viscoelastic if observed for sufficiently long periods of time, or at sufficiently high temperature. In other words, most real materials are viscoelastic.

For elastic materials, the compliance D is the inverse of the modulus E, both of which are constants, and they are related by

$$DE = 1 \tag{7.4}$$

For viscoelastic materials in the time domain, the compliance is called $D(t)$ and it is related to the time-varying relaxation $E(t)$ in a similar but not so simple way, as will be shown in Section 7.3. Note that the relaxation $E(t)$ takes the place of the modulus E. A brief derivation of the relationship between compliance and relaxation is presented next, in order to facilitate the presentation of viscoelastic models in Section 7.1. When both the compliance D and the relaxation E are functions of time, (7.4) simply becomes

$$D(t)E(t) = 1 \tag{7.5}$$

Both $D(t)$ and $E(t)$ are functions of time and thus it is not possible to operate algebraically on (7.5) to get either function explicitly in terms of the other. To find one from the other, take the Laplace transform (see Section 7.3) to get

$$s^2 D(s) E(s) = 1 \qquad (7.6)$$

Since both $D(s)$ and $E(s)$ are algebraic functions of s, and the time t is not involved, it is possible to operate algebraically to get

$$E(s) = \frac{1}{s^2 D(s)} \qquad (7.7)$$

Finally, the relaxation in the time domain is the inverse Laplace of (7.7) or

$$E(t) = L^{-1}[E(s)] \qquad (7.8)$$

Similarly, the compliance $D(t)$ can be obtained from the relaxation $E(t)$ as

$$D(t) = L^{-1}\left[\frac{1}{s^2 L[E(t)]}\right] \qquad (7.9)$$

where $L[\,]$ indicates the Laplace transform and $L^{-1}[\,]$ indicates the inverse Laplace transform.

7.1 Viscoelastic Models

The viscoelastic material models presented in this section are convenient curve fits of experimental data. In the time domain, the usual experiments are the creep and relaxation tests. In the *creep test*, a constant stress σ_0 is applied and the ensuing strain is measured. The ratio of measured strain to applied stress is the compliance $D(t) = \varepsilon(t)/\sigma_0$. In the *relaxation test*, a constant strain ε_0 is applied and the stress needed to maintain that strain is measured. The ratio of measured stress to applied strain is the relaxation $E(t) = \sigma(t)/\varepsilon_0$.

7.1.1 Maxwell Model

To derive the compliance of the *Maxwell* model [45], a creep test is performed under constant stress σ_0 applied at the ends of the model shown in Figure 7.1(a). The rate of strain is given by (7.2). Integrating with respect to time we get

$$\varepsilon(t) = \frac{1}{\tau E_0} \int_0^t \sigma_0 dt + \frac{\sigma_0}{E_0} \qquad (7.10)$$

where E_0 is the elastic constant of the spring, τ is the time constant of the material, and $\eta_0 = \tau E_0$ in Figure 7.1(a). The spring and dashpot are subject to the same load and to the same constant stress σ_0, so evaluating the integral yields

$$\varepsilon(t) = \frac{\sigma_0 \, t}{\tau E_0} + \frac{\sigma_0}{E_0} \qquad (7.11)$$

Table 7.1: Some common Laplace transforms.

Function	$f(t)=L^{-1}\{f(s)\}$	$f(s)=L\{f(t)\}$
constant	a	a/s
linear	at	a/s^2
derivative	df/dt	$sf(s)-f(0)$
exponential	$\exp(at)$	$1/(s-a)$
convolution integral	$\int_0^t f(t-\tau)g(\tau)d\tau$	$L\{f\}L\{g\}$

Then, the compliance is

$$D(t) = \frac{1}{E_0} + \frac{t}{\tau E_0} \tag{7.12}$$

To derive the relaxation of the Maxwell model, take the Laplace transform of (7.12) (using Table 7.1 or MATLAB) to get

$$D(s) = \frac{1}{sE_0} + \frac{1}{s^2\tau E_0} = \frac{s\tau+1}{s^2\tau E_0} \tag{7.13}$$

At $t = 0$, the dashpot does not move, so E_0 is also the initial elastic modulus of the material. Now, the relaxation in the Laplace domain is

$$E(s) = \frac{1}{s^2D(s)} = \frac{\tau E_0}{s\tau+1} \tag{7.14}$$

and the relaxation in the time domain is obtained by taking the inverse Laplace transform (using Table 7.1 or MATLAB) to get

$$E(t) = E_0\exp(-t/\tau) \tag{7.15}$$

Note that at $t = \tau$, the relaxation decays to 36.8% of its initial value, and thus τ is called the *time constant* of the material.

7.1.2   Kelvin Model

For the *Kelvin* model, only the creep test is possible, since a relaxation test would require an infinitely large stress to stretch the dashpot in Figure 7.1(b) to a constant value in no time. For a creep test, a constant stress $\sigma = \sigma_0$ is applied. Then, (7.3) is an ordinary differential equation (ODE) in $\varepsilon(t)$, which is satisfied by $\varepsilon(t) = (\sigma_0/E)[1 - exp(-t/\tau)]$. Therefore, the compliance $D(t) = \varepsilon(t)/\sigma_0$ is

$$D(t) = 1/E_0[1 - \exp(-t/\tau)] \tag{7.16}$$

Using (7.8), the relaxation function can be written with the aid of the Heaviside step function $H(t)$ and the Dirac delta function $\delta(t)$, as follows

$$E(t) = EH(t) + E\tau\delta(t) \tag{7.17}$$

where $\delta(t-t_0) = \infty$ *if* $t = t_0$ and zero for any other time. The following MATLAB code implements (7.17), as follows

```
syms s complex; syms Dt Et t E tau real;
Dt=expand((1-exp(-t/tau))/E)
Ds=laplace(Dt)
Es=1/Ds/s^2
Et=ilaplace(Es)
```

7.1.3 Standard Linear Solid

To have an initial compliance $1/E_0$, a spring is added to the Kelvin model (Figure 7.1.c). Then, the compliance is

$$D(t) = 1/E_0 + 1/E_2 \left[1 - \exp\left(\frac{-t}{\tau_2}\right) \right] \tag{7.18}$$

and

$$E(t) = E_\infty + (E_0 - E_\infty) \exp\left(\frac{-t(E_0 + E_2)}{\tau_2 E_2}\right) \tag{7.19}$$

where $E_\infty = (E_0^{-1} + E_2^{-1})^{-1}$ is the equilibrium modulus as time goes to infinity.

To obtain a better correlation, more spring-dashpot elements are added in series, as in

$$D(t) = D_0 + \sum_{j=1}^{n} D_j \left[1 - exp\left(-t/\tau_j\right) \right] \tag{7.20}$$

where τ_j are the retardation times. When $n \to \infty$,

$$D(t) = \int_0^\infty \Delta(\tau) \left[1 - exp\left(-t/\tau\right) d\tau \right] \tag{7.21}$$

where $\Delta(\tau)$ is the compliance spectrum.

7.1.4 Maxwell-Kelvin Model

A crude approximation of a liquid material is the *Maxwell-Kelvin* model, also called the four-parameter model, described by Figure 7.1(d). Since the Maxwell and Kelvin elements are placed in series, the compliance is found by adding the compliances of the two individual modes, as

$$D(t) = \frac{1}{E_0} + \frac{t}{\tau_1 E_0} + \frac{1}{E_2} \left[1 - exp\left(-t/\tau 2\right) \right] \tag{7.22}$$

where E_0 is the elastic modulus, τ_1 takes the place of τ in (7.12), and E_2, τ_2 take the place of E, τ in (7.16). The relaxation modulus is given by [45, page 28], as follows

$$E(t) = \left(P_1^2 - 4P_2\right)^{-1/2}\left[\left(q_1 - \frac{q_2}{T_1}\right)\exp\left(-t/T_1\right) - \left(q_1 - \frac{q_2}{T_2}\right)\exp\left(-t/T_2\right)\right]$$

$$\eta_1 = E_0\tau_1 \quad ; \quad \eta_2 = E_0\tau_2$$

$$q_1 = \eta_1 \quad ; \quad q_2 = \frac{\eta_1\,\eta_2}{E_2}$$

$$T_1 = \frac{1}{2P_2}\left[P_1 + \sqrt{P_1^2 - 4P_2}\right] \quad ; \quad T_2 = \frac{1}{2P_2}\left[P_1 - \sqrt{P_1^2 - 4P_2}\right]$$

$$P_1 = \frac{\eta_1}{E_0} + \frac{\eta_1}{E_2} + \frac{\eta_2}{E_2} \quad ; \quad P_2 = \frac{\eta_1\,\eta_2}{E_0E_2} \tag{7.23}$$

Another way to determine if a material is a liquid or a solid is to look at its long-term deformation. If the deformation is unbounded, then it is a liquid. If the deformation eventually stops, then it is a solid.

7.1.5 Power Law

Another model, which is popular to represent relatively short-term deformation of polymers, is the *power law*

$$E(t) = At^{-n} \tag{7.24}$$

The parameters A and n are adjusted with experimental data. The power law is popular because it fits well the short-time behavior of polymers and because fitting the data is very easy; just take logarithm on both sides of (7.24) so that the equation becomes that of a line, then fit the parameters using linear regression. The compliance is obtained by using (7.9) as

$$D(t) = D_0 + D_c(t)$$
$$D_c(t) = [A\Gamma(1-n)\Gamma(1+n)]^{-1}t^n \tag{7.25}$$

where Γ is the Gamma function [46], $D_0 = 1/E_0$ is the elastic compliance, and the subscript $()_c$ indicates the creep component of the relaxation and compliance functions.

7.1.6 Prony Series

Although the short-term creep and relaxation of polymers can be described well by the power law, as the time range becomes longer, a more refined model becomes necessary. One such model is the *Prony series*, which consists of a number n of decaying exponentials

$$E(t) = E_\infty + \sum_{i=1}^{n} E_i \exp(-t/\tau_i) \tag{7.26}$$

where τ_i are the relaxation times, E_i are the relaxation moduli, and E_∞ is the equilibrium modulus, if any exists. For example, a Maxwell material is a "liquid" and thus $E_\infty = 0$. The larger the τ_i the slower the decay is. Note that at $t = 0$, $E_0 = E_\infty + \sum E_i$. Equation (7.26) can be rewritten as

$$E(t) = E_\infty + \sum_1^n m_i E_0 \exp(-t/\tau_i) \tag{7.27}$$

where $m_i = E_i/E_0$.

The Prony series can be written in terms of the shear modulus and bulk modulus $(G, K, \text{ see Section } 1.12.5)$, as follows

$$G(t) = G_\infty + \sum_1^n g_i G_0 \exp(-t/\tau_i)$$

$$k(t) = k_\infty + \sum_1^n k_i k_0 \exp(-t/\tau_i) \tag{7.28}$$

where $g_i = G_i/G_0$ and $k_i = K_i/K_0$, and G_0, K_0 are the initial values of shear modulus and bulk modulus, respectively. Noting that at $t = 0$, $G_\infty = G_0(1-\sum_1^n g_i)$, and $k_\infty = k_0(1 - \sum_1^n k_i)$, the Prony series can be rewritten as

$$G(t) = G_0 \left(1 - \sum_1^n g_i\right) + \sum_1^n g_i G_0 \exp(-t/\tau_i)$$

$$k(t) = k_0 \left(1 - \sum_1^n k_i\right) + \sum_1^n k_i K_0 \exp(-t/\tau_i)$$

$$\tag{7.29}$$

Data are input to Abaqus CAE in **Edit Material** by specifying separately the elastic properties (E_0 and ν_0) and the viscous properties (g_i, k_i, τ_i) for the i-th element of the Prony series. Internally, Abaqus calculates G_0 and K_0 using the values entered for E_0 and ν and equations (1.74) and (1.76). For most polymers and composites it is usual to assume that the Poisson's ratio does not change with time, which according to (1.74) and (1.76) is achieved by setting $k_i = g_i$. Also, if ν is constant over time, $m_i = g_i$ in (7.27).

Note that the Young modulus entered in **Edit Material** represents E_0 if **Moduli time scale (for viscoelasticity)** is set to **Instantaneous**. Else, if **Moduli time scale (for viscoelasticity)** is set to **Long term**, then the Young modulus represents E_∞. Of course, one can be calculated from the other by using the relationship $E_0 = E_\infty/(1 - \sum m_i)$.

A Maxwell material (7.15) can be modeled by a one term Prony series by setting $g_i = k_i = 1$. However, Abaqus cannot calculate a material with $E_\infty = 0$, so the data must be such that $\sum g_i < 1$. Thus, to model a Maxwell material one has to use $g_1 = k_1 = 0.999$.

7.1.7 Standard Nonlinear Solid

While the Prony series can fit any material behavior if a large number of terms are used, other models are more efficient for fitting purposes, if harder to manipulate mathematically. For example the *Standard Nonlinear Solid* model

$$D(t) = D_0 + D_1'[1 - \exp(-t/\tau)^m] \tag{7.30}$$

can approximate well the long-term compliance in the α-region of polymer creep [47]. At room temperature, this is the region of interest to structural engineers since it spans the range of time from seconds to years. In contrast the β-region [47] is of interest to sound and vibration experts, among others, since it spans the sub-second range of times. In other words, for long-term modeling, all compliance occurring in the β-region can be lumped in the term D_0, with D_1' representing all the compliance that could ever be accumulated in the α-region. Equation (7.30) has four parameters. When the data span short times, it may be impossible to determine all four parameters because the material behavior cannot be distinguished from a 3-parameter power law (7.31). This can be easily understood if (7.30) is expanded in a power series, truncated after the first term, as follows [47]

$$D(t) = D_0 + D_1'(t/\tau)^m[1 - (t/\tau)^m + ...] \approx D_0 + D_1 t^m \ ; \ D_1 = D_1'/\tau \tag{7.31}$$

For short times, all higher order powers of t can be neglected. What remains is a modified power law with only three parameters. Note that for short times, the parameter τ is combined with D_1' to form D_1. If the data cover a short time, the fitting algorithm will not be able to adjust both τ and D_1' in (7.31); virtually any combination of τ and D_1' will work. That means that short-term data must be modeled by a smaller number of parameters, in this case three.

7.1.8 Nonlinear Power Law

All models described so far represent linear viscoelastic materials. In the context of viscoelasticity, linear means that the parameters in the model are not a function of stress (see Section 7.2.1). That means that the deformation at any fixed time can be made proportionally larger by increasing the stress. If any of the parameters are a function of stress, the material is nonlinear viscoelastic. For example, a *nonlinear power law* takes the form

$$\dot{\varepsilon} = At^B \sigma^D \tag{7.32}$$

Take logarithm to both sides of (7.32) to get a linear equation in two variables

$$y = \bar{A} + BX_1 + DX_2 \ ; \ \bar{A} = \log(A), X_1 = \log(t), X_2 = \log(\sigma) \tag{7.33}$$

that can be fitted with a multiple linear regression algorithm in MATLAB.

Although most materials are not linearly viscoelastic, they can be approximated as linear viscoelastic if the range of stress at which the structure operates is narrow.

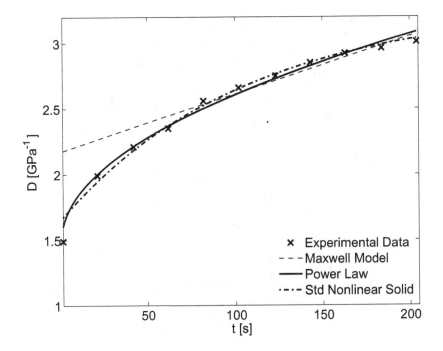

Fig. 7.2: Viscoelastic fit: Maxwell Model, Power Law, and Std Nonlinear Solid.

Example 7.1 *Fit the creep data in Table 7.2 with (a) Maxwell (7.12), (b) Power Law (7.31), and (c) Std Nonlinear Solid (7.30).*

Table 7.2: Creep data for Example 7.1.

time [sec]	1	21	42	62	82	102	123	143	163	184	204
$D(t)$ [GPa^{-1}]	1.49	1.99	2.21	2.35	2.56	2.66	2.75	2.85	2.92	2.96	3.01

Solution to Example 7.1 *To fit the Maxwell model, fit a line to the secondary creep data; that is, ignore the curvy portion for short times to get $E_0 = 0.460$ GPa, $\tau = 495$ s.*

To fit the Power Law, write (7.19) as

$$D(t) - D_0 = D_1 t^m$$

where $D_0 = 1.49$ GPa^{-1} is the first datum in Table 7.2 (see also (7.31)). Take logarithm to both sides of above equation and adjust a line using linear regression to get $D_0 = 1.49$ GPa^{-1}, $D_1 = 0.1117$ (GPa sec)$^{-1}$, and $m = 0.5$.

To fit the Standard Nonlinear Solid you need to use a nonlinear solver to minimize the error between the predicted (expected) values e_i and the experimental (observed) values o_i. Such error is defined as the sum over all the available data points: $\chi^2 = \sum (e_i - o_i)^2 / o_i^2$. In this way, the following are obtained: $D_0 = 1.657$ GPa^{-1}, $D_1' = 1.617$ GPa^{-1}, $\tau = 0.273$ sec, and $m = 0.0026$.

The experimental data and the fit functions are shown in Figure 7.2.

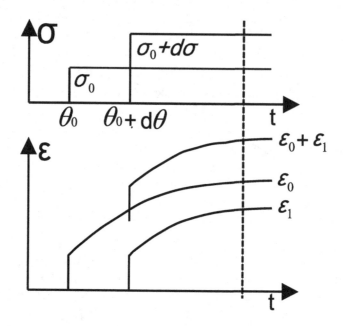

Fig. 7.3: Boltzmann superposition of strains.

7.2   Boltzmann Superposition

7.2.1   Linear Viscoelastic Material

A viscoelastic material is linear if superposition applies. That is, given a stress history

$$\sigma(t) = \sigma_1(t) + \sigma_2(t) \tag{7.34}$$

the strain is given by

$$\varepsilon(t) = \varepsilon_1(t) + \varepsilon_2(t) \tag{7.35}$$

where $\varepsilon_1(t)$, $\varepsilon_2(t)$ are the strain histories corresponding to $\sigma_1(t)$ and $\sigma_2(t)$, respectively. For linear materials, the creep compliance and relaxation modulus are independent of stress

$$D(t) = \frac{\varepsilon(t)}{\sigma_0}$$

$$E(t) = \frac{\sigma(t)}{\varepsilon_0} \tag{7.36}$$

For nonlinear materials, $D(t, \sigma)$ is a function of stress and $E(t, \varepsilon)$ is a function of strain.

For a linear material subjected to a stress σ_0 applied at time $t = \theta_0$ (Figure 7.3) we have

$$\varepsilon(t) = \sigma_0 \, D(t, \theta_0) \quad ; \quad t > \theta_0 \tag{7.37}$$

Adding an infinitesimal load step $d\sigma$ at time $\theta_0 + d\theta$ results in

$$\varepsilon(t) = \sigma_0 \, D(t, \theta_0) + d\sigma \, D(t, \theta_0 + d\theta) \quad ; \quad t > \theta_0 + d\theta \tag{7.38}$$

If stress changes continuously by $d\sigma$ over intervals $d\theta$, the summation (7.38) can be replaced by an integral to yield the accumulated strain as

$$\varepsilon(t) = \sigma_0\, D(t, \theta_0) + \int_{\theta_0}^{t} D(t, \theta) d\sigma = \sigma_0\, D(t, \theta_0) + \int_{\theta_0}^{t} D(t, \theta)\frac{d\sigma}{d\theta} d\theta \qquad (7.39)$$

where the discrete times θ_0, $\theta_0 + d\theta$, etc., are represented by the continuous function θ. Although aging effects are negligible over each infinitesimal $d\theta$, they are significant over time. Therefore, the compliance $D(t, \theta)$ is a function of the current time t and all the time-history represented by θ in $D(t, \theta)$.

7.2.2 Un-aging Viscoelastic Material

If the $\varepsilon_1(t, \theta)$ curve has the same shape as the $\varepsilon_1(t, \theta_0)$ curve, only translated horizontally, any curve can be shifted to the origin (Figure 7.3)

$$D(t, \theta) = D(t - \theta) \qquad (7.40)$$

Equation (7.40) is the definition of un-aging material. For a discussion of aging materials, see [45, 48]. Equation (7.40) means that all the curves have the same shape regardless of age θ, only shifted. Note θ in (7.40) is a continuous function $\theta < t$ that denotes the time of application of each load (σ_0, $d\sigma$, etc.).

The response $\varepsilon(t, \theta)$ at a fixed time t is a function of the response at all times $\theta < t$. Therefore, it is said that the response is hereditary. If the material is aging, t and θ are independent variables in $D(t, \theta)$. For un-aging materials, only one variable, $t - \theta$, is independent, so it does not matter how old the material is (t), it only matters for how long ($t - \theta$) it has been loaded with $d\sigma(\theta)$.

The creep compliance is the response of the material to stress and always starts when the stress is applied. If the change is gradual, from (7.39) we have

$$\varepsilon(t) = \int_{0}^{t} D(t - \theta)\, \dot{\sigma}(\theta)\, d\theta \qquad (7.41)$$

The relaxation is

$$\sigma(t) = \int_{0}^{t} E(t - \theta)\, \dot{\varepsilon}(\theta)\, d\theta \qquad (7.42)$$

The time-dependent behavior of linear viscoelastic materials is hereditary, meaning that the behavior at time t depends on what happened to the material since the beginning of loading at $t = 0$.

Example 7.2 *Consider an un-aging material represented by $D(t - \theta) = 1/E + (t - \theta)/\eta$ and loaded with (a) $\sigma_0 H(\theta)$ and (b) $\sigma_0 H(\theta - 1)$. Find $\varepsilon(t)$ in both cases and comment on the results.*

Solution to Example 7.2

(a) $\sigma = \sigma_0 H(\theta) \;\Rightarrow\; d\sigma/dt = \sigma_0 \delta(0)$

$$\varepsilon(t) = \int_0^t \left[\frac{1}{E} + \frac{(t-\theta)}{\eta} \right] \sigma_0 \delta(0) d\theta$$

$$\varepsilon(t) = \left[\frac{1}{E} + \frac{t}{\eta} \right] \sigma_0 \; ; \; t > 0$$

(b) $\sigma = \sigma_0 H(\theta - 1) \;\Rightarrow\; d\sigma/dt = \sigma_0 \delta(1)$

$$\varepsilon(t) = \int_0^t \left[\frac{1}{E} + \frac{(t-\theta)}{\eta} \right] \sigma_0 \delta(1) d\theta$$

$$\varepsilon(t) = \left[\frac{1}{E} + \frac{(t-1)}{\eta} \right] \sigma_0; \; t > 1$$

It can be seen that (b) is identical to (a), only shifted; meaning that there is no aging.

7.3   Correspondence Principle

The Laplace transform of a function $f(t)$ in the time domain (t-domain) maps to the Laplace domain (s-domain) as $f(s)$. The Laplace transform is defined as

$$L[f(t)] = f(s) = \int_0^\infty \exp(-st) f(t) dt \tag{7.43}$$

Most of the time, the Laplace transform can be obtained analytically, just using a table of transforms, such as Table 7.1. Taking the Laplace transform of (7.41, 7.42) yields

$$\varepsilon(s) = L[D(t)] \; L[\dot{\sigma}(t)] = sD(s)\sigma(s) \tag{7.44}$$

$$\sigma(s) = L[E(t)] \; L[\dot{\varepsilon}(t)] = sE(s)\varepsilon(s) \tag{7.45}$$

Multiplying (7.44) times (7.45) yields

$$s^2 \, D(s)E(s) = 1 \tag{7.46}$$

or

$$s \, D(s) = [s \, E(s)]^{-1} \tag{7.47}$$

where it can be seen that $sD(s)$ is the inverse of $sE(s)$. This is analogous to (7.4) for elastic materials.

The correspondence principle states that all the equations of elasticity, available for elastic materials, are valid for linearly viscoelastic materials in the Laplace domain. This principle is the basis, for example, of the determination of creep and relaxation of polymer matrix composites in terms of fiber and matrix properties using standard micromechanics methods, as shown in Section 7.6.

The inverse mapping from the Laplace domain to the time domain

$$f(t) = L^{-1}(f(s)) \tag{7.48}$$

is more difficult to compute. Decomposition in partial fractions [49] is a useful technique to break up $f(s)$ into simpler component functions for which the inverse Laplace can be found analytically. Another useful technique is the convolution theorem defined in Table 7.1. Also, the limiting value theorems

$$f(0) = \lim_{s \to \infty} [sF(s)]$$
$$f(\infty) = \lim_{s \to 0} [sF(s)] \qquad (7.49)$$

can be used to evaluate the initial and final response of a material in the time domain directly in the Laplace domain. Otherwise, the inverse Laplace can be found numerically using [50] or by the collocation method described in [9, Appendix D].

The Carson transform is defined as

$$\hat{f}(s) = sf(s) \qquad (7.50)$$

In the Carson domain, the constitutive equations (7.41-7.42) become

$$\varepsilon(s) = \hat{D}(s)\sigma(s)$$
$$\sigma(t) = \hat{E}(s)\varepsilon(s) \qquad (7.51)$$

which are analogous, in the Carson domain, to the stress-strain equations of elastic materials in the time domain. Furthermore, the relationship between compliance and relaxation becomes

$$\hat{D}(s) = 1/\hat{E}(s) \qquad (7.52)$$

7.4 Frequency Domain

The Fourier transform maps the time domain into the frequency domain. It is defined as

$$F[f(t)] = f(\omega) = \int_{-\infty}^{\infty} \exp(-i\omega t) f(t) \, dt \qquad (7.53)$$

and its inverse

$$f(t) = \frac{1}{\sqrt{2\pi}} \int_{-\infty}^{\infty} \exp(i\omega t) f(w) dw \qquad (7.54)$$

Applying the Fourier transform to (7.41-7.42) yields

$$\varepsilon(\omega) = D(\omega)\dot{\sigma}(\omega)$$
$$\sigma(\omega) = E(\omega)\dot{\varepsilon}(\omega) \qquad (7.55)$$

and

$$D(\omega) = -\frac{1}{\omega^2 E(\omega)} \tag{7.56}$$

where $D(\omega) = D' + iD''$ and $E(\omega) = E' + iE''$ are complex numbers. Here D', D'' are the storage and loss compliances, and E', E'' are the storage and loss moduli. Using standard complex analysis we get

$$D' = \frac{E'}{E'^2 + E''^2}$$

$$D'' = \frac{E''}{E'^2 + E''^2} \tag{7.57}$$

The frequency domain has a clear physical meaning. If a sinusoidal stress $\sigma(\omega, t) = \sigma_0 \exp(-i\omega t)$ is applied to a viscoelastic material, it responds with an out-of-phase sinusoidal strain $\varepsilon(\omega, t) = \varepsilon_0 \exp(-i\omega t + \phi)$. Furthermore, the complex compliance is $D(\omega) = \varepsilon(\omega, t)/\sigma(\omega, t)$, and the complex relaxation is simply the inverse of the complex compliance, $E(\omega) = \sigma(\omega, t)/\varepsilon(\omega, t)$.

7.5 Spectrum Representation

The Prony series (7.26) provides a physical interpretation of polymer behavior as a series of Maxwell models, each with its own decay time. In the limit, a real polymer has an infinite number of such models [51], so that

$$E(t) - E_\infty = \int_{-\infty}^{\infty} H(\theta) \exp(-t/\theta) d\ln\theta = \int_0^{\infty} \frac{H(\theta)}{\theta} \exp(-t/\theta) d\theta \tag{7.58}$$

where $H(\theta)$ is the relaxation spectrum [52]. In terms of compliance, we have

$$D(t) - D_0 = \frac{t}{\eta} + \int_{-\infty}^{\infty} \frac{L(\theta)}{\theta}[1 - \exp(-t/\theta)]d\theta \tag{7.59}$$

where $L(\theta)$ is the retardation spectrum [52], D_0 is the elastic compliance, η is the asymptotic viscosity of liquids, with $\eta \to \infty$ for solids (see also [53]).

7.6 Micromechanics of Viscoelastic Composites

7.6.1 One-Dimensional Case

Recall the constitutive equations (7.51) in the Carson domain. By the correspondence principle, all equations of micromechanics for elastic materials are valid in the Carson domain for linear viscoelastic materials. For example, the Reuss micromechanical model assumes uniform identical strain in the matrix and fiber (see discussion on page 222). Therefore, the stiffness of the composite $\mathbf{C}$ is a linear combination of the stiffness of the constituents (fiber and matrix) weighed by their respective volume fractions V_m, V_f

$$C = V_m C^m + V_f C^f \tag{7.60}$$

with $A^m = A^f = I$ in (6.1). Taking into account the correspondence principle for a viscoelastic material (Section 7.3), it is possible to write the stiffness tensor in the Carson domain by analogy with (7.60) simply as

$$\hat{C}(s) = V_m \hat{C}^m + V_f \hat{C}^f \tag{7.61}$$

From it, the stiffness tensor in the Laplace domain is (see (7.47))

$$C(s) = \frac{1}{s}\hat{C}(s) \tag{7.62}$$

Finally, the stiffness tensor in the time domain is obtained by finding the inverse Laplace transform (7.48) as

$$C(t) = L^{-1}[C(s)] \tag{7.63}$$

Example 7.3 *Derive the transverse compliance $D_2(t)$ in the time domain for a unidirectional composite with elastic fibers and a viscoelastic matrix represented by $D_m = 1/E_m + t/\eta_m$. Plot D_f, $D_m(t)$, and $D_2(t)$ for $0 < t < 0.1$, $E_f = 10$, $V_f = 0.5$, $E_m = 5$, $\eta_m = 0.05$. Use the Reuss model and discuss the results.*

Solution to Example 7.3 *The elastic behavior of the fiber and viscoelastic behavior of the matrix are defined as follows*

Fiber (elastic): $E_f = constant \rightarrow D_f = \dfrac{1}{E_f}$

Matrix (Maxwell model (7.12) with $E_m = E_0$, $\eta_m = \tau E_0$): $\dfrac{1}{E_m} = \dfrac{1}{E_m} + \dfrac{t}{\eta_m}$

Take the Laplace transform,

$D_f(s) = \dfrac{1}{sE_f}$ *because* $\dfrac{1}{E_f}$ *is constant.*

$D_m(s) = \dfrac{1}{sE_m} + \dfrac{1}{s^2\eta_m}$

Then, the Carson transform is

$\hat{D}_f(s) = s\,D_f(s) = \dfrac{1}{E_f}$

$\hat{D}_m(s) = s\,D_m(s) = \dfrac{1}{E_m} + \dfrac{t}{s\eta_m}$

Using the Reuss model (page 222) to compute the composite behavior

$\hat{D}_2 = V_f\hat{D}_f + V_m\hat{D}_m$

$\hat{D}_2 = V_f\dfrac{1}{E_f} + V_m\left(\dfrac{1}{E_m} + \dfrac{1}{s\eta_m}\right)$

Back to Laplace domain

$D_2(s) = \dfrac{V_f}{sE_f} + \dfrac{V_m}{sE_m} + \dfrac{V_m}{s^2\eta_m}$

Back to the time domain (inverse Laplace)

$D_2(t) = L^{-1}(D_2(s)) = \dfrac{V_f}{E_f} + \dfrac{V_m(E_m t + \eta_m)}{E_m\eta_m}$

To make a plot, take $E_f = 10$, $V_f = 0.5$, $E_m = 5$, $\eta_m = 0.05$, which results in

$$D_f = 0.1 = 1/10$$
$$D_m(t) = 0.2 + 20t$$
$$D_2(t) = 0.15 + 10t$$

Since $V_f = 0.5$, the initial compliance is halfway between those of the fiber and the matrix. The elastic fiber has constant compliance. The creep rate of the composite $1/\eta_c$ is $1/2$ of the creep rate of the matrix $1/\eta_m$.

7.6.2   Three-Dimensional Case

The constitutive equation for an elastic, isotropic material (1.78) can be written in terms of just two material parameters λ and $\mu = G$ as follows

$$\sigma = (\lambda \mathbf{I}^{(2)} \otimes \mathbf{I}^{(2)} + 2\mu \mathbf{I}^{(4)}) : \varepsilon \tag{7.64}$$

where $\mathbf{I}^{(2)}$ and $\mathbf{I}^{(4)}$ are the second- and fourth-order identity tensors[1] (see Appendix B). The constitutive equation for a viscoelastic, isotropic material can be written in terms of the viscoelastic Lamé constants $\lambda(s)$ and $\mu(s)$, as follows [53]

$$\sigma(t) = \int_0^t \lambda(t-\theta)\mathbf{I}^{(2)} \otimes \mathbf{I}^{(2)} : \dot{\varepsilon}(\theta)d\theta + \int_0^t 2\mu(t-\theta)\mathbf{I}^{(4)} : \dot{\varepsilon}(\theta)d\theta \tag{7.65}$$

Using the convolution theorem (Table 7.1), the Laplace transform of (7.65) is

$$\sigma(s) = s\lambda(s)\mathbf{I}^{(2)} \otimes \mathbf{I}^{(2)} : \varepsilon(s) + s\,2\mu(s)\mathbf{I}^{(4)} : \varepsilon(s) \tag{7.66}$$

or in terms of the Carson transform

$$\hat{\sigma}(s) = \hat{\mathbf{C}}(s) : \hat{\varepsilon}(s) \tag{7.67}$$

Assuming that the fiber is elastic, and the matrix is viscoelastic, the latter represented with a Maxwell model

$$D_m(t) = 1/E_m + t/\eta_m \tag{7.68}$$

the Carson transform is

$$\hat{D}_m = 1/E_m + 1/s\eta_m = \frac{E_m + s\eta_m}{s\eta_m E_m} \tag{7.69}$$

Using the correspondence principle yields

$$\hat{E}_m = 1/\hat{D}_m = \frac{s\eta_m E_m}{E_m + s\eta_m} = \frac{sE_m}{E_m/\eta_m + s} \tag{7.70}$$

----

[1]Tensors are indicated by boldface type, or by their components using index notation.

Using (1.75) and assuming the Poisson's ratio ν_m of the matrix to be constant, the Lamé constant of the matrix in the Carson domain is

$$\hat{\lambda}_m = \frac{\hat{E}_m \nu_m}{(1 + \nu_m)(1 - 2\nu_m)} \tag{7.71}$$

and the shear modulus of the matrix is

$$\hat{\mu}_m = \frac{\hat{E}_m}{2(1 + \nu_m)} \tag{7.72}$$

Barbero and Luciano [38] used the the Fourier expansion method to get the components of the relaxation tensor in the Carson domain for a composite with cylindrical fibers arranged in a square array with fiber volume fraction V_f. The elastic, transversely isotropic fibers are represented by the transversely isotropic stiffness tensor $\mathbf{C}'$ defined by (1.70,1.92) in terms of fiber properties in the axial and transverse (radial) directions E_A, E_T, G_A, G_T, and ν_T. Defining the matrix properties in the Laplace $\tilde{()}$ and Carson domain $\hat{()}$ as $\hat{\lambda}_m = s\tilde{\lambda}_m(s)$ and $\hat{\mu}_m = s\tilde{\mu}_m(s)$, the components of the relaxation tensor of the composite in the Carson domain $\hat{\mathbf{L}}^*$ become [38]

$$\hat{L}_{11}^*(s) = \hat{\lambda}_m + 2\hat{\mu}_m - V_f \left(-a_4^2 + a_3^2 \right)$$

$$\left(-\frac{\left(2\hat{\mu}_m + 2\hat{\lambda}_m - C_{33}' - C_{23}' \right)(a_4^2 - a_3^2)}{a_1} + \frac{2(a_4 - a_3)\left(\hat{\lambda}_m - C_{12}' \right)^2}{a_1^2} \right)^{-1}$$

$$\hat{L}_{12}^*(s) = \hat{\lambda}_m + V_f \left(\frac{\left(\hat{\lambda}_m - C_{12}' \right)(a_4 - a_3)}{a_1} \right)$$

$$\left(\frac{\left(2\hat{\mu}_m + 2\hat{\lambda}_m - C_{33}' - C_{23}' \right)(a_3^2 - a_4^2)}{a_1} + \frac{2(a_4 - a_3)\left(\hat{\lambda}_m - C_{12}' \right)^2}{a_1^2} \right)^{-1}$$

$$\hat{L}_{22}^*(s) = \hat{\lambda}_m + 2\hat{\mu}_m - V_f \left(\frac{\left(2\hat{\mu}_m + 2\hat{\lambda}_m - C_{33}' - C_{23}' \right)a_3}{a_1} - \frac{\left(\hat{\lambda}_m - C_{12}' \right)^2}{a_1^2} \right)$$

$$\left(\frac{\left(2\hat{\mu}_m + 2\hat{\lambda}_m - C_{33}' - C_{23}' \right)(a_3^2 - a_4^2)}{a_1} + \frac{2(a_4 - a_3)\left(\hat{\lambda}_m - C_{12}' \right)^2}{a_1^2} \right)^{-1}$$

$$\widehat{L}^*_{23}(s) = \widehat{\lambda}_m + V_f \left(\frac{\left(2\,\widehat{\mu}_m + 2\,\widehat{\lambda}_m - C'_{33} - C'_{23}\right) a_4}{a_1} - \frac{\left(\widehat{\lambda}_m - C'_{12}\right)^2}{a_1^2} \right)$$

$$\left(\frac{\left(2\,\widehat{\mu}_m + 2\,\widehat{\lambda}_m - C'_{33} - C'_{23}\right)(a_3^2 - a_4^2)}{a_1} + \frac{2\,(a_4 - a_3)\left(\widehat{\lambda}_m - C'_{12}\right)^2}{a_1^2} \right)^{-1}$$

$$\widehat{L}^*_{44}(s) = \widehat{\mu}_m - V_f \left(\frac{2}{2\,\widehat{\mu}_m - C'_{22} + C'_{23}} - \left(2\,S_3 - \frac{4\,S_7}{2 - 2\,\nu_m}\right)\widehat{\mu}_m^{-1} \right)^{-1}$$

$$\widehat{L}^*_{66}(s) = \widehat{\mu}_m - V_f \left(\left(\widehat{\mu}_m - C'_{66}\right)^{-1} - \frac{S_3}{\widehat{\mu}_m} \right)^{-1} \tag{7.73}$$

where

$$a_1 = 4\,\widehat{\mu}_m^2 - 2\,\widehat{\mu}_m\, C'_{33} + 6\,\widehat{\lambda}_m\,\widehat{\mu}_m - 2\,C'_{11}\,\widehat{\mu}_m - 2\,\widehat{\mu}_m\, C'_{23} + C'_{23}\,C'_{11} + 4\,\widehat{\lambda}_m\, C'_{12}$$
$$- 2\,C'^{\,2}_{12} - \widehat{\lambda}_m\, C'_{33} - 2\,C'_{11}\,\widehat{\lambda}_m + C'_{11}\, C'_{33} - \widehat{\lambda}_m\, C'_{23}$$

$$a_2 = 8\,\widehat{\mu}_m^3 - 8\,\widehat{\mu}_m^2 C'_{33} + 12\,\widehat{\mu}_m^2 \widehat{\lambda}_m - 4\,\widehat{\mu}_m^2 C'_{11}$$
$$- 2\,\widehat{\mu}_m\, C'^{\,2}_{23} + 4\,\widehat{\mu}_m\,\widehat{\lambda}_m\, C'_{23} + 4\,\widehat{\mu}_m\, C'_{11}\, C'_{33}$$
$$- 8\,\widehat{\mu}_m\,\widehat{\lambda}_m\, C'_{33} - 4\,\widehat{\mu}_m\, C'^{\,2}_{12} + 2\,\widehat{\mu}_m\, C'^{\,2}_{33} - 4\,\widehat{\mu}_m\, C'_{11}\,\widehat{\lambda}_m + 8\,\widehat{\mu}_m\,\widehat{\lambda}_m\, C'_{12}$$
$$+ 2\,\widehat{\lambda}_m\, C'_{11}\, C'_{33} + 4\,C'_{12}\, C'_{23}\,\widehat{\lambda}_m - 4\,C'_{12}\, C'_{33}\,\widehat{\lambda}_m - 2\,\widehat{\lambda}_m\, C'_{11}\, C'_{23}$$
$$- 2\,C'_{23}\, C'^{\,2}_{12} + C'^{\,2}_{23} C'_{11} + 2\,C'_{33}\, C'^{\,2}_{12} - C'_{11}\, C'^{\,2}_{33} + \widehat{\lambda}_m\, C'^{\,2}_{33} - \widehat{\lambda}_m\, C'^{\,2}_{23}$$

$$a_3 = \frac{4\,\widehat{\mu}_m^2 + 4\,\widehat{\lambda}_m\,\widehat{\mu}_m - 2\,C'_{11}\,\widehat{\mu}_m - 2\,\widehat{\mu}_m\, C'_{33} - C'_{11}\,\widehat{\lambda}_m - \widehat{\lambda}_m\, C'_{33} - C'^{\,2}_{12}}{a_2}$$
$$+ \frac{C'_{11}\, C'_{33} + 2\,\widehat{\lambda}_m\, C'_{12}}{a_2} - \frac{S_3 - \dfrac{S_6}{2 - 2\nu_m}}{\widehat{\mu}_m}$$

$$a_4 = -\frac{-2\,\widehat{\mu}_m\, C'_{23} + 2\,\widehat{\lambda}_m\,\widehat{\mu}_m - \widehat{\lambda}_m\, C'_{23} - C'_{11}\,\widehat{\lambda}_m - C'^{\,2}_{12} + 2\,\widehat{\lambda}_m\, C'_{12} + C'_{11}\, C'_{23}}{a_2}$$
$$+ \frac{S_7}{\widehat{\mu}_m\,(2 - 2\nu_m)} \tag{7.74}$$

The coefficients S_3, S_6, S_7 account for the geometry of the microstructure, including the geometry of the inclusions and their geometrical arrangement [35]. For cylindrical fibers arranged in a square array [36] we have

$$S_3 = 0.49247 - 0.47603V_f - 0.02748V_f^2$$
$$S_6 = 0.36844 - 0.14944V_f - 0.27152V_f^2 \qquad (7.75)$$
$$S_7 = 0.12346 - 0.32035V_f + 0.23517V_f^2$$

Note that (7.73) yields six independent components of the relaxation tensor. This is because (7.73) represents a composite with microstructure arranged in a square array. If the microstructure is random (Figure 1.12), the composite is transversely isotropic (Section 1.12.4) and only five components of the relaxation tensor are independent. When the axis x_1 is the axis of transverse isotropy for the composite, the averaging procedure (6.7) yields the relaxation tensor with transverse isotropy as

$$\hat{C}_{11} = \hat{L}_{11}^*$$
$$\hat{C}_{12} = \hat{L}_{12}^*$$
$$\hat{C}_{22} = \frac{3}{4}\hat{L}_{22}^* + \frac{1}{4}\hat{L}_{23}^* + \frac{1}{2}\hat{L}_{44}^*$$
$$\hat{C}_{23} = \frac{1}{4}\hat{L}_{22}^* + \frac{3}{4}\hat{L}_{23}^* - \frac{1}{2}\hat{L}_{44}^*$$
$$\hat{C}_{66} = \hat{L}_{66}^* \qquad (7.76)$$

where the remaining coefficients are found using (1.70) due to transverse isotropy of the material. For example, $\hat{C}_{44} = (\hat{C}_{22} - \hat{C}_{23})/2$. This completes the derivation of the relaxation tensor $\hat{\mathbf{C}}_{ij} = s\mathbf{C}_{ij}(s)$ in the Carson domain. Next, the inverse Laplace transform of each coefficient yields the coefficients of the stiffness tensor in the time domain as

$$\mathbf{C}_{ij}(t) = L^{-1}\left[\frac{1}{s}\hat{\mathbf{C}}_{ij}\right] \qquad (7.77)$$

A MATLAB code based on [50] is available in [2] to perform the inverse Laplace numerically. Another algorithm is provided in [9, Appendix D].

Example 7.4 *Consider a composite made with 60% by volume of transversely isotropic fibers with axial properties $E_A = 168.4$ GPa, $G_A = 44.1$ GPa, $\nu_A = 0.443$, and transverse properties $E_T = 24.8$ GPa and $\nu_T = 0.005$. The epoxy matrix is represented by a Maxwell model (7.12) with $E_0 = 4.08$ GPa, $\tau = 39.17$ min, and $\nu_m = 0.311$. Plot the relaxation $E_2(t)$ of the composite as a function of time for $0 < t < 100$ minutes, compared to the elastic value of the transverse modulus E_2.*

Solution to Example 7.4 *This example has been solved using MATLAB. The elastic and viscoelastic values of the transverse modulus E_2 are shown in Figure 7.4. The calculation procedure is explained next:*

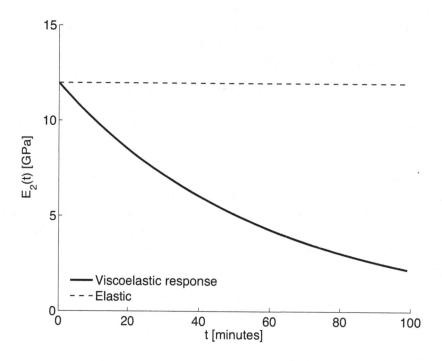

Fig. 7.4: Elastic and viscoelastic values of the transverse modulus E_2.

– *Program the equations of Section 7.6.2 and use them to calculate the elastic values of the composite's elastic properties such as E_2. These equations have been implemented in* PMMViscoMatrix.m.

– *Replace the elastic modulus of the matrix E_0 by the Maxwell model for the matrix Eq. (7.15) in the Carson domain $\hat{E}_0$ (see* PMMViscoMatrix.m), *as follows*

    i. *The output from the portion of the code implementing (7.73-7.76) are equations for the relaxation moduli in terms of s in the Carson domain. Note that it is necessary to declare the variable s as symbolic.*

    ii. *Divide them by s to go back to the Laplace domain.*

    iii. *Back transform to the time domain using the function* invlapFEAcomp, *which is derived from [50].*

    iv. *Finally, fit the numerical values of $E_2(t)$ with a viscoelastic model equation. Usually it is convenient to use the same model equation for the composite relaxation as that used for the matrix relaxation; in this case the Maxwell model. This step is implemented in* fitfunFEAcomp.m

    The MATLAB codes PMMViscoMatrix.m, invlapFEAcomp.m, *and* fitfunFEAcomp.m *are available in [2]. The results are shown in Figure 7.4. The complete set of Maxwell parameters for the composite are calculated in Example 7.7, p. 281.*

7.7 Macromechanics of Viscoelastic Composites

7.7.1 Balanced Symmetric Laminates

The in-plane viscoelastic behavior of a balanced symmetric laminate can be obtained using the procedure in Section 1.15 (Apparent Laminate Properties), but in the Carson domain. Start with the stiffness of the laminas (7.76) in lamina coordinates, in the Carson domain. Rotate each matrix to laminate coordinates. Then, average them using (1.102). Using (1.105), find the laminate engineering properties in the Carson domain and divide by s to go back to the Laplace domain. Finally, take the inverse Laplace transform to find the laminate stiffness in the time domain. Then, fit them with a model equation as it is done in Example 7.4, p. 275.

7.7.2 General Laminates

Thanks to the correspondence principle, the stress-resultant vs. strain-curvature equations from classical lamination theory (CLT, see Chapter 3) are valid for linearly viscoelastic laminated composites in the Carson domain. The A, B, D, H matrices of a laminate in the Carson domain can be computed by using the equations from first-order shear deformation theory (FSDT, Section 3.1.1). This methodology was used in [54].

7.8 FEA of Viscoelastic Composites

Most commercial codes include viscoelasticity (creep) for isotropic materials only. This is a severe limitation for users interested in the analysis of viscoelastic behavior of polymer matrix composites.

However, it is possible to take advantage of the user-programmable features of commercial software in order to implement the formulations presented in this chapter. This is relatively easy because the approach used in this chapter is not stress dependent, but a linear viscoelastic approach, and its implementation is not complicated. A UMAT subroutine is used in Example 7.7 (p. 281) to implement the viscoelastic formulation.

Example 7.5 *Compute the creep response of a $[0/90_8]_s$ laminate. The thickness of each lamina is $t_k = 1.25$ mm. The laminate width is $2b = 20$ mm and it has a length $2L = 80$ mm. Load the sample with a uniform stress σ_x by using coupling constraints on the free end at $x = L$ and a load $P = 1000$ N. Use solid elements on each lamina and symmetry conditions to model 1/8 of the laminate. Both laminas are isotropic with the properties $E_0 = 11975$ MPa, $\nu = 0.1886$, $g_1 = 0.999$ (Maxwell), $k_1 = 0.999$ (constant Poisson's ratio), and $\tau = 58.424$ min.*

Solution to Example 7.5 *Watch the video in [1].*
Most of the model can be obtained by modifying the model of Example 5.2 (p. 203) as illustrated by the following pseudo-code

i. Retrieve the cae *file of Example 5.2*

```
Menu:    File, Set Work Directory, [C:\SIMULIA\User\Ex_7.5], OK
Menu:    File, Open, [C:\SIMULIA\User\Ex_5.2\Ex_5.2.cae], OK
Menu:    File, Save As, [C:\SIMULIA\User\Ex_7.5\Ex_7.5.cae], OK
```

ii. Modifying the part

```
Module: Part
    # change thickness of 90-deg lamina editing the section solid extrude
    # on the .mdb tree, # expand: Parts(1), Part-1, Features
    # right-click: Solid extrude, Edit, Depth [11.25], OK
    # move the partition plane
    # right-click: Datum plane-1, Edit, Offset [10], OK
```

iii. Modifying materials and section properties. The material defined as Material-1 *in Example 5.2 must be removed because viscoelastic analysis in Abaqus only accepts isotropic materials.*

```
Module: Property
Menu:    Material, Manager
    # select: Material-1, Delete, Yes
    # create an isotropic viscoelastic material
    Create, Name [iso-visco], Mechanical, Elasticity, Elastic
    Type: Isotropic, [11975, 0.1866]
    Moduli time scale: Instantaneous
    # viscous part
    Mechanical, Elasticity, Viscoelastic, Domain: Time, Time: Prony
    g_1 [0.999], k_1 [0.999], tau_1 [58.242], OK
    # close Material Manager pop-up window
Menu:    Section, Edit, Section-1, Material: iso-visco, OK
```

iv. Modifying analysis step

```
Module: Step
Menu:    Step, Manager
    # select: Step-1, Delete, Yes, # this also deletes the BC applied
    Create, Name [Step-1], Procedure type: General, Visco, Cont
    Tab: Basic, Time period [150]
    Tab: Incrementation, Type: Fixed, Max number of increments [200]
    Increment size [1], OK, # close Step Manager pop-up window
```

v. Defining constraints and interactions

```
Module: Interaction
    # make surface X=40 into a rigid surface to apply a force
Menu:    Tools, Reference Point, [40,10,11.25], X # RP-1
Menu:    Constraint, Create
    Name [Constraint-1], Type: Coupling, Cont
    # pick: RP-1 # constant control point
    Surface, # pick the surfaces @ X=40, Done
    Constrained degrees of freedom: U1 # uncheckmark: all but U1, OK
```

vi. *Creating a set to request History Output*

```
Module: Step, Model: Model-1, Step: Step-1 # Step-1 must be selected
Menu:   Tools, Set, Create
    Name [Set-1], Cont, # pick: RP-1, Done
Menu:   Output, History Output Requests, Edit, H-Output-1
    Domain: Set : Set-1, Output Variables: Select from list below
    # expand: Displacement/Velocity/Acceleration, # checkmark: U, OK
```

vii. *Adding loads and boundary conditions (BC)*

```
Module: Load
    # restrain rotations UR2, UR3, of the reference point
Menu:   BC, Create
    Name [BC-RP-1], Step: Initial, Type: Disp/Rota, Cont
    # pick: RP-1, Done, # checkmark: UR2, UR3, OK
    # apply load at the reference point
Menu:   Load, Create
    Step: Step-1, Mechanical, Concentrated force, Cont
    # pick: RP-1, Done, CF1 [1000.0], OK
```

viii. *Remeshing the model*

```
Module: Mesh
Menu:   Mesh, Instance, Yes
```

ix. *Solving and visualizing the results*

```
Module: Job
Menu:   Job, Manager
    Create, Cont, OK, Submit, # when Completed, Results
Module: Visualization
    # plot displacement vs. time
Menu:   Result, History Output, U1, Plot, # close pop-up window
Menu:   Plot, Contours, On Deformed Shape
    Toolbox: Field Output, Primary, U, U1
Menu:   Result, Step/Frame, Index: 0, Apply
Menu:   Result, Step/Frame, Index: 1, Apply
Menu:   Result, Step/Frame, Index: 150, Apply
    # analogously visualize the results of E11 and S11
```

The results are shown in Table 7.3. Note that the contour plots of stress and strain display results in the local coordinate systems that were assigned in Example 5.2, and thus the results shown in Table 7.3 are extracted from the outer lamina. This has been done on purpose to emphasize that, by default, results are shown in the local coordinate system defined by CSYS for each lamina.

Example 7.6 *Compute the relaxation response of the laminate in Example 7.5. Load the sample with a uniform strain $\epsilon_x = 0.1$ by using coupling constraints on the free end at $x = L$ and imposing a displacement $U1 = 4$ mm.*

Solution to Example 7.6 *Watch the video in [1].*
Most of the model can be obtained by modifying the model of Example 7.5 as illustrated in

Table 7.3: Results Example 7.5.

Time [min]	U1 [mm]	ϵ_x [%]	σ_x [MPa]
0	0.0000	0.0000	0.0000
1	0.0299	0.0749	8.8889
150	0.1057	0.2643	8.8889

the following pseudo-code. Note how the initial displacement is applied over a short period of time (1 min) using automatic incrementation. Then, a second step is set up to compute the relaxation over 150 min with fixed intervals to facilitate data interpretation.

  i. Retrieve the CAE *file of Example 7.5*

```
Menu:   Set Work Directory, [C:\SIMULIA\User\Ex_7.6], OK
Menu:   File, Open, [C:\SIMULIA\User\Ex_7.5\Ex_7.5.cae], OK
Menu:   File, Save As, [C:\SIMULIA\User\Ex_7.6\Ex_7.6.cae], OK
```

 ii. Modifying analysis step

```
Module: Step
Menu:   Step, Edit, Step-1
    Tab: Basic, Time period [1]
    Tab: Incrementation, Type: Automatic
    Creep/swelling/viscoelastic strain error tolerance: [1E-6], OK
Menu:   Step, Create
    Name [Step-2], Procedure type: General, Visco, Cont
    Tab: Basic, Time period [150]
    Tab: Incrementation, Type: Fixed, Max number of increments [200]
    Increment Size [1], OK
```

iii. Creating a set to request History Output so that we can display it later

```
Module: Step, Model: Model-1, Step: Step-1 # Step-1 must be selected
Menu:   Tools, Set, Create
        Name [Set-elem-output],
        Type: Element, #pick smallest element, Done
Menu:   Output, History Output Requests, Edit, H-Output-1
        Domain: Set: Set-1, Output Variables: Select from list below
        # expand: Disp./Velocity/Acceleration, # checkmark: S, OK
        # expand: Disp./Velocity/Acceleration, # checkmark: E, OK
```

 iv. Modifying the applied loads and BC

```
Module: Load
    # delete the load at RP-1
Menu:   Load, Delete, Load-1, Yes
    # specify a displacement at RP-1
Menu:   BC, Create, Step: Step-1, Disp/Rota, Cont
    # pick: RP-1, Done, # checkmark: U1 [4.0], OK
```

Table 7.4: Results Example 7.6

Total Time [min]	Step Time [min]	Step	U1 [mm]	ϵ_x	σ_x [MPa]
0	0	Step-1	0.00	0.00	0.00
1	1	Step-1	4.00	0.10	1187.00
1	0	Step-2	4.00	0.10	1187.00
3	2	Step-2	4.00	0.10	1147.00
11	10	Step-2	4.00	0.10	1001.00
151	150	Step-2	4.00	0.10	91.48

Table 7.5: Lamina viscoelastic properties of the orthotropic material used in Example 7.7.

Young's Moduli	Shear Moduli	Poisson's Ratio
$(E_1)_0 = 102417$ MPa	$(G_{12})_0 = (G_{13})_0 = 5553.8$ MPa	$\nu_{12} = \nu_{13} = 0.4010$
$\tau_1 = 16551$ min	$\tau_{12} = \tau_{13} = 44.379$ min	
$(E_2)_0 = (E_3)_0 = 11975$ MPa	$(G_{23})_0 = 5037.3$ MPa	$\nu_{23} = 0.1886$
$\tau_2 = \tau_3 = 58.424$ min	$\tau_{23} = 54.445$ min	

v. Solving and visualizing the results

```
Module: Job
      # create new Job. Do not overwrite the results of the creep case
Menu:   Job, Manager
      Create, Cont, OK, Submit, # when Completed, Results
Module: Visualization
      # plot displacement vs. time
Menu:   Result, History Output, U1, Plot, # close pop-up window
      # visualize the results at different times as in Example 7.5
```

The results are shown in Table 7.4.

Example 7.7 *Compute the relaxation response of a $[0/90_8]_s$ laminate. The thickness of each lamina is $t_k = 1.25$ mm. The laminate width is $2b = 20$ mm and its length is $2L = 40$ mm. Load the sample with a uniform strain $\epsilon_x = 0.1$ by applying a uniform displacement at $x = L$. Use solid elements on each lamina and symmetry conditions. Plot the laminate stiffness $E_x(t)$ for $0 > t > 150$ minutes. Use the lamina material properties given in Table 7.5, which were computed with the procedure used in Example 7.4, p. 275.*

Solution to Example 7.7 *Watch the video in [1].*
The Maxwell parameters of an orthotropic lamina are shown in Table 7.5. Since the material is not isotropic, we cannot use Abaqus directly. Instead, we use a UMAT subroutine to implement the constitutive equation of an orthotropic material with the following time-dependent properties:

$E_1(t) = (E_1)_0 \exp(-t/\tau_1)$; $E_2(t) = E_3(t) = (E_2)_0 \exp(-t/\tau_2)$ *and*
$G_{12}(t) = G_{13}(t) = (G_{12})_0 \exp(-t/\tau_{12})$; $G_{23}(t) = (G_{23})_0 \exp(-t/\tau_{23})$.

The rest of the model can be constructed using a procedure similar to that used in Example 7.6, p. 279.

Details of the UMAT implementation are given in Section 12.3. The code is available in [2, umat3dvisco.for].

i. Retrieving cae *file of Example 7.6 and setting* Work Directory

```
Menu:    Set Work Directory, [C:\SIMULIA\User\Ex_7.7], OK
Menu:    File, Open, [C:\SIMULIA\User\Ex_7.6\Ex_7.6.cae], OK
Menu:    File, Save As, [C:\SIMULIA\User\Ex_7.7\Ex_7.7.cae], OK
```

ii. Creating a material to be used with the UMAT

```
Module: Property
Menu:    Material, Create
    Name [user], General, User Material, Type: Mechanical
    [102417 11975 0.401 0.1886 5553.8 5037.3 16551 58.424 44.379
     54.445]
    # E1o, E2o, nu12o, nu23o, G12o, G23o, tau1, tau2, tau12, tau23
    # entered as a column vector
    General, Depvar,
    Number of solution-dependent state variables [6], OK
```

iii. Modifying the analysis steps

```
Module: Step
Menu:    Step, Manager
    # pick: Step-1, Edit, Tab: Basic, Time period [0.001]
    Tab: Incrementation, Type: Automatic, Error tolerance [1E-6], OK
    # leave Step-2 as it is, # close Step Manager pop-up window
Step:    Step-1, # selected above WS
Menu:    Output, Field Output Requests, Edit, F-Output-1
         Output Variables: Edit variables [S,E,U,RF,SDV], OK
Menu:    Output, History Output Requests, Edit, H-Output-1
         Output Variables: Edit variables [S11,E11,U1,RF1,SDV], OK
```

iv. Remeshing the model with quadratic elements to avoid specifying hourglass stiffness

```
Module: Mesh
Menu:    Mesh, Element Type, # select all, Done
    Family: 3D Stress, Geometric Order: Quadratic
    # checkmark: Reduced integration, OK
Menu:    Mesh, Instance, Yes
```

v. Solving and visualizing the results

```
Module: Job
    # create a third Job
    # but do not to overwrite the results of previous one
Menu:    Job, Manager
    Create, Cont, Tab: General,
        User subroutine file [umat3dvisco.for], OK
    Submit, # when Completed, Results
Module: Visualization
    # visualize the reaction force to calculate laminate stiffness
    Toolbox: Field Output, Symbol, RF, RESULTANT
    # to save data to Excel
    # on the Left Tree
```

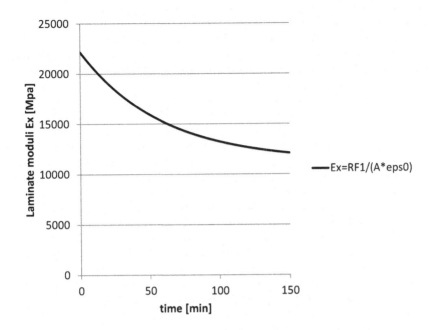

Fig. 7.5: Laminate stiffness $E_x(t)$ calculated in Example 7.7.

```
# expand: Output Databases, # expand: Job-4.odb (Job name used)
# expand: History Output, # right-click: Reaction Force
Save As, Name [XYData-RF1], OK
# expand: XYData, # right-click: XYData-RF1, Plot
# # right-click: XYData-RF1, Edit, # select: all data
# right-click: Copy, # paste in Excel
# close pop-up window
```

The results are shown in Figure 7.5.

Example 7.8 *Consider a composite made with 40% by volume of isotropic graphite fibers with properties $E_f = 168.4$ GPa, $\nu_f = 0.443$ and epoxy matrix represented by a Maxwell model with $E_0 = 4.082$ GPa, $\tau = 39.15$ min, and constant Poisson's ratio $\nu_m = 0.311$. Construct an FEA micromechanical model using hexagonal microstructure (see Example 6.3, p. 242), subject to shear strain $\gamma_4 = 0.02$ applied suddenly at $t = 0$. Tabulate the average stress σ_4 over the RVE at times $t = 0, 20, 40, 60, 80,$ and 100 minutes.*

Solution to Example 7.8 *Watch the video in [1].*
The fiber can be represented using elastic properties and the matrix using viscoelastic properties in Abaqus itself. The Part is very similar to Example 6.4, Figure 6.13, p. 250.
There are two ways to solve this example. One is to modify the Python script Ex_6.4.py, *as explained in Example 12.3. Another approach is to modify* Ex_6.4.cae, *as we do here, with the pseudo-code provided below.*
First, open Ex_6.4.cae *and save it as* Ex_7.8.cae. *Also, update the work directory.*

```
Menu: File, Open, C:\SIMULIA\USER\Chapter_6\Ex_6.4.cae
Menu: File, Save As, C:\SIMULIA\USER\Chapter_7\Ex_7.8.cae
Menu: File, Set Work Directory, C:\SIMULIA\User\Chapter_7\Ex_7.8
```

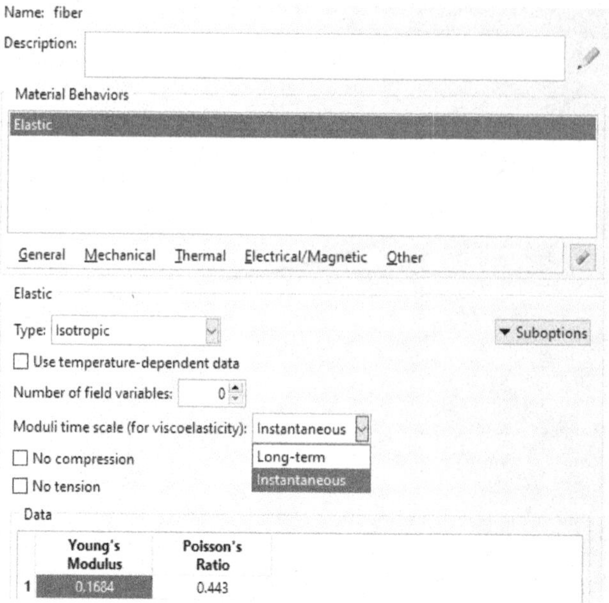

Fig. 7.6: Both fiber and matrix must have `Moduli time scale (for viscoelasticity):Instantaneous`.

Edit fiber material. See Figure 7.6 and pseudo-code below.

```
Module: Property
Menu:   Material, Manager,
        Name: "fiber", Edit, Mechanical, Elasticity, Elastic,
        Young: 0.1684 #TPa, Poisson: 0.443,
        Moduli time scale (for viscoelasticity): instantaneous,
        OK, Dismiss
```

Edit the matrix material and add viscoelastic properties. A Prony series with one term and $g_1 = k_1 = 1.0$ reduces to a Maxwell model but we cannot enter 1.0, we have to enter 0.999 (see Section 7.1.6). See Figure 7.7.

```
Module: Property
Menu:   Material, Manager,
        Name "matrix", Edit, Mechanical, Elasticity, Elastic,
        Moduli time scale (for viscoelasticity): Instantaneous, #important!
        Young: 0.004082 #TPa (instantaneous), Poisson: 0.311
        Name "matrix", Edit, Mechanical, Elasticity, Viscoelastic,
        Domain: Time, Time: Prony #reduces to Maxwell for g1=k1=0.999
        g_1: 0.999, k_1: 0.999, Tau_1: 39.15 #min
        OK, Dismiss
```

Define loading and relaxation steps for the viscoelastic analysis.

```
Module: Step
Menu:   Step, Manager, #delete Step-1, OK to delete dependent objects
        Create, Name: Step-1, After: Step-1, type: General, Visco, Cont
        Tab: Basic, Description: [sudden load], Time period: [0.001]
```

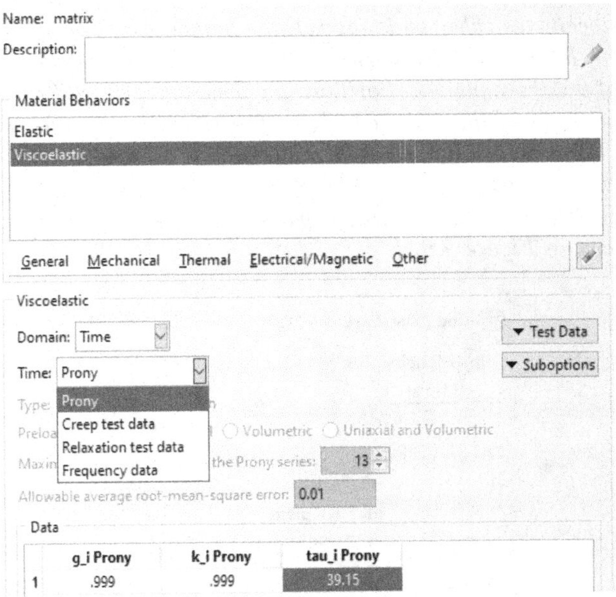

Fig. 7.7: Domain: Time, Time (data): Prony.

```
Tab: Incrementation, Type: Automatic, Tolerance: 1E-6, OK
Create, Name: Step-2, After: Step-1, type: General, Visco, Cont
Tab: Basic, Description: relaxation period, Time period: [100]
Tab: Incrementation, Type: Fixed, Maximum number increments: [200],
Increment size: [1], OK, Dismiss
```

Edit the Field Output Requests to include the element volume (IVOL). See Figure 7.8.

```
Module: Step
Menu:   Output, Field Output Requets, Edit, F-Output-1, Step-2
        Output variables, Edit variables, #add to the list: [IVOL], OK
```

In Example 6.4 we use constraint equations because there we have to apply two strains simultaneously, but in this example there is only shear γ_4 to apply. Even though the actual

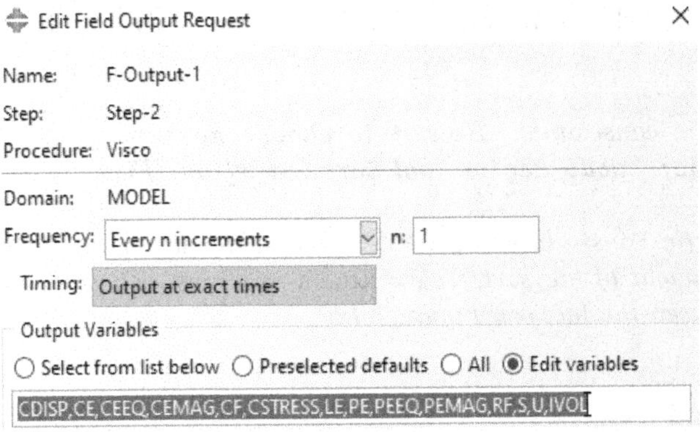

Fig. 7.8: Add IVOL to the list.

material is not homogeneous (fiber and matrix), the homogenized material is orthotropic but homogeneous. If the RVE is defined correctly [3, Section 4.1.2], as shown in Figure 6.13, pure shear in any of the three planes of orthotropy does not produce Poisson's deformations. Therefore, the periodicity of the RVE can be enforced with `tie` *constraints between pairs of opposite edges (top-bottom and left-right). Therefore, we can delete all the constraint equations (CE) from Example 6.4.*

First, delete the RP. On the left tree, expand `Parts`, `Part-1`, `Feature`, *right-click on* `RP`, `Delete`, *as shown in Figure 7.9.*

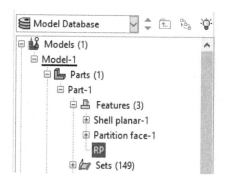

Fig. 7.9: Delete the RP.

Next, delete all the constraint equations. See Figure 7.10.

```
Module: Interaction
Menu:   Constraint, Manager
        #SHIFT-hold select all, Delete, OK
```

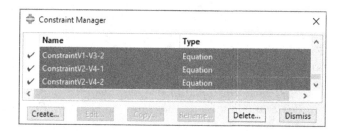

Fig. 7.10: Delete all the constraint equations. Use SHIFT-hold to select all, then Delete.

Now, create Tie constraints. Be sure to **choose master type: Node Region,** *and* **choose slave type: Node Region,** *not* `Surface` *region (Figure 7.11). See Section 2.4 for details.*

To define the tie constraints, we need to know the width of the part, $2\,a_2 = 10.54$ *microns, and the height of the part,* $2\,a_3 = 18.256$ *microns. These dimensions are use for the separation between the faces in Figure 7.12.*

```
Module: Interaction
Menu:   Constraint, Manager
        Create, Name: top, Tie, Cont.
        Choose master type: Node region, select bottom edge, Done, m_Set-1
```

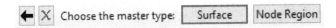

Fig. 7.11: Choose master type: Node Region. Also, Choose slave type: Node Region. Do not use Surface region.

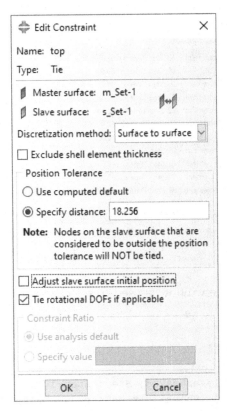

Fig. 7.12: Notice the master and slave surface sets, the discretization method, and how to specify the distance to be kept constant between the faces.

```
Choose slave type: Node region, select top edge, Done, s_Set-1
Discretization: surface to surface
Specify distance: 2a3=18.256 #microns
Adjust slave surface initial positions: un-check, OK
#
Create, Name: right, Tie, Cont.
Choose master type: Node region, select left edge, Done, m_Set-3
Choose slave type: Node region, select right edge, Done, s_Set-3
Discretization: surface to surface
Specify distance: 2a2=10.54 #microns
Adjust slave surface initial positions: un-check, OK
```

Create boundary conditions to apply only $\gamma_4 = 0.02$, with $\gamma_4 = \gamma_{12}$ in the c.s. of the model. Also, apply U1=0 to hold the bottom edge from sliding.

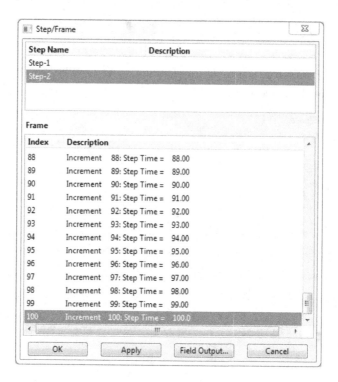

Fig. 7.13: The Step/Frame window is used to select the step and frame (increment) on the output database.

```
Module: Load
Menu:   BC, Manager
        Create, Name: U1=gamma*2a3, Step: Step-1, Type: Displacement,
            Set: s_Set-1, #highlight selection, Cont.,
            U1=gamma*2a3=0.36512 #microns
        Create, Name: U1=0, Step: Step-1, Type: Displacement,
            Set: s_Set-1, #highlight selection, Cont., U1=0, OK
```

Next, we submit a Job. When completed, click Results.

```
Module: Job
Menu:   Job, Manager,
        Create, Name: Job-1, Tab: parallelization,
            use multiple processors: checkmark, [8] # 2Xcores
            OK, Submit, #when completed, click Results
Module: Visualization
Menu:   Plot, Contours, On deformed shape
Menu:   Result, Step/Frame
```

Note how **Menu: Result, Step/Frame** *opens the* **Step/Frame** *window shown in Figure 7.13. Here, you can select the Step and Increment. In this example,* **Step-1** *is the loading and* **Step-2** *is the relaxation. Since* **Incrementation: Fixed** *was used in Step-2, increment numbers correspond to time in minutes. For a time-dependent analysis such as this example, frame and increment are synonymous.*

To calculate the average stress $\overline{\sigma}_4$ (S12 in the model) and average strain $\overline{\gamma}_4$ (E12 in the model), modify srecover2D.py, *as follows*

i. *Select the step. In this example, step-1 is the loading step and step-2 is the relaxation step. The step is selected in the script, as follows*

```
# In Ex. 7.8, Step-2 is the relaxation step
frameRepository = odb.steps['Step-2'].frames;
```

ii. *Parameterize the script to select the frame where the increment is stored*

```
i = 0
# Get the results for frame [i], where i is the increment number
frameS = frameRepository[i].fieldOutputs['S'].values;
frameE = frameRepository[i].fieldOutputs['E'].values;
frameIVOL = frameRepository[i].fieldOutputs['IVOL'].values;
```

The rest of the script is unchanged. The complete script is given below (also available [2, srecover2D.py]).

```
# srecover2D.py modified for Ex. 7.8
from visualization import *
# Open the output data base for the current Job
odb = openOdb(path='Job-1.odb');
myAssembly = odb.rootAssembly;

# Temporary variable to hold the frame repository speeds up the process
# In Ex. 7.8, Step-2 is the relaxation step
frameRepository = odb.steps['Step-2'].frames;

i = 0
# Get the results for frame [i], where i is the increment number
frameS = frameRepository[i].fieldOutputs['S'].values;
frameE = frameRepository[i].fieldOutputs['E'].values;
frameIVOL = frameRepository[i].fieldOutputs['IVOL'].values;

Tot_Vol=0.;          # Total Volume
Tot_Stress=0.;       # Stress Sum
Tot_Strain = 0.;     # Strain Sum

# Calculate Average
for II in range(0,len(frameS)):
    Tot_Vol+=frameIVOL[II].data;
    Tot_Stress+=frameS[II].data * frameIVOL[II].data;
    Tot_Strain+=frameE[II].data * frameIVOL[II].data;

Avg_Stress = Tot_Stress/Tot_Vol;
Avg_Strain = Tot_Strain/Tot_Vol;

# from Abaqus Analysis User's Manual - 1.2.2 Conventions
print '2D Abaqus/Standard Stress Tensor Order: 11-22-33-12'
print 'Average stresses Global CSYS: 11-22-33-12';
print Avg_Stress;
print 'Average strain Global CSYS: 11-22-33-12';
print Avg_Strain;
odb.close()
```

Table 7.6: Average stress σ_4 over time.

Frame: i	0	20	40	60	80	100
Time [min]	0	20	40	60	80	100
Average γ_4	0.02	0.02	0.02	0.02	0.02	0.02
Average σ_4 [MPa]	62.65	38.33	23.46	14.37	8.81	5.42

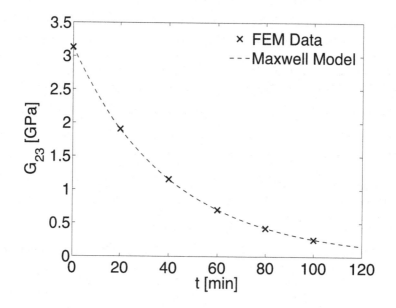

Fig. 7.14: Shear modulus G_{23} as a function of time.

The average values at step-1, increment/frame $i = 0$, are strain $\overline{\gamma}_4 = 2.0 \times 10^{-2}$ and average stress $\overline{\sigma}_4 = 6.26 \times 10^{-5}\ TPa$. In fact, the average strain is constant on a relaxation test. By changing the value of Step (Step-1 or Step-2) and frame (i) in the script, all the results can be obtained and tabulated as in Table 7.6.

Using an exponential regression it is possible to calculate the values of initial shear modulus $G_{23}^0 = 3.13\ GPa$ and relaxation time $\tau = 39.97\ min$ that represent the relaxation of the composite in the 23-shear direction using a Maxwell model (see Figure 7.14).

This example is solved in a different way in Example 12.3.

Suggested Problems

Problem 7.1 *Consider a composite made with 60% by volume of isotropic fibers with properties $E_f = 168.4\ GPa$ and $\nu_f = 0.443$, and epoxy matrix represented by a power law model (7.24) with $D_0 = 0.222\ GPa^{-1}$, $D_1 = 0.0135\ (GPa\ min)^{-1}$, $m = 0.17$, and $\nu_m = 0.311$. Plot the relaxation $C_{22}(t)$ of the composite as a function of time for $0 < t < 100$ minutes. Compare it to the elastic value of the stiffness C_{22} of the composite and the elastic stiffness C_{22} of the matrix.*

Problem 7.2 *Consider a composite made with 60% by volume of transversely isotropic graphite fibers with properties $E_A = 168.4\ GPa$, $E_T = 24.82\ GPa$, $\nu_A = 0.443$, $\nu_T = 0.005$, $G_A = 44.13\ GPa$, and epoxy matrix represented by a Maxwell model (7.15) with $E_0 = 4.082$*

GPa, $\tau = 39.15$ min, and $\nu_m = 0.311$. Plot the relaxation tensor stiffness components $C(t)$ of the composite as a function of time for $0 < t < 100$ minutes, compared to the elastic stiffness C of the composite and the elastic stiffness C_m of the matrix.

Problem 7.3 Compute the parameters in the Maxwell model for unidirectional lamina (see Section 1.14) of carbon/epoxy material used in Problem 7.2. Plot and compare the elastic and viscoelastic properties: $E_1(t)$, $E_2(t)$, and $G_{12}(t)$. Show all work in a report.

Problem 7.4 Use the user-programmable features to implement the Maxwell model constitutive equations for a transversely orthotropic lamina material under plane stress conditions. Using the viscoelastic materials properties obtained in Problem 7.3, compute the response of a $[\pm45/90_2]_s$ laminate. The thickness of each lamina is $t_k = 1.25$ mm. Load the sample with uniform edge load $N_x = 1$ N/mm. Plot the creep compliance $J_x(t)$ of the laminate as a function of time for $0 > t > 300$ minutes.

Problem 7.5 Compute the parameters in the Maxwell model for all the nine engineering properties of a $[0/90]_S$ laminate. Each lamina is 1.25 mm thick. The material is carbon T300 and epoxy 934(NR) with $V_f = 0.62$ and lamina thickness 1.25 mm. The epoxy is represented by a Maxwell model (7.15) with $E_0 = 4.082$ GPa, $\tau = 39.15$ min, and $\nu_m = 0.311$. Carbon T300 is transversely isotropic with axial modulus $E_A = 202.8$ GPa, transverse modulus $E_T = 25.3$ GPa, $G_A = 44.1$ GPa, $\nu_A = 0.443$, and $\nu_T = 0.005$, where the subscripts A and T indicate the axial and radial (transverse) directions of the fiber, respectively.

Chapter 8

Continuum Damage Mechanics

Many modes of damage can be observed in composite materials, including matrix cracks, fiber-reinforced breakage, fiber-matrix debonding, delaminations, and so on. Much work has been done trying to quantify each of these damage modes, their evolution with respect to load, strain, time, number of cycles, etc., and their effect on stiffness, remaining life, etc. *Continuum Damage Mechanics* (CDM) represents all these failure modes by the effect they have on the mesoscale behavior (lamina level) of the material. That is, CDM calculates the degraded moduli of the laminas and laminate in terms of continuum damage variables. Either strength or fracture mechanics failure criteria are used to predict damage initiation. Finally, damage evolution is predicted in terms of either measurable invariant material properties (Chapter 9) or empirical hardening equations set up in terms of additional parameters (this chapter) such as the hardening exponent used for metal plasticity. For example, a form of CDM is used in Chapter 10 to represent the degradation of the interface between laminas.

Hardening equations require non-standard experimentation to adjust the additional, empirical parameters. Since the parameters are adjusted to the model, some shortcomings of the model may be masked by the fitting of the parameters. From a thermodynamics point of view, damage variables are the state variables of the formulation, and they are not measurable. This is in contrast to micromechanics of damage models (Chapter 9) and metal plasticity where the state variables, i.e., crack density and plastic strain, are measurable. From a practical point of view, CDM's major shortcoming is the need for additional experimentation to determine parameters that are particular to each model. Furthermore, since the state variables are not measurable, the additional parameters need to be adjusted to the model through the loss of stiffness, which may not be sufficiently sensitive to damage [55].

One notable effect of damage is a reduction of stiffness, which can be used to define damage [56]. 1D models are used in Section 8.1 to introduce the concepts. The theoretical formulation for the general 3D case is developed in Sections 8.2-8.4.

8.1   One-Dimensional Damage Mechanics

The development of a 1D damage mechanics solution involves the definition of three major entities: 1) a suitable damage variable, 2) an appropriate damage activation function, and 3) a convenient damage evolution, or kinetic equation.

8.1.1   Damage Variable

Consider a composite rod of nominal area $\tilde{A}$, unloaded, and free of any damage (Figure 8.1.a). Upon application of a sufficiently large load P, damage appears (Figure 8.1.b). On a macroscopic level, damage can be detected by the loss of stiffness of the material. In CDM, damage is represented by a state variable D, called *damage variable*, which represents the loss of stiffness [56]

$$D = 1 - E/\tilde{E} \tag{8.1}$$

where $\tilde{E}$ is the initial (virgin) Young's modulus, and E is the modulus after damage.[1] Earlier work [57] conceptualized damage as the reduction of area due to accumulation of micro-cracks having the same effect as the actual damage

$$D = 1 - A/\tilde{A} \tag{8.2}$$

where $\tilde{A}$, A are the initial and remaining cross-sectional areas, respectively. The complement to damage is the integrity [58]

$$\Omega = 1 - D = A/\tilde{A} \tag{8.3}$$

which can be interpreted as the remaining cross-sectional area ratio, using the original area as basis. It is noted that, in principle, damage is a measurable parameter, which could be determined by measuring the damaged area, remaining area, or more practically measuring the initial and remaining moduli. Therefore, in thermodynamic terms, damage is a measurable state variable, in the same sense as the temperature is a measurable state variable that quantifies in macroscopic terms the random agitation of atoms, molecules, and other elementary particles. While it is possible, but extremely difficult, to track the agitation of atoms and molecules, it is very easy to measure the temperature with a thermometer or other device. The same holds true for damage in composite materials.

The analysis of a structural component is done in terms of the nominal area $\tilde{A}$, which is the only one known to the designer. The remaining area $A = (1-D)\tilde{A}$ is not known a priori. The nominal stress is $\sigma = P/\tilde{A}$. Neglecting stress concentrations at the tips of the fictitious cracks representing damage in the damaged configuration (Figure 8.1.b), the value of effective stress[2] acting on the remaining area A is $\tilde{\sigma} = P/A > P/\tilde{A}$.

---

[1]See also (8.10).

[2]Even taking into account stress concentrations, the volume average of the distribution of *effective stress* in the representative volume element (RVE, see Chapter 6) is still $\tilde{\sigma} = P/A$.

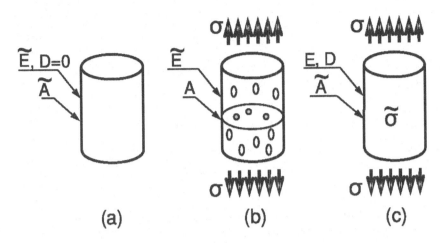

Fig. 8.1: (a) Unstressed material configuration, (b) stressed material configuration with distributed damage, (c) effective configuration.

Therefore, we can envision a configuration (Figure 8.1.c) free of damage, with nominal area $\tilde{A}$, loaded by nominal stress σ, but internally subjected to effective stress $\tilde{\sigma}$ and degraded stiffness E. Thus, the effective configuration allows us to perform structural analysis using the nominal geometry but effectively taking into account the increase of effective stress and the decrement of stiffness caused by damage.

In the undamaged configuration (a), $D = 0$, $\sigma = \tilde{\sigma}$, $\varepsilon = \tilde{\varepsilon}$, and Hooke's law is

$$\tilde{\sigma} = \tilde{E}\,\tilde{\varepsilon} \tag{8.4}$$

where $\tilde{E}$ is a constant.

In the effective configuration (c)

$$\sigma = E\,\varepsilon \tag{8.5}$$

where E is a function of D. The *principle of strain equivalence* assumes that the strain is the same in the configurations (b) and (c), or $\varepsilon = \tilde{\varepsilon}$. Starting with the nominal stress $\sigma = P/\tilde{A}$, multiplying by A/A and using (8.3), the relationship between effective stress $\tilde{\sigma}$ and nominal stress σ (under *strain equivalence*) is

$$\sigma = \tilde{\sigma}\,[1 - D] \tag{8.6}$$

Using (8.6), (8.4), and $\varepsilon = \tilde{\varepsilon}$ in (8.5), the apparent modulus is a function of the damage D

$$E(D) = \tilde{E}\,[1 - D] \tag{8.7}$$

The *principle of energy equivalence* [59] states that the elastic strain energy is identical in the configurations (b) and (c). That is, $\sigma : \varepsilon = \tilde{\sigma} : \tilde{\varepsilon}$, which is satisfied by

$$\sigma = \tilde{\sigma} \ [1 - D]$$
$$\tilde{\varepsilon} = \varepsilon \ [1 - D] \tag{8.8}$$

Substituting (8.8) in (8.5) yields

$$E(D) = \tilde{E} \ [1 - D]^2 \tag{8.9}$$

which redefines the damage variable as

$$D = 1 - \sqrt{E/\tilde{E}} \tag{8.10}$$

Every state variable has a conjugate thermodynamic force driving its growth. In plasticity, the measurable state variable is the plastic strain tensor ε^p, which is driven to grow by its conjugate thermodynamic force, the stress tensor σ. The thermodynamic damage force Y is defined as conjugate to the state variable D.

A kinetic equation $\dot{D}(Y)$ governs the growth of the state variable D as a function of its conjugate thermodynamic force Y. In principle, any relevant variable can be chosen as independent variable Y to define the kinetic equation $\dot{D}(Y)$, as long as it is independent of its conjugate state variable. When the damage D is a scalar and it is used to analyze 1D problems, various authors have chosen independent variables in the form of strain ε [60], effective stress $\tilde{\sigma}$ [61, 62], excess energy release rate $G - 2\gamma_c$ [63], and so on. However, the choice is better based on the appropriate form of the thermodynamic principle governing the problem, as it is shown in Section 8.3.

8.1.2   Damage Threshold and Activation Function

The elastic domain is defined by a threshold value for the thermodynamic force below which no damage occurs. When the load state is in the elastic domain, damage does not grow. When the load state reaches the limit of the elastic domain, additional damage occurs. Furthermore, the elastic domain modifies its size. Typical 1D responses of two materials are shown in Figure 8.2. Initially the elastic domain is defined by the initial threshold values, $\sigma \leq \sigma_0$ and $\varepsilon \leq \varepsilon_0$. While the load state is inside this domain, no damage occurs. When the load state is higher than the threshold, damage increases and the threshold changes. The elastic domain may evolve as hardening or softening. A stress threshold increases for materials with hardening (see Figure 8.2a), and it decreases for materials with softening (see Figure 8.2b). On the other hand, a strain or effective stress threshold always increases for hardening and softening behavior, as shown in Figure 8.2.

The elastic domain can be defined by the *damage activation function g* as

$$g = \hat{g} - \hat{\gamma} \leq 0 \tag{8.11}$$

where $\hat{g}$ is a positive function (norm) that depends on the independent variable (in a 1D case a scalar Y) and $\hat{\gamma}$ is the updated damage threshold for isotropic hardening. According to the positive dissipation principle (see Section 8.3 and (8.82),(8.97)), the updated damage threshold $\hat{\gamma}$ can be written as

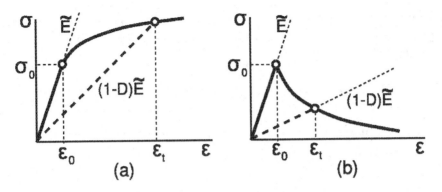

Fig. 8.2: (a) Hardening behavior and (b) softening behavior. No damage occurs until the strain reaches a threshold value ε_0, and no damage occurs during unloading.

$$\hat{\gamma} = \gamma(\delta) + \gamma_0 \qquad (8.12)$$

where γ_0 denotes the virgin damage threshold, and γ is a positive monotonic function, called *hardening (or softening) function*, that depends on the internal variable δ, called *damage hardening variable*.

8.1.3 Kinetic Equation

The rate of damage accumulation is represented by a kinetic equation. The evolution of damage and hardening are defined by

$$\dot{D} = \dot{\lambda}\frac{\partial g}{\partial Y} \quad ; \quad \dot{\delta} = \dot{\lambda}\frac{\partial g}{\partial \gamma} \qquad (8.13)$$

where Y is the independent variable and $\dot{\lambda} \geq 0$ is the damage multiplier that enforces consistency among the damage and hardening evolution as defined by (8.13). Furthermore, the values of $\dot{\lambda}$ and g allow us to distinguish among two possible situations, loading or unloading without damage growth, and loading with damage growth, according to the Kuhn-Tucker conditions [64]

$$\dot{\lambda} \geq 0 \quad ; \quad g \leq 0 \quad ; \quad \dot{\lambda}g = 0 \qquad (8.14)$$

In other words, the Kuhn-Tucker conditions allow us to differentiate among two different cases:

i. Un-damaging loading or unloading, in the elastic domain. The damage activation function is $g < 0$ and by condition (8.14.c) $\dot{\lambda} = 0$, and by (8.13.a) $\dot{D} = 0$.

ii. Damage loading. In this case $\dot{\lambda} > 0$ and condition (8.14.c) implies that $g = 0$. Then, the value of $\dot{\lambda}$ can be determined by the damage consistency condition

$$g = 0 \quad \text{and} \quad \dot{g} = 0 \qquad (8.15)$$

Example 8.1 *Compute $\dot{\lambda}$ for a 1D model under tensile load where the independent variable is the effective stress $Y = \tilde{\sigma}$, the activation function is defined by $\hat{g} = \tilde{\sigma}$, and the hardening function is defined by*

$$\hat{\gamma} = (F_0 - F_R)\delta + F_0$$

where F_0 and F_R are the initial threshold and the strength of the strongest microscopic element in the material, respectively.

Solution to Example 8.1 *The damage activation function g is defined as*

$$g = \hat{g} - \hat{\gamma} = \tilde{\sigma} - [(F_0 - F_R)\delta + F_0] \leq 0$$

Therefore,

$$\frac{\partial g}{\partial \tilde{\sigma}} = +1 \quad ; \qquad \frac{\partial g}{\partial \hat{\gamma}} = -1$$

Using (8.13), the kinetic equations can be written as

$$\dot{D} = \dot{\lambda}\frac{\partial g}{\partial \tilde{\sigma}} = \dot{\lambda} \quad ; \qquad \dot{\delta} = \dot{\lambda}\frac{\partial g}{\partial \hat{\gamma}} = -\dot{\lambda}$$

When new damage appears, the consistency conditions (8.15) yield

$$g = 0 \quad \Rightarrow \quad \hat{\gamma} = \tilde{\sigma}$$

and

$$\dot{g} = 0 \quad \Rightarrow \quad \dot{g} = \frac{\partial g}{\partial \tilde{\sigma}}\dot{\tilde{\sigma}} + \frac{\partial g}{\partial \hat{\gamma}}\dot{\hat{\gamma}} = \dot{\tilde{\sigma}} - \dot{\hat{\gamma}} = 0$$

where

$$\dot{\hat{\gamma}} = \frac{\partial \hat{\gamma}}{\partial \delta}\dot{\delta} = (F_0 - F_R)(-\dot{\lambda}) = (F_R - F_0)\dot{\lambda}$$

Substituting into the second consistency condition (8.15) we obtain $\dot{\lambda}$ as

$$\dot{\lambda} = \frac{1}{F_R - F_0}\dot{\tilde{\sigma}}$$

8.1.4   Statistical Interpretation of the Kinetic Equation

Let's assume that individual damaging events are caused by the failure of microscopic elements inside the material (e.g., fiber breaks, matrix cracks, fiber-matrix debond, etc.) Furthermore, assume each of these material points has a failure strength $\tilde{\sigma}$ and that the collection of failure strengths for all these points, i.e., elements failing at a certain stress $\tilde{\sigma}$ over the total number of elements available, is represented by a probability density $f(\tilde{\sigma})$ (Figure 8.3.b). The fraction of elements broken during an effective stress excursion from zero to $\tilde{\sigma}$ provides a measure of damage

$$D(\tilde{\sigma}) = \int_0^{\tilde{\sigma}} f(\sigma')d\sigma' = F(\tilde{\sigma}) \tag{8.16}$$

where $F(\tilde{\sigma})$ is the cumulative probability (Figure 8.3.b) corresponding to the probability density $f(\tilde{\sigma})$, and σ' is a dummy integration variable. Then, the kinetic equation in terms of effective stress $\tilde{\sigma}$ is

$$\dot{D} = \frac{dD}{d\tilde{\sigma}}\dot{\tilde{\sigma}} = f(\tilde{\sigma})\dot{\tilde{\sigma}} \tag{8.17}$$

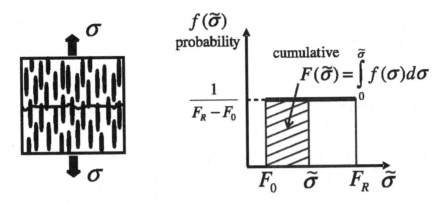

Fig. 8.3: One-dimensional random-strength model.

8.1.5 One-Dimensional Random-Strength Model

As explained in Section 8.1.3, the rate of damage accumulation is represented by a kinetic equation. Equation (8.17) represents a generic kinetic equation, which becomes specific once a particular probability density of failure is adopted.

Consider a loose bundle of short fibers embedded in a matrix and subjected to a uniform stress. The fiber-matrix interface strength is assumed to be identical for all fibers but the embedment length is random. The fiber pull out strength is therefore random. Random means that the probability of finding a fiber pulling out at any value of stress $F_0 < \tilde{\sigma} < F_R$ is constant. In other words, there is no stress level at which more fibers or less fibers pull out, because the probability of pull out is random. This is represented in Figure 8.3 and given by the equation $f(\tilde{\sigma}) = 1/(F_R - F_0)$. Substituting $\tilde{\varepsilon}$ for $\tilde{\sigma}$ as the independent variable in (8.17), and assuming strain equivalence $\varepsilon = \tilde{\varepsilon}$, we have

$$f(\tilde{\varepsilon}) = \frac{\tilde{E}}{F_R - F_0} \quad ; \quad F_0 \leq \tilde{\sigma} \leq F_R \tag{8.18}$$

Equation (8.18) yields the model proposed in [60], which represents well the damaging behavior of Haversian bone [65], concrete in tension [66], fiber composites when damage is controlled by fiber pull out [67], and transverse damage of unidirectional composites.

Damage Activation Function

For a 1D problem, choosing strain as the independent variable, it is possible to write $\hat{g} = \varepsilon$. Therefore, the damage activation function can be written as

$$g = \varepsilon - \hat{\gamma} \leq 0 \tag{8.19}$$

where $\hat{\gamma}$ is the updated damage threshold. Assuming that no damage occurs until the strain reaches a threshold value $\varepsilon_0 = F_0/\tilde{E}$, and applying the consistency conditions (8.15) and using (8.19), the updated damage threshold $\hat{\gamma}$ is given by the highest value of strain seen by the material, or

$$\hat{\gamma} = \max(\varepsilon_0, \varepsilon) \qquad (8.20)$$

Kinetic Equation

The kinetic equation (8.17) for the case of random strength (8.18) in terms of strains $\tilde{\varepsilon} = \varepsilon$ can be expressed as

$$\dot{D} = \frac{dD}{d\varepsilon}\dot{\varepsilon} = \begin{cases} \tilde{E}/(F_R - F_0)\dot{\varepsilon} & \text{when ; } \varepsilon > \hat{\gamma} \\ 0 & \text{otherwise} \end{cases} \qquad (8.21)$$

In this case, the independent variable is ε, and using (8.19), the kinetic equation (8.13) reduces to

$$\dot{D} = \dot{\lambda} \qquad (8.22)$$

Using the Kuhn-Tucker conditions and (8.21), the consistency condition (8.15) reduces to

$$\dot{\lambda} = \tilde{E}/(F_R - F_0)\dot{\varepsilon} \qquad (8.23)$$

when damage occurs and $\dot{\lambda} = 0$ otherwise. In this particular case, the kinetic equation is known explicitly (8.22-8.23). Therefore, it is not necessary to evaluate the evolution of hardening (8.12) because hardening is computed explicitly by (8.20). Note that (8.23) is identical to the solution of Example 8.1 because the hardening function was chosen deliberately to yield this result.

Secant Constitutive Equation

In this particular case, the damage variable is active when tensile load appears, and it can be obtained by integrating (8.21) as

$$D_t = \tilde{E}\frac{\hat{\gamma} - \varepsilon_0}{F_R - F_0} \quad \text{when } \varepsilon > 0 \qquad (8.24)$$

Note that the damage state does not depend on the actual load state ε; it only depends on the history of the load state $\hat{\gamma}$. In this example, crack closure is assumed in compression, damage becomes passive, and $D_c = 0$. Mathematically, damage under unilateral contact conditions can be defined by the following equation

$$D = D_t\frac{\langle\varepsilon\rangle}{|\varepsilon|} + D_c\frac{\langle-\varepsilon\rangle}{|\varepsilon|} \qquad (8.25)$$

where the Macaulay operator $\langle x \rangle$ is defined as $\langle x \rangle = \frac{1}{2}(x + |x|)$.

Substituting (8.24) into (8.5), and using strain equivalence, yields the following constitutive equation

$$\sigma = E(D)\,\varepsilon = \begin{cases} \left(1 - \tilde{E}\frac{\hat{\gamma} - \varepsilon_0}{F_R - F_0}\right)\tilde{E}\varepsilon & \text{when } \varepsilon > 0 \\ \tilde{E}\,\varepsilon & \text{when } \varepsilon < 0 \end{cases} \qquad (8.26)$$

Tangent Constitutive Equation

In a finite element formulation, it is necessary to provide the constitutive equation in rate form, where the rates of stress $\dot{\sigma}$ and strain $\dot{\varepsilon}$ are expressed as functions of pseudo-time. In this particular case, the tangent constitutive equation can be obtained by differentiation of the secant constitutive equation as

$$\dot{\sigma} = E(D)\dot{\varepsilon} + \dot{E}(D)\varepsilon \tag{8.27}$$

The term $\dot{E}(D)$ is zero when new damage does not appear, i.e., when there is elastic loading or unloading. When damaging behavior occurs (8.20) yields $\hat{\gamma} = \varepsilon$, and differentiating $E(D)$ in (8.26) we obtain

$$\dot{E}(D) = -\frac{\widetilde{E}^2}{F_R - F_0}\dot{\varepsilon} \tag{8.28}$$

Substituting (8.28) into (8.27) if damage occurs, or $\dot{E}(D) = 0$ if no damage occurs, the tangent constitutive equation can be written as

$$\dot{\sigma} = \begin{cases} \left(1 - \widetilde{E}\,\dfrac{2\hat{\gamma} - \varepsilon_0}{F_R - F_0}\right)\widetilde{E}\dot{\varepsilon} & \text{when } \varepsilon > \hat{\gamma} \\ E(D)\,\dot{\varepsilon} & \text{when } \varepsilon < \hat{\gamma} \end{cases} \tag{8.29}$$

Model Identification

The initial damage threshold ε_0 represents the minimum strain to initiate damage, and it is proportional to F_0, as follows

$$F_0 = \widetilde{E}\varepsilon_0 \tag{8.30}$$

Under load control, a tensile specimen breaks at $\varepsilon = \hat{\gamma} = \varepsilon_{cr}$ when $d\sigma/d\varepsilon = 0$. Then, using (8.29.a), the only unknown parameter in the model can be computed as

$$F_R = 2\widetilde{E}\varepsilon_{cr} \tag{8.31}$$

The material parameters F_0 and F_R can be calculated from the experimental data using (8.30) and (8.31), with $\widetilde{E}$ being the undamaged modulus of the material. The measurable values ε_0 and ε_{cr} can be obtained easily from material testing at the macroscopic level.

For the particular case $\varepsilon_0 = 0$, using (8.24) and (8.31) at $\varepsilon = \varepsilon_{cr}$, the critical damage at failure under tensile load is

$$D_{cr} = 0.5 \tag{8.32}$$

Therefore the critical effective stress is

$$\widetilde{\sigma}_{T\,cr} = \widetilde{E}\varepsilon_{cr} = 0.5F_R \tag{8.33}$$

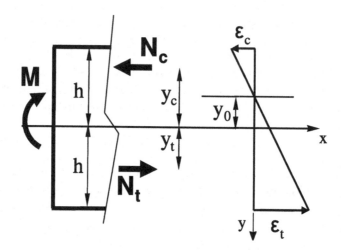

Fig. 8.4: One-dimensional random-strength model.[3]

and using (8.7), the critical applied stress is

$$\sigma_{T\,cr} = 0.25 F_R \tag{8.34}$$

Therefore, in a material with initial threshold $\varepsilon_0 = 0$, a tensile specimen under load control fails when $D = 1/2$ and applied stress $F_R/4$.

A conservative estimate of transverse tensile strength of a fiber reinforced lamina can be obtained, assuming that the fiber-matrix bond strength is negligible. In the limit, only the matrix between fibers carries the transverse load, with the fibers acting as holes. In this limit case, the matrix links can be assumed to have a random distribution of strength (8.18). Therefore, the random-strength model (8.29) applies, and the critical damage for transverse tensile loading of a unidirectional fiber-reinforced lamina can be estimated by (8.32) as $D_{2t}^{cr} = 0.5$. At the present time, there is no model available to estimate the critical transverse-direction compression damage D_{2c}^{cr}.

Example 8.2 *A beam of rectangular cross-section, width $b = 100$ mm and height $2h = 200$ mm, is subjected to pure bending. The bending moment at failure is 25.1 MN mm. The beam is made of carbon/epoxy composite with randomly oriented short fibers with undamaged Young's modulus $\widetilde{E} = 46$ GPa. Find the bending moment at failure in terms of F_R in (8.21). Assume that the material does not damage in compression and it has a random distribution of strength in tension, with the strongest material element having unknown strength $F_R > 0$ and $F_0 = 0$. Determine F_R using the data given.*

Solution to Example 8.2 *This problem was solved in [67]. With reference to Figure 8.4, M is the applied bending moment, and y_c, y_t are the distances from the neutral axis to the stress resultants N_c, N_t on the tensile and compression portions of the beam.*

Denoting by ε_t and ε_c the tension and compression strain on the outer surfaces of the beam, y_0 the distance from the mid-plane to the neutral surface, and assuming linear strain

---

[3]Reprinted from *Mechanics of Materials*, vol. 8, issue 2-3, D. Kracjcinovic, Damage Mechanics, Figure 2.11, p. 134, copyright (1989), with permission from Elsevier.

distribution through the thickness we have

$$\varepsilon(y) = \frac{y - y_0}{h - y_0}\varepsilon_t \quad or \quad \varepsilon(y) = \frac{-y + y_0}{h + y_0}\varepsilon_c$$

Since there is no damage in compression, the compression stress distribution is linear, and the resulting compression stress resultant is

$$N_c = \frac{1}{2}b(h + y_0)\tilde{E}\varepsilon_c$$

and the distance y_c is

$$y_c = \frac{1}{3}(y_0 - 2h)$$

As the tensile side of the beam damages, the neutral axis moves away from the mid-surface. The tensile stress resultant is obtained using (8.26) and integrating the stress between y_0 and h as

$$N_t = \int_{y_0}^{h} dN_t = b\int_{y_0}^{h} E(D)\varepsilon(y)dy = \frac{1}{6}(h - y_0)b\left(3 - 2(\tilde{E}/F_R)\varepsilon_t\right)\tilde{E}\varepsilon_t$$

where $\tilde{E}$ is the undamaged elastic moduli. The distance y_t is

$$y_t = \frac{1}{N_t}\int_{y_0}^{h} ydN_t = \frac{4h - 2y_0 - (\tilde{E}/F_R)\varepsilon_t\,(3h + y_0)}{6 - 4(\tilde{E}/F_R)\varepsilon_t}$$

The force and moment equations of equilibrium are

$$N_c + N_t = 0$$
$$N_c y_c + N_t y_t = M$$

Using the force equilibrium equation and assuming linear strain distribution through the thickness, it is possible to obtain the strains ε_t and ε_c in terms of y_0 as

$$\varepsilon_t = -\frac{6hy_0}{(h - y_0)^2}\frac{F_R}{\tilde{E}} \quad ; \quad \varepsilon_c = \frac{6hy_0(h + y_0)}{(h - y_0)^3}\frac{F_R}{\tilde{E}}$$

Using the above relation, it is possible to reduce the moment equilibrium equation to a single cubic equation in y_0

$$M = \frac{-y_0(4h^2 + 9hy_0 + 3y_0^2)}{(h - y_0)^3}\,bh^2 F_R$$

The ultimate bending moment can be determined by differentiation with respect to y_0

$$\frac{dM}{dy_0} = 0$$

that yields $y_{0\,cr} = -0.175\,h$ at beam failure. Therefore, the rupture bending moment is

$$M_{cr} = 0.2715\,bh^2\,F_R$$

A simple test (ASTM D790 or D6272) can be used to obtain the bending moment at failure; in this example $M_{cr} = 25.1\,10^6\,N\,mm$. Therefore, F_R can be estimated as $F_R = 92\,MPa$. As it is customary in structural engineering, the equivalent bending strength is defined as

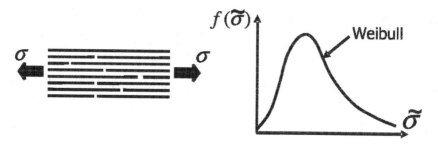

Fig. 8.5: One-dimensional Weibull-strength model.

$$\sigma_{Bcr} = \frac{M_{cr}}{S} = 0.407 \, F_R$$

where S is the section modulus (for a rectangular beam $S = \frac{2}{3}bh^2$). Note that according to (8.34), the tensile strength of the same material assuming the same kinetic equation (8.26) would be $\sigma_{T\ cr} = 0.25\ F_R$. This gives a ratio of equivalent bending to tensile strength equal to $\sigma_{B\ cr}/\sigma_{T\ cr} = 1.63$, which is in good agreement with experimental data $\sigma_{B\ cr}/\sigma_{T\ cr} = 1.6$ [68] obtained for unreinforced concrete and also with the value $\sigma_{B\ cr}/\sigma_{T\ cr} = 1.5$ recommended by the ACI Code [69].

8.1.6   Fiber Direction, Tension Damage

If a lamina is subjected to tensile stress in the fiber direction, it is reasonable to assume that the matrix carries only a small portion of the applied load and no damage is expected in the matrix during loading. Then, the ultimate tensile strength of the composite lamina can be accurately predicted by computing the strength of a bundle of fibers.

Fiber strength is a function of the gauge length used during fiber strength tests. The length scale that determines how much of the fiber strength is actually used in a composite is the ineffective length δ. Starting at a fiber break point, the ineffective length is that length over which a fiber recovers a large percentage of its load (say, 90%). Rosen [70] recognized this fact and proposed that the longitudinal ultimate strength of fibers embedded in a ductile-matrix can be accurately predicted by the strength of a dry bundle of fibers with length δ.

A dry bundle is defined as a number of parallel fibers of some given length and diameter which, if unbroken, carry the same load. After a fiber within a dry bundle fails, the load is shared equally by the remaining unbroken fibers. A dry bundle typically refers to fibers which have not yet been combined with a matrix. As tensile load is slowly applied to a dry bundle of fibers, the weaker fibers (with large flaw sizes) begin to fail according to a Weibull distribution (Figure 8.5) and the stress on the remaining unbroken fibers increases accordingly. The Weibull cumulative distribution [71] is often used to describe the cumulative probability $F(\widetilde{\sigma})$ that a fiber of length δ will fail at or below an effective stress $\widetilde{\sigma}$.

$$F(\widetilde{\sigma}) = 1 - \exp\left(-\frac{\delta}{L_0}\left(\frac{\widetilde{\sigma}}{\widetilde{\sigma}_0}\right)^m\right) \qquad (8.35)$$

The values of $\tilde{\sigma}_0$ and m, which represent the characteristic strength and dispersion of fiber strength, respectively, can be determined from fiber strength experiments performed with a gauge length L_0. Equation (8.35) can be simplified as

$$F(\tilde{\sigma}) = 1 - \exp\left(-\delta\alpha\tilde{\sigma}^m\right) \tag{8.36}$$

where

$$\alpha = \frac{1}{L_0\tilde{\sigma}_0^m} = \left[\frac{\Gamma(1+1/m)}{\tilde{\sigma}_{av}}\right]^m \frac{1}{L_0} \tag{8.37}$$

where $\Gamma(x)$ is the Gamma function and $\tilde{\sigma}_{av}$ is the average strength for a gauge length L_0. Equation (8.36) provides the percentage of fibers in a bundle which are broken as a function of the actual (or apparent) stress in the unbroken fibers. The percentage of fibers which are unbroken is $1 - F(\tilde{\sigma})$. The apparent stress, or bundle stress $\sigma = \sigma_b$, is equal to the applied load divided by the total fiber cross-sectional area. It is also equal to the product of the stress in unbroken fibers and the percentage of fibers which are unbroken

$$\sigma = \sigma_b = \tilde{\sigma}\exp(-\delta\alpha\tilde{\sigma}^m) \tag{8.38}$$

The value $\tilde{\sigma}_{\max}$, which maximizes (8.38), can be easily determined and is given by

$$\tilde{\sigma}_{\max} = (\delta\alpha m)^{-1/m} \tag{8.39}$$

The maximum (or critical) bundle strength σ_{cr} is determined by substituting (8.39) into (8.38)

$$\sigma_{cr} = (\delta\alpha m)^{-1/m}\exp(-1/m) = (\delta\alpha me)^{-1/m} \tag{8.40}$$

where e is the basis of the natural logarithms. The composite longitudinal tensile strength is [3, (4.68)]

$$F_{1t} = \left[V_f + \frac{E_m}{E_f}(1-V_f)\right]\sigma_{cr} \tag{8.41}$$

where V_f is the fiber volume fraction, and E_f and E_m are the fiber and matrix elastic Young's moduli, respectively.

Combining (8.36) and (8.39), we get

$$D_{1t}^{cr} = 1 - \exp(-1/m) \tag{8.42}$$

Therefore, the critical or maximum damage D_{1t}^{cr} for longitudinal tensile loading can be computed as the area fraction of broken fibers in the lamina prior to catastrophic failure [61, 62], which turns out to be a function of the Weibull shape modulus m only.

Example 8.3 *The data sheet of carbon fiber T300 from Toray^{TM} Carbon Fibers, Inc. gives average tensile strength of the fiber $\sigma_{av} = 3.53\ GPa$, and tensile modulus $E_f = 230\ GPa$. Also, the same data sheet provides results of tensile tests of a uni-directional (UD) composite with epoxy $E_m = 4.5\ GPa$ and fiber volume fraction $V_f = 0.6$. The tensile strength reported is $F_{1t} = 1860\ MPa$. Using this experimental data, and assuming a Weibull shape parameter $m = 8.9$, identify the damage model under tensile load. Then, formulate the damage model and implement it in Abaqus for a 1D bar element. Finally, obtain the strain vs. stress response of the UD composite.*

Solution to Example 8.3

MODEL IDENTIFICATION *From (8.41) and using the experimental data available, it is possible to obtain σ_{cr} as*

$$\sigma_{cr} = \frac{F_{1t}}{V_f + \frac{E_m}{E_f}(1 - V_f)} = 3060\ MPa$$

Then, the product $\delta\alpha$ can be obtained using (8.40) as

$$\delta\alpha = \frac{(\sigma_{cr})^{-m}}{me} = 3.92\ 10^{-33}$$

The properties $E_f = 230\ GPa$, $m = 8.9$, and $\delta\alpha = 3.92 \times 10^{-33}$ are sufficient for the identification of the model.

MODEL FORMULATION *Following a procedure similar to that shown in Section 8.1.5 to implement a damage model, the following items are needed*

Damage Activation Function *In this example, the effective stress is chosen as the independent variable. Therefore, the damage activation function can be written as*

$$g = \tilde{\sigma} - \hat{\gamma} \leq 0 \qquad (8.43)$$

where $\hat{\gamma}$ is the updated damage threshold. Assuming an initial threshold value $\sigma_0 = 0$, from the consistency conditions (8.15) and (8.19), $\hat{\gamma}$ is given by the highest value of effective stress seen by the material

$$\hat{\gamma} = \max(0, \tilde{\sigma}) \qquad (8.44)$$

Secant Constitutive Equation *In this example, the kinetic equation (8.1.3) is available in integral form and given explicitly by (8.36) as*

$$D = 1 - \exp\left(-\delta\alpha\hat{\gamma}^m\right) \quad when\ \tilde{\sigma} > 0;\ \varepsilon > 0 \qquad (8.45)$$

where the damage state does not depend on the actual load state $\tilde{\sigma}$; it only depends on the load history state $\hat{\gamma}$.

Substituting (8.45) into (8.5) and (8.7), and using strain equivalence, yields the constitutive equation

$$\sigma = E(D)\,\varepsilon = \exp\left(-\delta\alpha\hat{\gamma}^m\right) \tilde{E}\,\varepsilon \quad when\ \varepsilon > 0 \qquad (8.46)$$

Tangent Constitutive Equation *The tangent constitutive equation can be obtained by differentiating the secant constitutive equation as*

$$\dot{\sigma} = E(D)\dot{\varepsilon} + \dot{E}(D)\varepsilon \qquad (8.47)$$

The factor $\dot{E}(D)$ is zero when no new damage appears, i.e., during elastic loading or unloading. When damage occurs, (8.44) yields $\hat{\gamma} = \widetilde{E}\varepsilon$, and differentiating $E(D)$ in (8.46) we obtain

$$\dot{E}(D) = -\delta\alpha m \hat{\gamma}^{m-1} \exp\left(-\delta\alpha\hat{\gamma}^{m}\right) \widetilde{E}^{2}\dot{\varepsilon} \qquad (8.48)$$

The tangent constitutive equation is obtained by substituting (8.48) into (8.47) when damage occurs, or $\dot{E}(D) = 0$ when no new damage appears. Therefore, the tangent constitutive equation can be written as

$$\dot{\sigma} = \begin{cases} (1 - \delta\alpha m \hat{\gamma}^{m}) \exp\left(-\delta\alpha\hat{\gamma}^{m}\right) \widetilde{E}\dot{\varepsilon} & \text{when} \quad \varepsilon > \hat{\gamma}/\widetilde{E} \\ E(D)\,\dot{\varepsilon} & \text{when} \quad \varepsilon < \hat{\gamma}/\widetilde{E} \end{cases} \qquad (8.49)$$

NUMERICAL ALGORITHM *The 1D damage model is implemented in Abaqus using the UMAT subroutine* umat1d83.for, *available in [2]. The following items describe the procedure used to explicitly evaluate the damage constitutive equation.*[4]

   i. Read the strain at time t

$$\varepsilon_t$$

  ii. Compute the effective stress (assuming strain equivalence)

$$\widetilde{\sigma}_t = \widetilde{E}\varepsilon_t$$

 iii. Update the threshold value

$$\hat{\gamma}_t = \max(\hat{\gamma}_{t-1}, \widetilde{\sigma}_t)$$

  iv. Compute the damage variable

$$D_t = 1 - \exp\left(-\delta\alpha(\hat{\gamma}_t)^m\right)$$

   v. Compute the nominal stress

$$\sigma_t = (1 - D_t)\,\widetilde{E}\,\varepsilon_t$$

  vi. Compute the tangent stiffness

$$E_t^{dam} = \begin{cases} (1 - \delta\alpha m(\hat{\gamma}_t)^m) \exp\left(-\delta\alpha(\hat{\gamma}_t)^m\right) \widetilde{E} & \text{when } \hat{\gamma}_t > \hat{\gamma}_{t-1} \\ (1 - D_t)\,\widetilde{E} & \text{when } \hat{\gamma}_t = \hat{\gamma}_{t-1} \end{cases}$$

MODEL RESPONSE *The pseudo code below and user subroutine* umat1d83.for, *explained in Section 12.5.4 and available in [2], are used to model a 1D bar representative of a carbon fiber UD composite. The nominal stress-strain response is shown with a solid line in Figure 8.6. The UD composite fails at $\varepsilon_{cr} = 1.5\%$, in good agreement with the strain to failure reported by Toray.*

---

[4]See Section 8.4.1 for those cases for which it is not possible to integrate the constitutive equation explicitly.

    i. Retrieve **Ex_2.4.cae** *or use the instructions given in Example 2.4 (p. 76) to recreate the model*

```
Menu:    File, Open, [C:\SIMULIA\User\Ex_2.4\Ex_2.4.cae], OK
Menu:    File, Save As, [C:\SIMULIA\User\Ex_8.3\Ex_8.3.cae], OK
Menu:    File, Set Work Directory, [C:\SIMULIA\User\Ex_8.3], OK
```

   ii. Modify the geometry

```
Module: Part
    # on the left tree, expand (+) the following items:
    # expand: Model-1,  # expand: Parts (1)
    # expand: Part-1,   # expand: Features, # expand: Wire-1
    # right-click: Sketch, Edit
    # use dimension tool to change the length to [10]
Menu:    Add, Dimension
    # pick: Line, # create the dimension below the line, [10]
    X # to close the command, Done
Menu:    Feature, Regenerate
Module: Assembly
Menu:    Feature, Regenerate
```

  iii. Delete the existing Material-1 and create a new Material-1 with the properties and parameters required by the user material. Modify the cross-section.

```
Module: Property
Menu:    Material, Manager
    # select: Material-1, Delete, Yes
    Create, Name [Material-1], General, User Material
    Mechanical Constants [230.0E3  8.9 3.92E-33]
    # or right click Read from File, File: [props.txt]
    General, Depvar, Number of state variables [2],
    # SDV1: effective stress, see umat1d83.for
    # SDV2: damage, see umat1d83.for
    OK, # close Material Manager pop-up window
Menu:    Section, Edit, Section-1, Cross-sectional area: [1.0], OK
```

  iv. Delete the load and add a specified displacement

```
Module: Load
Menu:    Load, Delete, Load-1, Yes
Menu:    BC, Create
    Name [BC-2], Step: Step-1, Disp/Rota, Cont
    # pick: free-end node, Done, # checkmark: U1 [0.25], OK
```

   v. Remesh with five elements

```
Module: Mesh
    # to make the instance "independent", on the left tree
    # expand: Assembly, # expand: Instances
    # right-click: Part-1-1, Make Independent
Menu:    Seed, Instance, Approximate global size [2], Apply, OK
Menu:    Mesh, Element Type
    Element Library: Standard, Geometric Order: Linear
    Family: Trus # verify T2D2 is chosen, OK
```

```
Menu:    Mesh, Instance, Yes
Menu:    View, Assembly Display Options
    Tab: Mesh, # checkmark: Show node labels, OK
    # create a set to track the history of some variables
Menu:    Tools, Set, Create
    Name [Set-1], Type: Element, Cont
    # pick: element between nodes 5 and 6, Done
```

vi. *Modify the step to define the incrementation procedure, using 50 steps. A pseudo-time period of 50 is used but any time period could be used because it is not real time. Also, add solution dependent variables (SDV) to the Output Requests.*

```
Module: Step
Step:    Step-1
    # set up an incremental analysis over 50 units of pseudo time
Menu:    Step, Edit, Step-1
    Tab: Basic, Time Period [50]
    Tab: Incrementation, Type: Fixed, Maximum number of incr. [50]
    Increment size [1], OK
Menu:    Output, Field Output Requests, Edit, F-Output-1
    # expand: State/Field/User/Time, # checkmark: SDV, OK
Menu:    Output, History Output Requests, Edit, H-Output-1
    Domain: Set: Set-1
    # uncheck: all
    # expand: Stresses, # expand: S, # checkmark: S11
    # expand: Strains,  # expand: E, # checkmark: E11
    # expand: State/Field/User/Time, # checkmark: SDV, OK
```

vii. *Edit the Job to define the user material subroutine*

```
Module: Job
Menu:    Job, Manager
    # select: Job-1, Edit
    Tab: General, User Subroutine, Select, [umat1d83.for], OK, OK
    Submit, # when Completed, Results
```

viii. *Visualize the results. Failure occurs at Increment: 30, with strain $E = 1.5 \times 10^{-2}$, Cauchy stress $\sigma = 3059.0\ MPa$, effective stress SDV1=3450 MPa, and SDV2, damage, D=0.1132.*

```
Module: Visualization
Menu:    Plot, Contours, On Deformed Shape
Menu:    Result, Step/Frame
    # select: Increment 30 and verify your results
    # close Step/Frame pop-up window
    # plotting the stress vs. strain results
Menu:    Result, History Output
    # select: E11, Save As [e11], OK
    # select: S11, Save As [s11], OK
    # close History Output pop-up window
Menu:    Tools, XYData, Create
    Source: Operate on XY data, Cont, Operators: Combine
    # select: e11, Add to Expr., # select: s11, Add to Expr.
```

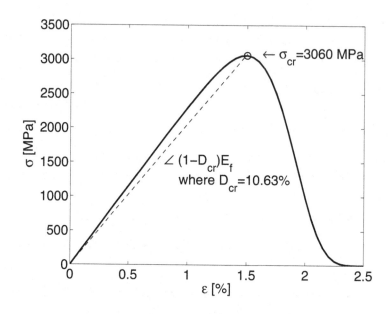

Fig. 8.6: Fiber tensile damage model response.

```
Save As [s11-vs-e11], OK, Plot Expression
```

Your results must be similar to those shown in Figure 8.6.

8.1.7   Fiber Direction, Compression Damage

Many models have been proposed trying to improve the prediction of compression strength of composites first introduced by Rosen [72]. The literature encompasses fiber buckling modes [20, 24, 73, 74], kink-band models [75], and kink-bands induced by buckling [76]. In fiber buckling models, it is assumed that buckling of the fibers initiates a process that leads to the collapse of the material [72]. Rosen's model has been refined with the addition of initial fiber misalignment and non-linear shear stiffness [73]. Experimental evidence suggests that fiber buckling of perfectly aligned fibers (Rosen's model) is an imperfection sensitive problem (see Section 4.1.1). Therefore, small amounts of imperfection (misalignment) cause large reductions in the buckling load, thus the reduction of the compression strength with respect to Rosen's prediction. Each fiber has a different value of fiber misalignment. The probability of finding a fiber with misalignment angle α is given by a Gaussian distribution [77, 78].

An optical technique [78] can be used to measure the misalignment angle of each fiber in the cross-section. The resulting distribution of fiber misalignment was shown to be Gaussian (Figure 8.7) by using the cumulative distribution function (CDF) plot and the probability plot [77]. Therefore, the probability density is

$$f(z) = \frac{\exp(-z^2)}{\Lambda\sqrt{2\pi}} \quad ; \quad z = \frac{\alpha}{\Lambda\sqrt{2}} \qquad (8.50)$$

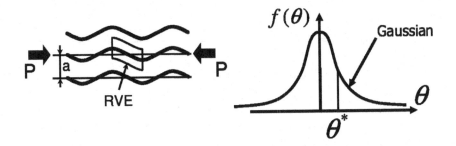

Fig. 8.7: Experimental distribution of misalignment angle in a unidirectional lamina [77].

where Λ is the standard deviation and α is the continuous random variable, in this case being equal to the misalignment angle. The CDF gives the probability of obtaining a value smaller than or equal to some value of α, as follows

$$F(z) = \text{erf}(z) = \frac{2}{\sqrt{\pi}} \int_0^z exp(-z'^2)dz' \qquad (8.51)$$

where $\text{erf}(z)$ is the error function.

The relationship between the buckling stress and the imperfection (misalignment) is known in stability theory as the imperfection sensitivity curve. Several models from the literature can be used to develop this type of curve. The deterministic model, similar to the one presented by Wang [73], is developed in [20] but using the representation of the shear response given by Equation (8.52).

The shear stress-strain response of polymer-matrix composites can be represented [24, 76] by

$$\sigma_6 = F_6 \tanh\left(\frac{G_{12}}{F_6}\gamma_6\right) \qquad (8.52)$$

where γ_6 is the in-plane shear strain. Furthermore, G_{12} is the initial shear stiffness and F_6 is the shear strength, which are obtained by fitting the stress-strain experimental data. Complete polynomial expansions [79] fit the experimental data well but they are not antisymmetric with respect to the origin. This introduces an artificial asymmetric bifurcation during the stability analysis [74]. Shear experimental data can be obtained by a variety of techniques including the ±45 coupon, 10° off-axis, rail shear, Iosipescu, Arcan, and torsion tests [80]. The nonlinear shear stress-strain curve should be measured for the actual composite being tested in compression.

Barbero [20] derived the equilibrium stress σ_{eq} as a function of the shear strain and the misalignment angle as

$$\widetilde{\sigma}_{eq}(\alpha, \gamma_6) = \frac{F_6}{2(\gamma_6 + \alpha)} \frac{(\sqrt{2} - 1)a + (\sqrt{2} + 1)(b - 1)}{1 - a + b}$$

$$\begin{aligned} a &= \exp(\sqrt{2}g) - \exp(2g) \\ b &= \exp(2g + \sqrt{2}g) \end{aligned} \qquad (8.53)$$

$$g = \frac{\gamma_6 G_{12}}{F_6}$$

with G_{12} and F_6 as parameters. Note that if the shear behavior is assumed to be linear $\widetilde{\sigma}_6 = G_{12}\gamma_6$ [81], then (8.53) does not have a maximum with respect to γ_6 and thus misaligned fibers embedded in a linear elastic matrix do not buckle. On the contrary, by using the hyperbolic tangent representation of shear (8.52), a maximum with respect to γ_6 is shown in (8.53). The maxima of the curves $\widetilde{\sigma}(\gamma_6)$ as a function of the misalignment angle α is the imperfection sensitivity curve, which represents the compression strength of a fiber (and surrounding matrix) as a function of its misalignment. For negative values of misalignment, it suffices to assume that the function is symmetric $\widetilde{\sigma}(-\alpha) = \widetilde{\sigma}(\alpha)$.

The stress carried by a fiber reduces rapidly after reaching its maximum because the load-carrying capacity of a buckled fiber is much lower than the applied load. Several models can be constructed depending on the assumed load that a fiber carries after buckling. A lower bound can be found assuming that buckled fibers carry no more load because they have no post-buckling strength. According to the imperfection sensitivity equation (8.53), fibers with large misalignment buckle under low applied stress. If the post-buckling strength is assumed to be zero, the applied stress is redistributed onto the remaining, unbuckled fibers, which then carry a higher effective stress $\widetilde{\sigma}(\alpha)$. At any time during loading of the specimen, the applied load σ (applied stress times initial fiber area) is equal to the effective stress times the area of fibers that remain unbuckled

$$\sigma = \bar{\sigma}(\alpha)[1 - D(\alpha)] \qquad (8.54)$$

where $0 \leq D(\alpha) \leq 1$ is the area of the buckled fibers per unit of initial fiber area. For any value of effective stress, all fibers having more than the corresponding value of misalignment have buckled. The area of buckled fibers $D(\alpha)$ is proportional to the area under the normal distribution located beyond the misalignment angle $\pm\alpha$.

Equation (8.54) has a maximum that corresponds to the maximum stress that can be applied to the composite. Therefore, the compression strength of the composite is found as

$$\sigma_c = \max \left[\bar{\sigma}(\alpha) \int_{-\alpha}^{\alpha} f(\alpha')d\alpha' \right] \qquad (8.55)$$

where $\bar{\sigma}(\alpha)$ is given by (8.53) and $f(\alpha')$ is given by (8.50). The maximum of (8.54), given by (8.55), is a unique value for the compression strength of the composite that incorporates both the imperfection sensitivity and the distribution of fiber misalignment. Note that the standard deviation Λ is a parameter that describes the actual, measured distribution of fiber misalignment, and it is not to be chosen arbitrarily as a representative value of fiber misalignment for all the fibers.

Since the distribution given in (8.50) cannot be integrated in closed form, (8.55) is evaluated numerically. However, it is advantageous to develop an explicit formula so that the compression strength can be easily predicted. Following the explicit formulation in [20], the compression strength of the unidirectional composite, explicitly in terms of the standard deviation of fiber misalignment Λ, the in-plane shear stiffness G_{12}, and the shear strength F_6 is

$$\frac{F_{1c}}{G_{12}} = (1 + 4.76 B_a)^{-0.69} \tag{8.56}$$

where 4.76 and -0.69 are two constants chosen to fit the numerical solution to the exact problem [20], with the dimensionless group B_a given by

$$B_a = \frac{G_{12}\Lambda}{F_6} \tag{8.57}$$

The misalignment angle of the fibers that buckle just prior to compression failure is given by [20, (23)]

$$
\begin{aligned}
\alpha_{cr} &= a/b \\
a &= 1019.011 G_{12} C_2^2 \Lambda^3 - 375.3162 C_2^3 \Lambda^4 - 845.7457 G_{12}^2 C_2 \Lambda^2 \\
&\quad + g \left(282.1113 G_{12} C_2 \Lambda^2 - 148.1863 G_{12}^2 \Lambda - 132.6943 C_2^2 \Lambda^3 \right) \\
b &= 457.3229 C_2^3 \Lambda^3 - 660.77 G_{12} C_2^2 \Lambda^2 - 22.43143 G_{12}^2 C_2 \Lambda \\
&\quad + g \left(161.6881 C_2^2 \Lambda^2 - 138.3753 G_{12} C_2 \Lambda - 61.38939 G_{12}^2 \right) \\
g &= \sqrt{C_2 \Lambda \left(8.0 C_2 \Lambda - 9.424778 G_{12} \right)} \\
C_2 &= -G_{12}^2/(4F_6)
\end{aligned}
\tag{8.58}
$$

Additionally, the shear strain at failure is

$$\gamma_{cr} = -\alpha_{cr} + \sqrt{\alpha_{cr}^2 + \frac{3}{2}\frac{\pi F_6 \alpha_{cr}}{G_{12}}} \tag{8.59}$$

In summary, when a fiber-reinforced lamina is compressed, the predominant damage mode is fiber buckling. However, the buckling load of the fibers is lower than that of the perfect system because of fiber misalignment, so much that a small amount of fiber misalignment could cause a large reduction in the buckling load. For each misalignment angle α, the composite area-fraction with buckled fibers $D(\alpha)$, corresponding to fibers with misalignment angle greater than α, can be taken as a measure of damage. If the fibers are assumed to have no post-buckling strength, then the applied stress is redistributed onto the remaining unbuckled fibers, which will be carrying a higher effective stress. The applied stress, which is lower than the effective stress by the factor $(1 - D)$, has a maximum, which corresponds to the compressive strength of the composite. Therefore, it is possible to compute the critical damage D_{1c} for longitudinal compressive loading as

$$D_{1c}^{cr} = 1 - \Omega_{1c} = 1 - erf\left(\frac{\alpha_{cr}}{\Lambda\sqrt{2}}\right) \tag{8.60}$$

where erf is the error function, Λ is the standard deviation of the actual Gaussian distribution of fiber misalignment (obtained experimentally [78]), and α_{cr} is the critical misalignment angle at failure. The three-dimensional theoretical formulation is developed in the next three sections.

8.2   Multidimensional Damage and Effective Spaces

The first step in the formulation of a general multidimensional damage model is to define the damage variable as well as the effective stress and strain spaces, as shown in this section. The second step is to define the form of either the Helmholtz free energy or the Gibbs energy and from them derive the thermodynamic forces conjugate to the state variables representing damage and hardening, as shown in Section 8.3. The third step is to derive the kinetic laws governing the rate of damage hardening, which are functions of the damage and hardening potentials, as shown in Section 8.4.

Experimental knowledge of the degradation and subsequent material response is used to guide the selection on the variable used to represent damage. A second-order damage tensor $\mathbf{D}$ can be used to represent damage of orthotropic fiber-reinforced composite materials, following Kachanov-Rabotnov's approach [58, 82]. For composite materials reinforced with stiff and strong fibers, damage can be accurately represented by a second-order tensor[5] with principal directions aligned with the material directions $(1, 2, 3)$ [83–87]. This is due to the fact that the dominant modes of damage are microcracks, fiber breaks, and fiber-matrix debond, all of which can be conceptualized as cracks either parallel or perpendicular to the fiber direction.[6] Therefore, the damage tensor can be written as

$$\mathbf{D} = D_{ij} = D_i \delta_{ij} \quad \text{no sum on i} \tag{8.61}$$

where D_i are the eigenvalues of $\mathbf{D}$, which represent the net stiffness reduction along the principal material directions n_i, and δ_{ij} is the Kronecker delta ($\delta_{ij} = 1$ if $i = j$, or zero otherwise). The integrity tensor is also diagonal, and using energy equivalence (8.8), we have

$$\mathbf{\Omega} = \Omega_{ij} = \sqrt{1 - D_i} \delta_{ij} \quad \text{no sum on i} \tag{8.62}$$

The integrity tensor is always symmetric and positive, because the net area reduction must be positive definite during damage evolution [88]. Both tensors are diagonal when represented in the principal system. Introducing a symmetric fourth-order tensor, $\mathbf{M}$, called the *damage effect tensor*, as

$$\mathbf{M} = M_{ijkl} = \frac{1}{2} \left(\Omega_{ik}\Omega_{jl} + \Omega_{il}\Omega_{jk} \right) \tag{8.63}$$

---

[5]Tensors are denoted by boldface type, or by their components with index notation.

[6]Strictly speaking, damage is transversely isotropic since cracks can be aligned along any direction in the 2-3 plane.

The transformation of stress and strain between the effective ($\tilde{\sigma}$) and damaged (σ) configurations is accomplished as follows

$$\tilde{\varepsilon} = \mathbf{M} : \varepsilon$$

$$\tilde{\sigma} = \mathbf{M}^{-1} : \sigma \qquad\qquad \tilde{\varepsilon}^e = \mathbf{M} : \varepsilon^e \qquad (8.64)$$

$$\tilde{\varepsilon}^p = \mathbf{M} : \varepsilon^p$$

where an over-bar indicates that the quantity is evaluated in the effective configuration and the superscripts e, p, denote quantities in the elastic and plastic domains, respectively.

By the energy equivalence hypothesis [59], it is possible to define the constitutive equation in the effective configuration (Figure 8.1.c) as

$$\tilde{\sigma} = \tilde{\mathbf{C}} : \tilde{\varepsilon}^e \quad ; \qquad \tilde{\varepsilon}^e = \tilde{\mathbf{C}}^{-1} : \tilde{\sigma} = \tilde{\mathbf{S}} : \tilde{\sigma} \qquad (8.65)$$

where the fourth-order tensors $\mathbf{C}$ and $\mathbf{S}$ denote the secant stiffness tensor and compliance tensor, respectively. The stress-strain equations in the damaged configuration (Figure 8.1.b) are obtained by substituting (8.65) into (8.64),

$$\begin{aligned}
\sigma &= \mathbf{M} : \tilde{\sigma} = \mathbf{M} : \tilde{\mathbf{C}} : \tilde{\varepsilon}^e, & \varepsilon^e &= \mathbf{M}^{-1} : \tilde{\varepsilon}^e = \mathbf{M}^{-1} : \tilde{\mathbf{S}} : \tilde{\sigma}, \\
\sigma &= \mathbf{M} : \tilde{\mathbf{C}} : \mathbf{M} : \varepsilon^e, & \varepsilon^e &= \mathbf{M}^{-1} : \tilde{\mathbf{S}} : \mathbf{M}^{-1} : \sigma, \qquad (8.66) \\
\sigma &= \mathbf{C} : \varepsilon^e & \varepsilon^e &= \mathbf{S} : \sigma
\end{aligned}$$

with

$$\mathbf{C} = \mathbf{M} : \tilde{\mathbf{C}} : \mathbf{M} \qquad\qquad \mathbf{S} = \mathbf{M}^{-1} : \tilde{\mathbf{S}} : \mathbf{M}^{-1} \qquad (8.67)$$

The explicit form of these tensors are presented in Appendix C. Given that the tensor $\mathbf{M}$ is symmetric, the secant stiffness and compliance tensors are also symmetric.

8.3 Thermodynamics Formulation

The damage processes considered in this chapter can be described by a series of equilibrium states reached while the system traverses a non-equilibrium path due to the irreversibility of damage and plasticity. In general, the current state of a system (e.g., stress, stiffness, compliance) depends on the current state (e.g., strain) as well as on the history experienced by the system. This is the case for viscoelastic materials discussed in Chapter 7. However, for damaging and elastoplastic materials, the current state can be described in terms of the current strain and the *effects of history* on the material, which in this chapter are characterized by the damage tensor $\mathbf{D}$ and the plastic strain tensor ε^p.

8.3.1   First Law

The *first law* of thermodynamics states that any increment of internal energy of the system is equal to the heat added to the system minus the work done by the system on its surroundings

$$\delta U = \delta Q - \delta W \tag{8.68}$$

The system under consideration in this section is a representative volume element (RVE), which is the smallest volume element that contains sufficient features of the microstructure and irreversible processes, such as damage and plasticity, to be representative of the material as a whole. Further discussions about the RVE can be found in Chapter 6.

In rate form, (8.68) is

$$\dot{U} = \dot{Q} - \dot{W} \tag{8.69}$$

where

$$\dot{U} = \frac{d}{dt} \int_{\Omega} \rho u \, dV \tag{8.70}$$

Here ρ is the density, Ω is the volume of the RVE, and u is the *internal energy density*, which is an internal variable and a potential function.[7]

For a deformable solid, the rate of work done *by* the system is minus the product of the stress *applied* on the system times the rate of strain

$$\dot{W} = - \int_{\Omega} \boldsymbol{\sigma} : \dot{\boldsymbol{\varepsilon}} \, dV \tag{8.71}$$

where $\boldsymbol{\varepsilon}$ is the *total* strain (see (8.126)).

The heat flow into the RVE is given by

$$\dot{Q} = \int_{\Omega} \rho r \, dV - \int_{\partial\Omega} \mathbf{q} \cdot \mathbf{n} \, dA \tag{8.72}$$

where r is the radiation heat per unit mass, $\mathbf{q}$ is the heat flow vector per unit area, and $\mathbf{n}$ is the outward normal vector to the surface $\partial\Omega$ enclosing the volume Ω. Since the volume Ω of the RVE does not change with time, and using the divergence theorem,[8] the *first law at the local level* becomes

$$\rho \dot{u} = \boldsymbol{\sigma} : \dot{\boldsymbol{\varepsilon}} + \rho r - \nabla \cdot \mathbf{q} \tag{8.73}$$

The internal energy accounts for all the energy stored into the system. For example, a system undergoing elastic deformation $\delta\varepsilon^e$, raising temperature δT, and damage in the form of cracks of area growing by δA_c undergoes a change of internal energy u given[9] by

$$\delta u = \boldsymbol{\sigma} : \delta\boldsymbol{\varepsilon}^e + C_p \delta T - (G - G_c)\delta A_c \tag{8.74}$$

---

[7]The values of the potential functions depend on the state and not on the path or process followed by the system to reach such state [89].

[8]$(\int_{\partial\Omega} \mathbf{q} \cdot \mathbf{n} dA = \int_{\Omega} \nabla \cdot \mathbf{q} dV)$; $div(\mathbf{q}) = \nabla \cdot \mathbf{q} = \partial q_i / \partial x_i$

[9]Thermodynamics custom and [90] are followed here in representing the internal energy with the letter u, not to be confused with the displacement vector $\mathbf{u}$ used elsewhere.

where G is the strain energy release rate, G_c is the surface energy needed to create the increment of the two surfaces of an advancing crack, and $C_p = C_v$ is the specific heat capacity of the solid.

In general $\varepsilon = \varepsilon(\sigma, u, s_\alpha)$ where s_α are internal variables. Let's assume for the time being that the system is adiabatic, i.e., $\rho r - \nabla \cdot \mathbf{q} = 0$. Further, if there are no dissipative effects or heat transfer, then u is a function of ε only, $u = u(\varepsilon^e)$, where ε^e is the elastic strain. In such case, the internal energy density reduces to the strain energy density, which in rate form is

$$\dot{\varphi}(\varepsilon) = \sigma : \dot{\varepsilon}^e \tag{8.75}$$

and the complementary strain energy density is

$$\dot{\varphi}^*(\varepsilon) = \sigma : \dot{\varepsilon}^e - \dot{\varphi} = \dot{\sigma} : \varepsilon^e \tag{8.76}$$

8.3.2 Second Law

The *second law* of thermodynamics formalizes the fact that heat flows from hot to cold. Mathematically, the heat flow $\mathbf{q}$ has opposite direction to the gradient[10] of temperature T, which is formally written as

$$\mathbf{q} \cdot \nabla T \leq 0 \tag{8.77}$$

where the equal sign holds true only for adiabatic processes, i.e., when there is no heat exchange and thus no thermal irreversibility.

Let's visualize a process of heat transfer from a hot reservoir to a cold reservoir, happening in such a way that no heat is lost to, and no work is exchanged with, the environment. Once heat has flowed to the cold reservoir, it is impossible to transfer it back to the hot reservoir without adding external work. That is, the process of heat transfer is irreversible even though, on account of the first law energy balance (8.73), no energy has been lost. For future use (8.77) can be written[11] as

$$\mathbf{q} \cdot \nabla T^{-1} \geq 0 \tag{8.78}$$

The second law justifies the introduction of a new internal variable, the entropy density $s = s(u, \varepsilon)$, which is also a potential function [91]. According to the second law, the entropy density rate is $\dot{s} \geq 0$, where the equal sign holds true only for adiabatic processes.

Assume the specific entropy $s = s(u, \varepsilon)$ is a potential function such that for a reversible process [91]

$$ds = \left(\frac{\delta q}{T} \right)_{rev} \tag{8.79}$$

with $\delta Q = \int_\Omega \rho \, \delta q \, d\Omega$, where $\delta q = r - \rho^{-1} \nabla \cdot \mathbf{q}$ is the heat input per unit mass, and $S = \int_\Omega s \, \rho \, d\Omega$ is the entropy. We use δ, not d, to emphasize that δq is not the differential (perfect or total) of any (potential) function.

---

[10]The gradient of a scalar yields a vector, $\nabla T = \partial T / \partial x_i$
[11]$\nabla T^{-1} = -T^{-2} \nabla T$

As a preamble to the definition of conjugate variables (see (8.86, 8.92, 8.99)), note that using (8.79), the first law can be rewritten for a *reversible* process on an ideal gas $(pv = RT)$, as the *Gibbs* equation for an ideal gas,

$$du = T\, ds - p\, dv \tag{8.80}$$

where v is the specific volume (volume per unit mass). It can be seen in (8.80) that v is conjugate to $-p$ for calculating work input for an ideal gas and s is conjugate to T for calculating thermal energy input.

For a *cyclic reversible* process returning to its initial state characterized by state variables (e.g., u, T, ε), by virtue of (8.79) we have $\oint ds = \oint \left(\frac{\delta q}{T}\right)_{rev} = 0$. Since s is a potential function but q is not, for an *irreversible* process we have $\oint ds = 0$ but $\oint \left(\frac{\delta q}{T}\right)_{irrev} < 0$, as corroborated by experiments. The heat δq entering at temperature T_i provides less entropy input $\delta q/T_i$ than the entropy output $\delta q/T_o$ leaving the same cycle at temperature $T_o < T_i$ (see also [92, Example 6-2]). Since entropy is a potential function, and therefore a state variable, it always satisfies $\oint ds = 0$. Therefore, a negative net entropy supply must be compensated by internal entropy production. The entropy of a system can be raised or lowered by adding or extracting heat (in the form of $\delta q/T$) *but it is always raised by internal irreversible processes such as crack formation and so on* (positive dissipation principle).

Adiabatic systems do not exchange heat with the surroundings ($\delta q = 0$), so the only change in entropy is due to internal irreversibility $\dot{s} \geq 0$, where the equal sign holds for reversible processes only. Note that any system and its surroundings can be made adiabatic by choosing sufficiently large surroundings, e.g., the universe. For an arbitrary system, the total entropy rate is greater than (or equal to) the net entropy input due to heat

$$\dot{s} \geq \frac{r}{T} - \frac{1}{\rho}\nabla \cdot \left(\frac{\mathbf{q}}{T}\right) \tag{8.81}$$

The left-hand side of (8.81) represents the total entropy rate of the system. The right-hand side of (8.81) represents the external entropy supply rate. The difference is the internal entropy production rate

$$\dot{\gamma}_s = \dot{s} - \frac{r}{T} + \frac{1}{\rho}\nabla \cdot \left(\frac{\mathbf{q}}{T}\right) \geq 0 \tag{8.82}$$

Equation 8.82 is called the *local Clausius-Duhem inequality*. Noting that $\nabla \cdot (T^{-1}\mathbf{q}) = T^{-1}\nabla \cdot \mathbf{q} + \mathbf{q}\nabla\, T^{-1}$ results in

$$\dot{\gamma}_s = \dot{s} - \frac{1}{\rho T}(\rho r - \nabla \cdot \mathbf{q}) + \frac{1}{\rho}\mathbf{q} \cdot \nabla\, T^{-1} \geq 0 \tag{8.83}$$

where the first two terms represent the local entropy production due to local dissipative phenomena, and the last term represents the entropy production due to heat conduction[12] [91]. Assuming it is possible to identify all local dissipative phenom-

---

[12]Even absent local dissipative phenomena, $\mathbf{q} \cdot \nabla T^{-1} \geq 0$ represents the well-known fact that heat flows opposite to the temperature gradient ∇T.

ena, their contributions can be written as products of conjugate variables $p_\alpha \dot{s}_\alpha \geq 0$, and (8.83) can be written as

$$\rho T \dot{\gamma}_s = p_\alpha \dot{s}_\alpha + T \mathbf{q} \cdot \nabla T^{-1} \geq 0 \qquad (8.84)$$

where $\alpha = 1 \ldots n$, spans the total number of dissipative phenomena considered. Note that the dissipation is a scalar given by the contracted product of a thermodynamic force p_α times the increment of a measurable state variable s_α. The state variable, also called thermodynamic flux, describes univocally the effects of history (e.g., yield, damage) on the material. Note that γ_s is defined as an entropy, not as a dissipation heat, so that it is a potential function, while q is not.

For the particular case of damage due to penny-shaped cracks growing self-similarly [63], the state variable is the crack area A_c and the thermodynamic force is the energy available to grow the cracks $p_c = G - G_c$, which is equal to the difference between the energy release rate (ERR) G and the critical ERR $G_c = 2\gamma_c$, the latter being equal to twice the surface energy because two surfaces must be created every time a crack appears (see Chapter 10). In this case, the dissipation (heat) is $\rho T \dot{\gamma} = p_c \dot{A}_c$.

From the first law (8.73), considering an adiabatic process $(\rho r - \nabla \cdot \mathbf{q})$ and using the chain rule $\dot{u} = \partial u / \partial \varepsilon : \dot{\varepsilon}$ we have

$$\left[\boldsymbol{\sigma} - \rho \frac{\partial u}{\partial \varepsilon} \right] : \dot{\varepsilon} = 0 \qquad (8.85)$$

Since $\dot{\varepsilon} = 0$ would be a trivial solution, the stress tensor, conjugate to strain, is defined as

$$\boldsymbol{\sigma} = \rho \frac{\partial u}{\partial \varepsilon} \qquad (8.86)$$

The Clausius-Duhem inequality (8.83) for an *isothermal* $(\nabla T = 0)$ system reduces to

$$\dot{\gamma}_s = \dot{s} - \frac{1}{\rho T} (\rho r - \nabla \cdot \mathbf{q}) \geq 0 \qquad (8.87)$$

and using the first law we get

$$\rho T \dot{\gamma}_s = \rho T \dot{s} - (\rho \dot{u} - \boldsymbol{\sigma} : \varepsilon) \geq 0 \qquad (8.88)$$

The *Helmholtz free energy (HFE) density* is defined as

$$\psi(T, \varepsilon, s_\alpha) = u - Ts \qquad (8.89)$$

which is also a potential function. The corresponding extensive function is the Helmholtz free energy[13] $A = \int_\Omega \rho \psi dV$. The rate of change of HFE density is

$$\dot{\psi} = \dot{u} - \dot{T}s - T\dot{s} \qquad (8.90)$$

---

[13]The nomenclature of [90] has been used.

and introducing (8.88), with $\dot{\gamma}_s = 0$ at an equilibrium state, we get

$$\rho\dot{\psi} = -\rho s\dot{T} + \boldsymbol{\sigma} : \dot{\boldsymbol{\varepsilon}} \tag{8.91}$$

from which an alternative definition of stress, conjugate to strain, is found as

$$\boldsymbol{\sigma} = \rho\frac{\partial\psi}{\partial\boldsymbol{\varepsilon}} = \mathbf{C} : \boldsymbol{\varepsilon} \tag{8.92}$$

where the secant elastic stiffness, which is affected by dissipative phenomena, including damage, is defined as

$$\mathbf{C}(s_\alpha) = \rho\frac{\partial^2\psi}{\partial\boldsymbol{\varepsilon}^2} \tag{8.93}$$

Using the first law (8.73) in the internal entropy production per unit volume, or local Clausius-Duhem inequality (8.83), and expanding $\nabla \cdot (\mathbf{q}T^{-1}) = T^{-1}\nabla \cdot \mathbf{q} + \mathbf{q} \cdot \nabla T^{-1}$, we get

$$\rho T\dot{\gamma}_s = \frac{\mathbf{q}}{T} \cdot \nabla T^{-1} - \rho\left(\dot{\psi} + s\dot{T} - \rho^{-1}\boldsymbol{\sigma} : \dot{\boldsymbol{\varepsilon}}\right) \geq 0 \tag{8.94}$$

Realizing that $\nabla T^{-1} = -\nabla T/T^2$, the Clausius-Duhem inequality becomes

$$T\rho\dot{\gamma}_s = \boldsymbol{\sigma} : \dot{\boldsymbol{\varepsilon}} - \rho\left(\dot{\psi} + s\dot{T}\right) - \frac{\mathbf{q}}{T} \cdot \nabla T \geq 0 \tag{8.95}$$

Since the HFE density is a function of the internal variables ε, T, s_α, we have

$$\dot{\psi} = \left.\frac{\partial\psi}{\partial\boldsymbol{\varepsilon}}\right|_{T,s_\alpha} : \dot{\boldsymbol{\varepsilon}} + \left.\frac{\partial\psi}{\partial T}\right|_{\boldsymbol{\varepsilon},s_\alpha}\dot{T} + \left.\frac{\partial\psi}{\partial s_\alpha}\right|_{\boldsymbol{\varepsilon},T}\dot{s}_\alpha \quad ; \quad \alpha = 1\ldots n \tag{8.96}$$

where $\left.\dfrac{\partial}{\partial y}\right|_x$ represents the partial derivative with respect to y at constant x.

Inserting (8.96) into (8.95), using (8.89), (8.92), and $\nabla T^{-1} = -\nabla T/T^2$, the second law can be written as follows

$$\dot{\gamma} = \rho T\dot{\gamma}_s = -\rho\frac{\partial\psi}{\partial s_\alpha}\dot{s}_\alpha + T\mathbf{q} \cdot \nabla T^{-1} \geq 0 \tag{8.97}$$

where $\dot{\gamma}$ is the heat dissipation rate per unit volume. Comparing (8.97) to (8.84) it becomes clear that $-\rho\partial\psi/\partial s_\alpha = p_\alpha$ are the thermodynamic forces conjugated to s_α, which provides a definition for the thermodynamic forces.

The complementary free-energy density, or *Gibbs energy density*, is defined as

$$\chi = \rho^{-1}\boldsymbol{\sigma} : \boldsymbol{\varepsilon} - \psi \tag{8.98}$$

which is also a potential function. The corresponding extensive function is the Gibbs energy[14] $G = \int_\Omega \rho\chi dV$. From (8.98) it follows the definition of strain, conjugate

---

[14]The nomenclature of [90] has been used.

to stress, and the definition of the thermodynamic forces, conjugate to the state variables s_α, as

$$\varepsilon = \rho\frac{\partial\chi}{\partial\sigma} \quad ; \quad p_\alpha = \rho\frac{\partial\chi}{\partial s_\alpha} = -\rho\frac{\partial\psi}{\partial s_\alpha} \tag{8.99}$$

where s_α includes the damage variables and consequently p_α includes the thermodynamic damage forces (see Example 8.4).

The secant elastic compliance, which is affected by dissipative phenomena, including damage, is defined by

$$S(s_\alpha) = \rho\frac{\partial^2\chi}{\partial\sigma^2} \tag{8.100}$$

Example 8.4 *The following Gibbs free energy is proposed to represent the onset and accumulation of transverse matrix cracks resulting from transverse tension and in-plane shear loads:*

$$\chi = \frac{1}{2\rho}\left[\frac{\sigma_1^2}{\tilde{E}_1} + \frac{\sigma_2^2}{(1-D_2)^2\,\tilde{E}_2} + \frac{\sigma_6^2}{(1-D_6)^2\,2\tilde{G}_{12}} - \left(\frac{\tilde{\nu}_{21}}{\tilde{E}_2} + \frac{\tilde{\nu}_{12}}{\tilde{E}_1}\right)\frac{\sigma_1\sigma_2}{1-D_2}\right]$$

where $\tilde{E}_1$, $\tilde{E}_2$, $\tilde{\nu}_{12}$, $\tilde{\nu}_{21}$, and $\tilde{G}_{12}$ are the undamaged in-plane elastic orthotropic properties of a unidirectional lamina where the subscript $()_1$ denotes the fiber direction and $()_2$ denotes the transverse direction. The damage variables D_2 and D_6 represent the effect of matrix cracks. The proposed formulation distinguishes between active (D_{2+}) and passive damage (D_{2-}), corresponding to the opening or closure of transverse matrix cracks, respectively. The determination of the active damage variable is based on the following equation:

$$D_2 = D_{2+}\frac{\langle\sigma_2\rangle}{|\sigma_2|} + D_{2-}\frac{\langle-\sigma_2\rangle}{|\sigma_2|}$$

where $\langle x\rangle$ is defined as $\langle x\rangle := \frac{1}{2}(x + |x|)$.

For a lamina in a state of plane stress, subjected to in-plane stress only, without fiber damage $(D_1 = 0)$, and using the energy equivalence principle (8.8), derive expressions for a) the secant stiffness tensor, b) the effective stress, and c) the thermodynamic forces associated to the model. Use tensor components of strain $(\varepsilon_1, \varepsilon_2, \varepsilon_6)$.

Solution to Example 8.4 *The constitutive model is defined as the derivative of the Gibbs free energy with respect to the stress tensor*

$$\varepsilon = \rho\frac{\partial\chi}{\partial\sigma} = \mathbf{S} : \sigma$$

whith $\varepsilon_6 = \gamma_6/2$ and the compliance tensor $\mathbf{S}$ defined as

$$\mathbf{S} = \rho\frac{\partial^2\chi}{\partial\sigma^2}$$

The compliance tensor for plane stress $\mathbf{S}$ in Voigt contracted notation is

$$\mathbf{S} = \begin{bmatrix} \dfrac{1}{\tilde{E}_1} & -\dfrac{\tilde{\nu}_{21}}{\tilde{E}_2\,(1-D_2)} & 0 \\[3mm] -\dfrac{\tilde{\nu}_{12}}{\tilde{E}_1\,(1-D_2)} & \dfrac{1}{\tilde{E}_2\,(1-D_2)^2} & 0 \\[3mm] 0 & 0 & \dfrac{1}{2\tilde{G}_{12}\,(1-D_6)^2} \end{bmatrix}$$

The damage variables appear in S_{12}, S_{21}, S_{22}, and S_{66}. To perform tensor products using matrix multiplications, see (B.14) and (B.20). Using the energy equivalence principle and (8.67), the compliance matrix can be written as

$$\mathbf{S} = \mathbf{M}^{-1} : \tilde{\mathbf{S}} : \mathbf{M}^{-1}$$

where the undamaged compliance is

$$\tilde{\mathbf{S}} = \begin{bmatrix} \dfrac{1}{\tilde{E}_1} & -\dfrac{\tilde{\nu}_{21}}{\tilde{E}_2} & 0 \\[2mm] -\dfrac{\tilde{\nu}_{12}}{\tilde{E}_1} & \dfrac{1}{\tilde{E}_2} & 0 \\[2mm] 0 & 0 & \dfrac{1}{2\tilde{G}_{12}} \end{bmatrix}$$

and where the effective damage tensor $\mathbf{M}$, written in contracted notation, multiplied by the 3×3 version of the Reuter matrix (1.38) is

$$\mathbf{M} = \begin{bmatrix} 1 & 0 & 0 \\ 0 & (1 - D_2) & 0 \\ 0 & 0 & (1 - D_6) \end{bmatrix}$$

The stiffness tensor $\mathbf{C}$ is obtained by

$$\mathbf{C} = \mathbf{M} : \tilde{\mathbf{C}} : \mathbf{M}$$

where, for this particular example, the secant stiffness tensor is

$$\mathbf{C} = \begin{bmatrix} \dfrac{\tilde{E}_1}{1 - \tilde{\nu}_{21}\tilde{\nu}_{12}} & \dfrac{\tilde{\nu}_{12}\tilde{E}_2(1 - D_2)}{1 - \tilde{\nu}_{21}\tilde{\nu}_{12}} & 0 \\[3mm] \dfrac{\tilde{\nu}_{21}\tilde{E}_1(1 - D_2)}{1 - \tilde{\nu}_{21}\tilde{\nu}_{12}} & \dfrac{\tilde{E}_2(1 - D_2)^2}{1 - \tilde{\nu}_{21}\tilde{\nu}_{12}} & 0 \\[3mm] 0 & 0 & 2\tilde{G}_{12}\left(1 - \tilde{D}_6\right)^2 \end{bmatrix}$$

The effective stress $\tilde{\sigma}$ is related to the nominal stress σ by the effective damage tensor $\mathbf{M}$ using $\tilde{\sigma} = \mathbf{M}^{-1} : \sigma$, which yields

$$\tilde{\sigma}^T = \left\{ \sigma_1, \frac{\sigma_2}{1 - D_2}, \frac{\sigma_6}{1 - D_6} \right\}$$

The thermodynamic forces are obtained by using $Y = \rho \partial \chi / \partial D$, which for this particular example yield

$$\mathbf{Y} = \left\{ \begin{array}{c} Y_1 \\ Y_2 \\ Y_6 \end{array} \right\} = \left\{ \begin{array}{c} 0 \\[2mm] \dfrac{\sigma_2{}^2}{(1 - D_2)^3\,\tilde{E}_2} - \dfrac{\sigma_1\sigma_2\tilde{\nu}_{12}}{(1 - D_2)^2\,\tilde{E}_1} \\[4mm] \dfrac{\sigma_6{}^2}{(1 - D_6)^3\,2\tilde{G}_{12}} \end{array} \right\}$$

8.4   Kinetic Law in Three-Dimensional Space

The damage variable $\mathbf{D}$ introduced in Section 8.2 is a state variable that represents the history of what happened to the material. Next, a kinetic equation is needed to predict the evolution of damage in terms of the thermodynamic forces. Kinetic equations can be written directly in terms of internal variables as in (8.21) or as derivatives of potential functions. For three-dimensional problems, it is convenient to derive the kinetic law from a potential function, similar to the flow potential used in plasticity theory.

Two functions are needed. A damage surface $g(\mathbf{Y}(\mathbf{D}), \gamma(\delta)) = 0$ and a convex damage potential $f(\mathbf{Y}(\mathbf{D}), \gamma(\delta)) = 0$ are postulated. The damage surface delimits a region in the space of thermodynamic forces $\mathbf{Y}$ where damage does not occur because the thermodynamic force $\mathbf{Y}$ is inside the g-surface. The function $\gamma(\delta)$ accomplishes the expansion of g and f needed to model hardening. The damage potential controls the direction of damage evolution (8.102).

If the damage surface and the damage potential are identical ($g = f$), the model is said to be associated, and the computational implementation is simplified significantly. For convenience, the damage surface is assumed to be separable in the variables $\mathbf{Y}$ and γ, and written as the sum (see (8.11)-(8.12))

$$g(\mathbf{Y}(\mathbf{D}), \gamma(\delta)) = \hat{g}(\mathbf{Y}(\mathbf{D})) - (\gamma(\delta) + \gamma_0) \tag{8.101}$$

where $\mathbf{Y}$ is the thermodynamic force tensor, $\gamma(\delta)$ is the hardening function, γ_0 is the damage threshold, and δ is the hardening variable.

As a result of damage, $\hat{g}$ may grow, but the condition $g < 0$ must be satisfied. This is possible by *increasing* the value of the hardening function $\gamma(\delta)$, effectively allowing $\hat{g}(\mathbf{Y}(\mathbf{D}))$ to grow. Formally, the hardening function $\gamma(\delta)$ can be derived from the dissipation potential as per (8.99), (8.125), provided the form of the potential can be inferred from knowledge about the hardening process. Alternatively, the form of the function (e.g., polynomial, Prony series, etc.) can be chosen so that the complete model fits adequately the experimental data available. The latter approach is more often followed in the literature.

When $g = 0$, damage occurs, and a kinetic equation is needed to determine the magnitude and components of the damage $\dot{\mathbf{D}}$. This is accomplished by

$$\dot{\mathbf{D}} = \frac{\partial \mathbf{D}}{\partial \mathbf{Y}} = \dot{\lambda} \frac{\partial f}{\partial \mathbf{Y}} \tag{8.102}$$

where $\dot{\lambda}$ yields the magnitude of the damage increment and $\partial f / \partial \mathbf{Y}$ is a direction in $\mathbf{Y}$-space. To find the damage multiplier $\dot{\lambda}$, it is postulated that $\dot{\lambda}$ is also involved in the determination of the rate of change of the hardening variable, as follows

$$\dot{\delta} = \dot{\lambda} \frac{\partial g}{\partial \gamma} \tag{8.103}$$

There are two possible situations regarding g and $\dot{\lambda}$:

i. If $g < 0$, damage is not growing and $\dot{\lambda} = 0$, so $\dot{\mathbf{D}} = 0$.

ii. If $g = 0$, damage occurs and $\dot{\lambda} > 0$, so $\dot{\mathbf{D}} > 0$.

These are summarized by the Kuhn-Tucker conditions

$$\dot{\lambda} \geq 0 \quad ; \quad g \leq 0 \quad ; \quad \dot{\lambda} g = 0 \tag{8.104}$$

The value of $\dot{\lambda}$ can be determined by the consistency condition, which leads to

$$\dot{g} = \frac{\partial g}{\partial \mathbf{Y}} : \dot{\mathbf{Y}} + \frac{\partial g}{\partial \gamma} \dot{\gamma} = 0 \; ; \quad g = 0 \tag{8.105}$$

On the other hand, the rates of thermodynamic forces and hardening function can be written as

$$\dot{\mathbf{Y}} = \frac{\partial \mathbf{Y}}{\partial \varepsilon} : \dot{\varepsilon} + \frac{\partial \mathbf{Y}}{\partial \mathbf{D}} : \dot{\mathbf{D}}$$

$$\dot{\gamma} = \frac{\partial \gamma}{\partial \delta} \dot{\delta} \tag{8.106}$$

or in function of $\dot{\lambda}$, introducing (8.103) and (8.104) into (8.106) as follows

$$\dot{\mathbf{Y}} = \frac{\partial \mathbf{Y}}{\partial \varepsilon} : \dot{\varepsilon} + \dot{\lambda} \frac{\partial \mathbf{Y}}{\partial \mathbf{D}} : \frac{\partial f}{\partial \mathbf{Y}}$$

$$\dot{\gamma} = \frac{\partial \gamma}{\partial \delta} \dot{\lambda} \frac{\partial g}{\partial \gamma} \tag{8.107}$$

Introducing (8.107) into (8.105) we obtain the following equation

$$\dot{g} = \frac{\partial g}{\partial \mathbf{Y}} : \left[\frac{\partial \mathbf{Y}}{\partial \varepsilon} : \dot{\varepsilon} + \dot{\lambda} \frac{\partial \mathbf{Y}}{\partial \mathbf{D}} : \frac{\partial f}{\partial \mathbf{Y}} \right] + \frac{\partial g}{\partial \gamma} \frac{\partial \gamma}{\partial \delta} \dot{\lambda} \frac{\partial g}{\partial \gamma} = 0 \tag{8.108}$$

Next, $\partial f / \partial \gamma = \partial g / \partial \gamma = -1$, (8.108) can be written as

$$\dot{g} = \frac{\partial g}{\partial \mathbf{Y}} : \frac{\partial \mathbf{Y}}{\partial \varepsilon} : \dot{\varepsilon} + \left[\frac{\partial g}{\partial \mathbf{Y}} : \frac{\partial \mathbf{Y}}{\partial \mathbf{D}} : \frac{\partial f}{\partial \mathbf{Y}} + \frac{\partial \gamma}{\partial \delta} \right] \dot{\lambda} = 0 \tag{8.109}$$

Therefore, the damage multiplier $\dot{\lambda}$ can be obtained as

$$\dot{\lambda} = \begin{cases} \mathbf{L}^d : \dot{\varepsilon} & \text{when} \quad g = 0 \\ 0 & \text{when} \quad g < 0 \end{cases} \tag{8.110}$$

where

$$\mathbf{L}^d = -\frac{\dfrac{\partial g}{\partial \mathbf{Y}} : \dfrac{\partial \mathbf{Y}}{\partial \varepsilon}}{\dfrac{\partial g}{\partial \mathbf{Y}} : \dfrac{\partial \mathbf{Y}}{\partial \mathbf{D}} : \dfrac{\partial f}{\partial \mathbf{Y}} + \dfrac{\partial \gamma}{\partial \delta}} \tag{8.111}$$

Equations (8.103), (8.104), and (8.110) yield the pair $\mathbf{D}, \delta$ in rate form as

$$\dot{\mathbf{D}} = \mathbf{L}^d : \frac{\partial f}{\partial \mathbf{Y}} : \dot{\varepsilon} \quad ; \quad \dot{\delta} = -\dot{\lambda} \tag{8.112}$$

The tangent constitutive equation can be obtained by differentiation of the constitutive equation $\boldsymbol{\sigma} = \mathbf{C} : \boldsymbol{\varepsilon}$, which yields

$$\dot{\boldsymbol{\sigma}} = \mathbf{C} : \dot{\boldsymbol{\varepsilon}} + \dot{\mathbf{C}} : \boldsymbol{\varepsilon} \tag{8.113}$$

where the last term represents the stiffness reduction. Next, the last term in (8.113) can be written as

$$\dot{C} : \boldsymbol{\varepsilon} = \frac{\partial \mathbf{C}}{\partial \mathbf{D}} : \dot{\mathbf{D}} : \boldsymbol{\varepsilon} \tag{8.114}$$

Introduce (8.112) and rearrange

$$\dot{C} : \boldsymbol{\varepsilon} = \frac{\partial \mathbf{C}}{\partial \mathbf{D}} : \boldsymbol{\varepsilon} : \mathbf{L}^d : \frac{\partial f}{\partial \mathbf{Y}} : \dot{\boldsymbol{\varepsilon}} \tag{8.115}$$

Since $(\boldsymbol{\varepsilon}, \mathbf{D})$ are state variables, and thus independent variables,

$$\frac{\partial \boldsymbol{\varepsilon}}{\partial \mathbf{D}} = 0 \tag{8.116}$$

Therefore,

$$\dot{\mathbf{C}} : \boldsymbol{\varepsilon} = \frac{\partial \boldsymbol{\sigma}}{\partial \mathbf{D}} : \mathbf{L}^d : \frac{\partial f}{\partial \mathbf{Y}} : \dot{\boldsymbol{\varepsilon}} \tag{8.117}$$

Finally, reintroduce the above into (8.113) to get

$$\dot{\boldsymbol{\sigma}} = \mathbf{C}^{ed} : \dot{\boldsymbol{\varepsilon}} \tag{8.118}$$

where the *damaged tangent constitutive tensor*, $\mathbf{C}^{ed}$, can be written as follows

$$\mathbf{C}^{ed} = \begin{cases} \mathbf{C} & \text{if } \dot{\mathbf{D}} \leq 0 \\ \mathbf{C} + \dfrac{\partial \boldsymbol{\sigma}}{\partial \mathbf{D}} : \mathbf{L}^d : \dfrac{\partial f}{\partial \mathbf{Y}} & \text{if } \dot{\mathbf{D}} \geq 0 \end{cases} \tag{8.119}$$

The internal variables $\mathbf{D}, \delta$, and related variables, are found using numerical integration, usually using a return-mapping algorithm as explained in Section 8.4.1.

As explained in Section 8.1.3 and 8.4, a number of internal material parameters are needed to define the damage surface, damage potential, and hardening functions. These parameters cannot be obtained directly from simple tests, but rather the model is *identified* by adjusting the internal parameters in such a way that model predictions fit well some observed behavior that can be quantified experimentally. Model identification is very specific to the particular model formulation, material, availability of experiments, and feasibility of conducting relevant experiments. Therefore, model identification can be explained only on a case by case basis, as it is done in Example 8.3, p. 306.

8.4.1   Return-Mapping Algorithm

A return-mapping algorithm [83, 85, 86] is used to solve for the variables $\dot{\lambda}$, $\dot{\delta}$, $\dot{\mathbf{D}}$, δ, and $\mathbf{D}$, in numerically approximated form.

The internal variables are updated by a linearized procedure between two consecutive iterations ($k - 1$ and k). The first-order linearization of (8.109) yields

$$(g)_k - (g)_{k-1} = \left(\frac{\partial g}{\partial \mathbf{Y}} : \frac{\partial \mathbf{Y}}{\partial \mathbf{D}} : \frac{\partial f}{\partial \mathbf{Y}} + \frac{\partial \gamma}{\partial \delta} \right)_{k-1} \Delta \lambda_k = 0 \qquad (8.120)$$

Successful iterations yield $[g]_k = 0$ and

$$\Delta \lambda_k = \frac{-(g)_{k-1}}{\left(\dfrac{\partial g}{\partial \mathbf{Y}} : \dfrac{\partial \mathbf{Y}}{\partial \mathbf{D}} : \dfrac{\partial f}{\partial \mathbf{Y}} + \dfrac{\partial \gamma}{\partial \delta} \right)_{k-1}} \qquad (8.121)$$

The complete algorithm used for a typical integration of constitutive equations is shown next:

i. Retrieve the strain $(\varepsilon)^{n-1}$ from the previous increment and the strain increment $(\Delta \varepsilon)^n$ for the current increment from the FEM code. The updated strain is calculated as

$$(\varepsilon)^n = (\varepsilon)^{n-1} + (\Delta \varepsilon)^n$$

ii. Retrieve the state variables from the previous step and start the return-mapping algorithm by setting the predictor iteration $k = 0$

$$(\mathbf{D})_0^n = (\mathbf{D})^{n-1}; \quad (\delta)_0^n = (\delta)^{n-1}$$

iii. Update the secant stiffness and Cauchy stress, which are used to calculate the thermodynamic forces and damage hardening at this point

$$(\mathbf{C})_k^n = (\mathbf{M})_k^n : \tilde{\mathbf{C}} : (\mathbf{M})_k^n$$
$$(\sigma)_k^n = (\mathbf{C})_k^n : (\epsilon)^n$$
$$(\mathbf{Y})_k^n \quad ; \quad (\gamma)_k^n$$

iv. The damage threshold is evaluated at this point

$$(g)_k = g \left((\mathbf{Y})_k^n, (\gamma(\delta))_k^n, \gamma_0 \right)$$

There are two possible cases:

(a) If $(g)_k \leq 0$, there is no damage, then $\Delta \lambda_k = 0$. Go to (viii).

(b) If $(g)_k > 0$, there is damage evolution, then $\Delta \lambda_k > 0$. Go to (v).

v. Damage evolution. Starting at iteration k, the damage multiplier is found from $(g)_k = 0$ as

$$\Delta\lambda_k = \frac{-(g)_{k-1}}{\left(\dfrac{\partial g}{\partial \mathbf{Y}}\right)_{k-1} : \left(\dfrac{\partial \mathbf{Y}}{\partial \mathbf{D}}\right)_{k-1} : \left(\dfrac{\partial f}{\partial \mathbf{Y}}\right)_{k-1} + \left(\dfrac{\partial \gamma}{\partial \delta}\right)_{k-1}}$$

vi. Update the state variables using $\Delta\lambda_k$

$$\left(D_{ij}\right)^n_k = \left(D_{ij}\right)^n_{k-1} + \Delta\lambda_k \left(\frac{\partial f}{\partial \mathbf{Y}}\right)_{k-1}$$

$$(\delta)^n_k = (\delta)^n_{k-1} + \Delta\lambda_k \left(\frac{\partial f}{\partial \gamma}\right)_{k-1} = (\delta)^n_{k-1} - \Delta\lambda_k$$

vii. End of linearized damage process. Go to (iii).

viii. Compute the tangent stiffness tensor

$$\left(\mathbf{C}^{ed}\right)^n = (\mathbf{C})^n + \left(\frac{\partial \sigma}{\partial \mathbf{D}}\right)^n : (\mathbf{L}^d)^n : \left(\frac{\partial f}{\partial \mathbf{Y}}\right)^n$$

ix. Store the stress and state variables to be used on the next load increment

$$(\sigma)^n = (\sigma)^n_k ; \quad (\mathbf{D})^n = (\mathbf{D})^n_k ; \quad (\delta)^n = (\delta)^n_k$$

x. End of the integration algorithm.

Example 8.5 *Implement the damage model developed in Example 8.4 (p. 321) into a user material subroutine, which is available in [2, Ex_8.5]. Additional details can be found in Section 12.6. We use a return mapping algorithm as shown in Section 8.4.1. Furthermore, use the following damage activation function*

$$g = \hat{g} - \hat{\gamma} = \sqrt{\left(1 - \frac{G^c_I}{G^c_{II}}\right)\frac{Y_2 \tilde{E}_2}{F_{2t}^2} + \frac{G^c_I}{G^c_{II}}\left(\frac{Y_2 \tilde{E}_2}{F_{2t}^2}\right)^2 + \left(\frac{Y_6 \tilde{G}_{12}}{F_6^2}\right)^2} - \hat{\gamma} \le 0$$

where G^c_I and G^c_{II} are the critical energy release in mode I and in mode II, respectively, and F_{2t} and F_6 are the transverse tensile strength and the shear strength, respectively. Also, use the following damage hardening function

$$\hat{\gamma} = \gamma + \gamma_0 = c_1 \left[\exp\left(\frac{\delta}{c_2}\right) - 1\right] + \gamma_0 \quad ; \quad \gamma_0 - c_1 \le \hat{\gamma} \le \gamma_0$$

where γ_0 defines the initial threshold value, c_1 and c_2 are material parameters and δ is the hardening variable (8.103) and (8.112)

$$\delta_i = \delta_{i-1} - \lambda_i \tag{8.122}$$

For this particular damage model, the model parameters for AS4/8852 carbon/epoxy are given in Tables 8.1 and 8.2.

Table 8.1: Elastic and strength properties for AS4/8852 unidirectional lamina.

$\tilde{E}_1$	$\tilde{E}_2$	$\tilde{G}_{12}$	$\tilde{\nu}_{12}$	F_{2t}	F_6
171.4 GPa	9.08 GPa	5.29 GPa	0.32	62.29 MPa	92.34 MPa

Table 8.2: Critical energy release, and hardening parameters for AS4/8852 unidirectional lamina.

G_I^c	G_{II}^c	γ_0	c_1	c_2
170 J/m^2	230 J/m^2	1.0	0.5	-1.8

Solution to Example 8.5 *Watch the video in [1].*
This model represents damage caused by both transverse tensile stress and in-plane shear stress. Longitudinal tension/compression have no effect. Therefore, the model is defined in the thermodynamic force space Y_2, Y_6. The shape of the damage surface for AS4/8852 lamina is shown in Figure 8.8.

To implement the return mapping algorithm shown in Section 8.4.1, expressions for $\partial f/\partial \mathbf{Y}$, $\partial g/\partial \mathbf{Y}$, $\partial f/\partial \gamma$, $\partial g/\partial \gamma$, $\partial \gamma/\partial \delta$, and $\partial \mathbf{Y}/\partial \mathbf{D}$ are needed.

Assuming $f = g$, the derivative of the potential function and the damage surface with respect to the thermodynamic forces is given by

$$\frac{\partial g}{\partial \mathbf{Y}} = \frac{\partial f}{\partial \mathbf{Y}} = \left\{ \begin{array}{c} 0 \\ \frac{1}{\hat{g}}\left(\left(1 - \frac{G_I^c}{G_{II}^c}\right)\frac{1}{4F_{2t}}\sqrt{\frac{2E_2}{Y_2}} + \frac{G_I^c}{G_{II}^c}\frac{E_2}{(F_{2t})^2}\right) \\ \frac{1}{\hat{g}}G_{12} \end{array} \right\}$$

and the derivative of the damage surface with respect to the damage hardening function is

$$\frac{\partial g}{\partial \gamma} = \frac{\partial f}{\partial \gamma} = -1$$

Also, the derivative of the hardening function γ w.r.t conjugate variable δ is needed

$$\frac{\partial \gamma}{\partial \delta} = \frac{c_1}{c_2}\exp\left(\frac{\delta}{c_2}\right)$$

Next, the derivative of the thermodynamic forces w.r.t the internal damage variables is written as

$$\frac{\partial \mathbf{Y}}{\partial \mathbf{D}} = \frac{\partial \mathbf{Y}}{\partial \mathbf{D}}\bigg|_{\sigma=const} + \frac{\partial \mathbf{Y}}{\partial \sigma} : \frac{\partial \sigma}{\partial \mathbf{D}}$$

Furthermore, the derivative of the thermodynamic forces w.r.t strain is written as

$$\frac{\partial \mathbf{Y}}{\partial \varepsilon} = \frac{\partial \mathbf{Y}}{\partial \sigma} : \frac{\partial \sigma}{\partial \varepsilon} = \frac{\partial \mathbf{Y}}{\partial \sigma} : \mathbf{C}$$

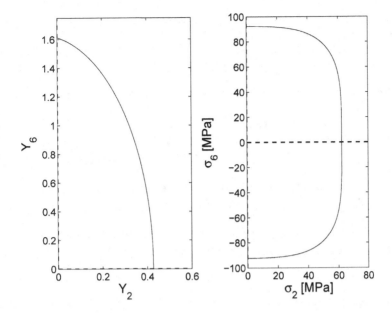

Fig. 8.8: Initial damage surface in thermodynamic and stress spaces.

The following are written in contracted notation as

$$\frac{\partial \mathbf{Y}}{\partial \boldsymbol{\sigma}} = \begin{bmatrix} 0 & 0 & 0 \\ \dfrac{-\sigma_2 \tilde{\nu}_{12}}{(1-D_2)^2\,\tilde{E}_1} & \dfrac{2\sigma_2}{(1-D_2)^3\,\tilde{E}_2} - \dfrac{\sigma_1 \tilde{\nu}_{12}}{(1-D_2)^2\,\tilde{E}_1} & 0 \\ 0 & 0 & \dfrac{\sigma_6}{(1-D_6)^3\,\tilde{G}_{12}} \end{bmatrix}$$

and

$$\frac{\partial \boldsymbol{\sigma}}{\partial \mathbf{D}} = \begin{bmatrix} 0 & -\dfrac{\tilde{E}_2 \tilde{\nu}_{12}}{1-\tilde{\nu}_{12}\tilde{\nu}_{21}}\varepsilon_2 & 0 \\ 0 & -\dfrac{\tilde{E}_1 \tilde{\nu}_{21}}{1-\tilde{\nu}_{12}\tilde{\nu}_{21}}\varepsilon_1 - \dfrac{2\,(1-D_2)\tilde{E}_2}{1-\tilde{\nu}_{12}\tilde{\nu}_{21}}\varepsilon_2 & 0 \\ 0 & 0 & -4(1-D_6)\,\tilde{G}_{12}\varepsilon_6 \end{bmatrix}$$

See Section 12.6 for implementation of these equations in the subroutine `umatps85.for`, *which is available in [2, Ex_8.5]. The material model subroutine can be used in conjunction with plain stress elements and laminated shells elements. The pseudo code is shown below.*

i. *Setting the* `Work Directory`

```
Menu:   File, Set Work Directory, [C:\SIMULIA\User\Ex_8.5], OK
Menu:   File, Save As, [C:\SIMULIA\User\Ex_8.5\ex8.5.cae], OK
```

ii. *Creating the part*

```
Module: Part
Menu:   Part, Create
```

```
        2D Planar, Deformable, Shell, Cont
Menu:    Add, Line, Rectangle, [0,0], [10,10], X, Done
```

iii. *Defining materials, sections, and assigning sections to parts*

```
Module: Property
Menu:    Material, Create
    General, User Material
    # read material properties from a file
    # right-click in Data area, Read from file, [props.txt], OK, OK
    General, Depvar
    Number of solution-dependent state variables [3], OK
Menu:    Section, Create
    Solid, Homogeneous, Cont
    # checkmark: Plane stress/strain thickness [1.0], OK
Menu:    Assign, Section, # pick: part, Done, OK
```

iv. *Creating the assembly*

```
Module: Assembly
Menu:    Instance, Create, Independent, OK
```

v. *Defining analysis steps*

```
Module: Step
Menu:    Step, Create
    Name [transverse], Procedure type: General, Static/General, Cont
    Tab: Basic, Time Period [50]
    Tab: Incrementation, Type: Fixed
    Maximum number of increments [48], Increment size [2.2], OK
Menu:    Output, Field Output Requests, Edit, F-Output-1
    Output Variables: Preselected Defaults, # add: [SDV], OK
```

vi. *Adding loads and BC*

```
Module: Load
Menu:    BC, Manager
    Create, Name [BC-1], Step: Initial, Symm/Anti/Enca, Cont
    # pick: node @ origin, Done, ENCASTRE, OK
    Create, Name [BC-2], Step: Initial, Symm/Anti/Enca, Cont
    # pick: line @ Y=0, Done, YSYMM, OK
    # transverse load imposed as displacement to create epsilon = 2%
    Create, Name [traction], Step: transverse, Disp/Rota, Cont
    # pick: line @ Y=10, Done, # checkmark: U2 [0.2], OK
    # close Boundary Condition Manager pop-up window
```

vii. *Meshing the model. Use only one element because you will have uniform stress/strain.*

```
Module: Mesh
Menu:    Seed, Instance, Approximate global size [10], OK
Menu:    Mesh, Element Type
    Geometric Order: Quadratic # CPS8R avoid hourglass, OK
Menu:    Mesh, Instance, Yes
```

viii. Creating a set to track the stresses and strains in the element

```
Module: Mesh
Menu:   Tools, Set, Create
     Type: Element, Cont, # pick: the only element in the mesh, Done
Module: Step
Menu:   Output, History Output Requests, Edit, H-Output-1
     Domain: Set: Set-1, # expand: Stresses, # checkmark: S
     # expand: Strains, # checkmark: E, OK
```

ix. Solving and visualizing the results

```
Module: Job
Menu:   Job, Manager
     Create, Cont
     Tab: General, User subroutine file [umatps85.for], OK, OK
     Submit, # when Completed, Results
Module: Visualization
Menu:   Result, History Output
     # select: E22 at Element 1 Int Point 1, Save As [e22], OK
     # select: S22 at Element 1 Int Point 1, Save As [s22], OK
     # close History Output pop-up window
Menu:   Tools, XYData, Create
     Source: Operate on XY data, Cont, Operations: Combine
     # select: e22, Add to Expr.
     # select: s22, Add to Expr.
     Save As [s22-vs-e22], OK, Plot Expression
     # save to external file [abaqus.rpt]
     Report, xy, [s22-vs-e22]
     # open file abaqus.rpt
# save the CAE file
Menu:   File, Save, [C:\SIMULIA\User\Ex_8.5\ex8.5.cae], OK
```

The model response is shown in Figure 8.9.

Example 8.6 *Apply shear ϵ_{12} to the model in Example 8.5.*

Solution to Example 8.6 *Watch the video in [1].*
A shear load can be applied easily modifying the cae *model from Example 8.5.*

i. Save the current cae *model into a new one into a diferent folder*

```
Menu:   File, Open, [C:\SIMULIA\User\Ex_8.5\ex8.5.cae], OK
Menu:   File, Save As, [C:\SIMULIA\User\Ex_8.6\ex8.6.cae], OK
```

ii. Defining analysis step

```
Module: Step
Menu:   Step, Manager
     # select: transverse, Rename, [shear], OK
     Edit, Tab: Incrementation, Type: Fixed
     Maximum number of increments [50], Increment size [3], OK
     # close Step Manager pop-up window, Dismiss
```

iii. Adding BC and loads

```
Module: Load
Menu:    BC, Manager
    # shear loading imposed as displacement U1 on step: shear
    # select: traction, Rename, [shear], OK
    Edit, # uncheck: U2, # checkmark: U1 [0.4], OK # epsilon_12=4%
    # close BC Manager pop-up window, Dismiss
```

iv. Adding constraint to guarantee pure shear loading

```
Module: Interaction
Menu:    Constraint, Create
    Type: Tie, Cont
    Node region, # pick: vertical line @ x=0, Done
    Node region, # pick: vertical line @ x=10, Done
    Discretization method: Surface to surface
    Position Tolerance: Specify distance [10]
    # uncheck: Adjust slave surface initial position, OK
    # check: Tie rotational DOF
```

v. Solving and visualizing the results

```
Module: Job
Menu:    Job, Manager,
    #select [Job-1], # verify the location of the UMAT
    Edit, Tab: General,
        [file: C:\SIMULIA\User\Chapter_8\Ex_8.5\umatps85.for]
    Submit, # when Completed, Results
Module: Visualization
Menu:    Result, History Output
    # select: E12 at Element 1 Int Point 1, Save As [e12], OK
    # select: S12 at Element 1 Int Point 1, Save As [s12], OK
    # close History Output pop-up window
Menu:    Tools, XYData, Create
    Source: Operate on XY data, Cont, Operators: Combine (X,X)
    # select: e12, Add to Expr. , <- do not forget the comma
    # select: s12, Add to Expr.
    Save As [s12-vs-e12], OK
    Plot Expression
```

The model response for the two cases, Example 8.5 and Example 8.6 are shown in Figure 8.9.

8.5   Damage and Plasticity

For polymer matrix composites reinforced by strong and stiff fibers, damage and its conjugate thermodynamic force can be described by second-order tensors $\mathbf{D}$ and $\mathbf{Y}$. Furthermore, the hardening processes that take place during plasticity and damage imply additional *dissipation*, so that

$$\rho\pi = T\rho\pi_s = \boldsymbol{\sigma} : \dot{\boldsymbol{\varepsilon}}^p + R\dot{p} + \mathbf{Y} : \dot{\mathbf{D}} + \gamma\dot{\delta} \qquad (8.123)$$

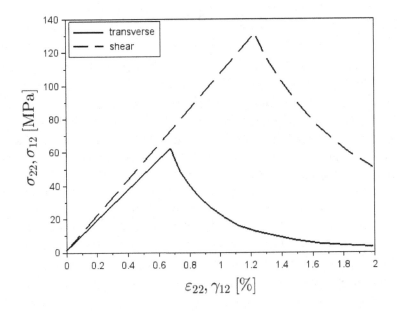

Fig. 8.9: Response to in-plane shear stress and transverse tensile stress.

where (R, p) is the thermodynamic force-flux pair associated to plastic hardening, and (γ, δ) is the thermodynamic force-flux pair associated to damage hardening, and $\rho\pi$ is the dissipation heat due to irreversible phenomena.

For the particular case of (8.123), from (8.92) and (8.99), the following definitions for the thermodynamic forces are obtained

$$\sigma = \rho\frac{\partial\psi}{\partial\varepsilon} = -\rho\frac{\partial\psi}{\partial\varepsilon^p} \quad \varepsilon = \rho\frac{\partial\chi}{\partial\sigma} \quad \mathbf{Y} = -\rho\frac{\partial\psi}{\partial\mathbf{D}} = \rho\frac{\partial\chi}{\partial\mathbf{D}} \tag{8.124}$$

as well as definitions for the hardening equations

$$\gamma = \rho\frac{\partial\chi}{\partial\delta} = -\rho\frac{\partial\psi}{\partial\delta} = \rho\frac{\partial\pi}{\partial\delta} \quad R = \rho\frac{\partial\chi}{\partial p} = -\rho\frac{\partial\psi}{\partial p} = \rho\frac{\partial\pi}{\partial p} \tag{8.125}$$

The additive decomposition [63]

$$\varepsilon = \varepsilon^e + \varepsilon^p \tag{8.126}$$

can be rewritten taking into account that the elastic component of strain can be calculated from stress and compliance, so that

$$\varepsilon = \mathbf{S} : \sigma + \varepsilon^p \tag{8.127}$$

Therefore, the strain-stress law in incremental and rate form is

$$\delta\varepsilon = \mathbf{S} : \delta\boldsymbol{\sigma} + \delta\mathbf{S} : \boldsymbol{\sigma} + \delta\varepsilon^p$$
$$\dot{\varepsilon} = \mathbf{S} : \dot{\boldsymbol{\sigma}} + \dot{\mathbf{S}} : \boldsymbol{\sigma} + \dot{\varepsilon}^p \qquad (8.128)$$

showing that an increment of strain has three contributions, elastic, damage, and plastic. The elastic strain occurs as a direct result of an increment in stress, the damage strain is caused by the increment in compliance as the material damages, and the plastic strain occurs at constant compliance. The elastic unloading stiffness does not change due to plasticity but it reduces due to damage. Following this argument, it is customary [93] to assume that the free energy and complementary free energy can be separated, as follows

$$\psi(\varepsilon, \varepsilon^p, p, \mathbf{D}, \delta) = \psi^e(\varepsilon^e, \mathbf{D}, \delta) + \psi^p(\varepsilon^p, p)$$
$$\chi(\boldsymbol{\sigma}, \varepsilon^p, p, \mathbf{D}, \delta) = \chi^e(\boldsymbol{\sigma}, \mathbf{D}, \delta) + \chi^p(\varepsilon^p, p) \qquad (8.129)$$

Suggested Problems

Problem 8.1 *Using the formulation and properties of Example 8.2 (p. 302), obtain a graphical representation of the evolution strain vs. nominal stress (ε vs. σ) and the evolution of strain vs. effective stress (ε vs. $\tilde{\sigma}$) for a point on the top surface of the beam and for another point on the bottom surface of the beam. Comment on the graphs obtained.*

Problem 8.2 *Implement a UMAT for a 1D CDM active in the x_1-direction only. Use 2D plane stress constitutive equations. Leave the x_2-direction, Poisson's, and shear terms as linear elastic with no damage. Verify the program by recomputing Example 8.2 and the plots obtained in Problem 8.1. Note that to obtain the same values, the Poisson's ratio should be set to zero.*

Problem 8.3 *The Gibbs free energy is defined in expanded form and using Voigt contracted notation as*

$$\chi = \frac{1}{2\rho} \left[\frac{\sigma_1^2}{(1-D_1)^2 \, \tilde{E}_1} + \frac{\sigma_2^2}{(1-D_2)^2 \, \tilde{E}_2} + \frac{\sigma_6^2}{(1-D_1)(1-D_2) \, \tilde{G}_{12}} \right. $$
$$\left. - \left(\frac{\tilde{\nu}_{21}}{\tilde{E}_2} + \frac{\tilde{\nu}_{12}}{\tilde{E}_1} \right) \frac{\sigma_1 \sigma_2}{(1-D_1)(1-D_2)} \right]$$

where $\tilde{E}_1$, $\tilde{E}_2$, $\tilde{\nu}_{12}$, $\tilde{\nu}_{21}$, and $\tilde{G}_{12}$ are the undamaged in-plane elastic orthotropic properties of a unidirectional lamina where the subindex $()_1$ denotes the fiber direction and $()_2$ denotes the transverse direction. (a) Obtain the secant constitutive equations, $\mathbf{C}$ and $\mathbf{S}$, using the given Gibbs free energy. (b) Obtain the thermodynamic forces Y_1 and Y_2 associated to D_1 and D_2. (c) If $\mathbf{M}$ is represented using Voigt contracted notation and multiplied by a Reuter matrix as

$$\mathbf{M} = \begin{bmatrix} (1-D_1) & 0 & 0 \\ 0 & (1-D_2) & 0 \\ 0 & 0 & \sqrt{1-D_1}\sqrt{1-D_2} \end{bmatrix}$$

check if this definition of $\mathbf{M}$ can be used as the damage effect tensor in a damage model using the principle of energy equivalence. Justify and comment on your conclusion.

Table 8.3: Elastic properties for composite material lamina.

$\widetilde{E}_1$	$\widetilde{E}_2$	$\widetilde{G}_{12}$	$\widetilde{\nu}_{12}$
171.4 GPa	9.08 GPa	5.29 GPa	0.32

Table 8.4: Identification model parameters for composite material lamina.

H_1	H_2	γ_0	c_1	c_2
0.024	8.36	1.0	1.5	-2.8

Problem 8.4 *The damage activation function, for the model shown in Problem 8.3, is defined as*

$$g := \hat{g} - \hat{\gamma} = \sqrt{Y_1^2 H_1 + Y_2^2 H_2} - (\gamma + \gamma_0)$$

where H_1 and H_2 are model parameters that depend on elastic and strength material properties, and Y_1 and Y_2 are the thermodynamic forces associated to the damage variables D_1 and D_2, respectively. The damage hardening depends on δ according to

$$\hat{\gamma} = \gamma + \gamma_0 = c_1 \left[\exp \left(\frac{\delta}{c_2} \right) - 1 \right] + \gamma_0$$

where γ_0 defines the initial threshold value, and c_1 and c_2 are material parameters. All necessary material parameters are shown in Tables 8.3 and 8.4.

a) *Using a flow chart diagram, describe the algorithm, with all necessary steps to implement it as a constitutive subroutine in a finite element package.*

b) *Compute the analytic expressions necessary to implement the model in a UMAT.*

c) *Program the algorithm using the UMAT capability for a plane stress constitutive equation.*

d) *Finally, using Abaqus, plot a single curve of apparent stress σ_2 vs. apparent strain ε_2 for an RVE loaded only with ε_2.*

e) *Using pseudo code, describe the process used to solve the problem in Abaqus CAE.*

Chapter 9

Discrete Damage Mechanics

Prediction of damage initiation and propagation is tackled in Chapter 8 using continuum damage mechanics (CDM). Alternatives to CDM include: micromechanics of damage, crack opening displacement methods, computational micromechanics, and synergistic methods. While CDM (Chapter 8) homogenizes the damage with a phenomenological approach, the alternative methods attempt to represent the actual geometry and characteristics of damage.

Prediction of transverse matrix cracking in laminated composites has been extensively studied for the particular case of symmetric $[0_m/90_n]_S$ laminates under membrane loads, for which matrix cracking is found in the 90° laminas (transverse laminas). Extensions to other laminate configurations, such as $[0/\pm\theta/0]_S$ and $[0/\theta_1/\theta_2]_S$, featuring cracks in the off-axis θ laminas, have been developed, but they are still limited to symmetric laminates subjected to in-plane loading.

Micromechanics of damage (MMD) models find an approximate elasticity solution for a laminate with a discrete crack or cracks [94–115]. The solutions are approximate because kinematic assumptions are made, such as a linear [116] or bilinear [117] distribution of interlaminar shear stress through the thickness of each lamina, as well as particular spatial distributions of in-plane displacement functions [114], stresses, and so on. The state variable is the crack density in the cracking lamina, defined as the number of cracks per unit distance perpendicular to the crack surface. Therefore, the state variable is measurable. One advantage of MMD is that the reduction of laminate moduli as a function of crack density is calculated without resorting to additional parameters as in the case of CDM. The main disadvantage of MMD is that most of the solutions available are limited to symmetric laminates under membrane loads with only one or two laminas cracking. A generalization to the case of multiple cracking laminas is presented in this chapter by resorting to the concept of synergistic methods, explained below.

Crack opening displacement (COD) methods [118–125] are based on the theory of elastic bodies with voids [126]. The distinct advantage of COD models is that the laminate stiffness can be calculated for any laminate configuration, even nonsymmetric laminate stacking sequence (LSS), subject to any deformation, including bending, featuring matrix cracking in any of its laminas [127].

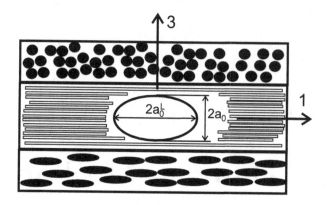

Fig. 9.1: Representative crack geometry.

Numerical solutions, such as FEA, provide 3D solutions without the kinematic simplifications of MMD and COD models [105, 121, 124, 128–131]. However, FEA solutions require a new mesh and boundary conditions for each LSS, crack orientation, and so on, making them too cumbersome for practical application. Another numerical approach is Monte Carlo simulation, where the probabilistic distribution of flaws in the material is considered [132–134]. Unfortunately, Monte Carlo simulations require additional parameters that have to be adjusted by fitting the results of the model to experimental damage evolution data. Such data are scarce.

Synergistic damage mechanics (SDM) methods combine elements of different modeling strategies such as CDM and MMD [44, 132, 133, 135–138], bringing the best features of each of the models involved. For example, in this chapter the laminate stiffness reduction is computed via MMD methods and the generalization to multiple cracking laminas is made via CDM concepts, but unlike CDM models, no additional parameters are needed.

9.1 Overview

In the following we describe how to use two material properties, the fracture toughness in modes I and II, G_I^c, G_{II}^c, to predict the damaging behavior and transverse tensile failure of a unidirectional fiber-reinforced lamina embedded in a laminate. The constraining effect of adjacent laminas is taken into account, leading to apparent transverse tensile strength F_{2t} being a function of ply thickness. The crack initiation strain, crack density evolution as a function of stress (strain) up to crack saturation, and stress redistribution to adjacent laminas are predicted accurately.

The physics of matrix cracking under transverse tension and in-plane shear is as follows. No matter how much care is taken during the production process, there are always defects in the material. These defects may be voids, microcracks, fiber–matrix debonding, and so on, but all of them can be represented by a typical matrix crack of representative size $2a_0$, as shown in Figure 9.1.

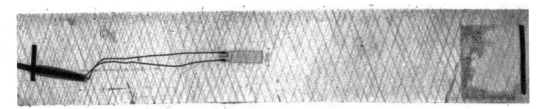

Fig. 9.2: Matrix cracks in the ± 70 laminas of a $[0/\pm 70_4/0_{1/2}]_S$ laminate loaded to 0.7% tensile strain along the $0°$ direction [139].

When subject to load, matrix cracks grow parallel to the fiber orientation, as shown in Figure 9.2, where it can be seen that cracks are aligned with the fiber direction in the $\pm 70°$ laminas. These sets of parallel cracks reduce the stiffness of the cracked lamina, which then sheds its share of the load onto the remaining laminas. In each lamina, the damage caused by this set of parallel cracks is represented by the crack density, defined as the inverse of the distance between two adjacent cracks $\lambda = 1/(2l)$, as shown in Figure 9.3. Therefore, the crack density is the only state variable needed to represent the state of damage in the cracked lamina. Note that the actual, discrete cracks are modeled by the theory, which is thus named *discrete damage mechanics* (DDM).

The basic ingredients of the DDM model for transverse tension and in-plane shear damage are listed below:

i. In each lamina i, the state variable is the crack density λ_i. Two damage variables $D_2(\lambda_i)$ and $D_6(\lambda_i)$ are defined for convenience but they are not independent variables; instead, they are computed in terms of the crack density. The set of crack densities for the laminate is denoted by $\lambda = \lambda_i$ with $i = 1...N$, where N is the number of laminas in the laminate.

ii. The independent variable is the midsurface[1] strain $\epsilon = \{\epsilon_1, \epsilon_2, \gamma_{12}\}^T$.

iii. The damage activation function separates the damage state from the undamaged state. Two equations have been used, yielding similar results.

$$g = \max\left[\frac{G_I(\lambda, \epsilon, \Delta T)}{G_I^c}, \frac{G_{II}(\lambda, \epsilon, \Delta T)}{G_{II}^c}\right] - 1 \leq 0 \qquad (9.1)$$

or

$$g = (1-r)\sqrt{\frac{G_I(\lambda, \epsilon)}{G_I^c}} + r\frac{G_I(\lambda, \epsilon)}{G_I^c} + \frac{G_{II}(\lambda, \epsilon)}{G_{II}^c} - 1 \leq 0 \qquad (9.2)$$

where $g \leq 0$ represents the un-damaging domain and $r = G_I^c/G_{II}^c$. The critical ERRs are not easily found in the literature but they can be fit to available experimental data using the methodology explained, for example in [55, 141].

---

[1] The analysis presented in this section is for symmetric laminates under membrane forces. For unsymmetric laminates and/or laminates under bending is being, see [127, 140].

Top View

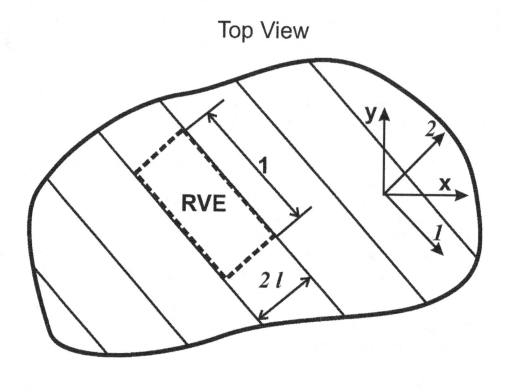

Side View

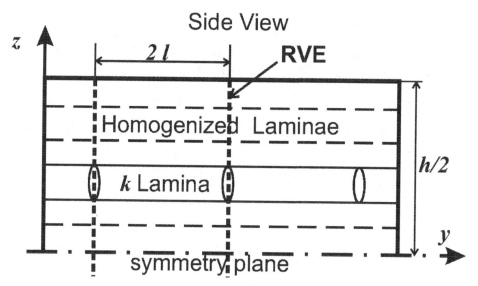

Fig. 9.3: Representative unit cell used in discrete damage mechanics.

iv. The damage threshold is embedded into g, and represented by the (invariant) material properties G_I^c, G_{II}^c. Before damage starts, $\lambda = 0$ and (9.1) is a damage initiation criterion, similar to [142], with or without mode interaction. With $\lambda = 0$, the strain for which $g = 0$ is the strain for crack initiation. Once damage starts, (9.1) becomes a damage activation function by virtue of the automatic hardening described below.

v. The hardening function is embedded into the damage activation function g. For a given value of strain, the calculated values of energy release rate $G_I(\lambda), G_{II}(\lambda)$ are monotonically decreasing functions of λ. Therefore, as soon as λ grows, $G_I(\lambda), G_{II}(\lambda)$ decrease, making $g < 0$ and thus stopping further damage until the *driving thermodynamic force*, i.e., the strain, is increased by the application of additional load.

vi. No damage evolution function needs to be postulated, with the advantage that no new empirical parameters are needed. Simply the crack density λ adjusts itself to a value that will set the laminate in equilibrium with the external loads for the current strain while satisfying $g = 0$. A return mapping algorithm (Section 8.4.1) achieves this by iterating until $g = 0$ and updating the crack density with iterative increments calculated as $\Delta\lambda = -g/\frac{\partial g}{\partial \lambda}$.

vii. The crack density grows until the lamina is saturated with cracks $(\lambda \to \infty)$. At that point, the lamina loses all of its transverse and shear stiffness $(D_2 \approx 1, D_6 \approx 1)$, at which point all of the load is already transferred to the remaining laminas in the laminate. The analysis of the cracked lamina is stopped when the crack density reaches $\lambda_{lim} = 1/h_k$, where h_k is the thickness of lamina k; i.e., when cracks are closely spaced at a distance equal to the lamina thickness.

Having described the ingredients of the model, it now remains to show how to calculate the various quantities. The solution begins by calculating the degraded[2] stiffness of the laminate $Q = [A]/h$ for a given crack density λ_k in a cracked lamina k, where $[A]$ is the in-plane laminate stiffness matrix,[3] and h is the thickness of the laminate.

The following conventions are used in this section:

- (i) denotes any lamina in the laminate.
- (k) denotes the cracking lamina.
- (m) denotes any lamina other than the cracking one $(m \neq k)$.
- A sub/superscript in parenthesis (i) denotes the lamina number; not a power or order of differentiation.
- x_j with $j = 1, 2, 3$ denote the coordinates x_1, x_2, x_3, or $j = 1, 2, 6$ for quantities expressed in Voigt contracted notation.
- $u(x_j), v(x_j), w(x_j)$ with $j = 1, 2, 3$ are the three components of the displacement.

---

[2] Also called "damaged", "reduced", or "homogenized".
[3] Not to be confused with the crack area A.

- *hat* $\hat{p}$ denotes the thickness average of quantity p, where the thickness is mentioned or it is obvious from context.
- *tilde* $\tilde{p}$ denotes the virgin value of quantity p.
- *over-line* $\bar{p}$ denotes the volume average of quantity p.

9.2   Approximations

Most practical laminates are symmetric and the most efficient use of them is by designing the structure to be loaded predominantly with membrane loads [3, Chapter 12]. Therefore, the solution presented here is for a symmetric laminate under membrane loads. In this case,

$$\frac{\partial w^{(i)}}{\partial x_1} = \frac{\partial w^{(i)}}{\partial x_2} = 0 \qquad (9.3)$$

where $u(x_j), v(x_j), w(x_j)$ with $j = 1, 2, 3$ are the displacements of a point in lamina i as a function of the coordinates x_j with $j = 1, 2, 3$. Furthermore, the thickness h_i of the laminas are assumed to be small, so that the plane stress assumption holds

$$\sigma_3^{(i)} = 0 \qquad (9.4)$$

Since all cracks are parallel to the fiber direction and practical designs avoid thick laminas, it can be expected that the cracks occupy the entire thickness of the lamina. Any crack smaller than the lamina thickness is unstable both through the lamina thickness and along the fiber direction [3, Section 7.2.1].

Since the objective is to calculate the laminate stiffness reduction due to cracks, it suffices to work with thickness averages of the variables. A thickness average is denoted by

$$\hat{\phi} = \frac{1}{h'} \int_{h'} \phi \, dx_3 \quad ; \quad h' = \int dx_3 \qquad (9.5)$$

where h' can be the lamina or laminate thickness, denoted by h_i, h, respectively. Specifically,

- $\hat{u}^{(i)}(x_j), \hat{v}^{(i)}(x_j), \hat{w}^{(i)}(x_j)$ are the thickness-average displacements in lamina i as a function of the in-plane coordinates x_j with $j = 1, 2$.

- $\hat{\epsilon}_1^{(i)}(x_j), \hat{\epsilon}_2^{(i)}(x_j), \hat{\gamma}_{12}^{(i)}(x_j)$ are the thickness-average strains in lamina i.

- $\hat{\sigma}_1^{(i)}(x_j), \hat{\sigma}_2^{(i)}(x_j), \hat{\tau}_{12}^{(i)}(x_j)$ are the thickness-average stress in lamina i.

Out-of-plane (intralaminar) shear stress components appear due to the perturbation of the displacement field caused by the crack. These are approximated by linear functions through the thickness of the lamina i, as follows

$$\tau_{13}^{(i)}(x_3) = \tau_{13}^{i-1,i} + \left(\tau_{13}^{i,i+1} - \tau_{13}^{i-1,i}\right) \frac{x_3 - x_3^{i-1,i}}{h_i}$$

$$\tau_{23}^{(i)}(x_3) = \tau_{23}^{i-1,i} + \left(\tau_{23}^{i,i+1} - \tau_{23}^{i-1,i}\right) \frac{x_3 - x_3^{i-1,i}}{h_i} \qquad (9.6)$$

where x_3^{i-1} is the thickness coordinate at the bottom of lamina i, i.e., at the interface between lamina $i-1$ and lamina i, and $\tau_{13}^{i-1,i}$ is the shear stress at the interface between the $i-1$ and the i lamina. This assumption, which is common to several other analytical models, is called the *shear lag assumption*. The linear approximation has been shown to yield accurate results [131].

The shear lag equations are obtained from the constitutive equations for out-of-plane shear strains and stresses using weighted averages [44, Appendix A],

$$
\left\{ \begin{array}{c} \hat{u}^{(i)} - \hat{u}^{(i-1)} \\ \hat{v}^{(i)} - \hat{v}^{(i-1)} \end{array} \right\} = \frac{h_{(i-1)}}{6} \left[\begin{array}{cc} S_{45} & S_{55} \\ S_{44} & S_{45} \end{array} \right]^{(i-1)} \left\{ \begin{array}{c} \tau_{23}^{i-2,i-1} \\ \tau_{13}^{i-2,i-1} \end{array} \right\}
$$
$$
+ \left[\frac{h_{(i-1)}}{3} \left[\begin{array}{cc} S_{45} & S_{55} \\ S_{44} & S_{45} \end{array} \right]^{(i-1)} + \frac{h_{(i)}}{3} \left[\begin{array}{cc} S_{45} & S_{55} \\ S_{44} & S_{45} \end{array} \right]^{(i)} \right] \left\{ \begin{array}{c} \tau_{23}^{i-1,i} \\ \tau_{13}^{i-1,i} \end{array} \right\}
$$
$$
+ \frac{h_{(i)}}{6} \left[\begin{array}{cc} S_{45} & S_{55} \\ S_{44} & S_{45} \end{array} \right]^{(i)} \left\{ \begin{array}{c} \tau_{23}^{i,i+1} \\ \tau_{13}^{i,i+1} \end{array} \right\} \tag{9.7}
$$

Inverting (9.7), the intralaminar stresses are written in terms of displacements at the interfaces, as follows

$$
\tau_{23}^{i,i+1} - \tau_{23}^{i-1,i} = \sum_{j=1}^{n-1} \left[[H]^{-1}{}_{2i-1,2j-1} - [H]^{-1}{}_{2i-3,2j-1} \right] \left\{ \hat{u}^{(j+1)} - \hat{u}^{(j)} \right\}
$$
$$
+ \left[[H]^{-1}{}_{2i-1,2j} - [H]^{-1}{}_{2i-3,2j} \right] \left\{ \hat{v}^{(j+1)} - \hat{v}^{(j)} \right\}
$$
$$
\tau_{13}^{i,i+1} - \tau_{13}^{i-1,i} = \sum_{j=1}^{n-1} \left[[H]^{-1}{}_{2i,2j-1} - [H]^{-1}{}_{2i-2,2j-1} \right] \left\{ \hat{u}^{(j+1)} - \hat{u}^{(j)} \right\}
$$
$$
+ \left[[H]^{-1}{}_{2i,2j} - [H]^{-1}{}_{2i-2,2j} \right] \left\{ \hat{v}^{(j+1)} - \hat{v}^{(j)} \right\} \tag{9.8}
$$

in terms of the $2(N-1)$ by $2(N-1)$ coefficient matrix H.

9.3 Lamina Constitutive Equation

The stress-strain law for the cracking lamina k is that of an intact material, i.e.,

$$
\hat{\sigma}_i^{(k)} = \tilde{Q}_{ij}^{(k)} \left(\hat{\epsilon}_j^{(k)} - \alpha_j^{(k)} \Delta T \right) \tag{9.9}
$$

where $\alpha^{(k)}$ is the CTE of lamina k, $\sigma_i^{(k)} = \left\{ \sigma_1^{(k)}, \sigma_2^{(k)}, \tau_{12}^{(k)} \right\}^T$, and *tilde* denotes a virgin property. The strain-displacement equations are

$$
\epsilon^{(k)} = \left\{ \begin{array}{l} \epsilon_1^{(k)} = u_{,1}^{(k)} \\ \epsilon_2^{(k)} = v_{,2}^{(k)} \\ \gamma_{12}^{(k)} = u_{,2}^{(k)} + v_{,1}^{(k)} \end{array} \right\} \tag{9.10}
$$

For the remaining laminas $(m \neq k)$, the constitutive equations can be obtained using (9.9) and the stiffness matrix $Q_{ij}^{(m)}$, written in terms of their previously calculated damage values $D_2^{(m)}, D_6^{(m)}$, defined in (9.33), and rotated to the k coordinate system using the usual transformation equations [3, Section 5.4]

$$
Q^{(m)} = T^{-1} \begin{bmatrix} \widetilde{Q}_{11}^{(m)} & \left(1 - D_2^{(m)}\right) \widetilde{Q}_{12}^{(m)} & 0 \\ \left(1 - D_2^{(m)}\right) \widetilde{Q}_{12}^{(m)} & \left(1 - D_2^{(m)}\right) \widetilde{Q}_{22}^{(m)} & 0 \\ 0 & 0 & \left(1 - D_6^{(m)}\right) \widetilde{Q}_{66}^{(m)} \end{bmatrix} T^{-T}
$$

$$(9.11)$$

where $T^{-1} = [T(-\theta)]$ and $T^{-T} = [T(-\theta)]^T$; see [3, (5.40)].

9.4   Displacement Field

The objective now is to solve for the average displacements $\hat{u}^{(i)}(x_j), \hat{v}^{(i)}(x_j); j = 1, 2$, in all laminas i for a given crack density set λ and applied strain ϵ. Taking into account that the intralaminar shear stresses are assumed to vary linearly through the thickness of each lamina, the equilibrium equations (1.15) for each lamina can be written as follows

$$
\hat{\sigma}_{1,1}^{(i)} + \hat{\tau}_{12,2}^{(i)} + \left(\hat{\tau}_{13}^{i,i+1} - \hat{\tau}_{13}^{i-1,i}\right) / h_i = 0 \tag{9.12}
$$

$$
\hat{\tau}_{12,1}^{(i)} + \hat{\sigma}_{2,2}^{(i)} + \left(\hat{\tau}_{23}^{i,i+1} - \hat{\tau}_{23}^{i-1,i}\right) / h_i = 0 \tag{9.13}
$$

Using the strain-displacement equations (9.10) and the constitutive equations (9.9) into the equilibrium equations (9.12)-(9.13) leads to a system of $2N$ partial differential equations (PDE) in $\hat{u}^{(i)}(x_j), \hat{v}^{(i)}(x_j)$. The PDE has particular solutions of the form

$$
\hat{u}^{(i)} = a_i \sinh\left(\eta_e x_2\right) + a\, x_1 + b\, x_2
$$

$$
\hat{v}^{(i)} = b_i \sinh\left(\eta_e x_2\right) + b\, x_1 + a^* x_2 \tag{9.14}
$$

where e is the eigenvalue number. The general solution can be written as

$$
\left\{
\begin{array}{c}
\hat{u}^{(1)} \\
\hat{u}^{(2)} \\
\cdot \\
\cdot \\
\cdot \\
\hat{u}^{(n)} \\
\hat{v}^{(1)} \\
\hat{v}^{(2)} \\
\cdot \\
\cdot \\
\cdot \\
\hat{v}^{(n)}
\end{array}
\right\}
= \sum_{e=1}^{2N} A_e
\left\{
\begin{array}{c}
a_1 \\
a_2 \\
\cdot \\
\cdot \\
\cdot \\
a_n \\
b_1 \\
b_2 \\
\cdot \\
\cdot \\
\cdot \\
b_n
\end{array}
\right\}_e
sinh\,(\eta_e x_2) +
\left\{
\begin{array}{c}
a \\
a \\
\cdot \\
\cdot \\
\cdot \\
a \\
b \\
b \\
\cdot \\
\cdot \\
\cdot \\
b
\end{array}
\right\} x_1 +
\left\{
\begin{array}{c}
b \\
b \\
\cdot \\
\cdot \\
\cdot \\
b \\
a^* \\
a^* \\
\cdot \\
\cdot \\
\cdot \\
a^*
\end{array}
\right\} x_2
\qquad (9.15)
$$

which substituted into the PDE leads to the eigenvalue problem

$$
\left[
\begin{array}{cc}
\alpha_1 & \beta_1 \\
\alpha_2 & \beta_2
\end{array}
\right]
\left\{
\begin{array}{c}
a_j \\
b_j
\end{array}
\right\}
+ \eta^2
\left[
\begin{array}{cc}
\zeta_{26} & \zeta_{22} \\
\zeta_{66} & \zeta_{26}
\end{array}
\right]
\left\{
\begin{array}{c}
a_j \\
b_j
\end{array}
\right\}
=
\left\{
\begin{array}{c}
0 \\
0
\end{array}
\right\}
\qquad (9.16)
$$

where $j = 1...2N$; η are the $2N$ eigenvalues and $\{a_j, b_j\}^T$ are the $2N$ eigenvectors of (9.16).

It turns out that two of the eigenvalues are always zero (corresponding to the linear terms in (9.15)), which can be taken to be the last two in the set, thus remaining only $2N - 2$ independent solutions. Then, the general solution of the PDE system is built as the linear combination of the $2N - 2$ independent solutions, as follows

$$
\left\{
\begin{array}{c}
\hat{u}^{(i)} \\
\hat{v}^{(i)}
\end{array}
\right\}
= \sum_{e=1}^{2N-2} A_e
\left\{
\begin{array}{c}
a_i \\
b_i
\end{array}
\right\}_e
sinh\,(\eta_e x_2) +
\left\{
\begin{array}{c}
a \\
b
\end{array}
\right\} x_1 +
\left\{
\begin{array}{c}
b \\
a^*
\end{array}
\right\} x_2
\qquad (9.17)
$$

where A_e are unknown coefficients in the linear combination. It can be seen that the general solution contains $2N + 1$ unknown coefficients, including the scalars a, b, a^* and the sets A_e with $e = 1...2N - 2$. To determine these coefficients, one needs $2N + 1$ boundary conditions on the boundary of the representative volume element (RVE) in Figure 9.3. Note that the RVE spans a unit length along the fiber direction x_1, a distance $2l$ between successive cracks (along x_2) and the whole thickness h of the symmetric laminate.

Two very important parameters are introduced through the boundary conditions, namely the crack density λ and the stress $\hat{\sigma} = \{N\}/h$ applied to the laminate, where $\{N\}$ represents the three components of the in-plane force per unit length.[4] The crack density enters through the dimension of the RVE, which has a width of $2l = 1/\lambda$. The applied stress (or strain) enters through the force equilibrium on the RVE. In summary, there are $2N + 1$ boundary conditions that lead to a system of $2N + 1$ algebraic equations that can be solved for the $2N + 1$ coefficients in (9.17). Therefore, the average displacements in all laminas are now known from (9.17) for given values of crack density λ and applied load $\hat{\sigma} = \{N\}/h$.

---

[4]Not to be confused with the number of laminas N.

9.4.1 Boundary Conditions for $\Delta T = 0$

First consider the case of mechanical loads and no thermal loads. To find the values of $A_e, a, a*, b$, the following boundary conditions are enforced: (a) stress-free at the crack surfaces, (b) external loads, and (c) homogeneous displacements. The boundary conditions are then assembled into an algebraic system, as follows

$$[B]\left\{A_e, a, a^*, b\right\}^T = \{F\} \tag{9.18}$$

where $[B]$ is the coefficient matrix of dimensions $2N+1$ by $2N+1$; $\left\{A_e, a, a^*, b\right\}^T$ represents the $2N+1$ unknown coefficients, and $\{F\}$ is the right hand side (RHS) force vector, also of dimension $2N+1$.

(a) Stress-free at the Crack Surfaces

The surfaces of the cracks are stress-free

$$\int_{-1/2}^{1/2} \hat{\sigma}_2^{(k)}(x_1, l)\, dx_1 = 0 \tag{9.19}$$

$$\int_{-1/2}^{1/2} \hat{\tau}_{12}^{(k)}(x_1, l)\, dx_1 = 0 \tag{9.20}$$

(b) External Loads

In the direction parallel to the surface of the cracks (fiber direction x_1) the load is supported by all the laminas

$$\frac{1}{2l} \sum_{i=1}^{N} h_i \int_{-l}^{l} \hat{\sigma}_1^{(i)}(1/2, x_2)\, dx_2 = h\hat{\sigma}_1 \tag{9.21}$$

In the direction normal to the crack surface (x_2 direction) only the uncracking (homogenized) laminas carry load

$$\sum_{m \neq k} h_m \int_{1/2}^{1/2} \hat{\sigma}_2^{(m)}(x_1, l)\, dx_1 = h\hat{\sigma}_2 \tag{9.22}$$

$$\sum_{m \neq k} h_m \int_{1/2}^{1/2} \hat{\tau}_{12}^{(m)}(x_1, l)\, dx_1 = h\hat{\tau}_{12} \tag{9.23}$$

(c) Homogeneous Displacements

For a homogenized symmetric laminate, membrane loads produce a uniform displacement field through the thickness, i.e., all the uncracking laminas are subjected to the same displacement

$$\hat{u}^{(m)}(x_1, l) = \hat{u}^{(r)}(x_1, l) \quad ; \quad \forall m \neq k \tag{9.24}$$

$$\hat{v}^{(m)}(x_1, l) = \hat{v}^{(r)}(x_1, l) \quad ; \quad \forall m \neq k \tag{9.25}$$

where r is an uncracked lamina taken as reference. In the computer implementation, lamina 1 is taken as reference unless lamina 1 is cracking, in which case lamina 2 is taken as reference.

9.4.2 Boundary Conditions for $\Delta T \neq 0$

Next, consider the case of thermal loads, which add a constant term to the boundary conditions. Constant terms do not affect the matrix $[B]$, but rather subtract from the forcing vector $\{F\}$, as follows

$$\{F\}_{\Delta T \neq 0} = \left\{ \begin{array}{c} \Delta T \sum\limits_{j=1,2,6} \bar{Q}_{1j}^{(k)} \bar{\alpha}_j^{(k)} \\ \Delta T \sum\limits_{j=1,2,6} \bar{Q}_{1j}^{(k)} \bar{\alpha}_j^{(k)} \\ \Delta T \sum\limits_{i \neq (k)} \sum\limits_{j=1,2,6} \bar{Q}_{1j}^{(i)} \bar{\alpha}_j^{(i)} \\ \Delta T \sum\limits_{i \neq k} \sum\limits_{j=1,2,6} \bar{Q}_{2j}^{(i)} \bar{\alpha}_j^{(i)} \\ \Delta T \sum\limits_{i \neq k} \sum\limits_{j=1,2,6} \bar{Q}_{6j}^{(i)} \bar{\alpha}_j^{(i)} \\ 0 \\ 0 \\ \cdots \\ \cdots \\ 0 \\ 0 \end{array} \right\} \tag{9.26}$$

In this way, the strain calculated for a unit thermal load ($\Delta T = 1$) is the degraded CTE of the laminate for the current crack density set λ.

9.5 Degraded Laminate Stiffness and CTE

In this section, we calculate the degraded stiffness of the laminate $Q = \mathbf{A}/h$ for a given crack density λ_k in a cracked lamina k, where $\mathbf{A}$ is the in-plane laminate stiffness matrix, and h is the thickness of the laminate. First, the thickness-averaged strain field in all laminas can be obtained by using the kinematic equations (9.10), namely by differentiating (9.17). Then, the compliance of the laminate S in the coordinate system of lamina k can be calculated one column at a time by solving for the strains (9.10) for three load cases, a, b, and c, all with $\Delta T = 0$, as follows

$$^a\hat{\sigma} = \left\{ \begin{array}{c} 1 \\ 0 \\ 0 \end{array} \right\} ; \quad ^b\hat{\sigma} = \left\{ \begin{array}{c} 0 \\ 1 \\ 0 \end{array} \right\} ; \quad ^c\hat{\sigma} = \left\{ \begin{array}{c} 0 \\ 0 \\ 1 \end{array} \right\} ; \quad \Delta T = 0 \tag{9.27}$$

Since the three applied stress states are unit values, for each case, a, b, c, the volume average of the strain (9.10) represents one column in the laminate compliance matrix

$$S = \begin{bmatrix} {}^a\epsilon_x & {}^b\epsilon_x & {}^c\epsilon_x \\ {}^a\epsilon_y & {}^b\epsilon_y & {}^c\epsilon_y \\ {}^a\gamma_{xy} & {}^b\gamma_{xy} & {}^c\gamma_{xy} \end{bmatrix} \tag{9.28}$$

where x, y are the coordinates of lamina k (Figure 9.3). Next, the laminate stiffness in the coordinate system of lamina k is

$$Q = S^{-1} \tag{9.29}$$

To get the degraded CTE of the laminate, one sets $\hat{\sigma} = \{0,0,0\}^T$ and $\Delta T = 1$. The resulting strain is equal to the CTE of the laminate, i.e., $\{\alpha_x, \alpha_y, \alpha_{xy}\}^T = \{\epsilon_x, \epsilon_y, \gamma_{xy}\}^T$.

9.6   Degraded Lamina Stiffness

The stiffness of lamina m, with $m \neq k$, in the coordinate system of lamina k (see Figure 9.3) is given by (9.11) in terms of the previously calculated values $D_2^{(m)}$, $D_6^{(m)}$, given by (9.33). The stiffness of the cracking lamina $Q^{(k)}$ is yet unknown. Note that all quantities are expressed in the coordinate system of lamina k.

The laminate stiffness is defined by the contribution of the cracking lamina k plus the contribution of the remaining $N-1$ laminas, as follows

$$Q = Q^{(k)}\frac{h_k}{h} + \sum_{m=1}^{n}(1 - \delta_{mk})Q^{(m)}\frac{h_m}{h} \tag{9.30}$$

where the *delta* Dirac is defined as $\delta_{mk} = 1$ if $m = k$, otherwise 0. The left-hand side (LHS) of (9.30) is known from (9.29) and all values of $Q^{(m)}$ can be easily calculated since the m laminas are not cracking at the moment. Therefore, one can calculate the degraded stiffness $Q^{(k)}$ of lamina k, as follows

$$Q^{(k)} = \frac{h}{h_k}\left[Q - \sum_{m=1}^{n}(1 - \delta_{mk})Q^{(m)}\frac{h_m}{h}\right] \tag{9.31}$$

where Q without a superscript is the stiffness of the laminate.

To facilitate later calculations, the stiffness $Q^{(k)}$ can be written using concepts of continuum damage mechanics (Section 8.2) in terms of the stiffness of the undamaged lamina and damage variables $D_2^{(k)}$, $D_6^{(k)}$, as follows

$$Q^{(k)} = \begin{bmatrix} \widetilde{Q}_{11}^{(k)} & (1 - D_2)\,\widetilde{Q}_{12}^{(k)} & 0 \\ (1 - D_2)\,\widetilde{Q}_{12}^{(k)} & (1 - D_2)\,\widetilde{Q}_{22}^{(k)} & 0 \\ 0 & 0 & (1 - D_6)\,\widetilde{Q}_{66}^{(k)} \end{bmatrix} \tag{9.32}$$

with $D_j^{(k)}$ calculated for a given crack density λ_k and applied strain ϵ^0, as follows

$$D_j^{(k)}(\lambda_k, \epsilon^0) = 1 - Q_{jj}^{(k)}/\widetilde{Q}_{jj}^{(k)} \quad ; \quad j = 2, 6; \text{ no sum on } j \tag{9.33}$$

where $\widetilde{Q}^{(k)}$ is the original value of the undamaged property and $Q^{(k)}$ is the degraded[5] value computed in (9.31), both expressed in the coordinate system of lamina k.

Although the lamina's crack density λ_k grows and its corresponding stiffness is degraded accordingly, the coefficient of thermal expansion of the cracking lamina remains unchanged, because damage is homogenized by a change of stiffness, not by a change of CTE. Only the the coefficient of thermal expansion of the laminate changes with damage and that is because the stiffness of the laminas decrease with damage. To calculate the laminate CTE, first we calculate the thermal stress [3, (6.71)], as follows

$$\{N^T\} = \sum_{k=1}^{N} [\bar{Q}]\{\bar{\alpha}\} t_k \tag{9.34}$$

where *overbar* means the quantity has been transformed to laminate c.s.; $[Q]$ is the damaged stiffness of the lamina k; α is the intact CTE of the lamina k; and t is the thickness of lamina k. To transform we use $\bar{\alpha}^k = [R][T^{-1}][R^{-1}]\alpha^k$ where $[R]$ is the Reuter matrix. With that, the laminate CTE is [3, (6.77)]

$$\{\alpha^o\} = [A]^{-1}\{N^T\} \tag{9.35}$$

9.7 Fracture Energy

Under displacement control, the energy release rate (ERR) is defined as the partial derivative of the strain energy U with respect to the crack area A (see (10.1)). According to experimental observations on laminated, brittle matrix composites (e.g., using most toughened epoxy matrices), cracks develop suddenly over a finite length, and thus are not infinitesimal. Then, Griffith's energy principle is applied on its discrete (finite) form in order to describe the observed, discrete (finite) behavior of crack growth, as follows.

The ERR values in modes I (opening) and II (sliding shear) are calculated as

$$G_I = \frac{U_I^a - U_I^b}{\Delta A} \quad ; \quad G_{II} = \frac{U_{II}^a - U_{II}^b}{\Delta A} \tag{9.36}$$

---

[5]Homogenized.

where $U_I^a, U_I^b, U_{II}^a, U_{II}^b$ are the elastic strain energies in mode I and II for crack densities $\lambda_a, \lambda_b = 2\lambda_a$, and ΔA is the newly created (finite) crack area, which is one half of the new crack surface. Counting crack area as one-half of crack surface is consistent with the classical fracture mechanics convention for which fracture toughness G_c is twice of Griffith's surface energy γ_c. In the current implementation of the model, which is used in Example 9.1 (p. 351), $\Delta A = h_k$ is the area of one new crack appearing halfway between two existing cracks. Alternative crack propagation strategies are considered in [127].

Mode decomposition is achieved by splitting the strain energy U into mode I (opening) and II (shear) and adding the contribution of each lamina $k = 1 \ldots n$, as follows

$$U_I = \frac{V_{RVE}}{2h} \sum_{k=1}^{n} t_k (\epsilon_2 - \tilde{\alpha}_2^{(k)} \Delta T) \, Q_{2j}^{(k)} (\epsilon_j - \tilde{\alpha}_j^{(k)} \Delta T) \qquad (9.37)$$

$$U_{II} = \frac{V_{RVE}}{2h} \sum_{k=1}^{n} t_k (\epsilon_6 - \tilde{\alpha}_6^{(k)} \Delta T) Q_{6j}^{(k)} (\epsilon_j - \tilde{\alpha}_j^{(k)} \Delta T) \qquad (9.38)$$

where $h = \sum_{k=1}^{n} t_k$, $\epsilon_6 = \gamma_{12}$, and $\tilde{\alpha}^{(k)}$ are the undamaged CTE of lamina k. Equations (9.37–9.38) are cast in the coordinate system of the cracking lamina k so that ϵ_2 is mode I (crack opening) and ϵ_6 is mode II (crack shearing). Laminate ultimate failure is predicted by a fiber damage and failure criterion [143].

The damage activation function (9.1) can now be calculated for any value of λ and applied strain $\epsilon_x, \epsilon_y, \gamma_{xy}$. Note that the computation of the ERR components derives directly from the displacement solution (9.17) for a discrete crack (Figure 9.3). When this formulation is used along with the finite element method (FEM), it does not display mesh dependency on the solution and does not require the arbitrary specification of a characteristic length [137], in contrast to formulations based on smeared crack approximations [55]. The effect of residual thermal stresses is incorporated into the formulation. The code is available as a user material for ANSYS [137], a shell user element for ANSYS [138], and a user general section (UGENS) for Abaqus [141]; see Example 9.1.

9.8   Solution Algorithm

The solution algorithm consists of (a) strain steps, (b) laminate-iterations, and (c) lamina-iterations. The state variables for the laminate are the array of crack densities for all laminas i and the membrane strain ϵ. At each load (strain) step, the strain on the laminate is increased and the laminas are checked for damage.

9.8.1   Lamina Iterations

When matrix cracking is detected in lamina k, a return mapping algorithm (RMA) (Section 8.4.1) is invoked to iterate and adjust the crack density λ_k in lamina k in such a way that g_k returns to zero while maintaining equilibrium between the

Table 9.1: Properties for Example 9.1. E-glass/epoxy (experimental).

Property	Value	Comment
E_1 [GPa]	44.7	Lamina longitudinal modulus
E_2 [GPa]	12.7	Lamina transverse modulus
G_{12} [GPa]	5.8	Lamina in-plane shear modulus
ν_{12}	0.297	Lamina in-plane Poisson's ratio
ν_{23}	0.41	Lamina out-of-plane Poisson's ratio
α_1 [1E-6/C]	3.7	Lamina longitudinal coefficient of thermal expansion
α_2 [1E-6/C]	30	Lamina transverse coefficient of thermal expansion
G_I^c [kJ/m^2]	0.254	Lamina critical ERR mode I
G_{II}^c [kJ/m^2]	10^6	Lamina critical ERR mode II
t [mm]	0.144	Ply thickness, used in the LSS

external forces and the internal forces in the laminas. The iterative procedure works as follows. At a given strain level ϵ for the laminate and given λ_k for lamina k, calculate the value of the damage activation function g_k and the damage variables, which are both functions of λ_k. The RMA calculates the increment (decrement) of crack density as

$$\Delta\lambda_k = -g_k / \frac{\partial g_k}{\partial \lambda} \tag{9.39}$$

until $g_k = 0$ is satisfied within a given tolerance, for all $k = 1...N$, where N is the number of laminas in the laminate. The analysis starts with a negligible value of crack density present in all laminas ($\lambda = 0.02$ cracks/mm were used in the examples).

9.8.2 Laminate Iterations

To calculate the stiffness reduction of a cracked lamina (k-lamina), all of the other laminas (m-laminas) in the laminate are considered not damaging during the course of lamina-iterations in lamina k, but with damaged properties calculated according to the current values of their damage variables $D_i^{(m)}$. Given a trial value of λ_k, the analytical solution provides g_k, $D_i^{(k)}$, for lamina k assuming all other laminas do not damage while performing lamina iterations in lamina k. Since the solution for lamina k depends on the stiffness of the remaining laminas, a converged iteration for lamina k does not guarantee convergence for the same lamina once the damage in the remaining laminas is updated. In other words, within a given strain step, the stiffness and damage of all the laminas are interrelated and they must all converge. This can be accomplished by laminate-iterations; that is, looping over all laminas repeatedly until all laminas converge to $g = 0$ for all k. Unlike classical RMA set up for plasticity, where the hardening parameter is monotonically increasing (Section 8.4.1), the crack density λ_k must be allowed to decrease if in the course of laminate iterations, other laminas sustain additional damage that makes the laminate more compliant and thus requires a reduction of λ_k.

Example 9.1 *Consider a* $[0/90_8/0_{1/2}]_S$ *laminate made of glass/epoxy with properties given in Table 9.1 subjected to a membrane strain* $\epsilon_x \neq 0, \epsilon_y = \gamma_{xy} = 0$. *Plot crack density in lamina* $k = 2$ *vs. strain* $0 < \epsilon_x < 2\%$. $N = 3$, $t_k = [1, 8, 0.5] * t$, *ply thickness* $t = 0.144$ *mm.*

Solution to Example 9.1 *Watch the video in [1].*

Example 9.1 requires using the UGENS code precompiled into **abaqusddm-std.obj**. *This plugin extends the capabilities of Abaqus to predict transverse damage of laminated composites. It is linked to Abaqus in* **Module: Job**. *The plugin is compiled for Abaqus 2020.*

Visual Studio and Intel Fortran need to be installed properly (see App. E) for Abaqus to link the .obj file to Abaqus itself, which is done automatically during **Job: Submit**.

First we build a model using CAE. The file **Ex_9.1_pre.cae** *is available in [2, Ex_9.1]. See the video in [1] or open and review* **Ex_9.1_pre.cae** *using CAE by following the description provided here. Alternatively, see Section* **Pseudo-code** *below in this example or Section 12.7.1 in Chapter 12.*

In **Module: Part**, *we create a part to represent a 10 by 10 mm shell and we call it* **Laminate**. *The names we assign are important because later we use a script that uses some of these names.*

Still in Module: Part, we use tools to create a set called **WholeLaminate**, *so that later we can refer to the part by its name.*

We do not define the material properties now, but rather wait to define them using a Python script, just before we set up the Job. Instead, we go directly to **Module: Assembly**, *to define and instance of the part.*

Next, we define a **Step** *to apply in-plane strain to the plate. We set the time period equal to the maximum strain we want to apply, so that time and strain are the same. In* **Tab: Incrementation**, *set the initial and maximum size to a small value and the minimum to a very small value. While in* **Module: Step**, *we edit the* **Field Output Request**, *to get the state variables (SDV) stored for visualization.*

In **Module: Load**, *we use the boundary conditions manager to set support conditions and the applied strain. First, symmetry boundary condition X on the left vertical edge of the plate. Next, symmetry boundary condition Y on the bottom horizontal edge of the plate. Next, restrict rigid body motion by clamping the plate at the origin with and* **Encastre** *boundary condition. Next, we apply a displacement* U_1 *on the right vertical edge of the plate to produce a set strain, thus loading the plate. To apply 1.873% strain, we need a displacement equal to the strain times the length of the plate,* $1.873/100 \times 10 = 0.1873$. *Finally, we restrict the out of plane displacements, to be sure we get pure inplane deformations.*

In **Module: Mesh**, *on the second toolbar (second below the menu), make sure that* **Part: Laminate** *is selected, because earlier we set the instance to dependent, so it will mesh the part but not the instance.*

For element type, we select S4R, which is a 4 node shell with reduced integration. Then we seed with a global size of 10, which is the length of the plate, to get just one element for the entire plate. At last we mesh.

Next, we need to enter the material properties into the PART block of the input file, like this,

```
*Shell General Section,elset=WholeLaminate,USER,PROPERTIES=30,VARIABLES=9
2.736
0.254,1e16,0,44700,12700,5800,.297,.41,
3.7,30.,0,0.144,44700,12700,5800,.297,
.41,3.7,30.,90,1.152,44700,12700,5800,
.297,.41,3.7,30.,0,0.072
```

```
*Transverse Shear Stiffness
1,1,0
```

but we cannot do it with CAE, because CAE does not know how many material properties we use in the software plug in, or what the variables mean. There is no CAE interface yet built in CAE to enter the material properties for the plugin. Although we could manually edit the .inp file, we prefer to do it with the Python script Ex-9-1-props.py.

In addition, we need to insert another line into the STEP block of the input file, like this,

```
*Initial Conditions, type=SOLUTION, USER
```

to tell Abaqus to initialize the state variables, which in this example are the crack densities, to a very small value. There is no CAE interface yet built in CAE to enter initial values for the plugin's state variables. Although we could manually edit the .inp file, we prefer to do it with the Python script Ex-9-1-props.py. Note that the script calls ugenKeyword.py. Both are available in [2, Ex_9.1].

On the bottom of the CAE display, switch to the python shell, by clicking on the 3 brackets >>>. *Then type Ex-9-1-props.py and hit Enter. That will execute the Python script which inserts the lines needed, then writes Job=ddm-exe.inp and creates a new Job based on the modified input file.*

Now we go back to normal use of CAE. In Module: Job, the first Job, based on Model1, does not have the lines inserted. You must use the second Job, based on the file job-ddm-exe.inp. Then, click Submit, and wait for it to be completed. If you Monitor the progress of the execution, look at the column called Increments (same as Frames) to see how the analysis progresses with more increments being completed.

When the Job is finished, click Results. In the Field Output toolbar, select SDV4 to see the crack density in the center lamina k=2. In the legend, you see SDV = 1.016, which is the crack density in the center lamina at frame 188 that corresponds to the final time, 1.873. Each SDV has a meaning (see Section 12.7.1 and Table 12.1).

For frame=0 (increment=0) at the beginning of the analysis, you can see that the initial crack density was set to SDV4=0.02 cracks per mm, that is one crack every 50 mm. With a ply thickness of 0.125 mm, that is one crack every 400 ply thicknesses. Indeed very small.

By advancing one frame at a time, you can pinpoint the frame, time, and strain at which crack initiation takes place, i.e., when the crack density of the center lamina (k=2, SDV=4) jumps from 0.02 to a higher value. Other laminas crack at different strain or do not crack at all. These predictions compare very well with crack density vs. strain data from experiments measuring crack density by optical or acoustic emission methods [139].

*Since conventional shell elements (not continuum shell elements) cannot extract the shell thickness from the mesh (the mesh has no thickness), a *Shell General Section needs to specify the total laminate thickness for the shell elements*

$$LT = \sum_{k=1}^{2N} t_k$$

where N is the number of laminas in the bottom half of the symmetric laminate. Therefore, the input file to Abaqus must have LT as its first parameter (2.736 in this example), as follows

```
*Shell General Section,elset=WholeLaminate,USER,PROPERTIES=30,VARIABLES=9
2.736
```

The DDM plugin requires $3+9N$ parameters and uses $3N$ state variables, where N is $1/2$ the number of laminas. The laminate is assumed to be symmetric, so we enter only $1/2$ of the LSS (laminate stacking sequence). In this example, $1/2$ of the LSS is $[0/90_8/0_{1/2}]$ and the full LSS is $[0/90_8/0_{1/2}]_S$; underscore S makes all the difference.

For each of the N laminas, the three state variables per lamina are described in Table 12.1.

Of the $3+9N$ parameters, the first 3 parameters are:

```
GIc, critical ERR mode I
GIIC, critical ERR mode II
dummy (for future use)
```

Then, 7 material properties plus 2 geometrical properties, 9 per lamina, total $9N$, as follows

$$E_1, E_2, G_{12}, \nu_{12}, \nu_{23}, \alpha_1, \alpha_2, \theta_k, t_k$$

where θ_k, t_k are the orientation and thickness of lamina k, with $k = 1 \ldots N$, and N is the number of laminas in the bottom half of the laminate.

In this example (using the Python scripts Ex-9-1-props.py and UgenKeyword.py) we use the same material properties for all laminas:

```
ModelParameters['E1']      = '44700'
ModelParameters['E2']      = '12700'
ModelParameters['G12']     =  '5800'
ModelParameters['Nu12']    = '.297'
ModelParameters['Nu23']    = '.41'
ModelParameters['Cte1']    = '3.7'
ModelParameters['Cte2']    = '30.'
```

*but you can modify the script to provide different materials properties for each lamina. These properties are used to assemble the *Shell General Section and the *Transverse Shear Stiffness that must be added to the Job-ddm-exe.inp file, as follows*

```
*Shell General Section,elset=WholeLaminate,USER,PROPERTIES=30,VARIABLES=9
2.736
0.254,1e16,0,44700,12700,5800,.297,.41,
3.7,30.,0,0.144,44700,12700,5800,.297,
.41,3.7,30.,90,1.152,44700,12700,5800,
.297,.41,3.7,30.,0,0.072
*Transverse Shear Stiffness
1,1,0
```

*However, if you wish to assign different materials properties to each lamina, you can do it by editing the *Shell General Section data above, as explained next.*

The first value in the data is the laminate thickness $LT=2.736$ mm.

The next 2 numbers are $GIc=0.254$ kJ/m^2 and $GIIc=1e16$ kJ/m^2.

Notice that $E_1=44700$ MPa appears 3 times for this example with $N=3$. E_1 is the first property in the set $E_1, E_2, G_{12}, \nu_{12}, \nu_{23}, \alpha_1, \alpha_2, \theta_k, t_k$.

The coefficients of thermal expansion need to be provided with units of $10^6/C$ (or $10^6/F$ if you are using US customary units). In this example they are $3.7\ 10^6/C$ and $30\ 10^6/C$.

Pseudo-code: *The following pseudo-code formally describes the procedure used to construct the model with CAE.*

```
Windows Start, Abaqus CAE

File, Set Work Directory
# navigate to folder \Chapter_9 (create if not available)
# select folder Ex_9.1 (create if not available)

Module: Part
Menu: Part, Create, [Laminate], 3D, Deformable, Shell, Planar,
        Approx. size: [20]
        Add, Lines, Rectangle, [0,0], [10,10], X, Done
Menu: Tools, Set, Create, [WholeLaminate], Cont, #select the part, Done

Module: Assembly
Menu: Instance, Create, from: Parts, Dependent, OK

Module: Step
Menu:  Step, Create, Name [ApplyStrain], after: Initial,
        General, Static-General, Cont.
        Time period [1.873] #highest strain to be applied,
        Incrementation, Automatic
        Max. number of increments [1000]
        Initial [0.01], Minimum [0.0001], Maximum [0.10], OK

Menu: Output, Field Output Requests, Edit, F-Output-1,
     # select displacements, reaction forces, state variables
     Select from list below [U, RF, SDV], OK

Module: Load
Menu: BC, Manager
        # symmetryX
        Create, Name: [SymmetryX], Step: Initial,
        Type: Symmetry, Cont,
        #select region: vertical edge at X=0, Done, XSYMM, OK
        # symmetryY
        Create, Name: [SymmetryY], Step: Initial,
        Type: Symmetry, Cont,
        #select region: horizontal edge at Y=0, Done, YSYMM, OK
        # restrict rigid body motion
        Create, Name: [NoRigidBodyMotion], Step: Initial,
        Type: Encastre, Cont,
        #select region: point at origin, Done, Encastre, OK
        # next, apply U1=ShellDimensionX*Strain/100
        Create, Name: [ApplyStrain], Step: ApplyStrain,
        Type: Displacement, Cont,
        #select region: vertical edge on right, Done,
        U1: [0.1873], OK
        # next, restrict Z motion
        Create, Name: [restrictZmotion], Step: Initial,
        Type: Displacement, Cont,
        #select region: whole face, Done,
        U3: [check], UR1: [check], UR2: [check], UR3: [check], OK
```

```
        Dismiss

Module: Mesh, Model: Model-1,
        Object: Part(not Assembly):Laminate
Menu:   Mesh, Element type, #select the whole face, Done,
        #pick: S4R, OK
Menu: Seed, Part, Approx. global size: [10]
        #single element mesh, OK, Done
Menu: Mesh, Part, OK to mesh the part? [Yes]
```

At this point we have the model meshed.

Next, toggle the bottom window to Python >>> *and copy paste* **Ex-9-1-props.py** *into the Python shell to be executed.*

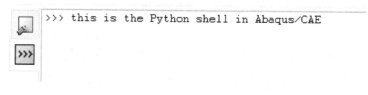

```
>>> this is the Python shell in Abaqus/CAE
```

Ex-9-1-props.py uses another script **UgenKeyword.pyc** *to write 8 values per line into the input file, as it is a requirement from Abaqus. Both codes are available in [2].*

The portion of the **Ex-9-1-props.py** *code that the user can modify is reproduced below.*

```
# filename: Ex-9-1-props.py
# use with: abaqusddm-std.obj
## Do not use until the Model is complete and meshed.
# Inserts material properties into the .inp file and defines the Job
from abaqusConstants import *
from mesh import *
from step import *
from regionToolset import Region
from multiprocessing import cpu_count
from visualization import openOdb
from abaqus import mdb
import csv                     # to write a .CSV file
from UgenKeyword import *    # to write UGENS parameters on the .inp file
session.journalOptions.setValues(replayGeometry=COORDINATE,
        recoverGeometry=COORDINATE)
# Enter your model parameters here:
NL = 3                             # number of laminas (scalar)
ModelParameters = {}               # define a dictionary data tye
ModelParameters['GIc']   = '0.254'
ModelParameters['GIIc']  = '1e16'
ModelParameters['dummy'] = '0'       # for future use
ModelParameters['E1']    = '44700'
ModelParameters['E2']    = '12700'
ModelParameters['G12']   = '5800'
ModelParameters['Nu12']  = '.297'
ModelParameters['Nu23']  = '.41'
ModelParameters['Cte1']  = '3.7'
ModelParameters['Cte2']  = '30.'
ModelParameters['tk']    = '.144'
#Next, [list] of (thickness,angle) pairs to describe the LSS
ModelParameters['Lss']   = [(1,0), (8,90), (.5,0)]
ModelParameters['Strain'] = '1.8735'  # % max. strain to run the model
# do not make any changes below
...
```

Next, in CAE, go to

```
Module: Job
Menu: Job Manager
```

and there you should see 2 Jobs (Figure 9.4. The second one is the one you should submit for execution.

```
Name: Job-ddmexe-1, File: Job-ddm-exe.inp
```

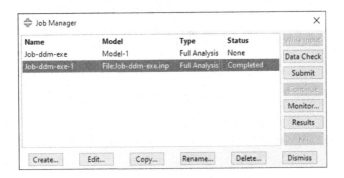

Fig. 9.4: Job-ddm-exe. The second one contains everything and it is ready for execution.

Once the Job is completed, click Results. It will take you to Module: Visualization. On the Field Output Toolbar, Select

```
Primary, SDV4
```

On the legend (Figure 9.5) you will see SDV4=1.016. That is the crack density in the 90_8 lamina (k=2) at Frame=188, Time=1.873, Strain=1.873. The meaning of the SDVs is explained in Section 12.7.1, Table 12.1.

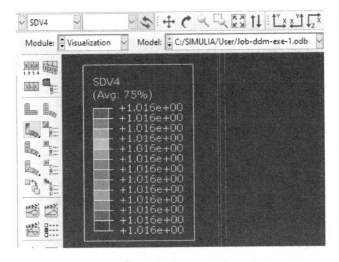

Fig. 9.5: Crack density in the 90_8 lamina (k=2) at Frame=188, Time=1.873, Strain=1.873.

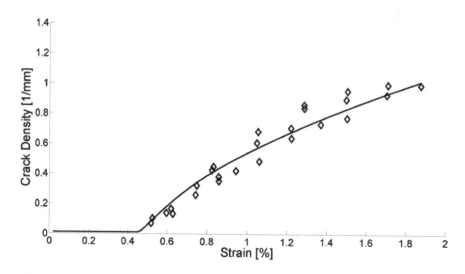

Fig. 9.6: Crack density vs. strain predicted by DDM and compared to experimental data from [144].

Next, selecting

Menu: Result, Step/Frame

we can see that Frame 188 is the last one, and Time=1.873 means Strain=1.873, that is the largest strain that we applied to the simulation.

Using the procedure described above, one may get a warning like this: "The following parts have some elements without any section assigned: Laminate", just ignore it.

Comparison between model predictions and experimental data from [144] is presented in Figure 9.6.

Suggested Problems

Problem 9.1 Calculate the critical laminate strain ϵ_x^c for which the first crack appears in laminates 1 to 3 in Table 9.2, all made of glass/epoxy with properties given in Table 9.1 (p. 351) and subjected to a membrane strain $\epsilon_x \neq 0, \epsilon_y = \gamma_{xy} = 0$. For each laminate, in what lamina does the first crack appear?

Problem 9.2 Using the results from Exercise 9.1, calculate the in-situ transverse strength F_{2t}^{is} of the laminas [3, Section 7.2.1].

Problem 9.3 Calculate the critical laminate strain ϵ_x^c for which the first crack appears in laminates 4 to 7 in Table 9.2, all made of glass/epoxy with properties given in Table 9.1 subjected to a membrane strain $\epsilon_x \neq 0, \epsilon_y = \gamma_{xy} = 0$. For each laminate, in what lamina does the first crack appear?

Problem 9.4 Using the results from Exercise 9.3, calculate the in-situ transverse strength F_{2t}^{is} of the laminas [3, Section 7.2.1].

Table 9.2: Laminates for Exercise 9.1.

Laminate Number	LSS
1	$[0/90_8/0/90_8/0]$
2	$[0/70_4/-70_4/0/-70_4/70_4/0]$
3	$[0/55_4/-55_4/0/-55_4/55_4/0]$
4	$[0_2/90_8/0_2]$
5	$[15/-15/90_8/-15/15]$
6	$[30/-30/90_8/-30/30]$
7	$[40/-40/90_8/-40/40]$

Chapter 10

Delaminations

A delamination is a frequent mode of failure affecting the structural performance of composite laminates. The interface between layers offers a low-resistance path for crack growth because the bonding between two adjacent layers depends only on matrix properties. A delamination may originate from manufacturing imperfections, cracks produced by fatigue or low velocity impact, stress concentration near geometrical/material discontinuity such as joints and free edges, or due to high interlaminar stresses.

In laminates loaded in compression, delaminated layers may buckle and cracks propagate due to interaction between delamination growth and buckling. The presence of delaminations may reduce drastically the buckling load and the compression strength of the composite laminates [145] (Figure 10.1). Delaminations may also be driven by buckling in laminates under transverse loading [146]. The analysis of delamination buckling requires the combination of geometrically nonlinear structural analysis with fracture mechanics.

According to its shape, delaminations are classified into through-the-width or strip [146–153], circular [153–159], elliptic [160], rectangular [161] or arbitrary [162, 163]. Depending on its location through the laminate thickness, delaminations are classified into thin film, symmetric split [145, 148, 149], and general [150, 153, 156, 157, 159]. In addition, analysis of combined buckling and growth for composite laminates containing multiple delaminations under in-plane compression loading have been carried out [164, 165]. Experimental results on delamination buckling are presented in [166, 167].

Other delamination configurations that have been investigated in the literature are the beam-type delamination specimens subjected to bending, axial, and shear loading [166–172] which form the basis for experimental methods used to measure interlaminar fracture toughness under pure mode I, mode II, and mixed-mode conditions in composites, adhesive joints, and other laminated materials (Figure 10.2).

In plates with piezoelectric sensors or actuators, an imperfect bonding between the piezoelectric lamina and the base plate may grow under mechanical and/or electrical loading. As a consequence, the adaptive properties of the smart system can be significantly reduced since debonding results in significant changes to the static or dynamic response [173, 174]. Finally, delamination growth may be caused

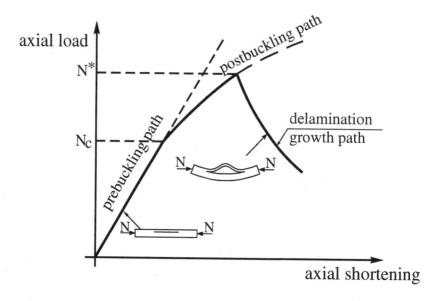

Fig. 10.1: Delamination buckling in a compressed laminate.

by dynamic effects, such as vibration and impact. For instance, the dynamics effects resulting from the inertia of the laminate on the growth process resulting from the buckling of the delamination have been investigated for a circular delamination and time-dependent loading [175].

Delaminations can be analyzed by using *cohesive damage models* (Section 10.1) and *fracture mechanics* (Section 10.2). A cohesive damage model implements interface constitutive laws defined in terms of damage variables and a damage evolution law. Cohesive damage elements are usually inserted between solid elements [176–179] or beam/shell elements [178].

In the fracture mechanics approach, the propagation of an existing delamination is analyzed by comparing the amount of energy release rate (ERR) with the fracture toughness of the interface. When mixed-mode conditions are involved, the decomposition of the total ERR into mode I, mode II, and mode III components becomes necessary due to the mixed-mode dependency of interface toughness [170, 180]. A number of fracture mechanics-based models have been proposed in the literature to study delamination, including three-dimensional models [181–183] and simplified beam-like models [145, 147, 172, 184, 185].

Fracture mechanics allows us to predict the growth of a preexisting crack or defect in brittle materials and in ductile materials as long as the plastic zone near the crack tip is small. In a homogeneous and isotropic body subjected to a generic loading condition, a crack tends to grow by kinking in a direction such that a pure mode I condition at its tip is maintained. On the contrary, delaminations in laminated composites are constrained to propagate in its own plane because the toughness of the interface is relatively low in comparison to that of the adjoining material. Since a delamination crack propagates with its advancing tip in mixed-mode condition, the analysis requires a fracture criterion including all three mode

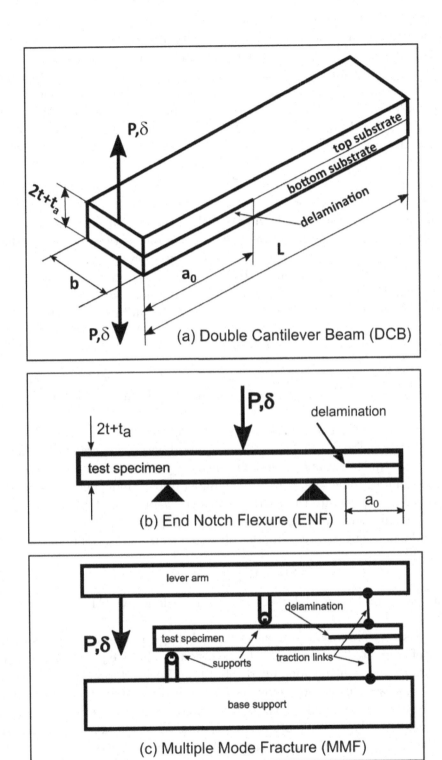

Fig. 10.2: Typical delamination specimens.

components (Section 10.1.2).

The elastic strain energy per unit volume (density, in J/m^3) is defined as $U_0 = 1/2\sigma_{ij}\epsilon_{ij}$. The strain energy (in Joules) is defined as the volume integral $U = \int_V U_0 dV$. The energy required to form, or to propagate, a crack is equal to the elastic energy released by the solid during crack formation. The energy released is the difference between the elastic strain energy available before and after the crack is formed, i.e., $-\Delta U = U_{after} - U_{before}$. The rate of energy released per unit of crack area A is given, in J/m^2, by

$$G = -\frac{\Delta U}{\Delta A} \qquad (10.1)$$

where A is the new crack area created, which is half of the new surface area created. The theory of crack growth may be developed by using one of two approaches due to Griffith and Irwin, respectively. The Griffith energy approach uses the concept of energy release rate G as the (quantifiable) energy available for fracture on one hand, and the material property G_c, which is the energy necessary for fracture, on the other hand. A crack does not grow as long as $G < G_c$ and it grows when

$$G = G_c \qquad (10.2)$$

where for completeness note that $G_c = 2\gamma_c$, where γ_c is the critical fracture energy per unit of new surface crack area created, i.e., the area of only one of the faces of the crack.

The Irwin (local) approach is based on the concept of stress intensity factor, which represents the energy stress field in the neighborhood of the crack tip. These two approaches are equivalent and, therefore, the energy criterion may be rewritten in terms of stress intensity factors. Further, a number of path independent integrals have been proposed to calculate the ERR, such as the J-integral [186].

The elastic strain energy released ΔU during crack propagation, and therefore used to create the new surface area, can be calculated as the work required to close the crack

$$\Delta U = W_{closure} \qquad (10.3)$$

providing the basis for the virtual crack closure technique (VCCT) described in Section 10.2.

When the plastic or process zone at the crack tip is large, as it often happens with delaminations of laminated composites and decohesion of adhesive joints, the energy dissipated contains a significant amount of plastic and other forms of decohesion dissipation. In this case, the amount of energy per unit of decohesion area required to propagate the crack, i.e., to fully separate the substrates, is called critical fracture energy, G_c. Similarly, the amount of energy per unit of decohesion area required to separate the substrates without reaching debonding is denoted by G. At full separation, (10.2) applies.

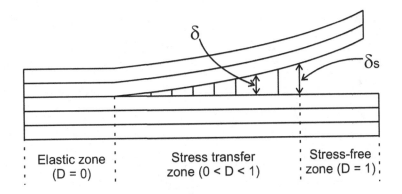

Fig. 10.3: Cohesive zone model to simulate crack propagation.

10.1 Cohesive Zone Model

The cohesive zone model (CZM) is based on the assumption that the stress transfer capacity between the two separating faces of a delamination is not lost completely at damage initiation, but rather is a progressive event governed by progressive stiffness reduction of the interface between the two separating faces (Figure 10.3).

The interface between two separating faces of the laminated material is modeled with *cohesive elements* [187]. There are two types of CZM elements in Abaqus:

- CZM elements with finite thickness, commonly used to simulate delamination behavior of adhesive bonds where the thickness of the interface is considerable.

- CZM elements with zero thickness, commonly used to simulate delamination behavior where the thickness of the interface is negligible, such as the interfaces between laminas in laminated composites.

The cohesive behavior is described in terms of a traction-separation equation (Figure 10.4). As the name implies, this approach replaces the engineering stress-strain ($\sigma - \epsilon$) equation with a traction-separation ($t - \delta$) equation, where t, ϵ, are traction and displacement, respectively. Tractions are forces per unit area in three directions n, s, t, depicted in Figure 10.5. The thickness of the element is set to zero by defining coincident opposite nodes of the cohesive element. However, even if the opposite nodes are initially coincident, they are still separate entities, and they can separate during the deformation of the laminated composite. The adjacent faces of the laminas can be regarded as being connected to each other through the stiffness of the cohesive element. During the deformation, the resulting separation between adjacent laminas is proportional to the stiffness of the cohesive element.

The CZM element with finite thickness COH2D4 (Example 10.1, p. 374) and the CZM element with zero thickness, which is implemented through *interaction cohesive behavior* (Example 10.2), are based on a traction-separation constitutive equation. The element stiffness matrix requires the stiffness $\tilde{K}$ of the interface material (called *penalty* stiffness in the Abaqus documentation), but the element stiffness matrix is not formulated as usual by integration over the volume of the element because the volume of the element is zero.

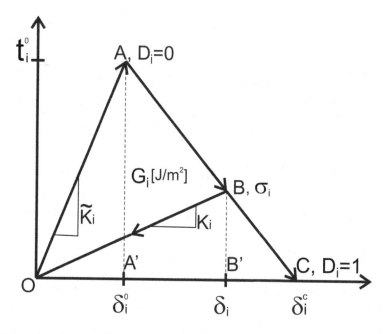

Fig. 10.4: Linear softening approximation for cohesive zone model. Other approximations include exponential, tabular, etc.

The CZM element can be visualized as a spring between the initially coincident nodes of the element. However, the stiffness of the CZM element is part of the structural stiffness, and the CZM element will undergo deformation during the loading of the laminate. The initially coincident nodes will open (mode n:opening) or slide (mode s:shear and t:tearing) relative to each other (Figure 10.6). The nodal separations of the CZM element are always known by solving the discretized structure.

Since the thickness of the CZM element is zero, the deformation state of the CZM element can not be described by the classical definition of strain. Instead, the measure of the deformation becomes the separation δ between the faces connected through the CZM element, and this makes possible the use of the $(t - \delta)$ traction-separation equation instead of the classical engineering $(\sigma - \epsilon)$ equation.[1]

10.1.1 Single Mode Cohesive Model

CZM is formulated assuming that the three propagation modes are uncoupled, even if multiple modes are active simultaneously, as described in Section 10.1.2. In this section we consider the case of a single-mode deformation at an interface of the laminated material, either mode n, s, or t (Figure 10.6).

The formulation is similar for any of the three modes. The surface tractions at the interface are t_i with $i = n, s, t$ denoting the three modes of crack propagation. The corresponding separations between the opposite faces of the CZM element are

---

[1]Abaqus documentation uses the word *strain* as synonymous to *separation* δ when describing the deformation of the CZM element based on traction-separation equation.

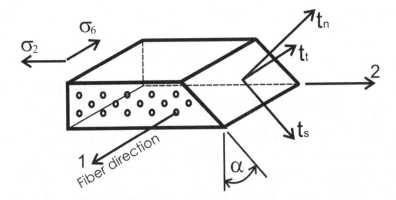

Fig. 10.5: Normal traction t_n (mode I, opening crack), sliding shear t_s (mode II), and tearing shear t_t (mode III).

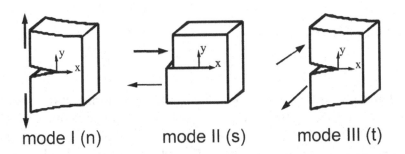

Fig. 10.6: Crack-propagation modes: opening, shear, and tearing.

denoted as δ_i, each related to the surface traction t_i through the interface stiffness K_i. The behavior of the material at the interface is assumed to be linear-elastic (OA in Figure 10.4) up to the onset of damage (A), then damaging with stress softening (AB), and elastic unloading (BO). Consequently, the stress-separation is described by

$$t_i = K_i \, \delta_i \tag{10.4}$$

and

$$K_i = (1 - D_i) \, \widetilde{K}_i \tag{10.5}$$

where D_i are the damage variables and $\widetilde{K}_i$ in $[N/mm^3]$ is the penalty stiffness of the *undamaged* material at the interface (segment OA), relating the tractions t_i to the relative separations δ_i between the opposite faces of the CZM element, with damage modes $i = n, s, t$. In CZM, these modes are called *damage modes* rather than *fracture modes* because the CZM replaces the fracture mechanics problem by a continuum damage mechanics problem (Chapter 8).

The stiffness values $\widetilde{K}_i$ are additional material properties needed for the CZM model, which are different from the Young's modulus E (traction/compression) and G for (shear) deformation. A discussion on how to choose numerical values for $\widetilde{K}_i$ is provided in Section 10.3.3.

For each mode, there is a critical level of traction t_i^o and separation δ_i^o, called *damage onset*, when the damage at the interface starts (point A in Figure 10.4). At damage onset, the two laminas do not separate completely into a physical crack but rather the interface material starts losing its stiffness. Here t_i^o represents the strength of the interface with one value of strength for each crack-propagation mode: t_n^o, t_s^o, t_t^o (normal, shearing, tearing). The interface strengths are therefore additional material parameters required by the CZM element. A discussion on how to choose numerical values for t_i^o is provided in Section 10.3.3.

The *damage initiation criteria* are

$$t_i = t_i^o \tag{10.6}$$

and the separation at damage onset are calculated as

$$\delta_i^o = t_i^o / \widetilde{K}_i \tag{10.7}$$

After damage onset, the interface material starts losing its stiffness (OB in Figure 10.4), according to (10.5). The damages D_n, D_s, D_t are state variables (8.61) with physical interpretation given by (10.5) as measures of stiffness degradation (see also Section 8.2). In Abaqus implementation of mixed-mode decohesion, they are replaced by a single, isotropic damage D_m (10.26) that captures the overall effect of damage.

The damage variables satisfy the following conditions:

- $D = 0$ up to damage onset (OA in Figure 10.4) while the interface material is undamaged, thus retaining its initial stiffness.

- $0 < D < 1$ during degradation of the interface material (AC in Figure 10.4), when the material is gradually losing its stiffness.

- $D = 1$ at fracture (point C in Figure 10.4), when there is no remaining stiffness for the interface material, which means no stress transfer capacity is provided by the interface. This corresponds to the fracture of the cohesive connection between the two faces of the CZM element.

With reference to Figure 10.4 we have

$$D_i = \begin{cases} 0, & \delta_i \leq \delta_i^o \\ 1, & \delta_i = \delta_i^c \end{cases} \tag{10.8}$$

In traction-separation space, the cohesive behavior depicted in Figure 10.4 displays *stress softening*. That is, after damage onset the traction t_i in the damaged interface have lower values than the peak values t_i^o.

The CZM element uses the classical assumption of *elastic damage* typical of continuum damage mechanics (CDM, Chapter 8). Therefore, unloading from any point such as B on the line AC in Figure 10.4 will return to the origin without permanent deformation upon unloading.

Eventually, there will be total fracture of the cohesive bond (point C in Figure 10.4) when the stiffness of the interface reduces to zero. Total loss of stiffness, and thus total loss of cohesive stress transfer does not takes place until point C. The area under the $(t - \delta)$ curve in Figure 10.4 is called *fracture energy*, which has the same units (energy per unit area, kJ/m^2) as the critical energy release rate (cERR) G_c in Griffith's principle (10.2). For linear softening (Figure 10.4), the separation at fracture δ_i^c can be calculated as

$$\delta_i^c = 2G_i^c/t_i^o \tag{10.9}$$

Since there is a critical fracture energy value for each damage mode $i = n, s, t$, similar to the linear elastic fracture mechanics (LEFM) modes $i = I, II, III$, there will be three separations at fracture–one per mode.

The three fracture energies G_i^c are material properties required by CZM, in addition to the three values of strength t_i^o and the three values of interface stiffness $\widetilde{K}_i$. A discussion on how to choose numerical values for G_i^c is provided in Section 10.3.3. From the point of view of the amount of experimental data needed to perform an analysis, this is a disadvantage of CZM because it requires nine experimental values. In comparison, discrete damage mechanics requires only the three values of ERR to predict both the onset and the evolution of damage (Chapter 9).

Substituting (10.4) into (10.5) and rearranging, we get

$$D_i = 1 - \frac{t_i \delta_i^o}{t_i^o \delta_i} \tag{10.10}$$

By similarity of triangles BB'C and AA'C, we have

$$\frac{t_i}{t_i^o} = \frac{\delta_i^c - \delta_i}{\delta_i^c - \delta_i^o} \tag{10.11}$$

which substituted into (10.10) yields

$$D_i = \frac{\delta_i^c(\delta_i - \delta_i^o)}{\delta_i(\delta_i^c - \delta_i^o)} \tag{10.12}$$

In this way, the damage variables D_i may be calculated as a function of the relative separation between the faces of the laminate δ_i, which is provided by the FE solution and the values δ_i^o, δ_i^c calculated previously.

In summary, there are four distinct stages in the material behavior described by Figure 10.4:

- Linear elastic undamaged material behavior (line OA), with the associated constitutive equation (10.4) and material property $\tilde{K}_i$.

- Damage initiation (point A), with the associated criterion (10.6) and material property t_i^o.

- Damage evolution (line AC), with the associated damage evolution equation (10.12) and material property G_i^c.

- Fracture (crack formation), with the associated crack formation criterion (10.9).

As previously stated, the formulation presented in this section works only for one mode of deformation, either $n, s,$ or t. The general case of the mixed-mode loading is presented in the following section.

10.1.2   Mixed-mode Cohesive Model

When the interface of the laminated material is under mixed modes, all three traction components t_n, t_s, t_t and all three separation components $\delta_n, \delta_s, \delta_t$ are active. In other words, a mixed mode implies that two or more pairs (t_i, δ_i) are non-vanishing, with $i = n, s, t$. Mixed-mode ratios can be defined between pairs of mode components. For example, in terms of separations

$$\beta_{\delta_s} = \frac{\delta_s}{\delta_n} \quad ; \quad \beta_{\delta_t} = \frac{\delta_t}{\delta_n} \tag{10.13}$$

or in terms of fracture energy

$$\beta_{G_s} = \frac{G_s}{\sum_1^3 G_i} \quad ; \quad \beta_{G_t} = \frac{G_t}{\sum_1^3 G_i} \tag{10.14}$$

Regardless of the definition used, mixed-mode ratios are just parameters characterizing the mixed-mode state, which allow for a simplification of the analysis by assuming that decohesion progresses at constant mixed-mode ratios. It is further assumed that the modes are uncoupled even though they occur simultaneously. That is, the stress-separation relationship for each uncoupled mode is again expressed by (10.4), separately for each mode.

Next, a mixed-mode separation is defined by the L^2 norm of the mode separations, i.e.,

$$\delta_m = \sqrt{\sum_{i=1}^{M} \delta_i^2} \tag{10.15}$$

where M is the number of modes involved (i.e., 2 or 3 modes). Next, the single-mode damage initiation criterion in (10.6) is replaced, for example, by a quadratic stress criterion

$$\sum_{i=1}^{M} \left(\frac{t_i}{t_i^o}\right)^2 = 1 \tag{10.16}$$

Other mixed-mode damage initiation criteria can be selected in Abaqus, such as *quadratic displacement criterion* or *maximum-stress/maximum-displacement criterion*. Regardless of the damage initiation criterion selected, the goal is to calculate the mixed-mode separation δ_m^o at damage onset under mixed-mode loading.

For the 2D case of mode n and s only, $M = 2$ and the *equivalent mixed-mode separation at damage onset* δ_m^o is found, as follows. First, rewrite the damage initiation criteria (10.16) in terms of separations using (10.7) and (10.4)-(10.5), taking into account that $D_i = 0$ up to damage initiation. Therefore, mixed-mode damage initiation is predicted by

$$\left(\frac{\delta_n}{\delta_n^o}\right)^2 + \left(\frac{\delta_s}{\delta_s^o}\right)^2 = 1 \tag{10.17}$$

Next, rewrite (10.15) using the first of (10.13) to get

$$\delta_n = \frac{\delta_m}{\sqrt{1 + \beta^2}} \tag{10.18}$$

and using (10.13) again

$$\delta_s = \beta \frac{\delta_m}{\sqrt{1 + \beta^2}} \tag{10.19}$$

Now, substitute (10.18) and (10.19) into (10.17), taking into account that since (10.17) represents damage initiation, one should write δ_m^o for δ_m. Therefore,

$$\delta_m^o = \sqrt{(\delta_n^o)^2 (\delta_s^o)^2 \frac{1 + \beta^2}{(\delta_s^o)^2 + \beta^2 (\delta_n^o)^2}} \tag{10.20}$$

The quantities δ_i^o in (10.20) represent the separations at damage onset during single-mode loading, calculated with (10.7), and β is the mixed-mode ratio, which is assumed to be constant during the damage process.

A mixed-mode crack-propagation criterion is now needed to replace the single-mode criterion (10.2). A possible choice is to use a power criterion, as follows

$$\sum_{i=1}^{3} \left(\frac{G_i}{G_i^c}\right)^{\alpha_i} = 1 \tag{10.21}$$

which attempts to predict fracture under mixed-mode conditions, similarly to point C in Figure 10.4 for the single-mode situation. To reduce the burden of experimentation, it is customary to assume that the exponents are the same for all modes, i.e., $\alpha_i = \alpha$.

Each of the single-mode components G_i can be calculated by one of two methodologies. Some authors [176, 178] calculate each single-mode component value G_i by considering the area OABB' in Figure 10.4, thus including the recoverable energy OBB' in the definition of G_i. This approach is indirectly related to linear elastic fracture mechanics.

Other authors [188] use a damage mechanics approach where each single-mode component value G_i is calculated by considering only the unrecoverable energy represented by the area OAB. Both approaches lead to the same results for a single-mode delamination since loss of adhesion occurs at δ_i^c where both approaches predict the same values for G_i (point B reaches point C). However, different results are obtained for mixed-mode delamination because the crack propagates when an interaction criterion is satisfied, the latter involving the ratios G_i/G_i^c.

In principle, the approach based on LEFM should lead to conservative predictions of the load carrying capacity for mixed-mode delamination. On the other hand, since the total energy dissipated during the delamination at each point is not released instantaneously as assumed in LEFM, the damage mechanics based definition appears appropriate especially when the size of the nonlinear fracture process zone ahead of the delamination front is not negligible as it may occur in the case of laminated composite materials where the damage zone may be comparable to or larger than the single-ply thickness, which generally scales with the near-tip stress field.

In the sequel, the damage mechanics approach is used. That is, for each single-mode separation δ_i, the single-mode component G_i is calculated as the area OAB in Figure 10.4, which represents dissipated energy, i.e.,

$$A_{OAB} = A_{OAF} - A_{OBF} \tag{10.22}$$

where A_{OAF} is the single-mode critical G_i^c, and A_{OBF} can be calculated based on the geometry in Figure 10.4 as

$$A_{OBF} = \frac{1}{2} BB' \times OF = \frac{1}{2} K_i \delta_i^o \frac{\delta_i^c - \delta_i}{\delta_i^c - \delta_i^o} \delta_i^c \tag{10.23}$$

where $\delta_i^c = OF$, $K_i \delta_i^o = t_i^o$, and t_i/t_i^o is given by (10.11). Based on (10.22), (10.23), the single-mode component G_i at the moment of mixed-mode fracture is calculated as [188, (14)]

$$G_i = G_i^c - \frac{1}{2} K_i \delta_i^o \delta_i^c \frac{\delta_i^c - \delta_i}{\delta_i^c - \delta_i^o} \tag{10.24}$$

The assumed mode decomposition (10.24) is necessary so that each single-mode component G_i corresponding to mixed-mode fracture can be expressed as a function of the single-mode separation δ_i, i.e., $G_i = G_i(\delta_i)$. All other quantities in (10.24) are known, as follows

– G_i^c is the single-mode critical ERR (material parameter).

– δ_i^o is the separation at damage onset under single-mode loading (point A in Figure 10.4), given by (10.7).

– δ_i^c is the separation at fracture under single-mode loading (point C in Figure 10.4), given by (10.9).

The single-mode G_i components of the mixed-mode condition in (10.24) have to satisfy the energy criterion (10.21) at the moment of fracture (crack propagation). For the case of two modes, and assuming $\alpha_i = \alpha = 2$ in (10.21), the mixed-mode separation at fracture (point C) is calculated in [188, (15)] as

$$\delta_m^c = \frac{\sqrt{1+\beta}}{\beta^2(\delta_n^{0F})^2 + (\delta_s^{0F})^2} \times$$
$$\left\{ \delta_n^o(\delta_s^{0F})^2 + \beta\delta_s^o(\delta_n^{0F})^2 \right.$$
$$\left. + \delta_n^{0F}\delta_s^{0F}\sqrt{(\delta_s^{0F})^2 - (\delta_s^o)^2 + 2\beta\delta_n^o\delta_s^o - \beta^2(\delta_n^o)^2 + \beta^2(\delta_n^{0F})^2} \right\} \tag{10.25}$$

where $\delta_n^{0F} = \delta_n^c - \delta_n^o, \delta_s^{0F} = \delta_s^c - \delta_s^o$.

There are three energy crack extension criteria implemented in Abaqus:

– The power equation criterion in (10.21).

– The BK criterion, which is another closed form function of (G_i, G_i^c).

– User input of the total mixed-mode $G_c = \sum_1^3 G_i^c$ at complete separation.

Once the mixed-mode separation at fracture (10.25) is calculated based on the selected criterion, and the mixed-mode separation at damage onset is known based on (10.20), the damage variable for mixed-mode conditions can be expressed in a similar manner as for the case of single-mode condition (10.12) by satisfying the requirement in (10.8), i.e.,

$$D_m = \frac{\delta_m^c(\delta_m - \delta_m^o)}{\delta_m(\delta_m^c - \delta_m^o)} \tag{10.26}$$

where the onset and fracture separations δ_m^o and δ_m^c are calculated based on (10.20) and (10.25), respectively; and δ_m is the current level of separation under mixed-mode conditions obtained using (10.15) in terms of the single-mode separations δ_i provided by the FE model. The stiffness degradation for the cohesive material is then calculated according with (10.5). The stress softening evolution of the cohesive material under mixed-mode conditions is similar to the one depicted in Figure 10.4, by substituting δ_m^o, δ_m^c for δ^o, δ^c.

Besides the energy approach to mixed-mode damage evolution as presented here, where the crack extension criterion is a function of fracture energy G, and δ_m^c is calculated based on this energy criterion, Abaqus also has implemented a displacement approach to damage evolution. The displacement approach requires the user to directly provide the mixed-mode separation at the moment of fracture δ_m^c, which is then directly used in (10.26).

Example 10.1 *A laminated double cantilever beam (DCB) 100 mm long and 20 mm wide is made up of two layers, each 1.5 mm thick, bonded by an adhesive layer of negligible thickness. In the parlance of adhesion science, the layers to be bonded are also called **substrates**. Apply a loading system to induce delamination and mode I crack growth through the adhesive layer. Assuming linear elastic behavior, create a 2D model of the DCB using CZM elements to represent the adhesive layer. Use the* element-based *cohesive behavior implemented in Abaqus to represent the adhesive.*

Solution to Example 10.1 *Watch the video in [1].*
To facilitate the solution of this example, run the Python script ws_composites_dcb.py *in your* Work Directory*. The script can be bound in [2] or fetched from the examples in the Abaqus documentation, as follows*

```
# open a Windows command prompt via the Windows Taskbar and Start Menu
Start, Run, Open [cmd], OK
# Navigate to the folder C:\SIMULIA\User\Ex_10.1
# you may need to create the directory using the command: mkdir Ex_10.1
# type the following command and press the Enter-key
abaqus fetch job=ws_composites_dcb.py
# this will copy the file ws_composites_dcb.py into the Work Directory
```

The script creates two models, of which only the 2D model named coh-els *is used in this example. The model contains the parts, assembly, materials, sections, sets, and surfaces needed to apply boundary conditions, constraints, and to define the output requests fields.*

During model set up, the cohesive elements representing the adhesive layer are defined with a finite thickness. This is done only to facilitate manipulation and the definition of constraints. At the end of the setup process, the thickness of the cohesive elements is adjusted to zero, as explained in Section vi, p. 377. That is, cohesive elements COH2D4 with zero thickness are used in this example.

Although the loading conditions applied in this example induce crack propagation in mode I only, Abaqus requires defining the parameters corresponding to all three propagation modes.

i. Select the Work Directory

> Menu: File, Set Work Directory, [C:\SIMULIA\User\Ex_10.1], OK

ii. Running the Python script

> Menu: File, Run Script, [ws_composites_dcb.py], OK
> # notice that the model has been named dcb.cae
> Menu: Save As, [Ex_10.1.cae], OK

iii. Defining materials, sections, and assigning sections to parts.
The coh-els *model is set up with units $[m, Pa]$. A material named 'bulk' contains the undamaged material properties of the laminas: $E_1 = 135.3 \times 10^9\ Pa; E_2 = E3 = 9 \times 10^9\ Pa; \nu_{12} = \nu_{13} = 0.24; \nu_{23} = 0.46; G_{12} = G_{23} = 4.5 \times 10^9\ Pa; G_{13} = 3.3 \times 10^9\ Pa.$*

To enter the properties of the adhesive, a new material called 'adhesive' must be defined using the pseudo code below. The penalty stiffness values used in this example are: $E_{nn} = E_{ss} = E_{tt} = 570 \times 10^{12}\ Pa/m$ along directions n, s, t.

In Abaqus CAE, the "/" in $E/E_{nn}, G1/E_{ss}, G2/E_{tt}$ means "or", not "quotient". When the user later creates a "section" for the adhesive/interface initial thickness, there are three choices: 1) use analysis default, 2) use nodal coordinates, 3) specify (see [1, Example 10.1] at time 2:29).

If the user selects "analysis default", the user-supplied values for E/Enn, $G1/Ess$, $G2/Ett$ are the values of penalty stiffness E_{nn}, E_{ss}, E_{tt}, calculated by the user (at time 1:18 in [1, Example 10.1]).

If the user selects "use nodal coordinates" or "specify", Abaqus calculates the penalty stiffness values as $E_{nn} = E/t_a$; $E_{ss} = G1/t_a$; $E_{tt} = G2/t_a$, where t_a thickness of the adhesive or interface, specified or calculated from the mesh, and $E, G1, G2$ are the Young modulus, sliding shear modulus, and tearing shear modulus of the adhesive (or interface), respectively. My preference is to calculate the penalty stiffness values myself.

The adhesive strength values are: $t_n^0 = t_s^0 = t_t^0 = 5.7 \ 10^7$ Pa. The critical fracture energies are $G_n^c = G_s^c = G_t^c = 280 \ J/m^2$ (10.2) and the BK exponent is $\alpha = 2.284$.

The "out-of-plane thickness" is the width of the beam, not the adhesive thickness.

```
Module: Property
Model:  coh-els
Part:   adhesive
Menu:   Material, Manager
    Create, Name [cohesive], Mechanical, Elasticity, Elastic
    Type: Traction, E/Enn [5.7E14], G1/Ett [5.7E14], G2/Ess [5.7E14]
    # damage onset criterion
    Mechanical, Damage for Traction Separation Laws, Quads Damage
    Nominal Stress Normal-only Mode [5.7E7]
    Nominal Stress First Direction  [5.7E7]
    Nominal Stress Second Direction [5.7E7]
    # crack extension criterion and required parameters
    # Benzeggagh-Kenane (BK) is used, see Abaqus documentation
    Suboptions, Damage Evolution
    Type: Energy, Mixed mode behavior: BK, # checkmark: Power [2.284]
    Normal Mode Critical Fracture Energy [280]               # Gnc
    Shear Mode Critical Fracture Energy First Direction  [280] # Gtc
    Shear Mode Critical Fracture Energy Second Direction [280] # Gsc
    OK, OK, # close Material Manager pop-up window
Menu:   Section, Manager
    # a section named 'bulk' defines the laminas
    # create a section for the adhesive
    Create, Name [cohesive], Category: Other, Type: Cohesive, Cont
    Material: cohesive, Response: Traction Separation
    # checkmark: Out-of-plane thickness [0.02], OK
    # close Section Manager pop-up window
Menu:   Assign, Section
    # pick the part, Done, Section: cohesive, OK
```

iv. Defining the analysis steps

```
Module: Step
Menu:   Step, Create, Cont
    # geometrical non-linearity for large displacements
    Tab: Basic, Nlgeom: On
    # parameters to setup the time increments for the solution
```

      Tab: Incrementation, Maximum number of increments [1000]
      Increment size: Initial [0.01], Minimum [1.0E-8], OK
   Menu: Output, Field Output Requests, Edit, F-Output-1
      Output Variables: # expand: State/Field/User/Time
      # checkmark: STATUS, OK
      # the STATUS variable is an on/off switch that indicates the CAE
      # postprocessor when to stop drawing the cohesive elements
      # a completely degraded (damaged) element will have STATUS 0 (off)
   Menu: Output, History Output Requests, Create
      Name [H-Output-2], Step: Step-1, Cont
      Domain: Set : top
      # expand: Displ/Veloc/Accel, # expand: U, # checkmark: U2
      # expand: Forces/Reactions, # expand: RF, # checkmark: RF2, OK
      # the values of U2 (imposed displacement) and RF2 (reaction force)
      # at the specified set (node) will be stored for postprocessing

 v. *Meshing the model*
   The sweep technique is used because the length (thickness) of the cohesive elements is
   orientation-dependent. For this case they need to be defined from the bottom to the
   top surface of the adhesive layer because their elongation will represent the separation
   of the laminas.

   A viscosity parameter is used along with the COH2D4 element to speed up the convergence of the iterative solution. You may change its value to see how it affects the
   solution.

   Module: Mesh
   Model: coh-els
   Object: Part: beam
   Menu: Mesh, Controls, Element Shape:Quad, Technique:Structured, OK
   Menu: Mesh, Element Type
      Family: Plane Strain, # checkmark: Incompatible modes
      # Notice element CPE4I has been assigned, OK
   Menu: Seed, Edges
      # pick the top horizontal line, Done
      Method: By number, Number of elements [400], OK
      # pick the left vertical line, Done
      Method: By number, Number of elements [2], OK
   Menu: Mesh, Part, Yes
   Object: Part: adhesive
   Menu: Mesh, Controls, Element Shape: Quad, Technique: Sweep
      Redefine Sweep Path
      # if the highlighted path coincides with the thickness direction
      Accept Highlighted
      # otherwise change accordingly to accomplish that, OK
   Menu: Mesh, Element Type
      Family: Cohesive, Viscosity: Specify [1.0E-5]
      # notice element COH2D4 has been assigned, OK
   Menu: Seed, Edges
      # pick the top horizontal line, Done
      Method: By number, Number of elements [280], OK
      # pick the left vertical line, Done

```
        Method: By number, Number of elements [1], OK
     Menu:    Mesh, Part, Yes
```

vi. *Creating model constraints.*
   The use of a finite-thickness part to represent the adhesive layer facilitated the defini-
   tion of the cohesive elements properties. The constraints applied here guarantee that
   the top and bottom surfaces of the cohesive elements are coincident once the solution
   process starts, effectively defining the initial length of the cohesive elements as zero.
   Although it is not visually shown, the procedure below adjusts the surfaces defined as
   slaves to follow those defined as masters.

```
Module: Interaction
Menu:    Constraint, Create, Name [top], Type: Tie, Cont
     Choose the master type: Surface, Surfaces, top, Cont
     Choose the slave type:  Surface, coh-top, Cont
     Position Tolerance: Specify distance [0.002], OK
Menu:    Constraint, Create, Name [bot], Type: Tie, Cont
     Choose the master type: Surface, bot, Cont
     Choose the slave type:  Surface, coh-bot, Cont
     Position Tolerance: Specify distance [0.002], OK
```

vii. *Applying boundary conditions (BC) and loads*

```
Module: Load
Menu:    BC, Create, Name [top], Step: Step-1, Disp/Rota, Cont
     Sets, top, Cont, #checkmark: U1 [0], #checkmark: U2 [0.006], OK
Menu:    BC, Create, Name [bot], Step: Step-1, Disp/Rota, Cont
     Sets, bot, Cont, #checkmark: U1 [0], #checkmark: U2 [-0.006], OK
     # due to the symmetry of the model no additional BCs are needed
```

viii. *Solving and visualizing the results*

```
Module: Job
Menu:    Job, Manager
     Create, Name [Ex-10-1],Cont,OK,Submit, #when completed, Results
Module: Visualization
Menu:    Plot, Deformed Shape # notice the beams are now separated
Menu:    Plot, Contours, On Deformed Shape
Menu:    Animate, Time History #see progression of delamination
```

To visualize the force–separation behavior in the model:

```
Module: Visualization
Menu:    Result, History Output
   Output Variables: Reaction force: RF2 PI,Save As [coh-els-RF2], OK
   Output Variables: Spatial displ: U2 PI,Save As [coh-els-U2], OK
   # close the History Output pop-up window
Menu:    Tools, XY Data, Create, Source: Operate on XY data, Cont
   Operators: Combine(X,X), # pick: coh-els-U2, Add to Expression
   # pick: coh-els-RF2, Add to Expression,
       Save As [cohesive elements], OK
   # close Operate on XY Data pop-up window
Menu:    Tools, XY Data, Plot, cohesive elements
```

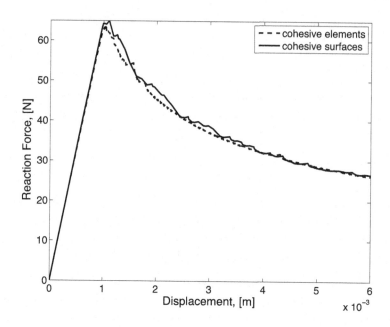

Fig. 10.7: Comparison of force–separation behavior for models using cohesive elements (Example 10.1) and cohesive surfaces (Example 10.2).

The resulting plot should look like the dashed line in Figure 10.7. The drop of the reaction force occurs when the cohesive elements start to degrade (damage). As the cohesive elements degrade, the reaction force reduces continuously. In the visualization module, zoom in close to the crack tip and identify the cohesive elements representing the adhesive layer.

Example 10.2 *Solve Example 10.1 using surface-based cohesive behavior instead of element-based cohesive method.*

Solution to Example 10.2 *Watch the video in [1].*
This example is solved modifying the .cae *file defined in Example 10.1. The part defined in Example 10.1 representing the adhesive layer is removed, then redefined as an interaction between the coincident surfaces of the two laminas. That is, element-based cohesive method features are replaced by surface-based cohesive method features.*

```
Menu: File, Set Work Directory, [C:\SIMULIA\User\Ex_10.2], OK
Menu: File, Open, [C:\SIMULIA\User\Ex_10.1\Ex_10.1.cae], OK
# on the Model-tree
# right-click: coh-els, Rename, [coh-surf], OK
Menu: File, Save As, [C:\SIMULIA\User\Ex_10.2\Ex_10.2.cae], OK
```

i. Deleting unnecessary features

```
Module: Assembly
Model:  coh-surf
Menu:   Tools, Surface, Manager
    # pick: coh-top, Delete, Yes
    # pick: coh-bot, Delete, Yes, # close Surface Manager
```

```
Module: Interaction
Menu:   Constraint, Manager
        # pick: bot, Delete, Yes, # pick: top, Delete, Yes
        # close Constraint Manager
        # on the model-tree
        # expand: coh-surf, # expand: Assembly, # expand: Instances
        # right-click: adhesive-1, Delete, Yes
Menu:   Feature, Regenerate
Module: Mesh
Model:  coh-surf
Object: Part:   beam
Menu:   Mesh, Delete Part Mesh, Yes
```

ii. Creating a partition.
At this point the model consists only of two stacked beams. Now, it is necessary to divide the interface between the beams to represent the region where the adhesive is applied. A partition will be created.

```
Menu:   Tools, Partition
        Type: Edge, Method: Enter parameter
        # pick bottom edge of beam, Done, Normalized edge parameter [0.7]
        Create Partition, Done, #close the Create Partition pop-up window
```

iii. Meshing the part

```
Menu:   Mesh, Part, Yes
```

iv. Creating sets and surfaces.
The surface called top in Example 10.1 will be redefined to be the whole interface between the beams. Since the two beams are stacked, selecting only one of the coincident surfaces is cumbersome. To facilitate selecting the desired surface, a useful visualization tool is introduced –Display Group Toolbar.

```
Module: Assembly
        # locate the Toolbar named Display Group, If it is not available
        # enable it using Menu: View, Toolbars, Display Group
        # click the first icon of Display Group Toolbar (Replace Selected)
        # pick the top beam, Done, # only the picked part will be visible
Menu:   Tools, Surface, Edit, top
        # pick the two lines forming the bottom edge of the beam, Done
Menu:   Tools, Set, Create
        Name [bond], Type: Geometry, Cont
        # pick the longest line at the bottom of the beam, Done
        # click on third icon of Display Group Toolbar (Replace All)
        # the full assembly will be visualized
```

The set named bond will represent the adhesive layer in the same way the part named adhesive-1 did in Example 10.1.

v. Defining interactions to represent the adhesive layer.
First the properties and characteristics of the contact are defined. Then the interacting surfaces are selected.

```
     Module: Interaction
     Menu:   Interaction, Property, Create
       Name [coh], Type: Contact, Cont
       Mechanical, Tangential Behavior, Friction formulation: Frictionless
       Mechanical, Normal Behavior, Pressure-Overclosure: Hard Contact
       Constraint enforcement method: Default
       Mechanical, Cohesive Behavior
       Eligible Slave Nodes: Specify the bounding node set in surf-to-surf
       Traction-separation Behavior: Specify stiffness coeffs, Uncoupled
       Knn [5.7e14], Kss [5.7e14], Ktt [5.7e14] # KI, KII, and KIII
       Mechanical, Geometric Properties,
           Out-of-plane surface thickness [0.02]
       Mechanical, Damage, Criterion: Quadratic traction
       Normal Only [5.7e7], Shear-1 Only [5.7e7], Shear-2 [5.7e7]
       # these define the onset of damage in the adhesive layer
       # checkmark: Specify damage evolution
       Type: Energy,
           # checkmark: Specify mixed-mode behavior: Benzeggagh-Kenane
       # checkmark: Specify power-law/BK exponent [2.284]
       Normal Critical Fracture Energy [280],   # GIc
       1st Shear Critical Fracture Energy [280] # GIIc
       2nd Shear Critical Fracture Energy [280] # GIIIc
       # checkmark: Specify damage stabilization, Viscosity coeff. [1e-5]
       # it plays the same role as the one used for cohesive elements
       OK # the properties of the adhesive layer have been defined
     Menu:   Interaction, Create
       Name [coh], Step: Initial, Type: Surface-to-surface(Standard), Cont
       Surfaces, # pick: bot, Cont # this is the master surface
       Surfaces, # pick: top, Cont # this is the slave surface
       Discretization method: Node to surface
       Tab: Bonding
       # checkmark: Limit bonding to slave nodes in subset: bond
       # This represents the adhesive layer bonding the beams, OK
```

vi. Solving and visualizing the results

```
     Module: Job
     Menu:   Job, Manager
       Create, Name [Ex-10-2], Cont, OK, Submit, #when completed, Results
     Module: Visualization
     Menu:   Plot, Deformed Shape # notice the beams have been separated
     Menu:   Plot, Contours, On Deformed Shape
     Menu:   Animate, Time History # progression of the delamination
```

To visualize the force-separation behavior in the model:

```
     Module: Visualization
     Menu:   Result, History Output
       Output Variables: Reaction force: RF2 PI,
           Save As [coh-surf-RF2], OK
       Output Variables: Spatial displ: U2 PI,
           Save As [coh-surf-U2],  OK
```

```
        # close the History Output pop-up window
    Menu:    Tools, XY Data, Create, Source: Operate on XY data, Cont
       Operators: Combine(X,X), # pick: coh-surf-U2, Add to Expression
       # pick: coh-surf-RF2, Add to Expression,
           Save As [cohesive surfaces], OK
       # close Operate on XY Data pop-up window
    Menu:    Tools, XY Data, Plot, cohesive surfaces
```

The resulting plot should look like the solid line in Figure 10.7. The drop-off of the reaction force occurs when the separation between the surfaces reaches a critical value, δ_i^0, which correspond to a critical stress value, t_i^0. Unlike Example 10.1, where cohesive elements are used, in this example the progressive damage of the interface is calculated by Abaqus without the use of particular elements. In the visualization module, zoom-in close to the crack-tip and notice that no elements are depicted to represent the adhesive layer.

10.2   Virtual Crack Closure Technique

The virtual crack closure technique (VCCT) can be used to analyze delaminations in laminated materials using a fracture mechanics approach. The method implements linear elastic fracture mechanics (LEFM). Only brittle crack-propagation is modeled. The energy dissipated by the formation of plastic zones at the crack-tip is not considered.

The condition for crack propagation is based on Griffith's principle (10.2). For the case of single-mode I deformation, the crack does not grow as long as $G_I < G_I^c$ and it grows when

$$G_I = G_I^c \tag{10.27}$$

where G_I is the ERR for mode I crack formation and G_I^c a material property representing the critical ERR for mode I crack formation.

The definition of ERR is given by (10.1). In VCCT, the Irwin principle (10.3) is used to calculate the change in strain energy ΔU, which is considered to be equal to the work required for crack closure $W_{closure}$.

By substituting (10.3) and (10.1) into (10.27), the condition for crack propagation under mode I loading becomes

$$\frac{W_{closure}/\Delta A}{G_I^c} = 1 \tag{10.28}$$

Abaqus [189] calculates $W_{closure}$ from the FE nodal displacements and forces as illustrated in Figure 10.8. Initially the crack surfaces are rigidly bonded. The nodal forces at the coincident nodes $2-5$ are calculated from the FE solution. The hypothesis of *self-similar crack propagation* is used, which says that during crack propagation, the crack configuration between nodes $2-3-4-5$ will be similar to the crack configuration between nodes $1-2-5-6$. This implies that the separation between nodes $2-5$ after crack propagation will be equal to the separation between nodes $1-6$ before crack propagation: $v_{2,5} = v_{1,6}$. If the nodes $2-5$ open (crack propagation), the elastic work required to close the crack is

$$W_{closure} = \frac{1}{2}F_{2,5}\, v_{2,5} = \frac{1}{2}F_{2,5}\, v_{1,6} \tag{10.29}$$

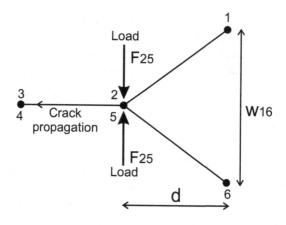

Fig. 10.8: VCCT

By substituting (10.29) in (10.28), the condition for crack propagation becomes

$$\frac{F_{2,5} \, v_{1,6}}{2\Delta A} \frac{1}{G_I^c} = 1 \tag{10.30}$$

The area ΔA of the newly formed crack is $\Delta A = d \times b$, where d is the length of the finite element undergoing crack propagation, and b is the width of the crack.

The VCCT method works similarly for the modes II or III, by considering the corresponding components of the separation and nodal forces. A simpler formulation called Jacobian derivative method (JDM) is available in [190].

The crack-propagation criterion (10.30) applies for single-mode loading only, as it is derived from the single-mode criterion in (10.2). For mixed-mode loading, the single-mode crack-propagation criterion (10.2) has to be replaced with a mixed-mode criterion. For example, one could use the power equation (10.21), where the critical ERRs G_i^c, with $i = I, II, III$, are material properties and the ERRs G_i are calculated similarly to (10.30) by using VCCT. Alternatively, one could use the BK equation [191] or the Reeder equation [180, 192].

Example 10.3 *Solve Example 10.1 (p. 374) using the virtual crack closure technique.*

Solution to Example 10.3 *Watch the video in [1].*
This example is solved modifying the **cae** *file defined in Example 10.2, p. 378. Initially, the contact definition must be changed deleting the properties associated with the surface-based cohesive formulation and redefining the contact characteristics. Then, the Output Requests must be modified to include the calculation of strain energy release rate (ERR) G, in the model. And finally, the Keyword Editor will be used to add the VCCT features before submitting the model to the solver.*

```
Menu: File, Set Work Directory, [C:\SIMULIA\User\Ex_10.3], OK
Menu: File, Open, [C:\SIMULIA\User\Ex_10.2\Ex_10.2.cae], OK
# on the Model-tree
# right-click: coh-surf, Rename, [vcct], OK
Menu: File, Save As, [C:\SIMULIA\User\Ex_10.3\Ex_10.3.cae], OK
```

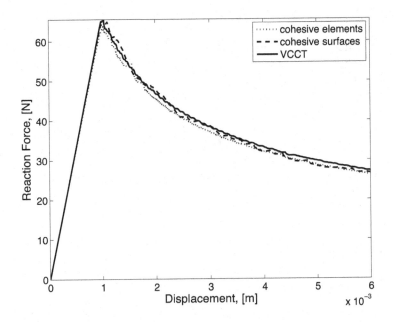

Fig. 10.9: Comparison of force–separation behavior using virtual crack closure technique (VCCT), cohesive elements, and cohesive surfaces.

i. Deleting unnecessary features

```
Module: Interaction
Menu:   Interaction, Property, Edit, coh
    # pick: Cohesive Behavior, Delete
    # pick: Damage, Delete, OK
```

ii. Redefining the contact definition

```
Menu:   Interaction, Edit, coh
    Sliding formulation: Small sliding
    Tab: Clearance
    Initial clearance: Uniform value across slave surface,
        [1.0e-7], OK
```

The use of an initial clearance will facilitate the solution of the model by preventing initial over-closures caused by the meshing algorithm. See Abaqus documentation for details.

iii. Modifying Field Output Requests

```
Module: Step
Step:   Step-1
Menu:   Output, Field Output Requests, Edit, F-Output-1
    Output Variables: # expand: Failure/Fracture
    # checkmark: ENRRT, # checkmark: BDSTAT, OK
    # ENRRT and BDSTAT allow tracking damage evolution
```

iv. Adding Keywords related to VCCT

```
Menu:    Model, Edit Keywords, vcct
     # click on the line that starts with the text: *Contact Pair
     Add After, # Type in the following text:
     [*Initial Conditions, type=CONTACT
     top, bot, bond]
     # click on the line that starts with the text: *Static
     Add After, # Type in the following text:
     [*Debond, slave=top, master=bot
     *Fracture criterion, type=VCCT, mixed-mode behavior=BK,
          tolerance=0.1
     280.0,280.0,280.0,2.284]
     OK
```

v. Solving and visualizing the results

```
Module: Job
Menu:    Job, Manager
   Create, Name [Ex-10-3], Cont, OK, Submit, # when completed, Results
Module: Visualization
Menu:    Plot, Deformed Shape # notice the beams have been separated
Menu:    Plot, Contours, On Deformed Shape
Menu:    Animate, Time History # see progression of the delamination
```

To visualize the force separation behavior in the model:

```
Module: Visualization
Menu:    Result, History Output
     Output Variables: Reaction force: RF2 PI, Save As [vcct-RF2], OK
     Output Variables: Spatial displ: U2 PI,  Save As [vcct-U2],  OK
     # close the History Output pop-up window
Menu:    Tools, XY Data, Create, Source: Operate on XY data, Cont
     Operators: Combine(X,X), # pick: vcct-U2, Add to Expression
     # pick: vcct-RF2, Add to Expression, Save As [vcct], OK
     # close Operate on XY Data pop-up window
Menu:    Tools, XY Data, Plot, vcct
```

The resulting plot should look like the solid line in Figure 10.9. The variable BDSTAT *stores the damage evolution associated with the bond between two nodes that were made coincident during the setup of the model. Its value ranges from 0, fully damaged bond, to 1, undamaged bond.* BDSTAT *can be used to find the location of the crack-tip at any stage of the delamination. In the* Visualization *Module, visualize* BDSTAT *and zoom-in to locate the crack-tip.*

10.3   Determination of CZM Parameters

In this section,[2] a methodology is presented to obtain the material parameters for cohesive zone model (CZM) in Abaqus using only load-deflection data from standard

---

[2]Before reading this section it is highly recommended that the user becomes familiarized with the three modalities used to interact with Abaqus, as explained in Section 2.3.1 p. 47.

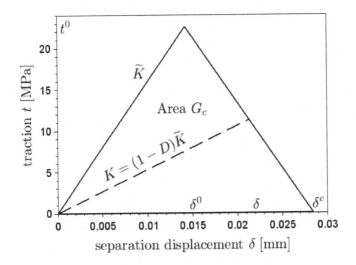

Fig. 10.10: Linear CZM with in-situ strength $t^o = 22.6$ MPa, undamaged penalty stiffness $\widetilde{K} = 1590$ N/mm^3, critical fracture energy $G_c = 1.56$ N/mm [195], and elastic unloading slope $K = (1 - D)\widetilde{K}$.

test methods such as double cantilever beam (DCB) [193] and end-notch flexure (ENF) [194]. Since these tests are designed to measure the critical fracture energy of a single mode, i.e., DCB for mode I and ENF for mode II, mixed-mode behavior cannot be inferred from these tests. However, for a practical application, one would have to choose a mixed-mode behavior and related parameters (Section 10.3.5).

10.3.1 Elastic Loading

To begin with, the simulation of the initial portion of the CZM diagram is linear elastic up to damage initiation. For example, see the portion of the diagram from the origin (0,0) to (t_n^o, δ_0) in Figure 10.10.

The elastic slope is controlled by the penalty stiffness $\widetilde{K}_n$ that is provided to Abaqus either directly as $\widetilde{K}_n$ or calculated by Abaqus as E_a/t_a from provided values of adhesive modulus and thickness E_a, t_a. The softening model only starts after the aforementioned damage initiation state. In Figure 10.10, softening is LINEAR. Figure 10.11 depicts a particular case of TABULAR softening, that including the elastic region, looks like a trapezoid.

In Abaqus CAE, define a cohesive material, as follows

```
Module: Property,
    Material Manager, Create "cohesive",
    Material Behaviors, Elastic, Elastic type: Traction
```

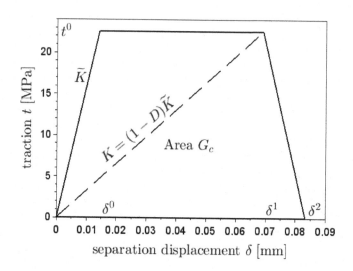

Fig. 10.11: TABULAR CZM with undamaged penalty stiffness $\widetilde{K} = 1590$ N/mm^3, in-situ strength $t^o = 22.6$ MPa, critical fracture energy $G_c = 1.56$ N/mm [195], and elastic unloading slope $K = (1 - D)\widetilde{K}$.[3]

To provide E/Enn, G1/Ess, G2/Ett, we use the Section Manager, as follows

Module: Property,
    Section Manager,Create "cohesive",Response "Traction Separation"

If we choose

initial thickness=analysis default

the values E/Enn, G1/Ess, G2/Ett are penalty stiffness $\widetilde{K}_n, \widetilde{K}_s, \widetilde{K}_t$.
    However, if we choose

initial thickness=specify

or

initial thickness=use nodal coordinates

Then, we specify the out-of-plane thickness t_a or let Abaqus calculate it from the mesh; then Abaqus calculates E/Enn=E_a/t_a, G1/Ess=G_a/t_a, and G2/Ett=G_a/t_a, where E_a, G_a, t_a are the Young's modulus, shear modulus, and thickness of the adhesive (or interface), respectively. I prefer to specify the penalty values myself.

---

[3]Superscripts ^{0},^{1},^{2},^{c}, denote instances of the variables δ, G, not exponentiation.

10.3.2   Softening Behavior

The mixed-mode deformation field in the cohesive zone, including opening (normal), sliding shear (s) and tearing shear (t), as shown in Figure 10.6, are represented by the mixed-mode ratios m_n, m_s, m_t. The mixed-mode ratios can be based on ENERGY, ACCUMULATED ENERGY, or TRACTION [196]. In this section we use MIXED MODE RATIO=ENERGY, defined as follows (see also section 10.3.5)

$$m_n = \frac{G_n}{G_m} \quad ; \quad m_s = \frac{G_s}{G_m} \quad ; \quad m_t = \frac{G_t}{G_m} \quad ; \quad G_m = G_n + G_s + G_t \qquad (10.31)$$

Along with the mixed-mode ratios, there are six types of *damage initiation criteria* available [196]:

QUADE Quadratic nominal strain

QUADS Quadratic nominal stress

MAXE Max. nominal strain

MAXS Max. nominal stress

MAXPE Max. principal strain, and

MAXPS Max. principal stress

This section is illustrated using the quadratic nominal stress criterion (QUADS)

$$\left\{ \frac{<t_n>}{t_n^o} \right\}^2 + \left\{ \frac{<t_s>}{t_s^o} \right\}^2 + \left\{ \frac{<t_t>}{t_t^o} \right\}^2 = 1 \qquad (10.32)$$

The evolution of damage at the process zone near the crack tip can be represented by either LINEAR (Figure 10.10), EXPONENTIAL, or TABULAR cohesive zone models (Figure 10.11).

To begin with, define the total damage up to failure, by one of two methods

TYPE=ENERGY You must provide the (total) fracture energy dissipated up to failure G_c, also called critical fracture energy, represented by the area under the diagram in Figure 10.10 or Figure 10.11, or

TYPE=DISPLACEMENT You must provide the displacement at failure (relative to the displacement at damage initiation, not to the origin of the $t - \delta$ diagram), e.g., $\delta^2 - \delta^o$ in Figure 10.11. For EXPONENTIAL softening, you must also provide the value of the exponent.

I prefer to use critical fracture energy because it has a clear physical meaning that is not apparent in the tabular description.

Next, specify the type of damage evolution, by one of two methods, using

LINEAR or EXPONENTIAL softening laws, or a

TABULAR description of the softening diagram by providing the displacements (with respect to the displacement at damage initiation) and the corresponding values of damage so that the $t - \delta$ diagram can be drawn, and the area G_c under the diagram can be calculated. For example, provide values for $\delta^j - \delta^o$ and D^j $(j = 0 \ldots 2)$ in Figure 10.11.

These combinations are possible [196]

```
DAMAGE EVOLUTION, TYPE=ENERGY, SOFTENING=LINEAR
DAMAGE EVOLUTION, TYPE=ENERGY, SOFTENING=EXPONENTIAL
DAMAGE EVOLUTION, TYPE=DISPLACEMENT, SOFTENING=LINEAR
DAMAGE EVOLUTION, TYPE=DISPLACEMENT, SOFTENING=EXPONENTIAL
DAMAGE EVOLUTION, TYPE=DISPLACEMENT, SOFTENING=TABULAR
```

Furthermore, if more than one mode of separation (crack opening and shear) are present, one must specify how to handle *damage evolution under mixed-mode conditions*. The available choices are

```
TYPE=DISPLACEMENT, SOFTENING=LINEAR
TYPE=DISPLACEMENT, SOFTENING=EXPONENTIAL
TYPE=DISPLACEMENT, SOFTENING=TABULAR
TYPE=DISPLACEMENT, MIXED MODE BEHAVIOR=MODE INDEPENDENT
TYPE=ENERGY, SOFTENING=LINEAR
TYPE=ENERGY, SOFTENING=EXPONENTIAL
TYPE=ENERGY, MIXED MODE BEHAVIOR=TABULAR
TYPE=ENERGY, MIXED MODE BEHAVIOR=POWER LAW, POWER=alpha
TYPE=ENERGY, MIXED MODE BEHAVIOR=BK, POWER=eta
TYPE=ENERGY, MIXED MODE BEHAVIOR=MODE INDEPENDENT
```

The default MIXED MODE BEHAVIOR is MODE INDEPENDENT.

Also note that there are restrictions regarding the types of *damage evolution* and *softening behavior* that can be used with each of the mixed-mode behaviors. Allowed choices can be discovered using **Abaqus CAE Module: Property, Material Manager, Edit Material, Damage Evolution**, as it can be seen in Figure 10.12.

For TYPE=ENERGY, SOFTENING=LINEAR, the damage is calculated by (10.26) where δ_m^o, δ_m^f are the initial and final CZM effective displacements, labeled δ^o, δ^c in Figure 10.10 for a single-mode situation. Furthermore, δ_m is the maximum effective displacement attained during the current loading history, where effective displacements are calculated, as follows [196]

$$\delta_m = \sqrt{< \delta_n >^2 + \delta_s^2 + \delta_t^2} \tag{10.33}$$

Since this section is illustrated with DCB data, only one mode (traction) is active. Thus, we chose TYPE=ENERGY, SOFTENING=LINEAR, MIXED MODE BEHAVIOR=MODE INDEPENDENT. Since the traction mode is dominant, it controls the simulation.

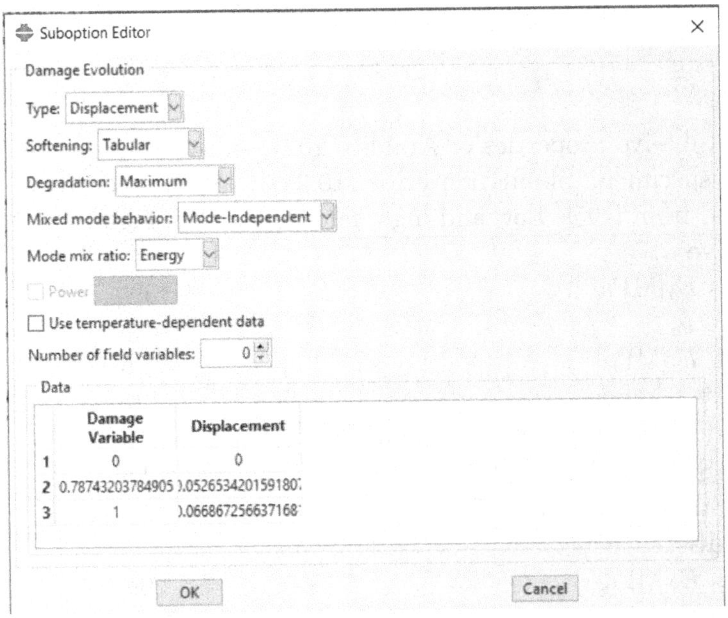

Fig. 10.12: CZM tabular data entry form.

10.3.3   LINEAR CZM

A schematic of the LINEAR traction-separation behavior is shown in Figure 10.10. It is described in terms of three parameters per deformation mode. The parameters are the penalty stiffness $\widetilde{K}_n$, the in-situ strength t_n^o and the critical fracture energy G_n^c. The adhesive stretches with traction-separation stiffness $\widetilde{K}_n$ until the stress t_n reaches the in-situ strength of the adhesive t_n^o. After that, further separation results in damage D so that material softening takes place in traction-separation space.

The material parameters can be adjusted to experimental data of load vs. crack opening displacement (COD), which is often readily available in the literature or easily obtained by experimentation. The percentage error between simulation and experiment is calculated as the difference between the areas under simulation and experimental curves (see Figure 10.13) normalized by the later, i.e.,

$$\%error = 100\frac{|\text{simulation area - experimental area}|}{\text{experimental area}} \tag{10.34}$$

Experimental data from double cantilever beam (DCB) tests often report COD, which for DCB is the total opening of the loading points. However, if both substrates are identical, it is more efficient to mesh only half of the specimen. Then, COD is twice the simulation displacement.

The error can be calculated with DCBCZMparams.py. Minimization of G_n^c is done with cost1d.m and minimize_Gnc.m, which are available in [2], used in Example 10.4, and further explained in Section 10.5. There are four parameters, $a_0, \widetilde{K}_n, t_n^o$, and G_n^c, that affect the error (10.34).

Table 10.1: Adhesive properties of Araldite 2021, substrate properties, and geometry of DCB specimen. Simulation error (10.34) with respect to two sets of data (low and high) from [195]. Low and high data depicted in Figures 10.13 and 10.14.

Adhesive		
Modulus E_a[MPa]	1590	
Poisson's ν_a	0.35	
Strength F_a[MPa]	22.6	
Thickness t_a[mm]	1.0	
Critical Fracture Energy G_n^c[N/mm] from [195]	1.56	
Penalty Stiffness $\widetilde{K} = E_a/t_a$	22.6	
In-situ Strength $t_n^o = F_a$[MPa]	22.6	
Steel substrate		
Modulus E_s[MPa]	200	
Poisson's ν_s	0.333	
Thickness t_s[mm]	2.0	
Geometry		
Length L[mm]	180	
Width B[mm]	25	
Initial crack length a_0[mm]	20	
Simulation error [%]	Low data	High data
LINEAR	3.57	18.6
EXPONENTIAL	2.66	17.8
TABULAR	22.1	3.06

Table 10.2: Critical fracture energy G_n^c of Araldite 2021 adjusted for each cohesive zone model using Load vs. COD data from [195] to minimize the error (10.34). Other properties from Table 10.1.

	G_n^c [N/mm]		Error [%]	
	Low data	High data	Low data	High data
LINEAR	1.648	2.158	$1.1005 \ 10^{-1}$	$5.6730 \ 10^{-3}$
EXPONENTIAL	1.625	2.111	$2.7737 \ 10^{-2}$	$2.1786 \ 10^{-2}$
TABULAR	1.267	1.511	$1.7879 \ 10^{-1}$	$1.0199 \ 10^{-1}$

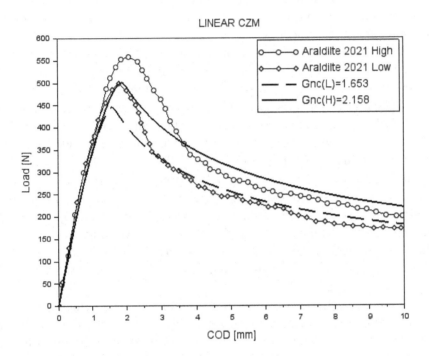

Fig. 10.13: Comparison between predicted and experimental Load vs. COD (high and low) for Araldite 2021 using Abaqus LINEAR CZM with $a_0, \widetilde{K}_n, t_n^o$ from (Table 10.1) and G_n^c adjusted to the experimental data (Table 10.2).

The analytical solution for the compliance of one of the substrates can be approximated by using beam theory as

$$C = \frac{\delta}{P} = \frac{8a_0^3}{E_x bh^3} + \frac{12a_0}{5bhG_{xz}} \tag{10.35}$$

where a_0 is the initial crack length, h, b are the thickness and width of the substrate (Figure 10.2 with $t = h$), and E_x, G_{xz}, are the axial and transverse shear moduli of the substrate. Except for the initial crack length a_0, all the parameters can be measured easily and accurately. Since the compliance is a cubic equation in terms of a_0, it is clear that any error measuring a_0 could lead to large errors in the estimation of the $P - \delta$ curve. Note that in (10.35) the deflection δ at the loading point is only that of the substrate being considered. If both substrates are identical, δ is one-half of the crack opening displacement (COD).

The initial crack length a_0 is very important but sometimes wrongly reported in the literature. If the simulation does not match the initial linear portion of the $P - \delta$ data, the value of a_0 is wrong. This situation is not due to rotation at the crack tip, because unlike analytical equations, the rotation is properly taken into account by FEA. First and foremost, one must make sure the simulation matches the initial slope of the data, as in Figure 10.13. If not, use the minimization code to adjust a_0 for the data in that range of COD (approximately 0–1 mm in Figure 10.13).

The penalty stiffness $\widetilde{K}_n = E_a/t_a$ is a function of the bulk modulus of the adhesive E_a and the adhesive thickness t_a. The effect of $\widetilde{K}_n$ on the initial slope of the $P - \delta$ data is minor. Therefore, do not try to match the initial slope of the data by adjusting the $\widetilde{K}_n$. It can be seen in Table 10.3 that varying $\widetilde{K}_n$ in the range 800–2400 has no appreciable effect on the response. The modulus E_a of the adhesive is available in the adhesive's materials specification literature and the thickness t_a of the adhesive can be measured.

The in-situ adhesive strength t_n^o does not normally has a significant impact on the simulation. Therefore, it can be approximated by the bulk adhesive strength F_a, which is available from the adhesive's literature. It can be seen in Table 10.3 that varying t_n^o in the range 15–30 has no significant effect on the response, but only a slight increase of peak load and error for higher values of t_n^o.

The most important parameter after a_0 is the critical fracture energy G_n^c. To find this value, we use the minimization code to adjust G_n^c for the data in the full range of COD. The results can be appreciated in Figure 10.13.

10.3.4   TABULAR CZM

A schematic of TABULAR traction-separation curve is shown in Figure 10.11. This is a particular case when the curve is described by only three points δ^i, D^i with $i = 0 \ldots 2$.

The separation displacements $\delta^0, \delta^1, \delta^2$ can be related to familiar quantities such as $\widetilde{K}_n, t_n^o$, and G_n^c by assuming that the downward slope at the end of the traction-separation curve has the same slope $\widetilde{K}_n$, albeit negative, as the linear elastic loading portion of the curve. In other words, $\delta^2 - \delta^1 = \delta^0$ and the area under the curve is

Table 10.3: Effect of varying $\widetilde{K}_n$ and t_n^o on error, peak load, and peak COD. Simulation using linear CZM.

G_n^c	t_n^o	$\widetilde{K}_n$	%error	peak COD	peak load
2.158	22.6	800	0.47900	1.9	501.4
2.158	**22.6**	**1590**	**0.00567**	**1.9**	**502.3**
2.158	22.6	2400	0.20000	1.9	502.4
2.158	15.0	1590	1.41750	2.1	476.0
2.158	30.0	1590	0.74450	1.8	518.4

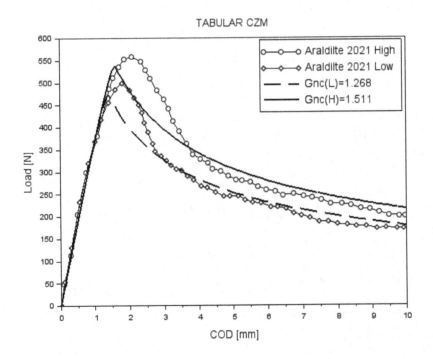

Fig. 10.14: Comparison between predicted and experimental load vs. COD (high and low) for Araldite 2021 using Abaqus TABULAR CZM with $a_0, \widetilde{K}_n, t_n^o$ from (Table 10.1) and G_n^c adjusted to the experimental data (Table 10.2).

$G_n^c = \delta^1 t_n^o$. Then,

$$\delta^0 = t_n^o / \widetilde{K}_n$$
$$\delta^1 = G_n^c / t_n^o$$
$$\delta^2 = \delta^1 + \delta^0 \tag{10.36}$$

and the damages are

$$D(\delta^0) = 0 \text{ ; no damage}$$
$$D(\delta^1) = 1 - t_n^o / (\widetilde{K}_n \delta^1)$$
$$D(\delta^2) = 1 \text{ ; completely damaged} \tag{10.37}$$

where the slope from the origin to (δ^1, t_n^o) in Figure 10.11 is $t_n^o / \delta^1 = (1 - D^1)\widetilde{K}_n$.

In Abaqus, the separation displacements δ and corresponding damage values D must be entered using a peculiar table (Figure 10.12), that must start at the damage initiation point (δ^0, t_n^o), not at the origin of the displacement–traction space shown in Figure 10.11. That is, D must be specified as a function of the effective displacement relative to the effective displacement at damage initiation [196]. Therefore, the separations found in (10.36) must be shifted left by $-\delta^0$, as follows

$$\delta^0 = 0$$
$$\delta^1 = G_n^c / t_n^o - t_n^o / \widetilde{K}_n$$
$$\delta^2 = G_n^c / t_n^o \tag{10.38}$$

and the damages remain as in (10.37). The resulting values are presented in Table 10.4. The calculations can be done with the following Python script

```
# LINEAR and EXPONENTIAL use tno, Gnc, Kn.
# For TABULAR, calculate tabular delta & D
delta = [0,0,0]# book notation [delta0, delta1, delta2]
delta[0] = tno/Kn# slope Kn linear region, D=0, tno=Kn*delta[1]
delta[1] = Gnc/tno# right and left triangles have same area
delta[2] = delta[1] + delta[0]# equal triangles
D = [0,0,0]# book notation
D[0] = 0.;# damage=0 from {0,0} to {delta[0],tno}
D[1] = 1.-tno/(Kn*delta[1]);# slope=tno/delta[1]=(1-D[1])*Kn
D[2] = 1.;# at {delta[1],0} damage reaches 1
print "before shift ", delta, D
# Abaqus table origin is at coordinates {delta[0],tno}, must shift
shift = delta[0]# shift this much
delta[0]=delta[0]-shift;# = 0
delta[1]=delta[1]-shift;
delta[2]=delta[2]-shift;
print "after shift ", delta, D
```

Example 10.4 *Using EXPONENTIAL CZM, calculate the critical fracture energy for Araldite 2021 using the experimental load–COD data in [195]. Use the data corresponding to the highest set of loads.*

Solution to Example 10.4 *The code used in this example is available in files that are listed in Section 10.5 and available in [2]. All the files must be placed in the same folder.*

Table 10.4: TABULAR CZM parameters for input to Abaqus when $t^o = 22.6$ MPa, $\widetilde{K} = 1590$ N/mm^3, $G_c = 1.56$ N/mm.

	Column 1	Column 2	Column 3
δ[mm]	0.0	0.0548	0.069
D	0.0	0.7940	1.0

From Table 10.1 we get the geometry and the initial crack length of the DCB specimen, and the material properties of the specimen and adhesive (except the critical fracture energy to be determined). The penalty stiffness is $\widetilde{K}_n = E_a/t_a$. The in-situ strength of the adhesive is $t_n^o = F_a$.

Digitizing the high-load vs. COD plot in [195] we get 46 data points

```
u2Exp=[0,0.099,0.28,0.493,0.757,1.086,1.431,1.546,1.711,1.859,
    2.072,2.336,2.484,2.615,2.796,2.993,3.125,3.257,3.437,
    3.651,3.832,4.046,4.26,4.408,4.638,4.852,5.066,5.247,
    5.493,5.707,5.954,6.217,6.431,6.743,7.023,7.27,7.533,
    7.796,8.125,8.372,8.651,8.914,9.194,9.474,9.786,9.942];# COD
p2Exp=[0,49.105,112.532,204.604,298.721,380.563,486.957,
    511.509,536.061,552.43,558.568,548.338,527.877,509.463,
    482.864,462.404,441.944,413.299,390.793,364.194,339.642,
    329.412,319.182,306.905,300.767,290.537,282.353,280.307,
    276.215,265.985,259.847,255.754,247.57,249.616,245.524,
    241.432,237.34,227.11,227.11,225.064,216.88,214.834,
    208.696,202.558,200.512,200.512];# P
```

The script DCBCZMparams.py, explained in Section 10.5, is capable of using LINEAR, EXPONENTIAL, and TABULAR CZM. For this example, we must select EXPONENTIAL CZM. Look at the heading of the script in Section 10.5 to find the line number where the model selection takes place. For this example we must select EXPONENTIAL, as follows

```
# Type: EXPONENTIAL
mdb.models['coh-els'].materials['cohesive'].quadsDamageInitiation.\
    DamageEvolution(
    softening=EXPONENTIAL, table=((Gnc, ), ), type=ENERGY,
    mixedModeBehavior=MODE_INDEPENDENT)
```

Next, we create the InitialState.txt file to include $a_0, \widetilde{K}_n, t_n^o$, as well as a guess value for G_n^c, and the maximum value of COD in the data

```
20.0 1590.0 22.6 1.56 10.0
```

The InitialState.txt will be read by the MATLAB code minimizeGnc.m, which uses the Nelder-Mead algorithm in fminsearch() to minimize the error (10.34) by changing the value of G_n^c.

```
% minimize_Gnc.m: find Gnc with a0 Kn tno rangeCOD constant
clc; dos('copy .\initialState.txt .\state.txt');% read by *.py script
[~, ~] = dos('clean.bat');% remove *.lck and other unnecessary files
is = fopen('initialState.txt','r');
A = fscanf(is, '%g %g %g %g %g' );
a0 = A(1);% adjust 1st if simulation doesn't match linear data
Kn = A(2);% use ~Ea/ta, do not adjust too much
tno = A(3);% use adhesive bulk strength, do not adjust too much
Gnc = A(4);% critical fracture energy, main property to adjust
```

```
rangeCOD = A(5); %COD = 2*simulationRange
disp([num2str([a0, Kn, tno, Gnc, rangeCOD])])
fclose(is);
% 'Display','iter' to see what's going on!
% 'Tolfun' = 1.0*%cost, 'TolX' = 0.01*Gnc
options=optimset('Display','iter','TolFun',1.0,'TolX',0.01,'MaxIter',100);
fun = @(x)cost1D(a0, Kn, tno, x, rangeCOD);% a0 Kn tno rangeCOD constant
x = fminsearch(fun, Gnc, options);% x is Gnc
disp('optimum = '+string(x))
```

Before executing *minimizeGnc.m*,

- *Close Abaqus CAE, because it locks the .odb*

- *Select the correct SOFTENING MODEL inside *.py (look at script's heading to find the line number where to make this selection).*

- *Select the correct experimental data inside *.py (look at script's heading to find the line number where to make this selection).*

- *Select the correct script (*.py) inside "cost1D.m" (reproduced below). For this example we use DCBCZMparams.py*

- *Set the initialState.txt*

- *If the value of cost does not change, check Abaqus, which maybe breaking with an error.*

- *To adjust a_0, use rangeCOD of the initial linear region of data.*

- *To adjust G_n^c, use rangeCOD of all experimental data.*

The cost *is calculated by* cost1D.m *by calling* DCBCZMparams.py. *The function is called* cost1D.m *because only one parameter, i.e.,* Gnc *is adjusted. It is possible to adjust a different parameter by slightly modifying* minimizeGnc.m *and* DCBCZMparams.py. *It is also possible to adjust several parameters simultaneously with further modifications, if needed.*

In this context, cost *is the error calculated in* DCBCZMparams.py *using (10.34). The script* DCBCZMparams.py *saves the cost, i.e., the error, in the* cost.txt *file. Thus, in this context, cost and error are identical.*

```
function cost = cost1D(a0, Kn, tno, Gnc, rangeCOD)
    % cost = 100*abs(area predicted-area experiment)/area experiment
    %           calculated in DCBCZMparams.py using Abaqus
    % Write the "full" state for DCBCZMparams.py to read it
    fs = fopen('state.txt','w');
    fprintf(fs, '%g %g %g %g %g\n', a0, Kn, tno, Gnc, rangeCOD );
    fclose(fs);% If Python error, use %.4f to avoid writing an integer
    % Execute "DCBCZMparams.py"
    [~, ~] = dos('abaqus cae -nogui DCBCZMparams.py');
    % Don't open until Abaqus is finished +> Job.waitForCompletion()
    fr = fopen('cost.txt','r');
    cost = fscanf(fr, '%g');% read the error calculated by DCBCZMparams.py
    disp([num2str([a0, Kn, tno, Gnc, cost])])
    fclose(fr);
end
```

Note that in `DCBCZMparams.py`*, the simulation displacement is half of the COD of the DCB specimen but the experimental data always reports COD.*

`DCBCZMparams.py` *reads the* `state.txt` *created by* `cost1D.m` *on each iteration of fminsearch(). With that information, the Python script uses Abaqus to calculate the error and writes it in* `cost.txt`*, which is then read by* `cost1D.m`*, and then used by* `fminsearch()`*.*

Once MATLAB converges with tolerances on the error (Tolfun) and the adjusted variable (TolX), we can read the adjusted value (MATLAB's optimum) and the error (MATLAB's min f(x)) for the last iteration displayed in the MATLAB Command Window, which may or may not be the last iteration calculated but rather the one with lowest error. See `log.txt` *for details. If in doubt, one can write the (just found) optimum value of G_n^c into* `state.txt`*, then execute*

```
abaqus cae -nogui DCBCZMparams.py
```

in a WINDOWS command shell or run the script `DCBCZMparams.py` *in Abaqus CAE. Then, the resulting error will be reported in* `cost.txt`*.*

10.3.5 Mixed-mode Parameter

The fracture energy G_i is the energy dissipated as a result of decohesion in mode i, with $i = n, s, t$. When only one mode of deformation is active, the fracture energy is the area *between* the traction separation law (either linear, exponential, or tabular) and the elastic return-to-origin line (see Figure 10.10). Thus, G_i can be calculated in terms of the adhesive separation δ_i in that particular mode by using the graph.

For each point on the graph, the slope of the return-to-origin line with slope $(1 - D_i)\widetilde{K}_i$ is a function of the undamaged penalty stiffness $\widetilde{K}_i$ and the damage D_i, from which the damage D_i can be calculated. Furthermore, the critical fracture energy G_i^c is the total area under the traction separation curve (all the way to δ^2 in Figure 10.10). But in CZM, even for a mixed-mode situation, the damage D is a scalar, that is, a single value. In a mixed-mode situation, a single value of mixed-mode fracture energy G_m can be calculated using either (10.26) or using a mixed-mode criterion. Two criteria are popular: power law and BK.

Power Law Criterion

The power law criterion [196]

$$\left(\frac{G_n}{G_n^c}\right)^\alpha + \left(\frac{G_s}{G_s^c}\right)^\alpha + \left(\frac{G_t}{G_t^c}\right)^\alpha \leq 1 \tag{10.39}$$

predicts total separation (failure) when the LHS equals one, otherwise there is damage but not failure. Here α is a material parameter to be determined by direct experimentation or indirectly adjusted to fit simulation results to experimental data. Furthermore, G_n^c, G_s^c, G_t^c are the three single-mode critical fracture energies that can be found, one at a time, by experimentation. Since tearing shear experiments are very difficult to perform and they are not standardized, it is often assumed that $G_t = G_s$, with G_s obtained from a ENF test [194].

The total mixed-mode energy dissipated at any stage of deformation is

$$G_m = G_n + G_s + G_t \tag{10.40}$$

and the mixed-mode ratios are defined as

$$m_i = \frac{G_i}{G_m} \qquad (i = n, s, t) \tag{10.41}$$

Substituting (10.41) into (10.39) we get the power law damage activation function

$$\left[\left(\frac{m_n}{G_n^c} \right)^\alpha + \left(\frac{m_s}{G_s^c} \right)^\alpha + \left(\frac{m_t}{G_t^c} \right)^\alpha \right]^{-1/\alpha} \le G_m^c \tag{10.42}$$

where G_m^c is the critical mixed-mode fracture energy at failure and the LHS is equal to G_m for a specific set of mixed-mode ratios. Failure occurs when the LHS equals G_m^c. As a corollary, (10.42) allows for the calculation of the mixed-mode fracture energy G_m when the LHS < RHS, and the critical mixed-mode fracture energy G_m^c when LHS=RHS, for a specific set of mixed-mode ratios and power exponent.

BK Criterion

The 3D BK damage activation criterion is defined by [196]

$$G_n^c + (G_s^c - G_n^c) \left(\frac{G_s + G_t}{G_m} \right)^\eta \le G_m^c$$

$$G_m = G_n + G_s + G_t \tag{10.43}$$

In 2D, $G_t = 0$. Therefore,

$$G_n^c + (G_s^c - G_n^c) \left(\frac{G_s}{G_n + G_s} \right)^\eta \le G_m^c \tag{10.44}$$

$$m = \frac{G_s}{G_n + G_s} \tag{10.45}$$

where G_n^c, G_s^c are single-mode critical fracture energies found experimentally by using DCB [193] and ENF [194] or similar tests. Furthermore, G_n, G_s, are the normal and shear fracture energies resulting from a two-mode (n,s) deformation test with mixed-mode ratio m. Failure occurs when a set of G_n, G_s, values allows the LHS of (10.44) to reach the critical mixed-mode fracture energy G_m^c.

Determination of the BK exponent relies on experimental data from three tests.

i. Using a double cantilever beam (DCB, [193]) or similar test, determine the critical fracture energy under pure normal separation G_n^c.

ii. Using an end notch flexure (ENF, [194]) or similar test, determine the critical fracture energy under pure shear deformation G_s^c.

iii. Finally, using a mixed-mode failure (MMF) test [197] determine the normal G_n and shear G_s fracture energies at a specific mixed-mode *failure* event with $G_m^c = G_n + G_s$ and 2D mixed-mode ratio m.

Then, η can be obtained when the LHS equals the RHS of (10.44), i.e.,

$$G_m^c = G_n + G_s$$
$$m = \frac{G_s}{G_m^c}$$
$$A = \frac{G_m^c - G_n^c}{G_s^c - G_n^c}$$
$$\eta = \frac{\log A}{\log m} \tag{10.46}$$

A plot of critical mixed-mode fracture energy G_m^c vs. 2D mixed-mode ratio m is shown in Figure 10.15 (plotting (10.44) with = sign). A BK value $\eta = 1$ produces a linear plot with a minimum of $G_m^c = G_n^c = 715$ N/m^2 at $m = 0$ (pure mode n, DCB test) and a maximum of $G_m^c = G_s^c = 1900$ N/m^2 at $m = 1$ (pure mode s, ENF test). Linear interaction is the usual choice when the value of η is unknown.

$\eta = 0.1$ produces an concave plot and $\eta > 1$ produces an convex plot between the same two extremes of G_m^c.

Using a single data point from a mixed-mode *failure* test with $G_n = 600, G_s = 180$ N/m^2, yields $\eta = 1.98$. Finally, the slightly convex curve passing through this experimental point provides the mixed-mode behavior predicted by the BK criterion. Furthermore, the value of η calculated in 2D is assumed to be valid for 3D modeling situations.

10.4 Modeling Considerations

In the softening region, you may encounter convergence problems or unjustifiable oscillation of the $P - \delta$ response. In that case, you may want to vary some parameters.

10.4.1 Damage Stabilization Cohesive

The Damage Stabilization Cohesive parameter is found in

```
Module:Property,
     Quads Damage Initiation, Damage Stabilization Cohesive
```

I my experience, I have not found any case where it had any effect.

10.4.2 Damping

The Damping parameter is found in

```
Module:Job,
     Stabilization Magnitude, Stabilization Method=DAMPING FACTOR.
```

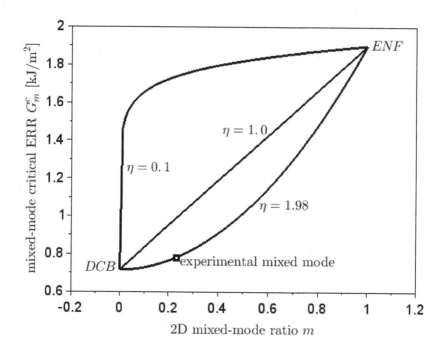

Fig. 10.15: Effect of BK parameter value η on the critical mixed-mode fracture energy G_m^c as a function of 2D mixed-mode ratio m. Value of η calculated with (10.46) using a single mixed-mode data point plus DCB and ENF data points.

Suggested value: 0.0002. Otherwise, you are likely to get oscillation and lack of convergence.

10.4.3 Symmetry

You cannot use symmetry boundary conditions on cohesive elements in the usual way. Instead, you must leave the cohesive elements unsupported and manually modify the .inp file, adding the keyword SYMMETRIC at the end of the line defining the section, as follows

```
*Cohesive Section, elset=Set-1, material=cohesive,
    response=TRACTION SEPARATION, SYMMETRIC, 15
```

where 15 is the width of the specimen. You must define the *Cohesive Section in CAE, then write the input file, then manually modify it, then create a new Job using the modified .inp file, and that will run correctly. The SYMMETRIC keyword is available from Abaqus release 2019 onward.

An alternative to manually modifying the .inp file is to program the modification in a Python script. For that we use the GetBlockPosition script:

```
import string
def GetBlockPosition(modelName, blockPrefix):
    #Finds block position in input file for modelName block
```

```
if blockPrefix == '':
    return len(mdb.models[modelName].keywordBlock.sieBlocks)-1
pos = 0
for block in mdb.models[modelName].keywordBlock.sieBlocks:
    if string.lower(
        block[0:len(blockPrefix)])==string.lower(blockPrefix):
        return pos
    pos=pos+1
return -1
```

that allows us to find, in the .inp file, the block number corresponding to the `blockPrefix` block that we want to modify (see Chapter 12).

10.4.4 ENCASTRE

You cannot use clamped boundary conditions on cohesive elements. One would think that ENCASTRE at the mid-surface of the adhesive should work to simulate symmetry of a DCB specimen because the longitudinal Poisson's effect on a beam is negligible. However, ENCASTRE will stiffen and delay the onset of separation.

10.4.5 Mesh Refinement

We cannot use more than one cohesive element through the thickness of the adhesive. They contract laterally even when the traction is vertical, and non-zero material properties are assigned to all three modes n, t, and s. As per Abaqus documentation [198] > Elements > Special-Purpose Elements > Cohesive elements > Modeling with cohesive elements "the cohesive zone must be discretized with a single layer of cohesive elements through the thickness."

Furthermore, [198] "If the cohesive zone represents an adhesive material with a finite thickness, the continuum macroscopic properties of this material (namely, the modulus of the bulk adhesive) can be used directly for modeling the constitutive response of the cohesive zone." That is, enter the modulus of the adhesive and either specify the thickness or let Abaqus calculate it from the mesh; then $\widetilde{K}_n = E_a/t_a$. This is not a good practice because the resulting penalty stiffness may be too high, causing poor convergence. We want to specify $\widetilde{K}_n$ directly so that we can reduce it if the convergence is poor.

Furthermore, both mesh density and element aspect ratio in both the adherent and adhesive have an effect on both convergence and the results of the simulation. Thus, [198] "if the cohesive zone represents an infinitesimally thin layer of adhesive at a bonded interface, it may be more relevant to define the response of the interface directly in terms of the traction at the interface versus the relative motion across the interface," i.e., enter the penalty stiffness directly, in which case Abaqus ignores the thickness of the adhesive that could be calculated from the mesh.

When $\widetilde{K}_n$ is specified, the adhesive mesh-thickness is irrelevant, and the aspect ratio of COH2D4 elements is irrelevant also. Only when the adhesive bulk modulus E_a is specified together with *thickness from mesh*$= t_a$ or *thickness specified*$= t_a$, then the thickness matters because Abaqus calculates $\widetilde{K}_n = E_a/t_a$.

Furthermore, we cannot use copy/mirror/merge operations because they may leave behind orphan segments at the ends of the adhesive, which forces a mesh with two elements through the thickness of the adhesive, which is not acceptable as was mentioned earlier. If we sketch a half-model to use symmetry, then mirror of the part, then merge it, even if we suppress or delete the two original parts, and the new part has no edge at the mid-surface of the adhesive, the two ends of the adhesive are still made up of two segments. Therefore, meshing produces two elements though the thickness of the adhesive, which is not acceptable.

10.4.6   Abaqus CAE and Input File

To understand the Abaqus documentation about cohesive elements [196, 198], one has to refer to the CAE interface and the INPUT FILE simultaneously. For example, CAE generates the following script to define a cohesive material

```
mdb.models['coh-els'].Material(name='cohesive')
mdb.models['coh-els'].materials['cohesive'].Elastic(table=((Kn, Ks,
    Kt), ), type=TRACTION)#2667., 1185.9, 1185.9
mdb.models['coh-els'].materials['cohesive'].QuadsDamageInitiation(
    table=((tno, tso, tto), ))#31.,31.,31.
mdb.models['coh-els'].materials['cohesive'].
    quadsDamageInitiation.DamageStabilizationCohesive(
    cohesiveCoeff=Vc1)#0.1
mdb.models['coh-els'].materials['cohesive'].
    quadsDamageInitiation.DamageEvolution(
    mixedModeBehavior=MODE_INDEPENDENT, table=(
    (Gnc, ), ), type=ENERGY)# 0.401,
```

The script results in the following section of the input file (.inp)

```
*Material, name=cohesive
*Damage Initiation, criterion=QUADS
31.,31.,31.
*Damage Evolution, type=ENERGY
0.401,
*Damage Stabilization
0.1
*Elastic, type=TRACTION
2667., 1185.9, 1185.9
```

That means that *Damage Initiation* is controlled by the quadratic stress criterion QUADS

$$\left(\frac{<t_n>}{t_n^o}\right)^2 + \left(\frac{t_s}{t_s^o}\right)^2 + \left(\frac{t_t}{t_t^o}\right)^2 = 1 \tag{10.47}$$

with values of strength t_n^o, t_s^o, t_t^o provided by *Elastic, type=TRACTION*.

Then, *Damage evolution* type=ENERGY with critical fracture energy $G_c = 0.401$ is controlled by a scalar damage variable D to calculate the (actual) damaged tractions (i.e., stresses)

$$t_n = (1 - D)\bar{t}_n \text{ if } \bar{t}_n \geq 0 \text{ else } = \bar{t}_n$$
$$t_s = (1 - D)\bar{t}_s$$
$$t_t = (1 - D)\bar{t}_t \tag{10.48}$$

where the damage D is computed from the ratio of dissipated energy G to (allowable) critical fracture energy G_c, and the (fictitious) undamaged traction $\bar{t}_n, \bar{t}_s, \bar{t}_t$ are computed from the current strains assuming $D = 0$.

10.5   Script for Example 10.4

Example 10.4 uses the Abaqus Python script `DCBCZMparams.py` to compute the **error** defined by (10.34) in terms of the state variables read from `state.txt`. The error is then written to the file `cost.txt`. A step by step description of the script is presented in Section 10.5.1. The files used to solve Example 10.4 are available in [2] and reproduced below

DCBCZMparams.py: Abaqus script used by `cost1D.m`. See Section 10.5.1.

minimizeGnc.m: root minimization code (MATLAB), reproduced below.

```
% minimizeGnc.m: find Gnc with a0 Kn tno rangeCOD constant
%! Remember to close Abaqus CAE, it locks the .odb!!!!!!!!!!!!!!!!
%! Remember to select the correct SOFTENING MODEL inside *.py!!!!
%! Remember to select the correct experimental data inside *.py!!
%! Remember to select the correct *.py inside "cost1D.m"!!!!!!!!!!
%! Remember to set the initialState.txt!!!!!!!!!!!!!!!!!!!!!!!!!!!
%! If "cost" does not change, check Abaqus breaking with an error
% To adjust a0 use rangeCOD of the initial linear region of data
% To adjust Gnc use rangeCOD of all experimental data
clc;
dos('copy .\initialState.txt .\state.txt');% read by *.py script
[~, ~] = dos('clean.bat');% remove *.lck and unnecessary files
is = fopen('initialState.txt','r');
A = fscanf(is, '%g %g %g %g %g' );
a0 = A(1);% adjust 1st if simulation doesn't match linear data
Kn = A(2);% use ~Ea/ta, do not adjust too much
tno = A(3);% use adhesive bulk strength, do not adjust too much
Gnc = A(4);% critical fracture energy, main property to adjust
rangeCOD = A(5); %COD = 2*simulationRange
disp([num2str([a0, Kn, tno, Gnc, rangeCOD])])
fclose(is);
% 'Display','iter' to see what's going on!
% 'Tolfun' = 1.0*%cost, 'TolX' = 0.01*Gnc
options = ...
  optimset('Display','iter','TolFun',1.0,'TolX',0.01,'MaxIter',100);
% a0 Kn tno rangeCOD are constant
fun = @(x)cost1D(a0, Kn, tno, x, rangeCOD);
x = fminsearch(fun, Gnc, options);% x is Gnc
disp('optimum = '+string(x))
```

initialState.txt: Initial state used by `minimizeGnc.m` (exemplified below). These are the initial values of a_0, $\widetilde{K}_n$, t_n^o, G_n^c, and max(COD) in the data.

```
20. 1590. 22.6 1.56 10.
```

state.txt: State variables used by `DCBCZMparams.py`. At the onset of the minimization process, `state.txt` is just a copy of `initialState.txt`. In subsequent iterations, the value of G_n^c is updated by `fminsearch()` in `minimize_Gnc.m`, and the whole set of values is written by `cost1D.m`.

cost1D.m: cost function used by `minimizeGnc.m`, reproduced below.

```
function cost = cost1D(a0, Kn, tno, Gnc, rangeCOD)
% cost = 100*abs(area predicted-area experiment)/area experiment
%        calculated in DCBCZMparams.py using Abaqus
% Write the "full" state for DCBCZMparams.py to read it
fs = fopen('state.txt','w');
```

```
    fprintf(fs, '%g %g %g %g %g\n', a0, Kn, tno, Gnc, rangeCOD );
    fclose(fs);% If Python error, use %.4f to avoid writing an integer
    % Execute "DCBCZMparams.py"
    [~, ~] = dos('abaqus cae -nogui DCBCZMparams.py');
    % Don't open until Abaqus is finished +> Job.waitForCompletion()
    fr = fopen('cost.txt','r');
    cost = fscanf(fr, '%g');% read error calc, by DCBCZMparams.py
    disp([num2str([a0, Kn, tno, Gnc, cost])])
    fclose(fr);
    end
```

cost.txt: Error calculated and written by `DCBCZMparams.py`

10.5.1   DCBCZMparams.py

`DCBCZMparams.py` is the Python script that calculates the difference (cost or error) between the predicted $P - \delta$ curve and the experimental data.

First, to keep track of how much time a simulation takes to complete, at the top of the script, include this

```
    # DCB CZM Parameters (c) 2022 Ever Barbero
    # select LINEAR, EXPONENTIAL, OR TABULAR at line ~221
    # Select H/L data at line ~364
    import time
    start_time = time.time()
    import os
```

and at the end of the script, include this

```
    minutes = (time.time()-start_time)/60
```

Next, provide all the geometrical, material, and simulation parameters, some of which may be changed later in the script, to accommodate the variables read from `state.txt`.

```
    # ***** Units [N, m] *****
    L1 = 180.;# [mm] 2*L in ENFdimensions.png
    t1 = 2.0;# [mm] Substrate thickness, each
    a0 = 20.0# initial crack length, re-read as state[0]
    La = L1-a0;# adhesive length, re-calculate below
    ta = 1.0;# adhesive thickness
    B = 25.0;# [mm] specimen width (vcct_width)
    u2target = 1.546;#[mm] COD total opening, FEA calc. 1/2 of this
    # Substrate properties
    E1 = 200.0E3; E2 = E1; E3 = E1 #Steel
    Nu12 = 0.33; Nu13 = Nu12; Nu23 = Nu12
    G12 = E1/2/(1+Nu12); G13 = G12; G23 = G12;
    # Adhesive moduli
    Ea = 1590.0 # MPa
    nua = 0.35   #
    Ess = Ea/2.0/(1.0+nua); Ett = Ess #[N/mm^2] shear
    # Calculate penalty stiffness here, do not let Abaqus calculate it
    Kn = Ea/ta; Ks = Ess/ta; Kt = Ett/ta; #re-read Kn from state below
    # Adhesive strengths
    tno = 22.6 #[N/mm^2] adhesive strength, re-read from state below
    tso = tno; tto = tno;#[N/mm^2]
    # Fracture energies
    Gnc = 1.56 #[N/mm], normal mode, re-read from state below
    Gsc = Gnc; Gtc = Gnc;#[N/mm]
    # BKpower = 0.284 #BK exponent
    Vc1 = 0.1     # DamageStabilizationCohesive has NO effect
    Vc2 = 0.0002 # DAMPING_FACTOR
    Vc3 = 0.0500 #adaptiveDampingRatio=max.0.05 else spurious results
    initialInc=0.01; maxNumInc=2000; minInc=1e-09; maxInc=0.01;
```

Next, read the current *state* from a file

```
import numpy
state = numpy.loadtxt("state.txt") # read state
a0  = state[0]# if adjusting a0, La = L1-a0 calculated below
Kn  = state[1]
tno = state[2]
Gnc = state[3]
u2target = state[4] # [mm}
u2target = u2target/2# 1/2 up, 1/2 down in the simulation
print state
```

For LINEAR and EXPONENTIAL CZM, that is sufficient data. For TABULAR CZM, we can use the following script to calculate the tabular data required by Abaqus. See Section 10.3.4 for details.

```
# LINEAR and EXPONENTIAL use tno, Gnc, Kn.
# For TABULAR, calculate tabular delta & D
delta = [0,0,0]# book notation [delta0, delta1, delta2]
delta[0] = tno/Kn# slope Kn linear region, D=0, tno=Kn*delta[1]
delta[1] = Gnc/tno# right and left triangles have same area
delta[2] = delta[1] + delta[0]# equal triangles
D = [0,0,0]# book notation
D[0] = 0.;# damage=0 from {0,0} to {delta[0],tno}
D[1] = 1.-tno/(Kn*delta[1]);# slope=tno/delta[1]=(1-D[1])*Kn
D[2] = 1.;# at {delta[1],0} damage reaches 1
print "before shift ", delta, D
# Abaqus table origin is at coordinates {delta[0],tno}, must shift
shift = delta[0]# shift this much
delta[0]=delta[0]-shift;# = 0
delta[1]=delta[1]-shift;
delta[2]=delta[2]-shift;
print "after shift ", delta, D
```

Meshing requires special treatment to make sure that the interaction between adhesive and substrate is correct

```
# Meshing (c) 2021 Ever Barbero
position_tolerance = 1.1*ta/2 # mm for interaction to occur
# Calculate element size (esize), number of elements neL1, neLa
t1Seeds = 2# number of elements thru thickness
t1Bias  = 1# bias, best is no bias
factor  = 4# multiplies L1 to get neX
neL1 = int(factor*L1)# elements over specimen length, L1 in [mm]
esize = L1/neL1# real value, element size along X
nea0 = int(a0/esize)# number of elements in the initial-crack
a0 = nea0*esize# recalculate a0 to fit exactly nea0 elements
La = L1-a0;# re-calculate the adhesive length
neLa = neL1 - nea0# number of elements in the adhesive
print ('nea0=%g neLa=%g neY=%g \n' % (nea0, neLa, t1Seeds))
# end properties input
```

Next, import the modules that are needed for parameterization

```
# Parameterized modeling script starts here
from abaqus import *
from abaqusConstants import *
session.viewports['Viewport: 1'].makeCurrent()
session.viewports['Viewport: 1'].maximize()
from caeModules import *
from driverUtils import executeOnCaeStartup
executeOnCaeStartup()
Mdb()
mdb.models.changeKey(fromName='Model-1', toName='coh-els')
```

Next, we parameterize Abaqus `ws_composites_dcb.py` so that we can create simulations for any set of geometrical and material parameters. Since the script provided by Abaqus is hardwired with set dimensions in meters, and materials properties with dimensions N/m^2, our first task is to replace every numerical dimension by a parameter, so that the script can be used for arbitrary dimensions L1, t1, La, ta, and properties E1, E2, E3, Nu12, Nu13, Nu23, G12, G13, G23. Extreme care must be taken when parameterizing the **findAt** methods to make sure that the script actually finds and selects the correct entities, points, lines, etc.

```python
# Parameterize ws_composites_dcb.py (c) 2021 Ever Barbero
# Module: Part
session.viewports['Viewport: 1'].setValues(displayedObject=None)
s = mdb.models['coh-els'].ConstrainedSketch(name='__profile__',
    sheetSize=0.5)
g, v, d, c = s.geometry, s.vertices, s.dimensions, s.constraints
s.setPrimaryObject(option=STANDALONE)
s.rectangle(
    point1=(0.0, 0.0),
    point2=(L1, t1))
p = mdb.models['coh-els'].Part(
    name='beam',
    dimensionality=TWO_D_PLANAR,
    type=DEFORMABLE_BODY)
p = mdb.models['coh-els'].parts['beam']
p.BaseShell(sketch=s)
s.unsetPrimaryObject()
session.viewports['Viewport: 1'].setValues(displayedObject=p)
del mdb.models['coh-els'].sketches['__profile__']
s1 = mdb.models['coh-els'].ConstrainedSketch(name='__profile__',
    sheetSize=0.5)
g, v, d, c = s1.geometry, s1.vertices, s1.dimensions, s1.constraints
s1.setPrimaryObject(option=STANDALONE)
s1.rectangle(
    point1=(0.0, 0.0),
    point2=(La, ta))
p = mdb.models['coh-els'].Part(
    name='adhesive',
    dimensionality=TWO_D_PLANAR,
    type=DEFORMABLE_BODY)
p = mdb.models['coh-els'].parts['adhesive']
p.BaseShell(sketch=s1)
s1.unsetPrimaryObject()
session.viewports['Viewport: 1'].setValues(displayedObject=p)
del mdb.models['coh-els'].sketches['__profile__']
#Create bulk material [MPa, mm]
mdb.models['coh-els'].Material(name='bulk')
mdb.models['coh-els'].materials['bulk'].Elastic(
    type=ENGINEERING_CONSTANTS,
    table= ((E1, E2, E3, Nu12, Nu13, Nu23, G12, G13, G23), ))
#Create and Assign Bulk Section
mdb.models['coh-els'].HomogeneousSolidSection(
    name='bulk',
    material='bulk',
    thickness=B)
p = mdb.models['coh-els'].parts['beam']
f = p.faces
faces = f
region = regionToolset.Region(faces=faces)
p.SectionAssignment(
    region=region,
    sectionName='bulk')
p.DatumCsysByThreePoints(
    name='Datum csys-1',
    coordSysType=CARTESIAN,
```

```
        origin=(0.0, 0.0, 0.0),
        line1=(1.0, 0.0, 0.0),
        line2=(0.0, 1.0, 0.0))
region = regionToolset.Region(faces=faces)
orientation = mdb.models['coh-els'].parts['beam'].datums[3]
mdb.models['coh-els'].parts['beam'].MaterialOrientation(
        region=region,
        orientationType=SYSTEM,
        localCsys=orientation,
        axis=AXIS_3,
        additionalRotationType=ROTATION_NONE,
        angle=0.0)
#Begin Assembly
a = mdb.models['coh-els'].rootAssembly
a.DatumCsysByDefault(CARTESIAN)
p = mdb.models['coh-els'].parts['beam']
a.Instance(name='beam-1', part=p, dependent=ON)
p = mdb.models['coh-els'].parts['beam']
a.Instance(name='beam-2', part=p, dependent=ON)
p1 = a.instances['beam-2']
p1.translate(vector=(0.0, -t1, 0.0))
p = mdb.models['coh-els'].parts['adhesive']
a.Instance(name='adhesive-1', part=p, dependent=ON)
p1 = a.instances['adhesive-1']
p1.translate(vector=(a0, -ta/2, 0.0))
v1 = a.instances['beam-1'].vertices
verts1 = v1.findAt(((0.0, t1, 0.0), ))
a.Set(vertices=verts1, name='top')
v1 = a.instances['beam-2'].vertices
verts1 = v1.findAt(((0.0, -t1, 0.0), ))
a.Set(vertices=verts1, name='bot')
s1 = a.instances['beam-1'].edges
side1Edges1 = s1.findAt(((L1/2, 0.0, 0.0), ))
a.Surface(side1Edges=side1Edges1, name='top')
s1 = a.instances['beam-2'].edges
side1Edges1 = s1.findAt(((L1/2, 0.0, 0.0), ))
a.Surface(side1Edges=side1Edges1, name='bot')
# All entities use same c.s., origing at left-end of crack.
s1 = a.instances['adhesive-1'].edges
side1Edges1 = s1.findAt(((a0+1, ta/2, 0.0), ))
a.Surface(side1Edges=side1Edges1, name='coh-top')
s1 = a.instances['adhesive-1'].edges
side1Edges1 = s1.findAt(((a0+1, -ta/2, 0.0), ))
a.Surface(side1Edges=side1Edges1, name='coh-bot')
mdb.saveAs('DCBCZMparams.cae')
#parameterized ws_composites_dcb.py ends here
```

Next we create and additional script, not available in `ws_composites_dcb.py`, to define the CZM material behavior,

```
# (c) 2021 Ever Barbero, not available in ws_composites_dcb.py
from part import *
from material import *
from section import *
from assembly import *
from step import *
from interaction import *
from load import *
from mesh import *
from optimization import *
from job import *
from sketch import *
from visualization import *
from connectorBehavior import *
# Module: Property
mdb.models['coh-els'].Material(name='cohesive')
mdb.models['coh-els'].materials['cohesive'].Elastic(table=((Kn, Ks,
```

```
        Kt), ), type=TRACTION)# see (*1) below
mdb.models['coh-els'].materials['cohesive'].QuadsDamageInitiation(
    table=((tno,tso, tto), ))#damage initiation stress tno, tso, tto
mdb.models['coh-els'].materials['cohesive'].quadsDamageInitiation.\
    DamageStabilizationCohesive(
    cohesiveCoeff=Vc1)#stabilization coefficient
# LINEAR SOFTENING (TRIANGULAR), default:MODE_INDEPENDENT
# mdb.models['coh-els'].materials['cohesive'].\
    # quadsDamageInitiation.DamageEvolution(
    # softening=LINEAR, table=((Gnc, ), ), type=ENERGY,
    # mixedModeBehavior=MODE_INDEPENDENT)
# EXPONENTIAL SOFTENING, default:MODE_INDEPENDENT
# mdb.models['coh-els'].materials['cohesive'].\
    # quadsDamageInitiation.DamageEvolution(
    # softening=EXPONENTIAL, table=((Gnc, ), ), type=ENERGY,
    # mixedModeBehavior=MODE_INDEPENDENT)
# TABULAR SOFTENING (TRAPEZOIDAL), default:MODE_INDEPENDENT
mdb.models['coh-els'].materials['cohesive'].\
    quadsDamageInitiation.DamageEvolution(softening=TABULAR,
    table=((D[0], delta[0]), (D[1], delta[1]), (D[2], delta[2])),
    type=DISPLACEMENT, mixedModeBehavior=MODE_INDEPENDENT)
# Menu: Section, Manager, works for linear, tabular, exponential.
# (*1) CAE section traction separation initial thickness
    # "analysis default", use Kn, Ks, Kt as calculated above
    # Kn : penalty stiffness
mdb.models['coh-els'].CohesiveSection(
    material='cohesive', name='cohesive',
    outOfPlaneThickness=B, response=TRACTION_SEPARATION)
# (*2) CAE section traction separation initial thickness
    # "specify" = ta, requires Elastic(table=((Ea, Ess,Ett),
    # confusing because LINEAR and TABULAR diagrams
    # use Kn,Ks,Kt, not Ea,Ess,Ett
# mdb.models['coh-els'].CohesiveSection(initialThickness=ta,
    # initialThicknessType=SPECIFY, material='cohesive',
    # name='cohesive',
    # outOfPlaneThickness=B, response=TRACTION_SEPARATION)
# Menu: Assign, Section
mdb.models['coh-els'].parts['adhesive'].Set(faces=
    mdb.models['coh-els'].parts['adhesive'].faces.findAt(((
    La/2, ta/3, 0.0), (0.0, 0.0, 1.0)), ), name='Set-1')
mdb.models['coh-els'].parts['adhesive'].SectionAssignment(
    offset=0.0, offsetField='', offsetType=MIDDLE_SURFACE, region=
    mdb.models['coh-els'].parts['adhesive'].sets['Set-1'],
    sectionName= 'cohesive', thicknessAssignment=FROM_SECTION)
# Module: Step
mdb.models['coh-els'].rootAssembly.regenerate()
# Create Set to apply the load
mdb.models['coh-els'].rootAssembly.Set(name='topLoad', vertices=
    mdb.models['coh-els'].rootAssembly.instances['beam-1'].\
    vertices.findAt(((0.0, t1, 0.0), ), ))#new top
# define Step
mdb.models['coh-els'].StaticStep(initialInc=initialInc,
    maxNumInc=maxNumInc, minInc=minInc, maxInc=maxInc, name='Step-1',
    nlgeom=ON, previous='Initial',
    adaptiveDampingRatio=Vc3, continueDampingFactors=False,
    stabilizationMagnitude=Vc2, stabilizationMethod=DAMPING_FACTOR)
mdb.models['coh-els'].fieldOutputRequests['F-Output-1'].setValues(
    variables=('S', 'LE', 'U', 'RF', 'STATUS'))
mdb.models['coh-els'].HistoryOutputRequest(createStepName='Step-1',
    name= 'H-Output-2', rebar=EXCLUDE, region=
    mdb.models['coh-els'].rootAssembly.sets['topLoad'],
    sectionPoints=DEFAULT, variables=('U2', 'RF2'))
# Module: Mesh
# Part: beam
mdb.models['coh-els'].parts['beam'].setMeshControls(elemShape=QUAD,
    regions=mdb.models['coh-els'].parts['beam'].faces.findAt(((
    L1/3, t1/4, 0.0), (0.0, 0.0, 1.0)), ), technique=STRUCTURED)
```

```
# plane stress
mdb.models['coh-els'].parts['beam'].setElementType(elemTypes=(
    ElemType(elemCode=CPS4I, elemLibrary=STANDARD),
    ElemType(elemCode=CPS3, elemLibrary=STANDARD)), regions=(
    mdb.models['coh-els'].parts['beam'].faces.findAt(((
    L1/3, t1/3, 0.0), (0.0, 0.0, 1.0)), ), ))
# seeds bulk
mdb.models['coh-els'].parts['beam'].seedEdgeByNumber(
    constraint=FINER, edges= mdb.models['coh-els'].parts['beam'].\
    edges.findAt(((L1/4, t1, 0.0), ), ),
    number=neL1) # neL1=400/100*L1 # w/L1 in [mm], L1 was 100 mm.
# mdb.models['coh-els'].parts['beam'].seedEdgeByNumber(
    #constraint=FINER,
    #edges=mdb.models['coh-els'].parts['beam'].edges.findAt(((
    0.0, t1/4, 0.0),),), number=neY)# elements thru thickness
mdb.models['coh-els'].parts['beam'].seedEdgeByBias(
    biasMethod=SINGLE, constraint=FINER,
    end1Edges=mdb.models['coh-els'].parts['beam'].edges.findAt(
    ((0.0, t1/2, 0.0),),), number=t1Seeds, ratio=t1Bias)#0.65=>t1/2
mdb.models['coh-els'].parts['beam'].generateMesh()
# Part: adhesive
mdb.models['coh-els'].parts['adhesive'].setMeshControls(
    elemShape=QUAD, regions=mdb.models['coh-els'].\
    parts['adhesive'].faces.findAt(((La/3, ta/3, 0.0),
    (0.0, 0.0, 1.0)), ), technique=SWEEP)
mdb.models['coh-els'].parts['adhesive'].setSweepPath(edge=
    mdb.models['coh-els'].parts['adhesive'].edges.findAt((
    La, ta/2, 0.0), ),
    region=mdb.models['coh-els'].parts['adhesive'].faces.findAt((
    La/3, ta/3, 0.0), (0.0, 0.0, 1.0)), sense=REVERSE)
mdb.models['coh-els'].parts['adhesive'].setElementType(elemTypes=(
    ElemType(
    elemCode=COH2D4, elemLibrary=STANDARD, viscosity=1.0E-05),
    ElemType(elemCode=UNKNOWN_TRI, elemLibrary=STANDARD)),
    regions=(mdb.models['coh-els'].parts['adhesive'].faces.findAt(
    ((La/3, ta/3, 0.0), (0.0, 0.0, 1.0)), ), ))
# seeds adhesive
mdb.models['coh-els'].parts['adhesive'].seedEdgeByNumber(
    constraint=FINER,
    edges=mdb.models['coh-els'].parts['adhesive'].edges.findAt(
    ((La/2,ta,0.0),),), number=neLa)#neLa=280/70*La #La was 70 mm.
mdb.models['coh-els'].parts['adhesive'].seedEdgeByNumber(
    constraint=FINER, edges=mdb.models['coh-els'].\
    parts['adhesive'].edges.findAt(((0.0,ta/4, 0.0),),),number=1)
mdb.models['coh-els'].parts['adhesive'].generateMesh()
# Module: Interaction
mdb.models['coh-els'].Tie(adjust=ON, master=
    mdb.models['coh-els'].rootAssembly.surfaces['top'], name='top',
    positionTolerance=position_tolerance,
    positionToleranceMethod=SPECIFIED,
    slave=mdb.models['coh-els'].rootAssembly.surfaces['coh-top'],
    thickness=ON, tieRotations=ON)
mdb.models['coh-els'].Tie(adjust=ON, master=
    mdb.models['coh-els'].rootAssembly.surfaces['bot'], name='bot',
    positionTolerance=position_tolerance,
    positionToleranceMethod=SPECIFIED,
    slave=mdb.models['coh-els'].rootAssembly.surfaces['coh-bot'],
    thickness=ON, tieRotations=ON)
# Module: Load
mdb.models['coh-els'].DisplacementBC(amplitude=UNSET,
    createStepName='Step-1', distributionType=UNIFORM,fieldName='',
    fixed=OFF,localCsys=None, name='topLoad',
    region=mdb.models['coh-els'].rootAssembly.sets['topLoad'],
    u1=0.0, u2=u2target, ur3=UNSET)#new top
mdb.models['coh-els'].DisplacementBC(amplitude=UNSET,
    createStepName='Step-1',
    distributionType=UNIFORM, fieldName='', fixed=OFF,
```

```
        localCsys=None, name='bot',
        region=mdb.models['coh-els'].rootAssembly.sets['bot'], u1=0.0,
        u2=-u2target, ur3=UNSET)# minus
# Module: Job
Job = mdb.Job(atTime=None, contactPrint=OFF, description='',
        echoPrint=OFF, explicitPrecision=SINGLE,
        getMemoryFromAnalysis=True, historyPrint=OFF,
        memory=90, memoryUnits=PERCENTAGE, model='coh-els',
        modelPrint=OFF,multiprocessingMode=DEFAULT,name='DCBCZMparams',
        nodalOutputPrecision=SINGLE,numCpus=8,numDomains=8,numGPUs=0,
        queue=None, resultsFormat=ODB, scratch='', type=ANALYSIS,
        userSubroutine='', waitHours=0, waitMinutes=0)
mdb.saveAs('DCBCZMparams.cae')    #save the model
# execution starts here
Job.submit(consistencyChecking=OFF)
Job.waitForCompletion()
```

Post-processing code

During post-processing, we calculate the **error** and write it into the **cost.txt** file
to be later read by MATLAB.

```
# Postprocessing starts here. (c) 2021 Ever Barbero
# calculate the error at experimental points, COD=2*u2
from odbAccess import *
import numpy
```

At this point, incorporate the experimental data

```
# Cabello's Araldite High
u2Exp=[0,0.099,0.28,0.493,0.757,1.086,1.431,1.546,1.711,1.859,2.072,2.336,
2.484,2.615,2.796,2.993,3.125,3.257,3.437,3.651,3.832,4.046,4.26,4.408,
4.638,4.852,5.066,5.247,5.493,5.707,5.954,6.217,6.431,6.743,7.023,7.27,
7.533,7.796,8.125,8.372,8.651,8.914,9.194,9.474,9.786,9.942];# COD
p2Exp=[0,49.105,112.532,204.604,298.721,380.563,486.957,511.509,536.061,
552.43,558.568,548.338,527.877,509.463,482.864,462.404,441.944,413.299,
390.793,364.194,339.642,329.412,319.182,306.905,300.767,290.537,282.353,
280.307,276.215,265.985,259.847,255.754,247.57,249.616,245.524,241.432,
237.34,227.11,227.11,225.064,216.88,214.834,208.696,202.558,200.512,
200.512];# P
# Cabello's Araldite Low (do not overwrite previous data)
u2Exp_low=[0,0.115,0.312,0.559,0.822,1.003,1.184,1.398,1.595,1.793,2.007,
2.138,2.27,2.368,2.434,2.533,2.664,2.747,2.911,3.059,3.191,3.355,3.503,
3.668,3.849,4.062,4.243,4.49,4.72,4.901,5.148,5.345,5.559,5.789,6.086,
6.283,6.513,6.743,6.957,7.204,7.451,7.73,7.993,8.257,8.52,8.816,9.128,
9.375,9.704,9.91];
p2Exp_low=[0,53.197,130.946,233.248,319.182,368.286,417.391,456.266,
484.91,499.233,482.864,466.496,450.128,433.76,411.253,388.747,364.194,
345.78,335.55,325.32,315.09,306.905,302.813,290.537,282.353,268.031,
263.939,253.708,245.524,245.524,245.524,237.34,231.202,229.156,220.972,
220.972,216.88,206.65,202.558,196.419,188.235,188.235,182.097,180.051,
178.005,178.005,171.867,173.913,171.867,171.867];
# Select data (low or high) to fit with Matlab fminsearch
# u2Exp = u2Exp_low
# p2Exp = p2Exp_low
```

Next, calculate the error and write it into **cost.txt**

```
# Calculate the error. All calculations using COD=2*u2 from FEA
odb = openOdb(path='DCBCZMparams.odb',readOnly=True)
step = odb.steps['Step-1']
# find region name with ">>> print step.historyRegions"
region = step.historyRegions['Node BEAM-1.1443']# mesh dependent!
# store u2, rf2, to lists
u2Data = region.historyOutputs['U2'].data
```

```
rf2Data = region.historyOutputs['RF2'].data
# find the peak load
rf2max = 0   #initialize the max. load
u2 = []      #initialize a list
rf2 = []     #initialize a list
u2AtPeekLoad = []   #initialize a list
# build the simulation results array
for i in range(len(u2Data)):#loop over simulation points
    u2.append(2*u2Data[i][1])    # COD: u2 = 2*U2
    rf2.append(rf2Data[i][1])    # Load: rf2 = RF2
    # print i, u2[i], rf2[i]
    if rf2[i] > rf2max:
        imax = i
        rf2max = rf2[i] #peek load
        u2AtPeekLoad = u2[i]
print ('increment %g, u2AtPeekLoad %g, rf2max %g,
    sim pts %g'%(imax, u2AtPeekLoad, rf2max, len(rf2Data)-1))
# Calculate area under experimental curve
areaExp = 0
for i in range(len(u2Exp)):# loop over experimental points
    if i == 0:
        continue # skip i=0
    if u2Exp[i] > 2*u2target:
        break # skip everything
    deltau2 = u2Exp[i] - u2Exp[i-1] # current minus previous u2Exp
    # experimental area
    areaExp = areaExp + deltau2 * (p2Exp[i-1] + p2Exp[i])/2
# Calculate area under simulation curve
areaSim = 0
for i in range(len(u2)):# loop over simulation points
    if i == 0:
        continue # skip i=0
    if u2[i] > 2*u2target:
        break # skip everything
    deltau2 = u2[i] - u2[i-1] # current minus previous u2
    # simulation area
    areaSim = areaSim + deltau2 * (rf2[i-1] + rf2[i])/2
# calculate error
error = 100*abs(areaSim-areaExp)/areaExp
print "areaExp=",areaExp, "areaSim=",areaSim, " %error=",error
#
simulation_data = open('simulation_data.txt','w')
for i in range(len(rf2Data)):#loop over simulation points
    simulation_data.write('%g %g\n' % (u2[i], rf2[i]))# COD, rf2
simulation_data.close()# to plot later
cost = open('cost.txt','w')# overwrite
cost.write('%10.4e \n' % error)# to be read by Matlab
cost.close()
log = open('log.txt','a+')
minutes = (time.time()-start_time)/60
log.write("Kn={} tno={} Gnc={} a0={} rangeCOD={}\n".format(
    Kn, tno, Gnc, a0, 2*u2target))
log.write("u2(p2max)={} p2max={} error={} time={}\n".format(
    u2AtPeekLoad, rf2max, error, minutes))
log.close()# useful info
odb.close()
# end
```

When the minimization of the error is achieved, the value of G_n^c and other state variables can be used to generate and plot the predicted load vs. COD path and to compare it with the experimental data that was used for the minimization, as shown in Figure 10.14.

Suggested Problems

Problem 10.1 *Retrieve the values of reaction force vs. displacement from Abaqus for Examples 10.1 and 10.2, pp. 374, 378. Plot them together and discuss the difference. Your plot should be similar to the one shown in Figure 10.7. To retrieve the data as a text file, proceed as follows*

```
Module: Visualization
Menu:   Report, XY
    # pick: cohesive surfaces, # cohesive elements for Example 10.1
    Tab: Setup, Name [coh_surf.rpt], # coh_ele.rpt for Example 10.1
    # un-checkmark: Append to file, OK
    # the report file will be saved in your Work Directory
```

Problem 10.2 *Retrieve the values of reaction force vs. displacement for Example 10.3 (p. 10.3) and plot them along those of Examples 10.1 and 10.2, p. 374, 378. Discuss the differences. The comparative plot should be similar to the one shown in Figure 10.9.*

Chapter 11

Fatigue

A methodology to predict fatigue damage in laminated composites is presented in this chapter. It is shown that limited experimental data can be used to characterize the material system by using appropriate micromechanics simulation. Furthermore, a small number of parameters are sufficient to characterize the fracture-controlled transverse cracking damage and the stress-driven fatigue damage due to defect nucleation and coalescence. A fatigue damage-initiation criterion and a kinetic equation for fatigue-damage growth are proposed. It is shown that *Paris law* applies for constant thermal ratio, which is further generalized into the *master Paris law*, applicable to arbitrary thermal ratio. These findings form the basis for a methodology to extrapolate available fatigue data for situations not considered by the experimental data, including other laminate stacking sequences, other thermal ratios, and other temperature ranges. Furthermore, a thermomechanical equivalence principle is developed to mimic thermal testing with surrogate mechanical testing at constant temperature.

11.1   Temperature Dependent Properties

Knowledge of the temperature-dependent properties of composites is crucial for accurate prediction of thermomechanical fatigue. Temperature-dependent properties, including moduli and coefficient of thermal expansion (CTE), of various fibers and matrices can be calculated from lamina data from the literature (Tables 11.1 and 11.2). First, temperature-independent properties of fibers and temperature-dependent properties of matrices can be back-calculated from lamina data using periodic microstructure micromechanics (PMM, Appendix D.1) and Levin's formula (Appendix D.2) as explained in [199]. Using such methodology, elastic properties of carbon fibers are calculated and reported in Table 11.1.

Temperature-dependent properties of polymer matrices are fitted with a quadratic polynomial

$$P(T) = a + b\,T + c\,T^2 \tag{11.1}$$

where P is the property, i.e., modulus, Poisson's ratio, or CTE, and a, b, c, are the coefficients of the quadratic polynomial. Values are reported in Table 11.2.

Table 11.1: Carbon fiber properties derived from lamina data in [200–203]. Reproduced from [199] © 2018 SAGE Publishing.

Property	AS4	T300	P75
E_A [GPa]	231.000	231.000	517.000
E_T [GPa]	23.453	26.864	11.158
G_A [GPa]	15.764	81.662	10.636
ν_A	0.253	0.156	0.269
ν_T	0.371	0.287	0.306
α_A [10^{-6}/ C]	-0.630	-0.600	-1.46
α_T [10^{-6}/ C]	5.997	11.086	12.500

Table 11.2: Temperature-dependent moduli and CTE of polymer matrices. Compiled from [199] © 2018 SAGE Publishing.

Property	a	b	c
Epoxy 3501-6, temperature range: $[180, -200]°C.$			
E_m [MPa]	4580.4836	-10.6103	0
ν_m	0.3812	$3.8564\ 10^{-5}$	0
α_m [10^{-6}/ C]	38.3445	0.1224	0
Epoxy 934, temperature range: $[-156, 120]°C.$			
E_m [MPa]	5032.7732	-16.7561	0.0251
ν_m	0.3659	$-1.1108\ 10^{-4}$	$-8.6080\ 10^{-7}$
α_m [10^{-6}/ C]	38.7655	0.1524	$-1.32553\ 10^{-4}$
Epoxy ERL 1962, temperature range: $[-156, 120]°C.$			
E_m [MPa]	5032.7732	-16.7561	0.0251
ν_m	0.3659	$-1.1108\ 10^{-4}$	$-8.6080\ 10^{-7}$
α_m [10^{-6}/ C]	49.3143	0.1594	$-4.5090\ 10^{-4}$
Epoxy 5208, temperature range: $[-156, 120]°C.$			
E_m [MPa]	4828.7124	-5.4846	$-5.2164\ 10^{-3}$
ν_m	0.4072	$-3.3332\ 10^{-4}$	$7.9119\ 10^{-7}$
α_m [10^{-6}/ C]	36.6598	0.1887	$-9.5441\ 10^{-5}$

Table 11.3: Temperature dependent fracture toughness of carbon/epoxy calculated from crack density data vs. temperature available in [204–206]. Compiled from [199] © 2018 SAGE Publishing.

Material	V_f	Temp. range	G_{IC} quadratic coefficients			Constant G_{IC}
			a	b	c	
P75/934	0.65	[-160,20]	50.0561	-4.300 10^{-2}	6.3749 10^{-4}	53.4050
P75/1962	0.52	[-160,-15]	77.8054	9.6211 10^{-2}	1.3948 10^{-3}	84.4808
AS4/3501-6	0.64	[-190,20]	61.9052	-1.610 10^{-1}	-5.041 10^{-4}	68.0664

Then, the temperature-dependent moduli and CTE of any composite material containing any combination of fiber and matrix, at any value of fiber volume fraction, can be calculated using PMM and Levin's formula (see Appendix D).

Finally, the temperature-dependent fracture toughness of the lamina can be calculated from experimental crack density vs. temperature data, as explained in Section 11.2. Values are reported in Table 11.3.

11.2 The Quasi-Static Problem

In this section we explain how quasi-static fracture mechanics, codified into discrete damage mechanics (DDM, Chapter 9), is used to obtain values for the temperature-dependent fracture toughness[1] G_{IC} in terms of available experimental data of crack density vs. temperature under quasi-static (no fatigue) conditions.

The bibliography on distributed damage in composite laminates is extensive [207]. Among the multitude of damage models available, DDM is attractive for this study because, in addition to the usual elastic properties, it requires at most two values of fracture toughness G_{IC} and G_{IIC} (usually just one needed) to predict both damage initiation and evolution due to transverse and in-plane shear strains for general laminates subjected to general loads [140]. Since for most materials $G_{IC} < G_{IIC}$, and for most laminates loaded under usual conditions the energy release rates (ERR) are such that $G_I > G_{II}$, only G_{IC} is usually necessary.

DDM is an objective (mesh independent) constitutive model, meaning that when implemented in FEM software, it does not require guessing a value for the characteristic length L_c in order to reduce mesh dependency [208]. DDM is available as a plugin for Abaqus [209] and ANSYS [210], and it has been extensively validated.

DDM is based on an analytical solution [211] of the displacement field inside a representative volume element (RVE, Figure 9.3) encompassing the laminate thickness t with N laminas, a unit length along the fiber direction x_1 of the cracking lamina $k = c$, and a distance $2\ell = 1/\lambda_c$ in the transverse direction x_2. Homogenization of the damaged stiffness of the N laminas, coupled with an iterative procedure allows all laminas $k = 1 \ldots N$ in a laminate to be cracking simultaneously with dif-

---

[1] Although G_{IC} is a critical energy release rate, with units [kJ/m^2], the abbreviated term "fracture toughness" is often used in the composites literature, not to be confused with the classical fracture toughness K_I^c used for isotropic materials.

ferent crack density λ_k values at a given time. For a given damage state $\lambda = \{\lambda_k\}^T$ and applied 2D thermomechanical field $\epsilon = \{\epsilon_1, \epsilon_2, \gamma_6, \Delta T\}^T$ at a Gauss integration point of a shell element, DDM determines the local 3D displacement field $u_i(x_j)$ analytically, with $i, j = 1 \ldots 3$, from which it calculates the local strain field and stress field including intralaminar stresses, as well as damaged lamina $[Q]$ and damaged laminate stiffness matrices, and ERRs (G_I, G_{II}) in modes I and II. The later are used in a damage initiation and evolution criterion (9.1–9.2).

Note that the fracture toughness G_{IC} and G_{IIC} are the only material properties needed to predict both initiation and evolution of crack density under quasi-static conditions, and usually only G_{IC} is necessary. Fatigue loading requires at most two additional parameters, which are the *defect nucleation rates* β_I, β_{II}, as explained later in this chapter. No hardening exponents or any other damage evolution material properties are needed to describe the kinetic evolution of damage.

The ERR values G_I, G_{II} in modes $i = I, II$ are calculated by (9.36). Mode decomposition is achieved by splitting the strain energy U into mode I (opening) and II (shear) and adding the contribution of each lamina $k = 1 \ldots n$ using (9.37–9.38).

DDM assumes local uniformity of crack spacing and linear distribution of intralaminar stresses in each lamina. Despite these assumptions, predicted results correlate extremely well with available data for a broad variety of material systems (carbon and glass reinforced composites), laminate stacking sequences (LSS) [212], and loading conditions including open hole tension data up to failure. [143,213–215]

Since the size of the RVE ($1 \times t \times 1/\lambda_k$) is dictated by the crack density λ_k, not by the element size, and the solution is in terms of displacements, not stress or strain, the predictions of the DDM constitutive model are mesh-density and element-type independent. The only effect of mesh density is on the quality of the stress/strain field, as it is well known for the finite element method. Mesh/element-type insensitivity (i.e., model objectivity) is a remarkable advantage with respect to cohesive zone models (CZM) and progressive damage analysis (PDA) that may be affected by mesh-density and element-type.

Thermomechanical fatigue produces intralaminar matrix cracking, resulting in stiffness reduction and stress redistribution that may induce fiber failure, delaminations, and loss of hermeticity. For each fatigue cycle, intralaminar matrix cracking can be predicted with DDM, which provides a solution for crack density λ [1/mm] in lamina k that can be summarized, as follows

$$\lambda = \lambda(G_I, G_{IC}) \quad ; \quad G_I = G_I(\epsilon, \Delta T, Q) \quad ; \quad Q = Q(\lambda) \qquad (11.2)$$

where the function $\lambda()$ calculates the crack density as a function of ERR G_I and fracture toughness G_{IC}; the function $G_I()$ calculates the ERR as a function mechanical strain ϵ (if applied), temperature range $\Delta T = T_{min} - T_{max}$, and laminate stiffness Q; and finally the function $Q()$ calculates the laminate stiffness as a function of crack density.

Although DDM can be summarized by (11.2), it is implemented in software, and thus it can also calculate G_I *implicitly* as a function of λ, for a given set of $G_{IC}, \Delta T, \epsilon$, which is useful for the formulation of the fatigue model explained later in this chapter. In other words, we can think of the DDM code as capable of solving (11.2) implicitly for any single variable in (11.2).

In DDM, an embedded iterative process finds λ for all laminae $k = 1 \ldots N$, as a function of applied strain and temperature. It solves for λ inside an RVE with volume $V = 2LH$, where $2L = L/\lambda$, and H is the laminate thickness. The quasi-static version of DDM is available as a plugin for Abaqus [209] and ANSYS [210], and it has been extensively validated. A plugin for the thermal fatigue version of discrete damage mechanics (DDM6TM) is available in [2] and used for Example 11.1.

DDM requires only one material property, the fracture toughness G_{IC} to predict initiation, accumulation, and saturation crack density in a laminate subjected to quasi-static thermal and/or mechanical loading. Thermal stress and temperature-dependent material properties are incorporated. Fatigue effects are explained in this chapter.

Matrix cracking of laminates is controlled by the in-situ fracture toughness G_{IC} of the material system. G_{IC} data can be obtained indirectly in terms of the effects of matrix cracking, such as modulus reduction or crack density. However, due to the high stiffness of carbon fibers, modulus reduction due to matrix cracks is small and difficult to measure because the modulus reduction in the direction normal to cracks is obscured by the high modulus of non-cracking laminae. Therefore, in this work we use crack density data, which is always measurable by optical, tomography, or acoustic emissions methods, allowing us to simply count the cracks as a function of applied mechanical or thermal strain [139].

The quasi-static fracture toughness G_{IC} is adjusted from crack density data *at each temperature* by minimizing the error D

$$D = \frac{1}{N} \sqrt{\sum_{j=1}^{N} (\lambda_j - \lambda_j^d)^2} \tag{11.3}$$

between crack density λ_j predicted with (11.2) and experimental λ_j^d crack density data *at each temperature*. The temperature-dependent results are then fitted as a function of temperature with a quadratic polynomial (11.1) with parameters a, b, c shown in Table 11.3.

A temperature-independent (average) value of G_{IC} can be obtained by using all the data for all temperatures at once in (11.3), with values shown on the last column of Table 11.3. The results are plotted in Figure 11.1, where it can be seen that the temperature-dependence of G_{IC} is not strong. This is due to the compensating effects of the material becoming more brittle but with a lower CTE at low temperature [199].

11.3 The First Cycle of a Fatigue Test

This section provides the phenomenological justification for the fatigue model presented in the rest of this chapter.

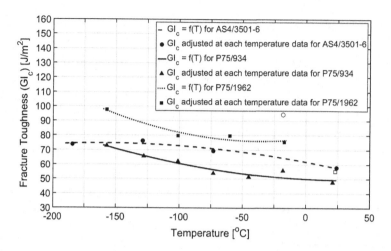

Fig. 11.1: G_{IC} vs. temperature for P75/934, P75/1962, and AS4/3501-6. Reproduced from [216] © 2018.

The Griffith/Irwin *fracture criterion* is used in this work as *damage-initiation criterion*, taking into account that a single intralaminar matrix crack does not by itself causes *fracture* of the whole laminate, but rather contributes to the deterioration of the laminate in the form of *damage*. The Griffith/Irwin fracture criterion states that when the Energy Release Rate (ERR) G_I exceeds the fracture toughness G_{IC}, a new crack appears, i.e.,

$$\xi = \frac{G_I}{G_{IC}} \geq 1 \rightarrow \text{new intralaminar crack} \qquad (11.4)$$

Inelastic dissipation at the crack tip is included in G_{IC} because the polymer is quasi-brittle and the plastic zone is small with respect to the crack dimensions, which is satisfied for this problem because the crack size is equal to the lamina thickness, and thus much larger than the plastic zone [3, Section 7.2.1]. Therefore, quasi-static DDM (Section 9) can be used to calculate crack density as a function of applied strain and/or temperature, or alternatively ERR as a function of crack density for fixed strain and temperature.

The stress intensity K_I and ERR can be written [217, Chapter 6] as $K_I = \beta \sigma \sqrt{\pi t_k}$ or

$$G_I(\lambda) = \beta^2 t_k (1 - \nu^2) E(\lambda) \epsilon^2 \qquad (11.5)$$

where t_k is the thickness of lamina k, ϵ is the applied thermomechanical strain, and β is a geometric coefficient that accounts for the geometry of the crack and the domain. DDM calculates the degraded material properties $\nu(\lambda), E(\lambda)$ and the geometric coefficient β. The thermal strain is $\alpha \Delta T$. Therefore the fracture problem is solved if we know the fracture toughness G_{IC}.

For the first cycle ($N = 1$) of a thermal fatigue test, we use the classical fracture toughness G_{IC} of the material (Table 11.3). Note that crack density is driven by energy considerations, specifically by energy release rate. If sufficient energy is available to propagate a crack, the Griffith/Irwin criterion states that the crack will actually propagate. It has nothing to do with stress. Furthermore, the energy required for crack formation (twice Griffith's surface energy plus Irwin's crack-tip dissipation effects) comes from the entire laminate, no just from the lamina that is cracking. This energy becomes available due to the relaxation caused by the propagation of the crack itself. This is in contrast to the fatigue problem, which is driven by stress, as we shall see in the next section. Also, there are no long-range dissipation effects such as fiber bridging because intralaminar cracks propagate mostly in the matrix.

For $N = 1$, cracks propagate at all locations where initial defects are large enough. Once those defects propagate into cracks, no more defects remain that are large enough to propagate into cracks, unless the mechanical or thermal strains are increased, or some other physical phenomenon, such as fatigue, produces new large enough defects that can grow into cracks.

Note that under constant-amplitude mechanical or thermal fatigue, the applied range of strain (mechanical or thermal) remains constant. All the defects that are large enough to grow into cracks are used during the first cycle, and on the next cycles there are no sites that can propagate into cracks, unless fatigue manages to nucleate small defects into large enough defects that can act as viable sites for additional crack propagation.

When crack density increases, the stiffness $E(\lambda)$ and coefficient of thermal expansion (CTE) $\alpha(\lambda)$ decrease [199]. The strain $\epsilon = \alpha \Delta T$ decreases because the CTE decreases while ΔT is constant, and the ERR G_I decreases as per (11.5) with both E and ϵ decreasing. Once the ERR and strain decreases, no more cracks can appear unless either the strain increases or the fracture toughness decreases. The strain cannot increase because the temperature range is constant, and the fracture toughness is an invariant material property (Table 11.3). For fatigue damage to occur when $N > 1$, the *fatigue fracture toughness* must decrease, as discussed next.

11.4 Fatigue Damage Criterion

In this section we propose a *fatigue damage criterion*

$$G_I(N) \geq G_{IC} \; f(N) \tag{11.6}$$

using separation of variables to describe fatigue-damage as the product of an energy-controlled fracture problem times a stress-controlled void and defect nucleation problem. Note that $G_I(N)$ and $f(N)$ are both functions of the number of cycles N but G_{IC} is only a function of temperature.

Thermal fatigue is caused by cyclic, repetitive oscillation of temperature T in the range $T_{min} < T < T_{max}$, with amplitude $\Delta T = T_{min} - T_{max} < 0$, thermal strain $\epsilon = \alpha \Delta T$ and thermal ratio defined as $R = T_{min}/T_{max}$.

Since the *quasi-static* fracture toughness G_{IC} is an invariant material property, the fatigue effect must be represented separately. In this chapter we use separation of variables to write the *fatigue fracture toughness* G'_{IC} as the product of the *quasi-*

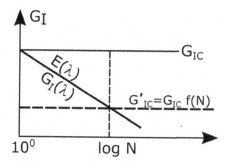

Fig. 11.2: As λ grows, $E(\lambda)$ decreases, $G_I(\lambda)$ decreases, and no new cracks can be propagated unless $f(N)$ decreases, which requires $N > 1$, thus fatigue loading.

static fracture toughness G_{IC} times the *defect nucleation function* $f(N)$, as follows

$$G'_{IC} = G_{IC}\, f(N) \quad ; \quad 0 < f(N) < 1 \tag{11.7}$$

Then, generalizing (11.2) for fatigue

$$\lambda = \lambda(G_I, G'_{IC}) \quad ; \quad G_I = G_I(\epsilon, \Delta T, Q) \quad ; \quad Q = Q(\lambda) \tag{11.8}$$

we can calculate crack density for every cycle with $N > 1$.

To explain the sequence of events during thermal fatigue loading, let's look at Figure 11.2. For $N = 1$, the material is pristine, thus $f(N) = 0$, and the Griffith/Irwin condition (11.4) means that ΔT must be large enough to produce enough thermal strain to propagate a crack from the largest defect in the material. If ΔT is not large enough, more cycles $N > 1$ are needed to grow the existing defects, for example by void/craze nucleation, so that the critical crack size is reached and the first crack can be propagated.

Once a crack is propagated, λ grows, $E(\lambda)$ decreases, and $G_I(\lambda)$ decreases. Since G_{IC} is constant, no new cracks can be propagated unless $f(N)$ decreases, which requires $N > 1$, thus fatigue loading.

With G_{IC} constant, as $f(N)$ decreases due to defect nucleation, fatigue toughness $G'_{IC} = G_{IC} f(N)$ can decrease sufficiently to catch up with decreasing ERR $G_I(\lambda)$ and thus, fatigue can take place.

The physical justification is as follows. In the absence of new cracks, cyclic load (mechanical or thermal) results in nucleation of voids and crazes in the polymer. Crazes are broken polymer branches and chains that occur due to stress [218, 219]. They multiply and coalesce into larger defects, driven by hydro-static stress [218, 219]. When the void or craze is large enough, a crack can propagate.

Thus, fatigue is a stress-driven phenomenon, not an energy-driven phenomenon. This provides justification for separating the fatigue phenomenon into an energy-driven quasi-static fracture toughness G_{IC} and a stress-driven defect nucleation function $f(N)$.

The defect nucleation function $f(N)$ can be characterized using low-cycle experimental data, as follows. Crack density vs. number of cycles can be obtained from experimental data (shown by symbols in Figure 11.3). For those cracks to be

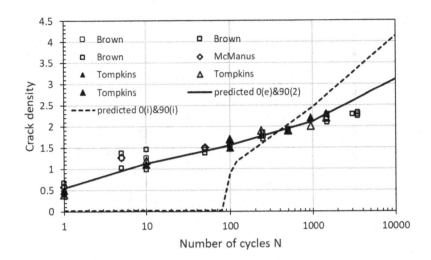

Fig. 11.3: Experimental crack density vs. number of cycles (N=1...3500) with thermal ratio $R = -156/121$ °C, for carbon/epoxy P75/1962 $[(0/90)_2]_S$, data from [205, 220, 221]. Reproduced from [216] © 2018.

present, the Griffith/Irwin criterion demands that $G_I = G'_{IC}$. DDM can calculate G_I given λ and ΔT. Therefore, DDM (represented by (11.8)) calculates the fatigue ERR G_I that corresponds to any set of known crack density λ, temperature range ΔT, and life N. From it and using the fracture criterion $G_I = G'_{IC}$, we immediately know the value of G'_{IC}.

Then, discrete values for the defect nucleation function $f(N)$ can be obtained from (11.7) dividing G'_{IC} by the constant G_{IC} (from Table 11.3). The resulting discrete values of $f(N)$ are displayed by square symbols Figure 11.4. The relationship between $f(N)$ and $\log(N)$ seen in Figure 11.4 can be approximated by

$$f(N) = 1 - \beta \log N \tag{11.9}$$

with $\beta > 0$. The defect nucleation function $f(N)$ is decreasing with N, so that with constant quasi-static fracture toughness G_{IC}, the fatigue fracture toughness G'_{IC} in (11.7) becomes a decreasing function of N, thus allowing for the fatigue phenomenon to be represented. At most there are two *defect nucleation rates*, β_I and β_{II}, for modes I and II, to be determined from experimental data as in Figure 11.3. However, β_{II} is seldom necessary.

11.5 Master Paris Law

In this section we propose a *kinetic equation for fatigue-damage growth* extending the modified Paris law [222][2] to define the master Paris law (MPL). This exercise

---

[2] *Modified* means that crack density and energy release rate substitute crack length and stress intensity K_I in the original Paris law.

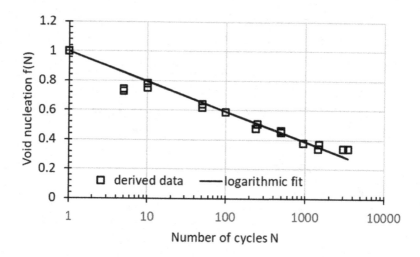

Fig. 11.4: Characterization of the defect nucleation function f(N) using experimental data from Figure 11.3. The linear fit has a slope $\beta = 2.04$. Reproduced from [216] © 2018.

provides further insight into the kinetics of fatigue damage in laminated composites, leading to a procedure that not only describes the data but also allows us to propose and extrapolation method for different LSS, thermal ratio R, and life N beyond the limitations of the experimental data. While only experimental data may someday prove or disprove the proposed method, numerical results agree with experimental data that is available for a modest range of number of cycles.

Using experimental data $\lambda(N)$, such as that available in Figure 11.3, we calculate the crack-growth rate $d\lambda/dN(N)$ as the local slope of the data at each data point. Then, use DDM (11.8) to calculate

$$\Delta G_I = G_I(T_{min}) - G_I(T_{max}) \tag{11.10}$$

for each value of N, and finally plot $d\lambda/dN$ vs. ΔG_I as in Figure 11.5.

The resulting data in Figure 11.5 can be approximated accurately by a linear equation, which is known as modified Paris law

$$\frac{d\lambda}{dN} = a\,[\Delta G_I(\lambda, \Delta T)]^b \tag{11.11}$$

with parameters a, b.

The outlier data (dark symbols) deviate from Paris law, thus suggesting particular events that can be attributed to damage initiation and crack saturation. Damage initiation is evident at the top-right of Figure 11.5, where both ΔG_I and $d\lambda/dN$ are large. Crack saturation is evident at the bottom-left of Figure 11.5, where ΔG_I and $d\lambda/dN$ are small. These two points bracket the range of applicability of Paris law.

On Figure 11.5, the thermal ratio R is the same for all data. But data with different thermal ratio does not fit Paris law, as shown by the dark symbols in

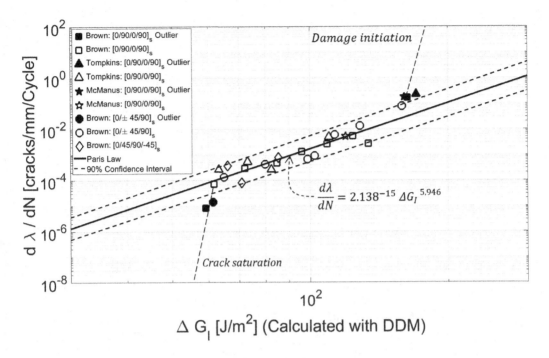

Fig. 11.5: Paris law. All data has the same thermal ratio $R = -156/121$ °C and $T_{min} = -156$°C. Reproduced from [216] © 2018.

Figure 11.6. To solve this problem, we note that a constant slope in log scale suggest Arrhenius phenomena. Therefore, taking a clue from the time-temperature superposition principle [223], we propose to normalize (shift) ΔG_I by $G_{IC}(T_{min})$ as shown in Figure 11.6, which can be described by the proposed *master Paris law* equation

$$\frac{d\lambda}{dN} = A \left[\frac{\Delta G_I(\lambda, \Delta T)}{G_{IC}(T_{min})} \right]^b \tag{11.12}$$

with A calculated in (11.13). For a given data set, the parameter b is unaffected by the shift. Thus, equation (11.12) provides a kinetic equation for damage growth-rate during thermal fatigue of laminated composites.

Note that in log scale, a quotient is a shift. The quasi-static fracture toughness at the lowest temperature $G_{IC}(T_{min})$ is chosen as the shift factor because maximum crack propagation takes place at temperature T_{min}, where the thermal stress is maximum and the polymer is most brittle. After the shift, all data fits in the 90% confidence interval regardless of thermal ratio.

Master Paris law (MPL) works as the time-temperature superposition principle (TTSP). We can find the parameters a and b by testing with low thermal ratio, such as $R = -46/10$ and $R = -101/66$, which is are easy tests, then shift the data (dark symbols in Figure 11.6) to $T_{min} = -156$ and have a Paris law for any situation with $T_{min} = -156$°C.

Once the shifting procedure is proven, the MPL line can be shifted and then used for analysis at any temperature. Note that T_{max} does not participate in the

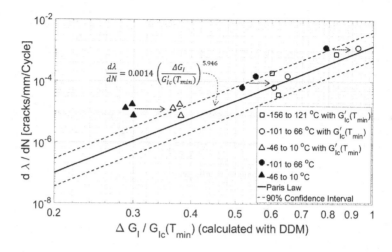

Fig. 11.6: Master Paris law for multiple thermal ratios $R = -156/121, -101/66$, and $-46/10^oC$, all shifted with their value of T_{min}. Reproduced from [216] © 2018.

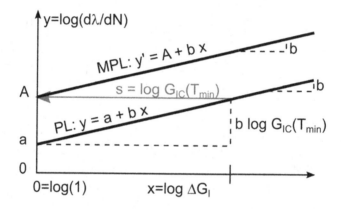

Fig. 11.7: Determination of constant A in master Paris law.

shifting process, but instead is taken into account by DDM during calculation of ΔG_I.

Since ΔG_I is a function of life N and temperature range T_{min}, T_{max}, we can calculate damage rate $d\lambda/dN$ with (11.12) for any set of values N, T_{min}, T_{max}, regardless of what temperature range was used to obtain the material properties a, b.

Since a is a function of the shift temperature T_{min} and the slope b is a constant, independent of shift temperature, we can derive a formula for the constant A in (11.12). With reference to Figure 11.7 we have

$$A = a + b \cdot s \quad \text{with shift } s = \log G_{IC}(T_{min}) \tag{11.13}$$

The only constraints on the applicability of shifting are damage initiation (DI) and crack saturation (CS), as illustrated in 11.5, where the applicable region is bracketed by DI on the top-right and CS on the bottom-left. The proposed formulation accounts for this bracketing implicitly.

Indeed, for a situation characterized by N, T_{min}, T_{max}, all of which are independent variables for the analysis, DDM will correctly calculate damage initiation, so that the portion of the MPL on the right of *damage initiation* is unused (see Figure 11.5). If damage initiation were not to occur at $N = 1$, the software tries for higher N until $f(N)$ decreases enough to allow the first crack to propagate. In either case, the portion of the MPL on the right of *damage initiation* is unused.

Similarly, for sufficiently large values of N, DDM predicts that $\Delta G_I \to 0$, because $E(\lambda) \to 0$, so the left portion of the MPL will not be used (see Figure 11.5). This can be seen in Figure 11.8 at $n = 10^5$, where it is evident that after crack density reaches a maximum, the material is damaged so much that the ERR G_I that would be released as a result of another crack is insufficient to exceed the *fatigue fracture toughness* $G'_{IC} = G_{IC}f(N)$, even though G'_{IC} is decreasing because $f(N)$ is decreasing with increasing N. In Figure 11.8, at $N = 10^5$ the predicted crack density drops, but in reality damage is irreversible, so the software detects it and keeps the crack density constant at the maximum value.

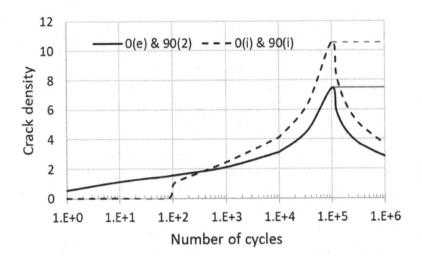

Fig. 11.8: Predicted crack saturation at $N \approx 10^5$.

11.6 Thermal and Fatigue Damage Prediction

The current implementation of DDM, called DDM6TM (version 6 thermomechanical) can analyze two situations, as follows

Monotonic Cooling simulates cooling of the material from stress free temperature (SFT) down to the coldest analysis temperature (T_{min}) for cycle one ($N = 1$); that is, no fatigue phenomenon is apparent ($f(N) = 1$). The total temperature excursion $\Delta T = T_{min} - SFT$ is divided in 1°C increments. Crack density is calculated at the end of each increment. The temperature-dependent properties at the beginning of the increment are used. Ply-by-ply

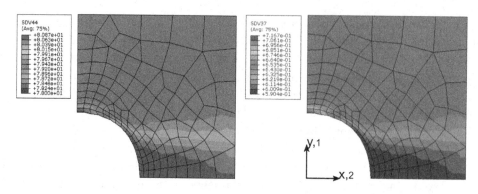

Fig. 11.9: Left: SDV44=σ_{22} at $T_{min} = 57^{o}C$ in the center lamina of a $[(0/90)_2]_S$ laminate clamped all around, SFT=$177^{o}C$. Right: SVD37=crack density. 1/4 model shown. σ_{22} is smaller where the damage is higher.

stress and strain components, ERR, and crack density are calculated at each temperature.

For example, crack density at T_{min} in the center ply of a $[(0/90)_2]_S$ laminate, clamped all around, is shown in Figure 11.9. The fiber direction is vertical for the 90_2 center lamina. When the material cools and shrinks, thermal contraction imposes tensile stress across the fibers. The stress is maximum at the edge of the hole (blue color, left figure), causing maximum damage at the edge of the hole (red color, right figure).

Thermal Fatigue Damage occurs when the temperature is cycled with constant thermal ratio $R = T_{min}/T_{max}$ with $T_{min} < T_{max} \leq SFT$ and $\Delta T = T_{min} - T_{max}$. Since the maximum crack density in each cycle occurs at T_{min}, the crack density is calculated at temperature T_{min}.

The reduction of fatigue toughness G'_{IC} with number of cycles and temperature is accounted for by two parameters, the quasi-static fracture toughness G_{IC} and the defect nucleation rate β_I. The temperature-dependent, quasi-static fracture toughness is a material property that can be obtained from quasi-static testing (at $N = 1$). The defect nucleation rate can inferred from experimental fatigue data. For every cycle N, the software calculates crack density, ply-by-ply stress and strain distributions, and energy release rate (ERR). Crack density vs. life N is compared to experimental data in Figure 11.3 (solid line).

For a laminate with n laminas, the proposed methodology finds the crack density that is compatible with the Griffith/Irwin condition $G_I = G'_{IC}$ with $G'_{IC} = G_{IC}f(N)$, G_{IC} constant, and $f(N)$ decreasing with N as shown in Figure 11.4.

Comparison of crack density vs. number of cycles predicted with the proposed methodology and experimental data [205, 220, 221] for $[(0/90)_2]_S$ is shown in Figure 11.3. Data is only available for exterior [0] and interior $[90_2]$ laminas. From quasi-static studies, these laminas are known to crack earlier and proceed with slower crack growth rate than thinner, interior laminas [3, Figure 7.9].

Data is not available for the interior [0] and [90] single-ply laminas, but the predicted values for interior laminas (dashed line) are consistent with previous experimental and analytical results for laminates subjected to quasi-static, strain-controlled loading, that show that interior laminas crack later and then proceed with faster crack-growth rate. Thus, the fatigue crack-growth rates shown in Figure 11.3 are consistent with previous quasi-static data. For the exterior [0] and interior [90_2] laminas, the agreement between prediction and data is good.

Agreement with experimental data is demonstrated in Figure 11.3 up to 3500 cycles. Beyond that, there is no experimental data available, but the proposed methodology is able to predict the response as shown in Figure 11.8. A crack density peak is predicted at approximately $N = 10^5$ when the *crack saturation* is reached. Since crack density is so high, ERR is very small, and the only way to satisfy the Griffith/Irwin condition is for the software to unrealistically lower the crack density. Thus, the onset of a negative rate of crack density evolution can be used to detect CS. This type of behavior is consistent with experimental observations under mechanical fatigue loads for a different LSS and material system [224, Sect. 6.1] but thermal loads seem to tolerate higher values of crack density. Note however that the current formulation does not account for delaminations, which may lower the CS significantly.

11.6.1 Implementation

The independent variables for the analysis are N, T_{min}, T_{max}. The range for N and values for T_{min}, T_{max} are to be provided as input to the software.

The dependent variables are the crack density $\lambda(k)$ in each lamina k, and from that, it is possible to calculate the ERRs G_I, G_{II}, that produced such crack density, as well stress components in each lamina. These variables constitute the output of the software.

The material properties are temperature dependent. The quasi-static fracture toughness G_{IC}, G_{IIC}, are quadratic in temperature, with three coefficients a, b, c, for each property, as in (11.1). They can be adjusted to crack density data for monotonic cooling (N=1) as described in [199].

The elastic properties $E_1, E_2, G_{12}, \nu_{12}, \nu_{23}$ are quadratic in temperature, with three coefficients a, b, c, for each property, representing the constant, linear, and quadratic terms, respectively. They can be predicted from available data as described in Section 11.1 and further explained in [199].

Each cycle of a thermal fatigue problem starts at SFT, cooling the material down to T_{max}. Then, further cooling to T_{min} and back up to T_{max}, thus completing one cycle. The analysis must begin at SFT so that residual thermal stresses are accounted for. Therefore, three temperatures must be specified at the onset of the analysis: SFT, T_{max}, T_{min}.

The CTE values α_1, α_2, are cubic functions of temperature, with four coefficients a, b, c, d, for each property, representing the constant, linear, quadratic, and cubic terms, respectively. They can be adjusted to available data as described in Section 11.1 and further explained in [199].

Fiber failure is monitored using the model in [143], which requires the lamina tensile and compression strength in the fiber direction F_{1T}, F_{2C}, as well as the Weibull modulus of the fiber [3, Tables 2.5–2.6]. The lamina compression strength in the transverse direction F_{2C} is included in the script in anticipation of future work.

Finally, the *defect nucleation function* (11.9) for transverse mode I and shear mode II are represented by one coefficient each, β_I, β_{II}, which are called *defect nucleation rates*. They can be calculated using low-cycle experimental data as explained in Section 11.4, Figures 11.3 and 11.4.

Since mode II ERR is small for thermal loads [199, Figure 5.5], it is assumed that the mode II component of ERR does not play a significant role. Therefore, only G_{IC} data is usually needed, but the software implements the mode II as well in case it becomes necessary in the future. The parameters β_I, β_{II}, are not necessary for thermomechanical analysis during the first cycle ($N = 1$) but they are necessary for thermal fatigue predictions. Similarly to G_{IIC}, the defect nucleation rate β_{II} is usually not needed when the laminate is subjected to thermal loads.

The input parameters can be provided to Abaqus using a Python script (see Section 12.7.2) or through the .inp file (see Example 11.1). The script includes additional code for tasks that are usually performed with the graphical user interface (CAE), such as mesh generation and specification of boundary conditions, so that all the code to execute and example is available in a single file. The script is set up to perform either monotonic cooling or fatigue calculations.

Figure 11.9 is made with Abaqus CAE using the results saved by the DDM6TM plugin via SDVs. In Abaqus, SDVs are meant to store state variables (e.g., crack densities for all laminas, at each Gauss point), but we also use them to store derived quantities, such as stress and ERRs, to allow for visualization.

For each lamina we save 12 values:

SDV1 Crack density λ_k in lamina k with $k = 1 \ldots n/2$ for a symmetric laminate, with $k = 1$ at the bottom of the laminate.

SDV2 Longitudinal tensile damage activation g_{1t} [143].

SDV3 Longitudinal compression damage activation g_{1c} (not implemented).

SDV4 Longitudinal tensile damage D_{1t}, calculated from SDV2 [143].

SDV5 Transverse damage D_2, calculated from SDV1 (9.33).

SDV6 Shear damage D_6, calculated from SDV1 (9.33).

SDV7 Longitudinal stress in lamina *coordinate-system* (c.s.).

SDV8 Transverse stress in lamina c.s.

SDV9 Shear stress in lamina c.s.

SDV10 Transverse damage activation function g_{2t} (9.2).

SDV11 Energy release rate mode I, $G_I(\lambda)$ (9.36).

SDV12 Energy release rate mode II, $G_{II}(\lambda)$ (9.36).

Upon completion of the Abaqus Job, the SDVs can be used for visualization within Abaqus CAE.

Example 11.1 *Compute the crack density in the external layer ($k = 1$) of a $[0/90/0/90_{1/2}]_S$ flat laminate, dimensions 100 by 100 mm, when subjected to a temperature excursion from stress free temperature $SFT = 177$ C to minimum temperature $T_{min} = -156$ C. Ply thickness is $t = 0.127$ mm. The elastic properties are described with $P = a + bT + cT^2$ as a function of the temperature T. The thermal properties are described with $P = a + bT + cT^2 + dT^3$. The fracture properties are described with $P = a + bT + cT^2$. The coefficients are given in Table 11.4. Elastic and thermal properties are known in the range $-156 \leq T \leq 120$, and thus they are assumed to be constant outside that range.*

Property	a	b	c	d
E_1	271270.586	-8.10997	1.19E-02	
E_2	6554.2638	-11.6689	4.93E-04	
G_{12}	3998.0213	-8.84364	6.12E-03	
ν_{12}	0.3147	-6.97E-05	-4.05E-07	
ν_{23}	0.5557	-1.01E-04	-1.14E-06	
α_1	-0.97667	-9.0549E-05	-7.67E-06	2.30E-08
α_2	38.4684	8.95E-02	-3.65E-04	0
G_{IC}	0.18187			
G_{IIC}	1.0			
F_{1t}	1900.87			
F_{1c}	441.2			
F_{2c}	57.23			
Weibull m	8.9			

Table 11.4: Material properties for Example 11.1. Elastic and thermal property data in the range $-156 < T < 120$ C. Assumed constant outside those ranges.

Solution to Example 11.1 *The Abaqus model can be built using the Python script described in Section 12.7.2 or by using CAE and modifying the input file to insert the material properties and initial conditions for the analysis.*

The Ex_11.1.cae file is available in [2]. We now look at this file using CAE, module by module, to observe the following:

In Module: Part, we define a flat shell with dimensions 100 by 100 mm, 3D deformable, named 'Laminate'.

In Module: Assembly, we define a dependent instance called 'Laminate-1'. Immediately we define a set called 'WholeLaminate' to be referenced later during the definition of the material properties.

In Module: Property, we create a dummy 'Material-1' that will be written to the .inp file for the sole purpose of later locating the material properties block in the .inp file, and thus allowing us to modify it to include the material properties required by the plugin.

   Still in Module: Property, we create a section, called 'Section-1', Type: Shell, Homogeneous, tied to 'Material-1'. Here we set three Simpson points per layer to get results at the top, bottom, and mid-point of each section. Further we specify the laminate thickness

$$H = 2 \sum_{k=1}^{N/2} t_k$$

where H is the whole laminate thickness of a symmetric laminate with N laminas. The transverse shear coefficients are set to $k11 = 1.0, k12 = 0.0, k22 = 1.0$, not actual values, because they are not used in the in-plane analysis that we shall perform in this example, but should be carefully calculated if bending of the laminate is expected (see discussion following (3.9)).

   Next, we assign the instance to the set 'WholeLaminate'.

   The first fatigue cycle, $N = 1$, is a purely thermal cycle (no fatigue) that runs in increments of -1 C from the stress free temperature (beginning of the analysis) to the minimum temperature Tmin. Therefore, at this point we define Step-1, with time period $SFT - Tmin$. We set both the initial and the maximum increments to 1.0 so that the analysis advances 1 C per increment. Also we set the Field Output Requests to save U, RF, and SDV so that we can visualize the reaction forces, displacements, and all the state variables and derived quantities calculated by the plugin.

   In Module: Load, we set symmetryX on the left edge, symmetryY on the bottom edge, restrain the vertex at (0,0) to prevent rigid body motion, and set $U3 = 0$ everywhere to ensure membrane deformation. The initial and final temperatures are applied as predefined fields (see Example 3.13).

   In Module: Mesh, we seed so that the entire domain is modeled with a 4×4 mesh of S4R elements.

   In Module: Job we define a Job that uses the user subroutine 'ddm6tm-2017-std.obj'. Using the UGENS compiled with Abaqus 2017 in Abaqus 2020 is OK. Since Abaqus does not allow UGENS to run with with multiple CPUS, do not use numCpus=cpu_count(). Instead use numCpus=1.

   From this point forward it becomes easier to work with a script as explained in Section 12.7.2 but it is possible to go ahead manually. In Module: Job, do not submit the Job, but instead, write the input file as 'Job-1.inp'. Then, edit the input file in three places (as explained next) and save it as a new file named 'Job.inp'.

   *In the input file, at the end of the block Instance, we need to substitute the dummy material properties for the properties used by the plugin. This is done by replacing *Shell Section from the following dummy material*

```
** Section: Section-1
*Shell Section, elset=WholeLaminate, material=Material-1
0.889, 3
```

*by *Shell General Section, to represent the damaging material, as follows*

```
** Section: Section-1
*Shell General Section, elset=WholeLaminate, USER, PROPERTIES=53, VARIABLES=48
1.016, 3.0
271270.586,-8.10997,1.18938E-02,6554.2638,-11.6689,4.9329E-04,3998.0213,-8.84364
6.1187E-03,0.3147,-6.9707E-05,-4.0521E-07,0.5557,-1.0089E-04,-1.1402E-06,-0.9766711
-9.0549E-05,-7.6732E-06,2.3038E-08,38.4684,8.9456E-02,-3.6454E-04,0.0,0.18187
0.0,0.0,24,-184,1,0.0,0.0,24
-184, 177,120,-156,120,-156,1900.87,441.2
57.23,8.9,0.0,0.0,1,0, 0.127, 90
0.127, 0, 0.127, 90, 0.0635
```

Then, remove the reference to the dummy material

```
*Material, name=Material-1
```

*Then, before *Step, insert this line to force Abaqus to set the initial crack density to a very small value, which is set internally in the plugin to 0.02.*

```
*Initial Conditions, type=SOLUTION, USER
```

If in doubt, compare the dummy input file 'Job-1.inp' with the modified file 'Job-2.inp', both available in [2, Ex_11.1].

Next, in Module: Job, create a new Job based not on 'Model-1' but on 'Job-2.inp'.

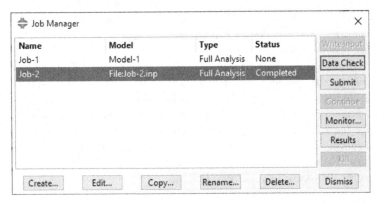

Make sure that Job-2 uses the user subroutine 'ddm6tm-2017-std.obj' and submit it.

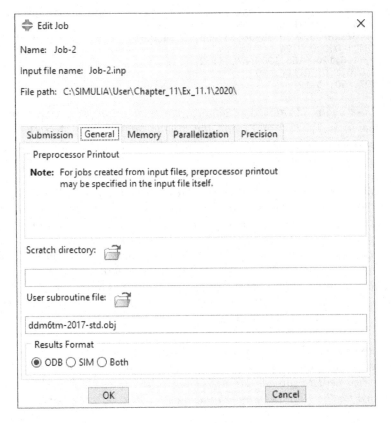

It will take a while to complete SFT-Tmin frames while advancing in -1 C increments. In Module: Job you can click 'Monitor' to monitor the progress.

For post-processing you can use the last part of the script in Section 12.7.2 or do the following. In Module: Job, click 'Results' so that the results are loaded from the output database .odb and the focus is switched to Module: Visualization. In the Field Output Toolbar, select primary, SDV1

SDV1, 13, 25, and 37 are the crack densities in laminas $k = 1, 2, 3, 4$. In a laminated composite with LSS $[0, 90, 0, 90_{1/2}]_s$. The outside lamina ($k = 1$) damages first because it is an external lamina. See Section 11.6.1 for the meaning of the twelve SDV per layer that are available for visualization.

Now select SDV1, then click Menu: Results, Step/frame. The current visualization should be for increment (frame) 333, which is the last frame. Double-click on increment 332 and SDV1 changes to 0.02, which is the initial damage. That means that the laminate does not start damaging until the last increment at temperature $T = -156$ C.

If we change the LLS to $[0, 90, 0, 90]_s$ the center lamina would damage at the same time as the external lamina because that would make the center lamina twice as thick, thus prone to damage early. To make this change, just edit the last value on the last line of *Shell General Section in Job-2.inp (see p.430), like this

```
** Section: Section-1
*Shell General Section, elset=WholeLaminate, USER, PROPERTIES=53, VARIABLES=48
1.016, 3.0
271270.586,-8.10997,1.18938E-02,6554.2638,-11.6689,4.9329E-04,3998.0213,-8.84364
6.1187E-03,0.3147,-6.9707E-05,-4.0521E-07,0.5557,-1.0089E-04,-1.1402E-06,-0.9766711
-9.0549E-05,-7.6732E-06,2.3038E-08,38.4684,8.9456E-02,-3.6454E-04,0.0,0.18187
0.0,0.0,24,-184,1,0.0,0.0,24
-184,  177,120,-156,120,-156,1900.87,441.2
57.23,8.9,0.0,0.0,1,0,  0.127,  90
0.127,  0,  0.127,  90,  0.127
```

and submit Job-2 again. In this case damage starts at frame 329, simultaneously in the exterior lamina ($k = 1, SDV = 1$) and the middle lamina ($k = 4, SDV = 37$) and in subsequent frames they both keep growing at the same rate. The other two laminas ($k = 2, 3, SDV = 13, 25$) do not damage.

Example 11.2 Compute the crack density in the external layer ($k = 1$) of a $[0/90/0/90_{1/2}]_s$ flat laminate, dimensions 100 by 100 mm, when subjected to $N = 10$ cycles of temperature excursion from stress free temperature $SFT = 177$ C to minimum temperature $T_{min} = -156$ C. Ply thickness is $t = 0.127$ mm. The elastic properties are described with $P = a + bT + cT^2$ as a function of the temperature T. The thermal properties are described with $P = a + bT + cT^2 + dT^3$. The fracture properties are described with $P = a + bT + cT^2$. The coefficients are given in Table 11.4. The defect nucleation rates are $\beta_I = \beta_{II} = -0.204$. Elastic and thermal properties are known in the range $-156 \leq T \leq 120$, and thus they are assumed to be constant outside that range.

Solution to Example 11.2 The solution to this example is similar to the solution of Example 11.1 with some data changes. Specifically, the number of cycles is $N = 10$, and because $N > 1$ we need to provide the defect nucleation rates β_I, β_{II}. Then, the analysis will be performed for the 10th thermomechanical cycle. The defect nucleation function $f(N)$

is calculated internally with (11.9) and with that the fatigue fracture toughness is calculated with (11.7).

As far as modeling is concerned, the procedure is similar to Example 11.1. We can either modify Job-2.inp from Example 11.1 and re-submit it, or use the script Ex_11.2.py, or look at Ex_11.2.cae, all of which are available in [2].

The only difference between Example 11.1 and 11.2 is the next to last line in the *Shell General Section. Below is the data in Example 11.1, where some lines have been omitted because they are identical in both examples

```
*Shell General Section,elset=WholeLaminate,USER,PROPERTIES=53,VARIABLES=48
0.889, 3.0
... omitted
... omitted
... omitted
... omitted
... omitted
57.23,8.9, 0, 0, 0, 0, 0.127, 90
0.127, 0, 0.127, 90, 0.0635
```

In the data for Example 11.2 below, notice that the numbers $-0.204, -0.204, 10$, appear in the next to last line, representing β_I, β_{II}, N

```
*Shell General Section,elset=WholeLaminate,USER,PROPERTIES=53,VARIABLES=48
0.889, 3.0
... omitted
... omitted
... omitted
... omitted
... omitted
57.23,8.9,-0.204,-0.204,10,0, 0.127, 90
0.127, 0, 0.127, 90, 0.0635
```

In CAE, once the Job is completed, click 'Results' to load the ODB and switch the focus to Module: Visualization. Now select SDV1, then click Menu: Results, Step/frame. The current visualization should be for increment (frame) 333, which is the last frame. Double-click on increment 285 and SDV1 changes to 0.02, which is the initial damage.

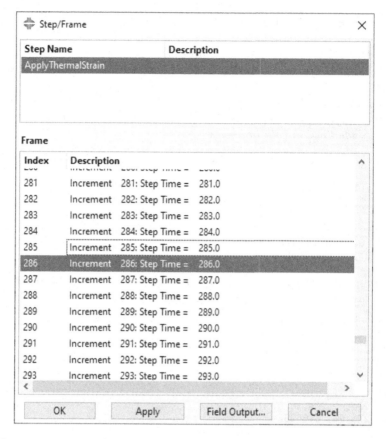

Now advance to increment 286, and SDV1 changes to 0.3594. That means that the laminate does not start damaging until increment 286. Since each increment represents 1 C, starting from $SFT = 177$, increment 286 represent $T = 177 - 286 = -109$ C. This means that at $T = -109$ C with $N = 10$ cycles the exterior lamina ($k = 1$) starts damaging. As you move forward with

the damage accumulates until it reaches $SDV1 = 1.102$ at frame 333 ($T = -156$).

11.7 Thermomechanical Equivalence

Thermomechanical equivalence (TME) means that there is a mechanical strain function of time that produces the same fatigue damage accumulation at room temperature ambient (RTA) as a thermal cycling program. Thermal fatigue can be done at about 4-5 cycles/hr (10^{-3} Hz). Mechanical fatigue test can be 1000 times faster. Thus, TME [216, Ch. 6] may be the only practical way to collect high-cycle thermal fatigue data. For a thermal fatigue test program with fixed $R = T_{min}/T_{max}$, the objective of thermomechanical equivalence is to calculate the equivalent mechanical strain amplitudes at RTA that produce the same fatigue damage as thermal fatigue. Then, a mechanical fatigue program can be designed so that crack density under thermal cyclic load is equal to crack density under equivalent mechanical strains for all cycles N in all laminas $k = 1 \ldots M$ in the laminate, where N, M, are the number of cycles and laminas, respectively.

11.7.1 Damage Equivalence

The objective of thermomechanical equivalence is to load a specimen with a fluctuating mechanical strain $\epsilon^M(N, t)$, where $0 < t < \tau$ is the elapsed time during cycle N, with period $\tau = 1/f$, of frequency f in Hertz. Mechanical testing would be done at constant temperature T_R, such as room temperature at ambient conditions (RTA). With this mechanical strain program applied, one tries to obtain fatigue damage evolution $\lambda^M(N, t)$ similar to that obtained with an alternating thermal load $T(t)$, with $T_L < T < T_H < SFT$, where T_L, T_H, are the minimum and maximum temperatures of each cycle, and SFT is the stress free temperature, usually approximated by the glass transition temperature of the polymer matrix T_g.

The best outcome would be to obtain the same damage evolution with either thermal or mechanical load, i.e.,

$$\lambda^M(N, \epsilon^M(t)) = \lambda^T(N, T) \qquad \forall N \qquad (11.14)$$

where superscripts T, M, denote thermal and mechanical loading, respectively. According to the fatigue damage criterion proposed in (11.6), cracks appear when the Griffith-Irwin criterion is satisfied, where $G_I, G'_{IC}, f(N)$, are the strain energy release rate ERR, fatigue fracture toughness, and defect nucleation functions, respectively. Since (11.6) is a necessary condition for calculating the damages λ^M and λ^T, and $G_I(N)$ is easier to calculate than λ, we propose to replace (11.14) by

$$G_I^M(N, \epsilon^M(t)) = G_I^T(N, T) \qquad \forall N \qquad (11.15)$$

and use (11.15) as part of the cost function (residual) in a minimization algorithm to find the equivalent mechanical strain ϵ^M.

11.7.2 Defect Nucleation Equivalence

Thermomechanical fatigue of laminated composites is controlled by two phenomena: energy and stress. On one hand, ERR, provided by the laminate while undergoing relaxation due to the crack formation, must exceed the fracture toughness of the material for the crack to propagate. On the other hand, cyclic stress precipitate nucleation of smaller defects into larger defects that become seed sites for new cracks. Defect nucleation, which is prerequisite for fatigue damage, is driven by stress [218, 219].

According to (11.7), the fatigue fracture toughness can be decomposed into quasi-static fracture toughness G_{IC} and defect nucleation function $f(N)$. The quasi-static fracture toughness G_{IC} can be approximated as a constant over the range of temperatures $[T_L, T_H]$ [199, Figure 3], but the temperature dependence of moduli and CTE must be included in the analysis.

On the left-hand side of the fracture criterion (11.6), the ERR G_I is forced by (11.15) to be the same on the thermal and mechanical cycles. On the right hand side, the quasi-static fracture toughness G_{IC} is an invariant material property. Therefore, to assure thermomechanical equivalence, it only remains to enforce

$$f^M(N) = f^T(N) \tag{11.16}$$

i.e., that the value of the defect nucleation function $f(N)$ be similar for thermal and mechanical loading. Since defect nucleation is driven by stress, (11.16) can be achieved if the stress history $\sigma_2(N,t)$ under thermal and mechanical load are similar, i.e.,

$$\sigma_2^M(N,t) = \sigma_2^T(N,T) \qquad \forall N \tag{11.17}$$

where $\sigma_2^M(N,t)$ is produced by a fluctuating strain program $\epsilon^M(N,t)$.

The defect nucleation function can be approximated by a linear equation in semi-log space (11.9), where β is the rate of defect nucleation with respect to $\log N$.

Since G_{IC} is constant, and the resulting G'_{IC} is the determinant of crack propagation in (11.6), it follows that $f(N)$ is effectively responsible for the rate of fatigue damage accumulation as per (11.9). Small values of β imply slow decrement of $G'_{IC}(N)$ and vice versa. Therefore, while (11.6) signals damage onset, (11.9) controls the rate of damage.

Consequently, (11.15) must be enforced together with (11.17) in order to achieve thermomechanical equivalence, i.e., to obtain a strain program $\epsilon^M(N,t)$ that produces similar damage accumulation as a thermal fatigue program.

11.8   Software Implementation

To simultaneously enforce (11.15) and (11.17), we propose to use minimization algorithm with a cost function (residual) defined by

$$R(N,t) = \left[\sum_{j=1}^{2N_L} w_j \left(\frac{\mu_j(N,t) - \theta_j(N,T)}{\theta_j(N,T)} \right)^2 \right]^{1/2} \tag{11.18}$$

where t is the time inside cycle N and $2N_L$ is the number of laminas. The component array for mechanical effects is

$$\mu_j = \left\{ \begin{array}{c} G_I^M(N, \epsilon(t), \lambda_{j-1}) \\ \sigma_2^M(N, \epsilon(t), \lambda_{j-1}) \end{array} \right\} \tag{11.19}$$

and for the thermal effects

$$\theta_j = \left\{ \begin{array}{c} G_I^T(N, T, \lambda_{j-1}) \\ \sigma_2^T(N, T, \lambda_{j-1}) \end{array} \right\} \tag{11.20}$$

from which the optimizer, by minimizing the residual $R(N,t)$, finds the optimal equivalent strain $\epsilon(N,t)$. Minimizing the difference between mechanical (11.19) and thermal response (11.20) is a multi-objective optimization with $2N_L$ objectives. Equation (11.18) reduces the multi-objective problem to a single-objective problem with a single residual R.

The weights w_j are normalized so that $\sum_{j=1}^{2N_L} w_j = 1$. Furthermore, all the weights for each class of response, ERR G_I or stress σ_2, are equal. For ERRs, $w_j = w_{j-1}$; $j = 2 \ldots N_L$, and for Stress, $w_j = w_{j-1}$; $j = N_L+1 \ldots 2N_L$. Therefore, all the weights are related by $w_j = N_L^{-1} - w_{j+N_L}$; $j = 1 \ldots N_L$. That means that only one weight, say w_0 needs to be minimized together with $\epsilon(N,t)$ using the Nelder-Mead algorithm [225]. The residual is minimized for each cycle $N = 1 \ldots N_{max}$.

If only the center lamina undergoes fatigue damage (as in next section), the optimization process is much more efficient because $N_L = 1$. Then, only two objective functions μ_j, θ_j, and two weights $w_j, j = 1 \ldots 2N_L$ need to be considered.

11.9 Uniaxial Mechanical Test

TME cannot be achieved simultaneously in all laminas because thermal cooling shrinks the laminate in all directions (controlled by the laminate CTE) while uniaxial mechanical loading introduces a uniaxial strain along the load direction, and Poisson's strains in the transverse direction, which cannot mimic the thermal strains. Therefore, we design the specimen so that fatigue damage is predominant in only one lamina, namely in the center lamina of a symmetric laminate. In this way, we can aim at achieving TME in the only damaging lamina.

For $N = 1$, a mechanical cycle $\epsilon^M(1,t)$ can be obtained that produces good comparison between mechanical and thermal damage $\lambda(1,T)$ in Figure 11.10, ERR $G_I(1,T)$ in Figure 11.11, and satisfactory stress $\sigma_2(1,T)$ in Figure 11.12. All graphs are shown in terms of temperature due to the fact that for every temperature T on the thermal cycle, there is a corresponding time t on the mechanical cycle, even if the actual period of those are different, with $\tau^T \gg \tau^M$ due to heat transfer being much slower than mechanical loading.

Note that if ERR agree perfectly, as in Figure 11.11, then crack density agree perfectly as well, as in Figure 11.10, thus validating our approach of using equality of ERRs (11.15) to enforce equality of crack densities (11.14). Usually, the shape of the thermal cycle is bi-linear, with a cooling ramp followed by a heating ramp. Therefore, the mechanical cycle should be bi-linear as well, albeit of much higher frequency.

In Figure 11.12, the comparison of thermally-induced vs. mechanically-induced stress is not perfect because the material properties are temperature dependent. Once crack densities, and thus ERRs, have been set to be equal for every temperature $T(t)$ and its corresponding mechanical strain $\epsilon(t)$, the transverse stress σ_2 on the thermal and mechanical cycles cannot be equal because the material properties are only equal at the temperature at which the mechanical test is conducted, namely T_R. To solve this problem, we realize that stress equality (11.17) is most important at the minimum temperature T_L where the maximum damage occurs at every cycle. Therefore, we propose to adjust the thickness h^M of the damaging lamina in the mechanically-loaded specimen to a different value than the thickness h^T of the same lamina in the thermally-loaded specimen. Assuming that h^T is known, we only have to adjust h^M by adding it as an unknown to the mechanical component

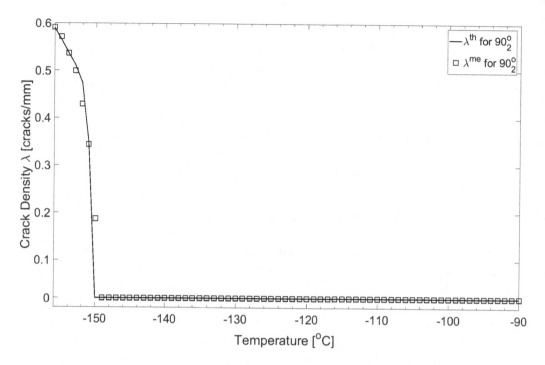

Fig. 11.10: Comparison between crack density $\lambda_{90_2}^T$ at $R_T = -156/121$ and crack density $\lambda_{90_2}^M$ subjected to uniaxial mechanical strain ϵ^M at RTA for P75/1962 $[(0/90)_2]_S$ with $h^T = t$. Reproduced from [216] © 2018.

of the residual (11.19), as follows

$$\mu_j = \left\{ \begin{array}{c} G_I^M(N, \epsilon(t), h^M, \lambda_{j-1}) \\ \sigma_2^M(N, \epsilon(t), h^M, \lambda_{j-1}) \end{array} \right\} \tag{11.21}$$

from which the optimizer, by minimizing the residual (11.18), finds the optimal equivalent strain $\epsilon(N, t)$ and equivalent thickness h^M.

For a thermally loaded specimen with LSS $[(0/90)_2]_S$, we have $h^T = 2h$, where h is the ply thickness. The optimizer finds the thickness of the middle lamina for the mechanically loaded specimen to be $0.7\,h$. The improvement in thermomechanical stress is evident in Figure 11.13, compared to Figure 11.12 and the calculation of ERR remains accurate as in Figure 11.11.

The equivalent mechanical strain $\epsilon^M(N, t)$ can be represented as $\epsilon^M(N, T)$ as well, because there is a univocal relationship between time t and temperature T, i.e., $t = T f^T f^{M-1}$, where f^t, f^M, are the thermal and mechanical excitation frequencies, respectively. Practically, f^M/f^T could be as high as 10^3, which is the *acceleration factor* for the proposed thermomechanical equivalence process. Due to this relationship, it is possible to plot the calculated mechanical strain program as a function of temperature and number of cycles, as in Figure 11.14.

As the temperature decreases (from right to left in Figure 11.14), the mechanically imposed longitudinal tensile strain ϵ_x on the specimen must increase to induce a transverse tensile strain ϵ_2 on the $90_{2h\,M}$ center lamina that is not equal, but

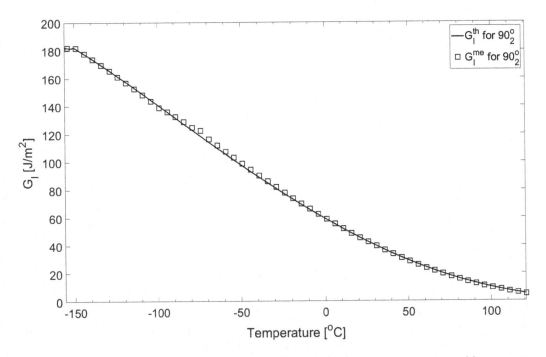

Fig. 11.11: Comparison between ERR G_I^T at $R_T = -156/121$ and G_I^M at RTA for middle 90_2 lamina subjected to equivalent mechanical strain ϵ^M for P75/1962 $[(0/90)_2]_s$ with $h^T = t$. Reproduced from [216] © 2018.

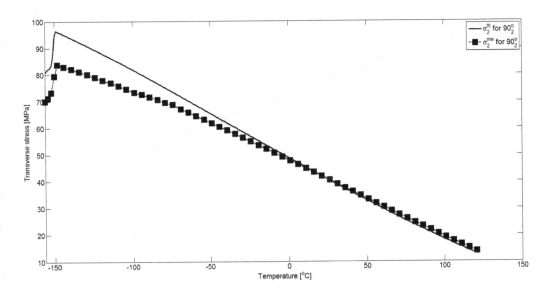

Fig. 11.12: Comparison between transverse stress σ_2^T at $R_T = -156/121$ and σ_2^M at RTA for middle 90_2 lamina subjected to equivalent uniaxial mechanical strain ϵ^M for P75/1962 $[(0/90)_2]_S$ with $h^T = t$. Reproduced from [216] © 2018.

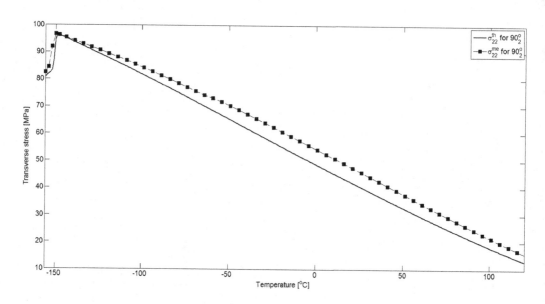

Fig. 11.13: Comparison between transverse stress σ_2^T at $R_T = -156/121$ and σ_2^{me} subjected to uniaxial equivalent mechanical strain ϵ^M at RTA for P75/1962 $[(0/90)_2]_S$ with $h^T = 0.70\,t$. Reproduced from [216] © 2018.

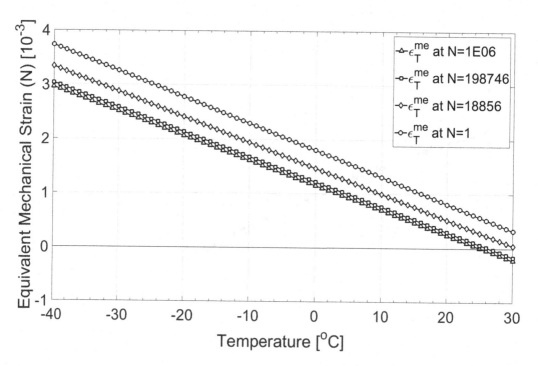

Fig. 11.14: Equivalent mechanical strain ϵ^M at discrete number between cycles N in the range $[-40, 30^oC]$ for P75/1962 $[0/90/0/90_{0.87t}]_S$, reference temperature $T_r = 30^oC$. Reproduced from [216] © 2018.

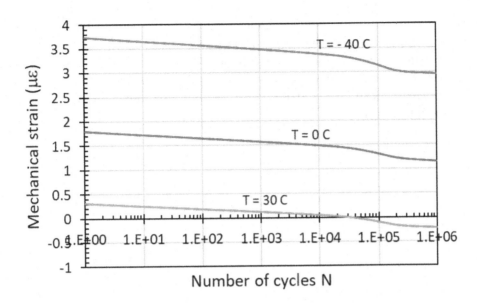

Fig. 11.15: Equivalent mechanical strain ϵ^M at discrete temperatures T for P75/1962 $[0/90/0/90_{0.87t}]_S$, reference temperature $T_r = 30^\circ C$.

thermomechanically equivalent to the tensile thermal strain produced by cooling on the 90_{2h^T} center lamina, when the material with CTE $\alpha_2 > \alpha_1$ cools down in a thermal cycle. The same data is shown in Figure 11.15 to highlight the dependence of the mechanical strain with the number of cycles (log scale), for three values of temperature. For each of the constant-temperature curves, the mechanical strain necessary to simulate thermal fatigue decreases with the number of cycles N to account for the stiffness reduction caused by damage accumulation.

Sometimes we may not be able to fabricate the reduced-thickness laminate using pre-preg that has constant ply thickness. That would require special design of the two laminates so that they can be fabricated, one to be subjected to thermal load and the other to mechanical load. For example, if the optimizer calculates, say, $h^M = 0.75 \, h^T$, we could change the LSS from $[(0/90)_2]_S$ to $[0_2/90_2]_S = [0_2/90_4/0_2]_T$, so that $h^T = 4\,h$, then $h^M = 4 \times 0.75h = 3\,h$, and the mechanical laminate would be $[0_2/90_3/0_2]_T$.

While designing the experimental program one must be aware that the mechanical strain $\epsilon^M(N)$ is calculated to satisfy TME, and thus some values of strain may be largely positive (possibly exceeding the longitudinal tensile strength of the 0° laminas) or negative (possibly exceeding the compression strength of the 0° laminas). Therefore, one must try not to exceed the tensile and/or longitudinal compression strength of the longitudinal laminas (along the loading direction), which may occur at T_L and T_H, respectively. If that happens, selecting a lower value of thermal ratio $R_T = T_L/T_H$ can solve the problem. Then, MPL (11.12) can be used to transform the findings to different temperatures than those used for the experiments.

Another issue to be aware of is that compression strains may be needed to simulate thermal cooling. If $T_H > RTA$ in the thermal test, compression strains will be needed in the equivalent mechanical test, if performed at RTA conditions. That could complicate the design of the specimen, due to potential buckling. The solution is to minimize the need for compression strain, which can be accomplished by raising the reference temperature from RTA to T_H. This may require performing the mechanical tests at elevated temperature, but virtually eliminates potential buckling of the specimen.

Chapter 12

Abaqus Programmable Features

Several examples in previous chapters are solved with the aid of Abaqus programmable features such as UMAT and UGENS. Because of the complexity of Fortran and Python coding, the programming required by those examples is deferred to this chapter. Programmable features require additional software such as compilers and libraries, necessitating detailed installation instructions that are included in Appendix E. In this chapter, detailed descriptions are given on how to write the Fortran code for UMAT and UGENS in various flavors depending on the type of elements required (1D, plane stress, 3D solid) and implementing new capabilities that native Abaqus does not have.

Furthermore, details about how to provide the augmented Abaqus solver with the data needed by these new capabilities and how to visualize their novel results are provided. All of this is illustrated by examples including custom-coded failure criteria, orthotropic viscoelasticity, continuum damage mechanics (CDM), non-local constitutive models (DDM), and fatigue.

In addition, interaction between Abaqus CAE and MATLAB is best done by using Python scripts that CAE can execute. This is illustrated with examples on how to write constraint equations (CE), calculate volume average of stresses and strains, simulate discrete damage mechanics, and evaluate thermal fatigue.

12.1   User Materials in Abaqus Standard

A user material subroutine (UMAT) represents a local constitutive equation. That is, the constitutive equation has no way of knowing or taking into account what happens to the material at any other point in model other than at the exact coordinates of the Gauss point at which the constitutive equation is being used. This is in contrast to UGENS where the constitutive model at a Gauss point in a conventional shell element does take into account what happens to all the laminas through the thickness of the shell (Section 12.7).

12.2   UMAT for Linear Elastic Shells

Failure criteria for a unidirectional lamina apply only in the linear elastic range. Once the lamina starts damaging, subsequent damage cannot be predicted using failure criteria (Chapter 9). For a laminate, the so-called *failure criteria* do not predict *failure* but rather *damage initiation*, because for a well designed LSS, damage in one lamina leads to stress redistribution to other laminas and the laminate does not *fail* [3, Chapter 5].

Even though Abaqus provides computation of several failure criteria, we develop UMATS8R5.for [2] to compute Tsai-Wu and truncated maximum strain (TMS) failure criteria for Ex. 3.8 (p. 158)[1] as an illustration on how to code failure criteria. The elastic moduli and strength values for the lamina are given in Table 3.1, p. 108. A failure index is the inverse of the safety factor R (called strength ratio in [3]). The load capacity P_c of a structure that is analyzed with a nominal load P_n is

$$P_c = R \, P_n = P_n/F_I \tag{12.1}$$

where R is the so called safety factor and F_I is the failure index.

In allowable stress design (ASD), a design is considered adequate if $R > 1$ or $F_I < 1$. In terms of damage, the non-damaging domain is described by $R > 1$ or $F_I < 1$. Likewise, the damage domain (or plastic domain for yielding materials) is described by $R < 1$ or $F_I > 1$. In Chapters 8 and 9, the damage surface is at $g = 0$, the non-damaging domain is at $g < 0$, and the region $g > 0$ is forbidden. With this in mind, we develop a UMAT to calculate both R_I and F_I.

12.2.1   User Subroutine Interface

The most important thing about writing a UMAT is to respect the user subroutine interface, which has to match exactly what Abaqus expects. The following is taken from the documentation for Release 2020.

```
      SUBROUTINE UMAT(STRESS,STATEV,DDSDDE,SSE,SPD,SCD,
     1 RPL,DDSDDT,DRPLDE,DRPLDT,
     2 STRAN,DSTRAN,TIME,DTIME,TEMP,DTEMP,PREDEF,DPRED,CMNAME,
     3 NDI,NSHR,NTENS,NSTATV,PROPS,NPROPS,COORDS,DROT,PNEWDT,
     4 CELENT,DFGRD0,DFGRD1,NOEL,NPT,LAYER,KSPT,JSTEP,KINC)
C
      INCLUDE 'ABA_PARAM.INC'
C
      CHARACTER*80 CMNAME
      DIMENSION STRESS(NTENS),STATEV(NSTATV),
     1 DDSDDE(NTENS,NTENS),DDSDDT(NTENS),DRPLDE(NTENS),
     2 STRAN(NTENS),DSTRAN(NTENS),TIME(2),PREDEF(1),DPRED(1),
     3 PROPS(NPROPS),COORDS(3),DROT(3,3),DFGRD0(3,3),DFGRD1(3,3),
     4 JSTEP(4)
...
      user coding to define DDSDDE, STRESS, STATEV, SSE, SPD, SCD
      and, if necessary, RPL, DDSDDT, DRPLDE, DRPLDT, PNEWDT
...
      RETURN
      END
```

---

[1]Compare the results with the native-Abaqus implementation of the same. However, Tsai-Hill, Azzi-Tsai-Hill, and Tsai-Wu are not recommended because they overemphasize the interaction between fiber and transverse matrix damage modes.

Beware that Abaqus may alter this interface in future releases, and if so, your UMAT will no longer work. Inside your UMAT, you can change the names of variables but you cannot change their type (REAL, REAL*8, INTEGER, CHARACTER, etc.), their dimension, the shape of arrays, etc. Everything has to match what the documentation shows.

12.2.2   State Variables and Constants

You can add your own variables to your version of UMAT, which will be local to your UMAT, thus unknown to the rest of Abaqus, and their content will be lost every time the execution exits through the RETURN statement. The only way to keep information live for the next time the execution comes back to the UMAT (for the same Gauss point) is to store it into *state variables* STATEV. Abaqus allocates storage for the STATEV for each Gauss point in the model. The number of STATEV is what you declare for DEPVAR in Figure 12.1 or what you declare

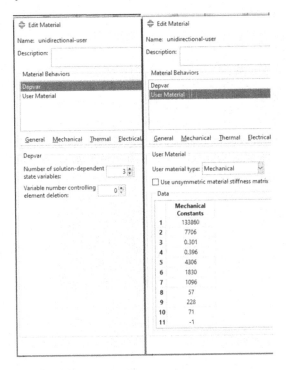

Fig. 12.1: Edit material for UMAT.

in Module: Property, inside the input file, as follows

```
*Material, name=unidirectional-user
*Depvar
      3,
```

In this example DEPVAR = 3. Besides state variables, we also need *user material parameters*. The values of the user material parameters that one may need to code an UMAT come into the UMAT via the array PROPS(NPROPS). These are read by Abaqus from the input file, as follows

```
*User Material, constants=11
133860., 7706., 0.301, 0.396, 4306., 1830., 1096.,   57.
   228.,    71.,    -1.
```

NPROPS is the value of **constants** and the NPROPS constants go into PROPS. In this case **constants** = 11. The constants may include material properties, geometrical properties, etc. In this case, they are five moduli (E1, E2, PR12, PR23, and G12) and six values of strength (F1T, F1C, F2T, F2C, F6, F12), as you can see in the full code described in Section 12.2.4.

12.2.3   User's Code

Write your code using the minimum number of variables needed to accomplish your objective. In this case, we need to calculate the index failure F_I. The Tsai-Wu criterion is defined by equations (3.34–3.36) p.173. Those equations are coded, as follows

```
!     TSAI-WU FAILURE criterion
!
      S_1   = STRESS(1)
      S_2   = STRESS(2)
      S_6   = STRESS(3) !PLANE STRESS
      A = S_1**2./(F1T*F1C)+S_2**2./(F2T*F2C)+&
   &    S_6**2./(F6)**2+ F12*S_1*S_2/SQRT(F1T*F1C*F2T*F2C)
      B = (1./F1T-1/F1C)*S_1 + (1./F2T-1/F2C)*S_2
      R = HUGE(R) !If R->infty, F_I->0
      IF (A.GT.EPS) R = -B/2./A + SQRT((B/2./A)**2.+1./A)
      F_I   = 1/R
!
!     STORE FAILURE INDEX & STRENGTH RATIO IN STATE VARIABLE ARRAY
!
      STATEV(1) = F_I
      STATEV(2) = R
```

We can see above that the results, F_I and R, are stored into state variables. These are called **STATEV** in the notation used in the UMAT interface, but called SDV in Module: Visualization. For example, the contour plot of failure index F_I, which is STATEV(1) in the code, is shown in Figure 12.2 by selecting SVD1 in the Field Output Toolbar

The calculation needs the in-plane stress components **S_1,S_2,S_6**, which are local to your UMAT. The STRESS values are also calculated by the code in terms of DSTRAN (the increment of strain) provided by Abaqus. Since this UMAT is used when the material is intact (no damage) the analysis consist of a single linear step and the only strain is DSTRAN. The elastic stiffness matrix DDSDDE is calculated in terms of material properties, as follows

```
!     ELASTIC STIFFNESS
!
      DO 20 K1=1,NTENS
         DO 10 K2=1,NTENS
            DDSDDE(K2,K1)=0.0D0
10       CONTINUE
```

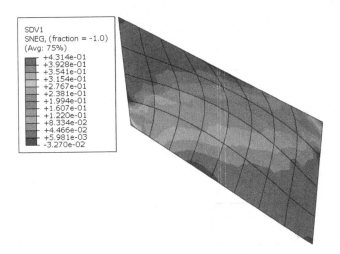

Fig. 12.2: Contour plot of failure index F_I (SDV1). The extreme values are exaggerated by extrapolation to the nodes. For better results, increase mesh refinement or use a quilted contour plot.

```
20      CONTINUE
!       PR21  = PR12*E2/E1
        IF (NDI.EQ.2) THEN
           DDSDDE(1,1) = -1/(-E1+PR12**2.*E2)*E1**2.
           DDSDDE(1,2) = -PR12*E1/(-E1+PR12**2.*E2)*E2
           DDSDDE(2,1) = DDSDDE(1,2)
           DDSDDE(2,2) = -E1/(-E1+PR12**2.*E2)*E2
           DDSDDE(3,3) = G12
           IF (NSHR.GT.1) THEN
              DDSDDE(4,4) = G13
              DDSDDE(5,5) = G23
           ENDIF
        ELSE
           WRITE (6,3)
        ENDIF
3       FORMAT(1x,'ERROR-UMAT: ONLY PLANE STRESS IMPLEMENTED')
!
!       CALCULATE STRESS FROM ELASTIC STRAINS
!
        DO 70 K1=1,NTENS
           DO 60 K2=1,NTENS
              STRESS(K2)=STRESS(K2)+DDSDDE(K2,K1)*DSTRAN(K1)
60         CONTINUE
70      CONTINUE
```

To know what values are provided by Abaqus and what values are to be calculated by the UMAT, all we have to do is look at the documentation, which for this example is shown in Section 12.2.1, from which we highlight the following

```
user coding to define DDSDDE, STRESS, STATEV, SSE, SPD, SCD
and, if necessary, RPL, DDSDDT, DRPLDE, DRPLDT, PNEWDT
```

12.2.4   UMATS8R5.FOR

UMATS8R5.FOR is called from Module: Job, tab General, as shown in Figure 12.3 and used in Example 12.1.

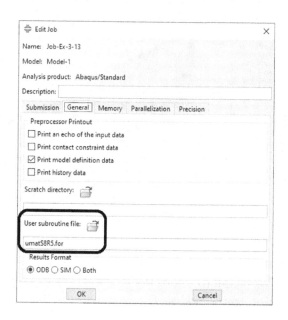

Fig. 12.3: Link to the user subroutine file.

The complete Fortran code is reproduced below

```
      SUBROUTINE UMAT(STRESS,STATEV,DDSDDE,SSE,SPD,SCD,&
     & RPL,DDSDDT,DRPLDE,DRPLDT,&
     & STRAN,DSTRAN,TIME,DTIME,TEMP,DTEMP,PREDEF,DPRED,CMNAME,&
     & NDI,NSHR,NTENS,NSTATV,PROPS,NPROPS,COORDS,DROT,PNEWDT,&
     & CELENT,DFGRD0,DFGRD1,NOEL,NPT,LAYER,KSPT,JSTEP,KINC)
!     COPYRIGHT (2012,2020,2021) EVER J. BARBERO
!     Ex. 12.1 Finite Element Analysis of Composite Materials
      INCLUDE 'ABA_PARAM.INC'
      PARAMETER (EPS=2.22D-16) !SMALLEST NUMBER REAL*8 CAN STORE
!
      CHARACTER*80 CMNAME    !CMNAME<=MATERL in R10
      DIMENSION STRESS(NTENS),STATEV(NSTATV),&
     & DDSDDE(NTENS,NTENS),DDSDDT(NTENS),DRPLDE(NTENS),&
     & STRAN(NTENS),DSTRAN(NTENS),TIME(2),PREDEF(1),DPRED(1),&
     & PROPS(NPROPS),COORDS(3),DROT(3,3),DFGRD0(3,3),DFGRD1(3,3),&
     & JSTEP(4)
!
! -----------------------------------------------------------
!     UMAT FOR SHELL ELEMENTS S8R5, F95 FREE FORMATTED CODE
!     F77 IMPLICIT NAME CONVENTION,
!     SINGLE/DOUBLE CONTROLLED BY ABAQUS
! -----------------------------------------------------------
!     NDI: # of direct components (11,...)
!        in DDSDDE, DDSDDT, and DRPLDE
!     NSHR: # of engineering shear components (12,...)
!        in DDSDDE, DDSDDT, and DRPLDE
!     NTENS = NDI + NSHR: Size of the stress or strain matrix
! -----------------------------------------------------------
!
      E1         = PROPS(1)
      E2         = PROPS(2)
      E3         = E2
      PR12       = PROPS(3)
      PR13       = PR12
      PR23       = PROPS(4)
      G12        = PROPS(5)
      G13        = G12
```

```
        G23       = E2/2/(1.+PR23)
!
        F1T       = PROPS(6)    ! Strength 1 tension
        F1C       = PROPS(7)    ! Strength 1 compresion positive value
        F2T       = PROPS(8)    ! Strength 2 tension
        F2C       = PROPS(9)    ! Strength 2 compresion positive value
        F6        = PROPS(10)   ! Strength 12 shear
        F12       = PROPS(11)   ! Tsai-wu interaction coefficient
!
!       ELASTIC STIFFNESS
!
        DO 20 K1=1,NTENS
          DO 10 K2=1,NTENS
            DDSDDE(K2,K1)=0.0D0
10        CONTINUE
20      CONTINUE
!       PR21 = PR12*E2/E1
        IF (NDI.EQ.2) THEN
          DDSDDE(1,1) = -1/(-E1+PR12**2.*E2)*E1**2.
          DDSDDE(1,2) = -PR12*E1/(-E1+PR12**2.*E2)*E2
          DDSDDE(2,1) = DDSDDE(1,2)
          DDSDDE(2,2) = -E1/(-E1+PR12**2.*E2)*E2
          DDSDDE(3,3) = G12
          IF (NSHR.GT.1) THEN
            DDSDDE(4,4) = G13
            DDSDDE(5,5) = G23
          ENDIF
        ELSE
          WRITE (6,3)
        ENDIF
3       FORMAT(1x,'ERROR-UMAT: ONLY PLANE STRESS IMPLEMENTED')
!
!       CALCULATE STRESS FROM ELASTIC STRAINS
!
        DO 70 K1=1,NTENS
          DO 60 K2=1,NTENS
            STRESS(K2)=STRESS(K2)+DDSDDE(K2,K1)*DSTRAN(K1)
60        CONTINUE
70      CONTINUE
!
!       TSAI-WU FAILURE criterion
!
        S_1   = STRESS(1)
        S_2   = STRESS(2)
        S_6   = STRESS(3) !PLANE STRESS
        A = S_1**2./(F1T*F1C)+S_2**2./(F2T*F2C)+&
   &        S_6**2./(F6)**2+ F12*S_1*S_2/SQRT(F1T*F1C*F2T*F2C)
        B = (1./F1T-1/F1C)*S_1 + (1./F2T-1/F2C)*S_2
        R = HUGE(R) !If R->infty, F_I->0
        IF (A.GT.EPS) R = -B/2./A + SQRT((B/2./A)**2.+1./A)
        F_I   = 1/R
!
!       STORE FAILURE INDEX AND STRENGTH RATIO IN STATE VARIABLES
!
        STATEV(1) = F_I
        STATEV(2) = R
!
        RETURN
        END
```

Example 12.1 *Compute the Tsai-Wu failure index I_F for Example 3.8 (p. 158) using a UMAT subroutine, available in [2, umatS8R5.for]. Details of the UMAT implementation are given in Section 12.2. The elastic moduli and strength values for the lamina are given in Table 3.1, p. 108. The shell consists of a single orthotropic lamina oriented at 30° with respect to the X-axis.*

Solution to Example 12.1 *Environment variables must be set properly before executing a job with user-programmable features (see Appendix E). To solve this example, first open* Ex_3.8.cae *and save it as* Ex_12.1.cae. *Also, set the Work Directory to the local directory so that all the files are easily accessed.*

Menu: File, Set Work Directory, [C:\SIMULIA\User\Ex_12.1], OK
Menu: File, Open, [C:\SIMULIA\User\Ex_3.8\Ex_3.8.cae], OK
Menu: File, Save As, [C:\SIMULIA\User\Ex_12.1\Ex_12.1.cae], OK

Only a few things need to be modified in the model database (.cae). Remember that in Abaqus the stress/strain components for shell elements are ordered as follows: 11, 22, 12, 13, 23. Then, the section shear stiffness *can be calculated using (3.9). Alternatively, for a single ply shell of thickness t, use*

$$H_{44} = (5/6) \, t \, G_{23}$$
$$H_{45} = 0$$
$$H_{55} = (5/6) \, t \, G_{13}$$

or for a laminate use [3, (6.20,6.22)], or use Abaqus to calculate the values and read them from the .dat *file as illustrated by item (ii) next.*

i. *Modifying the material definition to add failure parameters*

Module: Property
unidirectional material from Ex. 3.8 used to calculate H44, H45, H55
Menu: Material, Create
    Name [unidirectional], Mechanical,
    Elasticity, Elastic, Type: Lamina
    [133860   7706   0.301   4306   4306   2760]
    # or right-click on Data field, Read from File
    Select, [uni_lam_prop.txt], OK, OK
    # add Fail Stress values
    Suboptions, Fail Stress
    [1830   -1096   57   -228   71   -1], OK
    # or right-click on Data field, Read from File
    Select, [uni_fail_stress.txt], OK, OK, OK,
    OK # to close Edit Material
Menu: Section, Edit, Section-1
    Tab: Basic, Material: unidirectional, OK

ii. *Submitting the job and visualizing the results*

Module: Job
Menu: Job, Manager
    Edit, Tab: General, # checkmark: Print model definition data, OK
    Submit, # when completed, Results
    Field Output Toolbar: Primary, U, U3, # U3max = 8.704 mm

Use a text editor to open the file Job-1.dat. *The file is stored in the* Work Directory *(C:\SIMULIA\User\Ex_12.1). Read the values corresponding to the Transverse Shear*

Stiffness for the Section. They should be

$$Kts11 = K11 = H55 = 35883.0$$
$$Kts22 = K22 = H44 = 23000.0$$
$$Kts12 = K12 = H45 = 0.0$$

To run the user material subroutine, do the following

i. *Create a user material to be used with the UMAT subroutine and edit the corresponding section*

```
Module: Property
Menu:   Material, Create
    Name [ud-user], General, User Material,
    User material type: Mechanical
    Mechanical constants, [133860   7706   0.301   0.396
    4306   1830   1096   57   228   71   -1]
    # or right-click on Data field, Read from File
    Select, [user_mat_props.txt], OK, OK
    # see umats8r5.for for interpretation of Mechanical constants
    General, Depvar,
    Number of solution-dependent state variables [2], OK
Menu:   Section, Edit, Section-1
    Tab: Basic
    Section integration: During analysis
    Thickness: Shell thickness: Value [10]
    Material: ud-user
    Thickness integration rule: Simpson
    Thickness integration points [3]
    Tab: Advanced
    Transverse Shear Stiffness: # checkmark: Specify values
    K11: [35883], K12: [0], K22: [23000], OK
```

ii. *Define output variables for the analysis step*

```
Module: Step
Step:   Step-1
Menu:   Output, Field Output Requests, Edit, F-Output-1
    Output variables: Edit variables: [S,E,U,RF,SDV], OK
```

iii. *Submit the job and visualize the results.*
Before submitting the job make sure your system is properly setup to run a UMAT (see Appendix E). Notice that if your system is not properly setup to run UMATs you will get an error after submitting the job for execution. See also Figure 12.3.

```
Module: Job
Menu:   Job, Manager
    Edit, Tab: General,
    User subroutine file: Select, [umatS8R5.for], OK, OK
    Submit, # when completed, Results
Module: Visualization
Menu:   Options, Contour, Contour Type: Quilt, OK
```

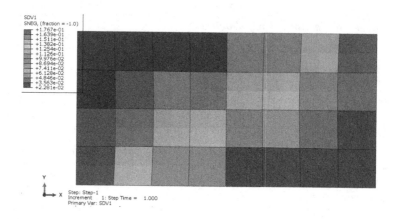

Fig. 12.4: Quilted contour of SDV1 in Example 12.1.

```
Menu:    Plot, Contours on Deformed Shape
      Field Output Toolbar: Primary, SDV1
Menu:    Result, Section Points, Selection method: Plies
      Ply result location: Bottommost, Apply
      Ply result location: Topmost, Apply
      OK # close the Section Points pop-up window
```

In case of an error, check that C:\SIMULIA\Abaqus\Commands\abq2020.bat (configuration file) is correct. Also, check that the following is included in the .inp file:

```
** Section: Section-1
*Shell Section, elset=_PickedSet2, material=uni-user, orientation=Ori-1
10., 3
*Transverse Shear
35883., 23000., 0.
```

and

```
*Material, name=uni-user
*Depvar
2,
*User Material, constants=11
133860., 7706., 0.301, 0.396, 4306., 1830., 1096.,    57.
228.,    71.,    -1.
```

*From the visualization of F_I (**SDV1**) on a quilted contour (Figure 12.4)*

```
Menu: Options, Contour, tab: Basic, Quilt, OK
```

the bottom-left of the laminate reaches $I_F = 1.767 \times 10^{-1}$ and the bottom-right reaches $I_F = 2.281 \times 10^{-2}$.

Example 12.2 *Compute the failure indexes I_F of the truncated maximum strain failure criterion for Example 3.8 (p. 158) using a UMAT subroutine, available in [2, umatTMS.for]. Details of a similar UMAT implementation are given in Section 12.2. The elastic moduli and strength values for the lamina are given in Table 3.1, p. 108. The shell consists of a single orthotropic lamina oriented at $30°$ with respect to the X-axis.*

Solution to Example 12.2 *Watch the video in* `https://youtu.be/JXYmeLeU9RU` *or [1, How to write an Abaqus UMAT]. The UMAT code is available in [2, Ex_12.2].*

The four failure indexes I_F of the truncated maximum strain failure criterion are:

SDV(1) : IF1 (fiber tensile failure index). Inverse of safety factor.

SDV(2) : IF2 (fiber compression failure index). Inverse of safety factor.

SDV(3) : IF3 (fiber shear failure index). Inverse of safety factor.

SDV(4) : IF4 (matrix cutoff index). Inverse of safety factor.

In the UMAT implementation presented in Section 12.2.4, replace the last portion of the code, starting with

```
!      TSAI-WU FAILURE criterion
```

by this code

```
!      Calculate strains to failure from strength values
!
       stran = stran + dstran
       call tmsfc (stran, F1t/E1, F1c/E1, F2t/E2, PR12, statev)
!
       RETURN
       END SUBROUTINE
```

and the following subroutine to actually perform the calculations of TMS.

```
subroutine tmsfc (strain, xet, xec, yet, nu12, SDV)
! evaluates the failure index (FI) for truncated max. strain
! [ref.] Section 7.3.2 in ISBN: 978-1-138-19680-3, CRC Press (2018)
! NOTE THE ORDER 1...4 TO INTERPRET RESULTS IN CAE LATER
! SDV (same as STATEV): failure index = 1 / safety factor
! SDV(1) : IF1 (fiber tensile failure index). Inverse of SF
! SDV(2) : IF2 (fiber compression failure index). Inverse of SF
! SDV(3) : IF3 (fiber shear failure index). Inverse of SF
! SDV(4) : IF4 (matrix cutoff index). Inverse of SF
! strain : is the current strain
! trunc. max. strain, no interlaminar stresses/strains
! ht : hygro-thermal strains
! xet : fiber tensile strain to failure
! xec : fiber compression strain to failure
! yet : lamina transverse tensile strain
implicit none
double precision strain(5), xet, xec, yet, nu12
double precision SDV(4)
double precision s, tmp, ht(4)
double precision, parameter :: zero=0.0D0
! hygro-thermal strains discounted from ultimate values
ht = zero        !but set them to zero for now
SDV = zero       !initialize when calculating I_F (like Abaqus)
! fiber direction f.c.
    if ( strain(1).gt.zero.and.(xet-ht(1)).ne.zero ) &
        SDV(1)= strain(1)/(xet-ht(1))
        !fiber tension, when strain is positive
    if ( strain(1).lt.zero.and.(xec-ht(1)).ne.zero ) &
        SDV(2)=-strain(1)/(xec+ht(1))
        !fiber compression, when strain is negative
! fiber shear cut-off, it is not a matrix cut off,
! but part of fiber failure
    s = (1+nu12)*max((xet-ht(1)),(xec+ht(1)))
    tmp = dabs(strain(1)-strain(2))
    if( tmp.gt.zero.and.s.ne.zero ) SDV(3)=tmp/s
    !fiber shear, on 2nd and 4th quadrants only
```

```
! matrix cut off on current layer
if ( strain(2).gt.zero.and.(yet-ht(1)).ne.zero )  &
    SDV(4)=strain(2)/(yet-ht(2))
    !matrix failure, when the transverse strain is positive
end subroutine
```

12.3   UMAT for Orthotropic Viscoelasticity

For linear viscoelasticity, if the material is isotropic you can use Abaqus directly, as in Examples 7.5 and 7.6 (p. 277-279). If not, we can develop a UMAT to implement orthotropic viscoelasticity [38].

12.3.1   User Subroutine Interface

The most important thing about writing a UMAT is to respect the user subroutine interface, which has to match exactly what Abaqus expects. The following is taken from the documentation for Release 2020.

```
      SUBROUTINE UMAT(STRESS,STATEV,DDSDDE,SSE,SPD,SCD,
     1 RPL,DDSDDT,DRPLDE,DRPLDT,
     2 STRAN,DSTRAN,TIME,DTIME,TEMP,DTEMP,PREDEF,DPRED,CMNAME,
     3 NDI,NSHR,NTENS,NSTATV,PROPS,NPROPS,COORDS,DROT,PNEWDT,
     4 CELENT,DFGRD0,DFGRD1,NOEL,NPT,LAYER,KSPT,JSTEP,KINC)
C
      INCLUDE 'ABA_PARAM.INC'
C
      CHARACTER*80 CMNAME
      DIMENSION STRESS(NTENS),STATEV(NSTATV),
     1 DDSDDE(NTENS,NTENS),DDSDDT(NTENS),DRPLDE(NTENS),
     2 STRAN(NTENS),DSTRAN(NTENS),TIME(2),PREDEF(1),DPRED(1),
     3 PROPS(NPROPS),COORDS(3),DROT(3,3),DFGRD0(3,3),DFGRD1(3,3),
     4 JSTEP(4)
...
      user coding to define DDSDDE, STRESS, STATEV, SSE, SPD, SCD
      and, if necessary, RPL, DDSDDT, DRPLDE, DRPLDT, PNEWDT
...
      RETURN
      END
```

In Abaqus, an analysis is often divided into steps $s = 1...ns$. Further, each step is often divided into increments $i = 1...ni$. Abaqus maintains a record of the (total) *time* elapsed t since the beginning of the analysis, including restart steps. In addition, Abaqus maintains a record of the time elapsed since the beginning of the step, called *step time* t_s.

With reference to Figure 12.5, the step time t_s at the beginning of the current increment (within the current step) is passed to the user subroutines in the variable TIME(1). The (total) time t at the beginning of the current increment is TIME(2). The current time increment ΔT is DTIME.

In this way the current step time can be calculated as

$$t_s = \text{TIME(1)} + \text{DTIME}$$

and the current (total) time can be calculated as

$$t = \text{TIME(2)} + \text{DTIME}$$

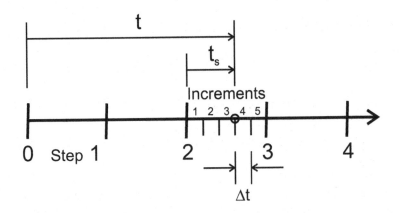

Fig. 12.5: Definition of step time t_s, total time t, and time increment Δt.

Linear viscoelastic analysis is linear but time-dependent, thus incremental. An increment of strain DSTRAN is calculated for each increment of time (Figure 12.5), and when the current increment is completed (or converged if the analysis were non-linear) the DSTRAN is added to the accumulated strain STRAN.

12.3.2 State Variables and Constants

Abaqus allocates storage for the state variables (STATEV) for each Gauss point in the model. The number of STATEV is what you declare for DEPVAR in Module: Property, or inside the input file:

```
*Material, name=user
*Depvar
      6,
*User Material, constants=10
102417., 11975.,  0.401, 0.1886, 5553.8, 5037.3, 16551., 58.424
 44.379, 54.445
```

Actually, no *true* state variables are needed for viscoelastic materials. Here we use state variables only to recover the resulting moduli vs. time, back to the Visualization module. This is a common practice in Abaqus. From a data management point of view, in Abaqus there is no difference between true state variables and placeholders for user results.

The material constants, 10 in this example, are read by Abaqus from the input file and stored inside PROPS(NPROPS). PROPS is then available to the UMAT code.

The 10 constants in this example are E1o, E2o, nu12o, nu23o, G12o, G23o, tau1, tau2, tau12, tau23. Actually, we need only nine properties because for an orthotropic material only three of nu12o, nu23o, G12o, G23o are needed, but we enter 10 because that is the way the CAE interface is built. You can see how these properties are assigned in Section 12.3.4.

12.3.3   User's Code

To implement the UMAT, first we recalculate the material properties using the Maxwell model:

```
!      Maxwell model
       E1 = E1o*dexp(-(Time(2)+DTime)/tau1)
       E2 = E2o*dexp(-(Time(2)+DTime)/tau2)
       G12 = G12o*dexp(-(Time(2)+DTime)/tau12)
       G23 = G23o*dexp(-(Time(2)+DTime)/tau23)
       PR12 = PR12o
       PR23 = PR23o
```

Then, update the stiffness matrix, with $NDI = 2$ for plane stress/strain and $NDI = 3$ for solids.

```
       IF (NDI.EQ.2) THEN
!          PR21 = PR12*E2/E1
           DDSDDE(1,1) = -1/(-E1+PR12**2.*E2)*E1**2.
    DDSDDE(1,2) = -PR12*E1/(-E1+PR12**2.*E2)*E2
           DDSDDE(2,1) = DDSDDE(1,2)
           DDSDDE(2,2) = -E1/(-E1+PR12**2.*E2)*E2
           DDSDDE(3,3) = G12
           IF (NSHR.GT.1) THEN
              DDSDDE(4,4) = G13
              DDSDDE(5,5) = G23
           ENDIF
         ELSEIF (NDI.EQ.3) THEN
!           calculate 3D stiffness
    DDSDDE(1,1) = E1**2*(PR23-1)/(E1*PR23-E1+2*PR12**2*E2)
    DDSDDE(1,2) = -E2*E1*PR12/(E1*PR23-E1+2*PR12**2*E2)
    DDSDDE(1,3) = -E2*E1*PR12/(E1*PR23-E1+2*PR12**2*E2)
    DDSDDE(2,1) = -E2*E1*PR12/(E1*PR23-E1+2*PR12**2*E2)
    DDSDDE(2,2) = E2*(-E1+PR12**2*E2)/(-E1+E1*PR23**2+2*PR12**2*E2 &
    & +2*PR12**2*E2*PR23)
    DDSDDE(2,3) = -E2*(E1*PR23+PR12**2*E2)/(-E1+E1*PR23**2        &
    &   +2*PR12**2*E2+2*PR12**2*E2*PR23)
    DDSDDE(3,1) = -E2*E1*PR12/(E1*PR23-E1+2*PR12**2*E2)
    DDSDDE(3,2) = -E2*(E1*PR23+PR12**2*E2)/(-E1+E1*PR23**2        &
    & +2*PR12**2*E2+2*PR12**2*E2*PR23)
    DDSDDE(3,3) = E2*(-E1+PR12**2*E2)/(-E1+E1*PR23**2+2*PR12**2*E2 &
    & +2*PR12**2*E2*PR23)
    DDSDDE(4,4) = G12    !Abaqus notation
    DDSDDE(5,5) = G12    !G13
    DDSDDE(6,6) = G23    !Abaqus notation
         ELSE
           WRITE (6,3)
         ENDIF
3      FORMAT(1x,'NDI VALUE NOT IMPLEMENTED')
```

Then, calculate the stress using STRAN+DSTRAN because the Maxwell model is viscous but elastic, meaning that the moduli are slopes with respect to the origin of strain. Beware that we cannot unload to the origin instantly because the model is viscous but still the slopes are elastic slopes.

```
!      CALCULATE STRESS
       DO K1=1,NTENS
         STRESS(K1)=0.0D0
         DO K2=1,NTENS
           STRESS(K1)=STRESS(K1)+DDSDDE(K2,K1)*(STRAN(K2)+DSTRAN(K2))
         ENDDO
       ENDDO
```

12.3.4   UMAT3DVISCO.FOR

UMAT3DVISCO.FOR is called from the Job, tab General, as shown in Figure 12.6
and used in Example 7.7.

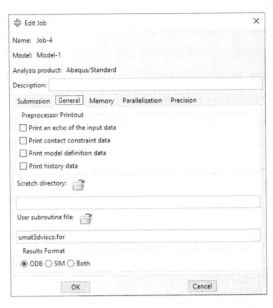

Fig. 12.6: Link to the user subroutine file.

The Fortran code is reproduced below:

```fortran
      SUBROUTINE UMAT(STRESS,STATEV,DDSDDE,SSE,SPD,SCD,          &
     & RPL,DDSDDT,DRPLDE,DRPLDT,STRAN,DSTRAN,                    &
     & TIME,DTIME,TEMP,DTEMP,PREDEF,DPRED,MATERL,NDI,NSHR,NTENS, &
     & NSTATV,PROPS,NPROPS,COORDS,DROT,PNEWDT,CELENT,            &
     & DFGRD0,DFGRD1,NOEL,NPT,KSLAY,KSPT,KSTEP,KINC)
!     COPYRIGHT (2012,2021) EVER J. BARBERO, ALL RIGHTS RESERVED
!     Example 7.7
      INCLUDE 'ABA_PARAM.INC'
      PARAMETER (EPS=2.22D-16) !SMALLEST NUMBER REAL*8 CAN STORE
      CHARACTER*80 MATERL
      DIMENSION STRESS(NTENS),STATEV(NSTATV),               &
     & DDSDDE(NTENS,NTENS),DDSDDT(NTENS),DRPLDE(NTENS),     &
     & STRAN(NTENS),DSTRAN(NTENS),TIME(2),PREDEF(1),DPRED(1), &
     & PROPS(NPROPS),COORDS(3),DROT(3,3),                   &
     & DFGRD0(3,3),DFGRD1(3,3)
! -------------------------------------------------------------
!     UMAT FOR 3D SOLID ELEMENTS
!     F77 IMPLICIT NAME CONVENTION, F95 FREE FORMAT TYPESETING
! -------------------------------------------------------------
!        NDI: # of direct components (11,...)
!             in DDSDDE, DDSDDT, and DRPLDE
!        NSHR: # of engineering shear components (12,...)
!             in DDSDDE, DDSDDT, and DRPLDE
!        NTENS = NDI + NSHR: Size of the stress or strain matrix
!        TIME(1):    step time beginning current increment
!        TIME(2):    total time beginning current increment
!        DTIME:      Time increment.
!        STRESS(NTENS): passed in as the stress tensor
!                       must be updated before return.
! -------------------------------------------------------------
      E1o       = PROPS(1)
```

```fortran
      E2o = PROPS(2)
      E3o       = E2o
      PR12o     = PROPS(3)
      PR13o     = PR12o
      PR23o     = PROPS(4)
      G12o      = PROPS(5)
      G13o      = G12o
      G23o      = E2o/2/(1.+PR23o)
      tau1 = props(7)
      tau2 = props(8)
      tau12 = props(9)
      tau23 = props(10)
!     Maxwell model
      E1 = E1o*dexp(-(Time(2)+DTime)/tau1)
      E2 = E2o*dexp(-(Time(2)+DTime)/tau2)
      G12 = G12o*dexp(-(Time(2)+DTime)/tau12)
      G23 = G23o*dexp(-(Time(2)+DTime)/tau23)
      PR12 = PR12o
      PR23 = PR23o
!     ELASTIC STIFFNESS
      DO K1=1,NTENS
        DO K2=1,NTENS
          DDSDDE(K1,K2)=0.0D0
        ENDDO
      ENDDO
      IF (NDI.EQ.2) THEN
!        PR21 = PR12*E2/E1
         DDSDDE(1,1) = -1/(-E1+PR12**2.*E2)*E1**2.
      DDSDDE(1,2) = -PR12*E1/(-E1+PR12**2.*E2)*E2
         DDSDDE(2,1) = DDSDDE(1,2)
         DDSDDE(2,2) = -E1/(-E1+PR12**2.*E2)*E2
         DDSDDE(3,3) = G12
         IF (NSHR.GT.1) THEN
           DDSDDE(4,4) = G13
           DDSDDE(5,5) = G23
         ENDIF
      ELSEIF (NDI.EQ.3) THEN
!        calculate stiffness 3D stiffness
      DDSDDE(1,1) = E1**2*(PR23-1)/(E1*PR23-E1+2*PR12**2*E2)
      DDSDDE(1,2) = -E2*E1*PR12/(E1*PR23-E1+2*PR12**2*E2)
      DDSDDE(1,3) = -E2*E1*PR12/(E1*PR23-E1+2*PR12**2*E2)
      DDSDDE(2,1) = -E2*E1*PR12/(E1*PR23-E1+2*PR12**2*E2)
      DDSDDE(2,2) = E2*(-E1+PR12**2*E2)/(-E1+E1*PR23**2+2*PR12**2*E2 &
     & +2*PR12**2*E2*PR23)
      DDSDDE(2,3) = -E2*(E1*PR23+PR12**2*E2)/(-E1+E1*PR23**2       &
     & +2*PR12**2*E2+2*PR12**2*E2*PR23)
      DDSDDE(3,1) = -E2*E1*PR12/(E1*PR23-E1+2*PR12**2*E2)
      DDSDDE(3,2) = -E2*(E1*PR23+PR12**2*E2)/(-E1+E1*PR23**2       &
     & +2*PR12**2*E2+2*PR12**2*E2*PR23)
      DDSDDE(3,3) = E2*(-E1+PR12**2*E2)/(-E1+E1*PR23**2+2*PR12**2*E2 &
     & +2*PR12**2*E2*PR23)
      DDSDDE(4,4) = G12    !Abaqus notation
      DDSDDE(5,5) = G12    !G13
      DDSDDE(6,6) = G23    !Abaqus notation
      ELSE
        WRITE (6,3)
      ENDIF
3     FORMAT(1x,'NDI VALUE NOT IMPLEMENTED')
!     CALCULATE STRESS
      DO K1=1,NTENS
        STRESS(K1)=0.0D0
        DO K2=1,NTENS
         STRESS(K1)=STRESS(K1)+DDSDDE(K2,K1)*(STRAN(K2)+DSTRAN(K2))
        ENDDO
      ENDDO
!     HOURGLASS CONTROL
      rFG = 0.005*(2*G12+G23)/3
```

```
!     STATE VARIABLES USED ONLY TO PLOT : moduli(t)
      STATEV(1) = E1
      STATEV(2) = E2
      STATEV(3) = G12
      STATEV(4) = G23
      STATEV(5) = PR12
      STATEV(6) = PR23
      RETURN
      END
```

12.4   Constraint Equations and Periodicity

In Chapter 6, Example 6.4 (p. 249) is briefly explained by using three Python scripts: `Ex_6.4.py`, `PBC_2D.py`, and `srecover2D.py`. The scripts are explained in this section.

The script `Ex_PBC_2D.py` creates the constraint equations (CE) to replicate periodic boundary conditions (PBC) as in (6.42–6.48).

The script `Ex_6.4.py` sets up the simulation of a 3D representative volume element (RVE).

The script `srecover2D.py` calculates the average stress and strain in the volume of the RVE. These can be used, for example, to calculate the homogenized effective properties of the composite material described by the RVE.

12.4.1   Ex_6.4.py

This script does all the work that in Example 6.4 is done by pseudo-code, or in the videos, using CAE [1, Example 6.4]. This script will be interpreted by CAE when it is provided to CAE in one of three ways.

Execute from a Powershell

The easiest but less friendly way to execute the script is to provide the entire script to CAE at the command line, as follows

Type `CRTL F S R` to Open a Powershell in the current folder (C:\Simulia\User-\Chapter6\Ex_6.4).

Type `abaqus cae -nogui .\Ex_6.4.py`

Execution happens in the background. It is very difficult to debug any errors that might exist in the script. You know when the job is completed when the lock file (.lck) disappears from the list of files in File Explorer, and the log file (.log) will say COMPETED. If everything went well, you can open the output database (.odb) with CAE.

Execute from CAE

Alternatively, you can open CAE, then do

Menu: `File, Run Script [Chapter6\Ex_6.4\Ex_6.4.py], OK`

You can watch the messages in the Progress window (below the workspace), and if all goes well you can even read the results right there, as follows

```
Average stresses Global CSYS: 11-22-33-12
[2.1511565e-05 1.1815099e-05 9.8874971e-06 2.4130654e-06]   TPa
```

If there is any error, the script will stop at the error, and you will get an error message. Once the error is corrected, you can execute the rest of the script using the method explained next.

Execute in CAE's Python Shell

This is the best method to understand and debug a script.

At the bottom of the display, below the Workspace, switch from *Progress Window* to the *Python Shell >>>*, as shown below.

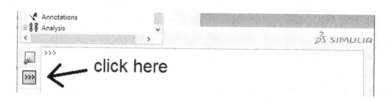

Then, copy/paste portions of the script at the Python shell >>> and monitor the development of the model in CAE itself.

The whole script is available in [2, Ex_6.4] and reproduced below with commentary.

First, close the model database (.cae) in case one is open, if not, no harm done. Immediately, read all the modules that Python needs for this problem, as follows

```
""" Example 6.4 using Constraint Equations CE, 2D Geometry """
mdb.close()
from part import *
from material import *
from section import *
from assembly import *
from step import *
from interaction import *
from load import *
from mesh import *
from job import *
from sketch import *
from visualization import *
from connectorBehavior import *
```

We do not need the next line of code to run the script, but we would need it if we wanted to parameterize the script in terms of geometry, material properties, load, seed spacing for meshing, etc.

```
## recoverGeometry for easy parameterization
session.journalOptions.setValues(recoverGeometry=COORDINATE)
```

Now, set the background of the Workspace to white color (optional) and set the work directory.

```
## session colors
session.viewports['Viewport: 1'].viewportAnnotationOptions.setValues(
    title=OFF)
```

```
session.graphicsOptions.setValues(backgroundStyle=SOLID,
    backgroundColor='#FFFFFF', translucencyMode=2)
# set the work directory
import os
os.chdir(r'c:\SIMULIA\User\Chapter_6\Ex_6.4')
```

Next, add regular Python code to define the geometry, material properties, load, and seed spacing for meshing:

```
# Geometry
rf = 3.5
a2 = 5.270
a3 = 9.128
a1 = a2/4.
# Materials
Ef, nuf = 0.241, 0.2  # TPa
Em, num = 0.00312, 0.38
# Load
strain = [0.002, 0.0, 0.001]     # epsilon_11, epsilon_22, gamma_12
# Seed spacing
MeshingSize = 0.4
```

Next, define 'Part-1'

```
mdb.models['Model-1'].ConstrainedSketch(
    name='__profile__', sheetSize=20.0)
mdb.models['Model-1'].sketches['__profile__'].rectangle(
    point1=(-a2, -a3), point2=(a2, a3))
mdb.models['Model-1'].Part(dimensionality=TWO_D_PLANAR,
    name='Part-1', type=DEFORMABLE_BODY)
```

Next, use the sketcher to define Part-1. *You may be asking yourself, how does he knows how to write all of this? I don't. I use CAE and capture the code from the record file (.rec). Then, I replace numerical values by variables (parameters).*

```
mdb.models['Model-1'].parts['Part-1'].BaseShell(sketch=
    mdb.models['Model-1'].sketches['__profile__'])
del mdb.models['Model-1'].sketches['__profile__']
mdb.models['Model-1'].ConstrainedSketch(gridSpacing=1.0, name='__profile__',
    sheetSize=40, transform=
    mdb.models['Model-1'].parts['Part-1'].MakeSketchTransform(
    sketchPlane=mdb.models['Model-1'].parts['Part-1'].faces.findAt((0.,
    0., 0.0), (0.0, 0.0, 1.0)), sketchPlaneSide=SIDE1,
    sketchOrientation=RIGHT, origin=(0.0, 0.0, 0.0)))
mdb.models['Model-1'].parts['Part-1'].projectReferencesOntoSketch(filter=
    COPLANAR_EDGES, sketch=mdb.models['Model-1'].sketches['__profile__'])
mdb.models['Model-1'].sketches['__profile__'].CircleByCenterPerimeter(center=(
    0.0, 0.0), point1=(0.0, rf))
mdb.models['Model-1'].sketches['__profile__'].ArcByCenterEnds(center=(-a2,
    a3), direction=CLOCKWISE, point1=(-(a2-rf), a3), point2=(-a2, a3-rf))
mdb.models['Model-1'].sketches['__profile__'].ArcByCenterEnds(center=(a2,
    a3), direction=COUNTERCLOCKWISE, point1=(a2-rf, a3), point2=(a2,
    a3-rf))
mdb.models['Model-1'].sketches['__profile__'].ArcByCenterEnds(center=(-a2,
    -a3), direction=COUNTERCLOCKWISE, point1=(-(a2-rf), -a3), point2=(-a2,
    -(a3-rf)))
mdb.models['Model-1'].sketches['__profile__'].ArcByCenterEnds(center=(a2,
    -a3), direction=CLOCKWISE, point1=(a2-rf, -a3), point2=(a2, -(a3-rf)))
mdb.models['Model-1'].sketches['__profile__'].Line(point1=(0.0, a3), point2=
    (0.0, -a3))
mdb.models['Model-1'].sketches['__profile__'].Line(point1=(-a2, 0.0), point2=
    (a2, 0.0))
mdb.models['Model-1'].sketches['__profile__'].Line(point1=(-a2, a3/2.),
    point2=(a2, a3/2.))
mdb.models['Model-1'].sketches['__profile__'].Line(point1=(-a2, -a3/2.),
```

```
        point2=(a2, -a3/2.))
mdb.models['Model-1'].parts['Part-1'].PartitionFaceBySketch(faces=
    mdb.models['Model-1'].parts['Part-1'].faces.findAt((((0.0, 0.0,
    0.0), (0.0, 0.0, 1.0)), ), sketch=
    mdb.models['Model-1'].sketches['__profile__'])
del mdb.models['Model-1'].sketches['__profile__']
```

To see the part in CAE, select Part: Part-1 in the drop-down box below.

Next, in Module: Property, define materials and sections for fiber and matrix, parametrically in therms of Ef, nuf, Em, num.

```
# Materials
mdb.models['Model-1'].Material(name='fiber')
mdb.models['Model-1'].materials['fiber'].Elastic(table=((Ef,nuf),))
mdb.models['Model-1'].Material(name='matrix')
mdb.models['Model-1'].materials['matrix'].Elastic(table=((Em,num),))
# Sections
mdb.models['Model-1'].HomogeneousSolidSection(material='fiber',
    name='fiber-sec', thickness=None)
mdb.models['Model-1'].HomogeneousSolidSection(material='matrix',
    name='matrix-sec', thickness=None)
```

We can check the material properties in CAE by selecting Module: Property, Menu: Material, Manager.

Still in Module: Property, assign sections to regions in Part-1. Note that the code is parameterized so that the script is able to 'find' the regions for any set of parameter values rf,a1,a2,a3.

```
# Section Assigment
mdb.models['Model-1'].parts['Part-1'].SectionAssignment(offset=0.0,
    offsetField='', offsetType=MIDDLE_SURFACE, region=Region(
    faces=mdb.models['Model-1'].parts['Part-1'].faces.findAt(((2.0**0.5/2.*rf,
    2.0**0.5/2.*rf, 0.0), (0.0, 0.0, 1.0)), ((-2.0**0.5/2.*rf, 2.0**0.5/2.*rf,0.0),
    (0.0, 0.0, 1.0)), ((-2.0**0.5/2.*rf, -2.0**0.5/2.*rf, 0.0), (0.0, 0.0, 1.0)),
    ((2.0**0.5/2.*rf, -2.0**0.5/2.*rf, 0.0), (0.0, 0.0, 1.0)),
    ((a2-2.0**0.5/4.*rf, a3-2.0**0.5/4.*rf, 0.0), (0.0, 0.0, 1.0)),
    ((-a2+2.0**0.5/4.*rf, a3-2.0**0.5/4.*rf, 0.0), (0.0, 0.0, 1.0)),
    ((-a2+2.0**0.5/4.*rf, -a3+2.0**0.5/4.*rf, 0.0), (0.0, 0.0, 1.0)),
    ((a2-2.0**0.5/4.*rf, -a3+2.0**0.5/4.*rf, 0.0), (0.0, 0.0, 1.0)), )),
    sectionName='fiber-sec', thicknessAssignment=FROM_SECTION)
mdb.models['Model-1'].parts['Part-1'].SectionAssignment(offset=0.0,
    offsetField='', offsetType=MIDDLE_SURFACE, region=Region(
    faces=mdb.models['Model-1'].parts['Part-1'].faces.findAt(((-(a2-rf)/2.,
    a3-a3/4.0, 0.0), (0.0, 0.0, 1.0)), (((a2-rf)/2., a3-a3/4.0, 0.0), (0.0, 0.0,
    1.0)), ((-a2+(a2-rf)/2., a3/4., 0.0), (0.0, 0.0, 1.0)), ((-(a2-rf)/2.,
    a3/4.0, 0.0), (0.0, 0.0, 1.0)), ((-(a2-rf)/2., -(a3-a3/4.0), 0.0), (0.0, 0.0,
    1.0)), (((a2-rf)/2., -(a3-a3/4.0), 0.0), (0.0, 0.0, 1.0)), ((-a2+(a2-rf)/2.,
    -a3/4., 0.0), (0.0, 0.0, 1.0)), ((-(-a2+(a2-rf)/2.), -a3/4.0,0.0), (0.0, 0.0,
    1.0)), )),
    sectionName='matrix-sec', thicknessAssignment=FROM_SECTION)
```

Next, create sets for the faces and vertices for later applying periodicity conditions using constraint equations (CE, see Section 2.4 and Example 6.4). To write CE between two faces, it is necessary to have a single part named **Part-1**. The part must be meshed with an equal number of nodes at both faces. Pairs of nodes must be located so that they have the same in-plane coordinates in both faces. The faces must be separated by a constant distance. Furthermore, the part must have the following sets predefined.

Face1 is the right face

Face2 is the top face

Face3 is the left face

Face4 is the bottom face

Vertex1 is the vertex intersection between Face1 and Face2

Vertex2 is the vertex intersection between Face2 and Face3

Vertex3 is the vertex intersection between Face3 and Face4

Vertex4 is the vertex intersection between Face4 and Face1

The sets can be defined by the following script.

```
# Create Sets:  Face1, Face2, Face3, Face4, Vertex1, Vertex2, Vertex3, Vertex4
mdb.models['Model-1'].parts['Part-1'].Set(edges=
    mdb.models['Model-1'].parts['Part-1'].edges.findAt(((a2, a3-(rf/2.0), 0.0), ),
    ((a2, a3-rf-(a3/2.-rf)/2., 0.0), ), ((a2, a3/4., 0.0), ), ((a2,-a3/4., 0.0), ),
    ((a2, -(a3-rf-(a3/2.-rf)/2.), 0.0), ), ((a2, -(a3-(rf/2.0)), 0.0), ), ),
    name='Face1')
mdb.models['Model-1'].parts['Part-1'].Set(edges=
    mdb.models['Model-1'].parts['Part-1'].edges.findAt(((a2-rf/2., a3, 0.0), ),
    ((-(a2-rf)/2., a3, 0.0), ), (((a2-rf)/2, a3, 0.0), ), ((-(a2-rf/2.), a3, 0.0),
    ), ), name='Face2')
mdb.models['Model-1'].parts['Part-1'].Set(edges=
    mdb.models['Model-1'].parts['Part-1'].edges.findAt(((-a2, a3-(rf/2.0), 0.0), ),
    ((-a2, a3-rf-(a3/2.-rf)/2., 0.0), ), ((-a2, a3/4., 0.0), ), ((-a2,-a3/4., 0.0),
    ), ((-a2, -(a3-rf-(a3/2.-rf)/2.), 0.0), ), ((-a2, -(a3-(rf/2.0)), 0.0), ), ),
    name='Face3')
mdb.models['Model-1'].parts['Part-1'].Set(edges=
    mdb.models['Model-1'].parts['Part-1'].edges.findAt(((a2-rf/2., -a3, 0.0), ),
    ((-(a2-rf)/2., -a3, 0.0), ), (((a2-rf)/2, -a3, 0.0), ), ((-(a2-rf/2.),
    -a3, 0.0), ), ), name='Face4')
mdb.models['Model-1'].parts['Part-1'].Set(name='Vertex1', vertices=
    mdb.models['Model-1'].parts['Part-1'].vertices.findAt(((a2, a3, 0.0), ), ))
mdb.models['Model-1'].parts['Part-1'].Set(name='Vertex2', vertices=
    mdb.models['Model-1'].parts['Part-1'].vertices.findAt(((-a2, a3, 0.0), ), ))
mdb.models['Model-1'].parts['Part-1'].Set(name='Vertex3', vertices=
    mdb.models['Model-1'].parts['Part-1'].vertices.findAt(((-a2, -a3, 0.0), ), ))
mdb.models['Model-1'].parts['Part-1'].Set(name='Vertex4', vertices=
    mdb.models['Model-1'].parts['Part-1'].vertices.findAt(((a2, -a3, 0.0), ), ))
```

In Module: Assembly, define an instance of the part. If we specify **dependent=ON**, meshing must be on the Part, not the the Instance.

```
# Instance
mdb.models['Model-1'].rootAssembly.DatumCsysByDefault(CARTESIAN)
mdb.models['Model-1'].rootAssembly.Instance(dependent=ON, name='Part-1-1',
    part=mdb.models['Model-1'].parts['Part-1'])
```

To see the assembly in CAE, just switch to module **Mesh**.

The periodicity conditions along the fiber direction, i.e., at faces $\pm a_1$ (Z direction in Figure 6.13) are enforced by a plane strain condition, which is implemented using plane strain elements CPE3 and CPE4.

```
# Mesh of plane strain elements CPE4
mdb.models['Model-1'].parts['Part-1'].seedPart(deviationFactor=0.1,size=MeshingSize)
mdb.models['Model-1'].parts['Part-1'].setElementType(elemTypes=(ElemType(
    elemCode=CPE4, elemLibrary=STANDARD), ElemType(elemCode=CPE3,
    elemLibrary=STANDARD)), regions=(
    mdb.models['Model-1'].parts['Part-1'].faces.findAt(((2.0**0.5/2.*rf,
    2.0**0.5/2.*rf, 0.0), (0.0, 0.0, 1.0)), ((-2.0**0.5/2.*rf, 2.0**0.5/2.*rf,0.0),
    (0.0, 0.0, 1.0)), ((-2.0**0.5/2.*rf, -2.0**0.5/2.*rf, 0.0), (0.0, 0.0, 1.0)),
    ((2.0**0.5/2.*rf, -2.0**0.5/2.*rf, 0.0), (0.0, 0.0, 1.0)), ((a2-2.0**0.5/4.*rf,
    a3-2.0**0.5/4.*rf, 0.0), (0.0, 0.0, 1.0)), ((-a2+2.0**0.5/4.*rf,
    a3-2.0**0.5/4.*rf, 0.0), (0.0, 0.0, 1.0)), ((-a2+2.0**0.5/4.*rf,
    -a3+2.0**0.5/4.*rf, 0.0), (0.0, 0.0, 1.0)), ((a2-2.0**0.5/4.*rf,
    -a3+2.0**0.5/4.*rf, 0.0), (0.0, 0.0, 1.0)),((-(a2-rf)/2., a3-a3/4.0, 0.0),
    (0.0, 0.0, 1.0)), (((a2-rf)/2., a3-a3/4.0, 0.0), (0.0, 0.0, 1.0)),
    ((-a2+(a2-rf)/2., a3/4., 0.0), (0.0, 0.0, 1.0)), ((-(a2+(a2-rf)/2.), a3/4.0,
    0.0), (0.0, 0.0, 1.0)), ((-(a2-rf)/2., -(a3-a3/4.0), 0.0), (0.0, 0.0, 1.0)),
    (((a2-rf)/2., -(a3-a3/4.0), 0.0), (0.0, 0.0, 1.0)), ((-a2+(a2-rf)/2., -a3/4.,
    0.0),(0.0, 0.0, 1.0)),((-(-a2+(a2-rf)/2.),-a3/4.0,0.0),(0.0, 0.0, 1.0)),),))
mdb.models['Model-1'].parts['Part-1'].generateMesh()
```

To see the mesh in CAE, just switch to module Mesh.

Now, define a step to apply external strains ϵ_2^0 and γ_4^0, as this example requires.

```
# Step
mdb.models['Model-1'].StaticLinearPerturbationStep(
    name='Step-1', previous='Initial')
```

The periodicity equations not only enforce periodicity on the RVE but simultaneously impose the applied strains, as explained in Section 6.3. All of this is accomplished by PBC_2D.py, which is executed simply by typing the following in CAE's Python shell.

```
# BC and periodicity
execfile('PBC_2D.py')
```

PBC_2D.py is explained in Section 12.4.2.

To see the boundary conditions in CAE, switch to Module:Load,BC,Manager. BC-1 is a unit displacement $U1 = 1$ that is necessary to write the CEs.

To see the periodicity conditions in CAE, switch to Module: Interaction, Constraint, Manager.

Next, modify the Field Output Request to make sure the output database (.odb) contains what we need for post-processing, including the stress S and element volume IVOL that are used to calculate the average stress in the RVE, with an equation similar to (6.18), but without restriction about the number of components of strain that can be applied to the model, i.e.,

$$\bar{\sigma}_\alpha = \frac{1}{V} \int_V \sigma_\alpha \left(x_1, x_2, x_3 \right) \ dV \tag{12.2}$$

and save the model database (.cae) before executing the Job.

```
# Field Output Request
mdb.models['Model-1'].fieldOutputRequests['F-Output-1'].setValues(variables=(
    'S', 'LE', 'U', 'RF', 'CF', 'IVOL', 'STH'))
# Save the .mdb and .cae
mdb.saveAs(pathName='c:/SIMULIA/User/Chapter_6/Ex_6.4/Ex_6.4.cae')
```

The Job is executed as follows

```
# Job
mdb.Job(atTime=None, contactPrint=OFF, description='', echoPrint=OFF,
    explicitPrecision=SINGLE, getMemoryFromAnalysis=True, historyPrint=OFF,
    memory=90, memoryUnits=PERCENTAGE, model='Model-1', modelPrint=OFF,
    multiprocessingMode=DEFAULT, name='Job-1', nodalOutputPrecision=SINGLE,
    numCpus=1, queue=None, scratch='', type=ANALYSIS, userSubroutine='',
    waitHours=0, waitMinutes=0)
mdb.jobs['Job-1'].setValues(numCpus=8, numDomains=8)
mdb.jobs['Job-1'].submit(consistencyChecking=OFF)
mdb.jobs['Job-1'].waitForCompletion()
```

Note the `waitForCompletion()` statement, which is used to wait for the Job to finish before trying to open the output database (.odb) to work on post-processing. *To take full advantage of your hardware you should use numCpus and numCpus equal to the number of processors you have, which usually is double the number of cores.*

Post-processing is performed with another script, `srecover2D.py`, explained in Section 12.4.3.

```
# Calculate Stresses and Strains
execfile('srecover2D.py')
```

You can use the following code to request visualization. Of course you can do the same using CAE graphical user interface (GUI).

```
# visualize
o3 = session.openOdb(
    name='c:/SIMULIA/User/Chapter_6/Ex_6.4/Job-1.odb')
session.viewports['Viewport: 1'].setValues(displayedObject=o3)
```

12.4.2 PBC_2d.py

The script `Ex_PBC_2D.py`, available in [2, Ex_6.4], creates the constraint equations to replicate periodic boundary conditions as in (6.42)–(6.48). Additional details, such how to deal with vertices, are provided in Section 6.3.

The script requires a single part, named `Part-1`, which could be partitioned if needed. The part must have the following sets predefined.

Face1 is the right face

Face2 is the top face

Face3 is the left face

Face4 is the bottom face

Vertex1 is the vertex intersection between Face1 and Face2

Vertex2 is the vertex intersection between Face2 and Face3

Vertex3 is the vertex intersection between Face3 and Face4

Vertex4 is the vertex intersection between Face4 and Face1

The faces must be meshed with identical meshes, with equal number of nodes placed at the same locations on both faces. If these requirements are met, we can execute `srecover2D.py` to find the effective strains and stresses.

12.4.3   Mesoscale Effective Stress from FEA-RVE Solution

The script `srecover.py`, available in [2, Ex_6.4], implements (12.2) to calculate the mesoscale effective stress and strain using the FEA solution for an RVE, as well as the maximum strain in the matrix.

```
# srecover2D.py
from visualization import *
# Open the output data base for the current Job
odb = openOdb(path='Job-1.odb');
myAssembly = odb.rootAssembly;
# Creating a temporary variable to hold the frame repository
# In Ex. 7.8, Step-2 is the relaxation step
frameRepository = odb.steps['Step-2'].frames;
# Get the results for frame [i], where i is the increment number
i = 100
frameS = frameRepository[i].fieldOutputs['S'].values;
frameE = frameRepository[i].fieldOutputs['E'].values;
frameIVOL = frameRepository[i].fieldOutputs['IVOL'].values;
Tot_Vol=0.;          # Total Volume
Tot_Stress=0.;       # Stress Sum
Tot_Strain = 0.;     # Strain Sum
# Calculate Average
for II in range(0,len(frameS)):
    Tot_Vol+=frameIVOL[II].data;
    Tot_Stress+=frameS[II].data * frameIVOL[II].data;
    Tot_Strain+=frameE[II].data * frameIVOL[II].data;
Avg_Stress = Tot_Stress/Tot_Vol;
Avg_Strain = Tot_Strain/Tot_Vol;
# from Abaqus Analysis User's Manual - 1.2.2 Conventions -
# Convention used for stress and strain components
print '2D Abaqus/Standard Stress Tensor Order: 11-22-33-12'
print 'Average stresses Global CSYS: 11-22-33-12';
print Avg_Stress;
print 'Average strain Global CSYS: 11-22-33-12';
print Avg_Strain;
odb.close()
```

Example 12.3 *Consider a composite made with 40% by volume of isotropic graphite fibers with properties $E_f = 168.4$ GPa, $\nu_f = 0.443$ and epoxy matrix represented by a Maxwell model with $E_0 = 4.082$ GPa, $\tau = 39.15$ min, and constant Poisson's ratio $\nu_m = 0.311$. Construct an FEA micromechanics model using hexagonal microstructure (see Example 6.3, p. 242), subject to shear strain $\gamma_4 = 0.02$ applied suddenly at $t = 0$. Tabulate the average stress σ_4 over the RVE at times $t = 0, 20, 40, 60, 80,$ and 100 minutes.*

Solution to Example 12.3 *Here is an alternative method to solve Example 7.8 by modifying the Python script Ex_6.4.py used in Example 6.4.*

Copy Ex_6.4\Ex_6.4.py as Ex_12.3.py in a different folder, such as

```
C:\SIMULIA\User\Chapter_12\Ex\_12.3
```

Copy PBC_2D.py and srecover2D.py to the local directory for this example.

Open Ex_12.3.py with Notepad++ for editing.

Start by updating the work directory to the current one, by modifying the corresponding line in Ex_12.3.py, as follows

```
os.chdir(r'C:\SIMULIA\User\Chapter_12\Ex_12.3')
```

Then, modify the material properties with new properties for the fibers and to include all the elastic and viscoelastic parameters for the matrix. The elastic moduli are given in TPa because the dimensions are given in micrometers.

```
# Materials
Ef, nuf = 168.4E-3, 0.443  # TPa
Em, num, tau, g_1, k_1 = 4.082E-3, 0.311, 39.15, 0.999, 0.999 # TPa,,min,,,
```

Then, modify the loading, keeping only $\gamma_4 = 0.02$ *(γ_{12} in the coordinate system of the model):*

```
# Load
strain = [0.000, 0.0, 0.02]     # epsilon_11, epslion_22, gamma_12
```

> *The best method to write Abaqus Python code is to use CAE's graphical user interface to do everything. CAE will write the code in the* `.rec` *file, but if the model database is saved, the* `.rec` *file will be erased. Therefore, before saving the* `.cae` *file, open the* `.rec` *file to get the script's code and save it in another file with extension* `.py`.

To create a Python script describing the definition of a viscoelastic material, open Abaqus CAE, and save the current session as Ex_12.3.cae. *Then, define a viscoelastic material exactly as it was done in Example 7.5, p. 277. The pseudo-code below describes the actions to be performed in CAE.*

```
Module: Property
Menu:   Material, Create
    Name [matrix], Mechanical, Elasticity, Elastic, Type: Isotropic
    Moduli time scale: Instantaneous, Data [4.082E-3, 0.311]
    Mechanical, Elasticity, Viscoelastic, Domain: Time, Time: Prony
    g_1, k_1, tau_1 [0.999, 0.999, 39.15], OK
```

Before saving the model, copy the following lines from Ex_12.3.rec *to* Ex_12.3.py, *replacing the two lines defining the properties of the matrix material in the section named* # Materials.

```
mdb.models['Model-1'].Material(name='matrix')
mdb.models['Model-1'].materials['matrix'].Elastic(moduli=INSTANTANEOUS
    , table=((0.004082, 0.311), ))
mdb.models['Model-1'].materials['matrix'].Viscoelastic(domain=TIME,
    table=((0.999, 0.999, 39.15), ), time=PRONY)
```

and immediately replace numerical values by variable names (to parameterize the script), as follows

```
mdb.models['Model-1'].Material(name='matrix')
mdb.models['Model-1'].materials['matrix'].Elastic(moduli=INSTANTANEOUS
    , table=((Em, num), ))
mdb.models['Model-1'].materials['matrix'].Viscoelastic(domain=TIME,
    table=((g_1, k_1, tau), ), time=PRONY)
```

Next save the model to empty the `.rec` *file.*

Next, create a viscoelastic step-1 to apply the strain, and a step-2 to track the ensuing relaxation, exactly as it was done in Example 7.6. Here is the pseudo-code detailing the actions to be done in Abaqus CAE.

```
Module: Step
Menu:   Step, Create
    Name [Step-1], Procedure type: General, Visco, Cont
    Tab: Basic, Time period [0.001]
    Tab: Incrementation, Type: Automatic, viscoelastic tolerance [1E-6], OK
Menu:   Step, Create
    Name [Step-2], Procedure type: General, Visco, Cont
    Tab: Basic, Time period [100]
    Tab: Incrementation, Type: Fixed, Maximum number of increments: [200]
    Increment Size [1], OK
```

Before saving the model, copy the following lines from Ex_12.3.rec to Ex_12.3.py, replacing the line defining the Step-1 in the section named # Step.

```
mdb.models['Model-1'].ViscoStep(cetol=1e-06, initialInc=0.001, maxInc=0.001,
    minInc=1e-08, name='Step-1', previous='Initial', timePeriod=0.001)
mdb.models['Model-1'].ViscoStep(cetol=0.0, initialInc=1.0, maxNumInc=200,
    name='Step-2', previous='Step-1', timeIncrementationMethod=FIXED,
    timePeriod=100.0)
```

Next, update the name of the file used to add periodicity conditions.

```
execfile('PBC_2D.py')
```

Next, update the name of the file in

```
mdb.saveAs(pathName='Ex_12.3.cae')
```

At this point, your updated script should look like the one available in [2, Ex_12.3.py]. Next, reset the Abaqus CAE session to start a new model, and then run the script, as follows

```
Menu:   File, New Model Database, With Standard/Explicit Model
Menu:   File, Run Script [Ex_12.3.py], OK
```

12.5   UMAT 1D

A 1D model for longitudinal damage of unidirectional composites is developed in Section 8.1.6. Model identification, i.e., determination of model parameters from experimental data is presented in Example 8.3. The model parameters are fiber modulus $E_f = 230$ GPa, Weibull shape parameter $m = 8.9$, and the product $\delta\alpha = 3.92 \times 10^{-33}$. For details, see Section 8.1.6 and the first section of the solution for Example 8.3. Once the parameters are known, the mathematical formulation of the damage model is presented in the second section of the solution for Example 8.3. Those equations and algorithms are implemented into a UMAT in this section, to be used along a 1D bar element to predict the material's response as in Figure 8.5, p. 304.

12.5.1   User Subroutine Interface

Your UMAT code has to match the user subroutine interface described in the documentation for Release 2020, as follows

```
      SUBROUTINE UMAT(STRESS,STATEV,DDSDDE,SSE,SPD,SCD,
     1 RPL,DDSDDT,DRPLDE,DRPLDT,
     2 STRAN,DSTRAN,TIME,DTIME,TEMP,DTEMP,PREDEF,DPRED,CMNAME,
     3 NDI,NSHR,NTENS,NSTATV,PROPS,NPROPS,COORDS,DROT,PNEWDT,
     4 CELENT,DFGRD0,DFGRD1,NOEL,NPT,LAYER,KSPT,JSTEP,KINC)
C
      INCLUDE 'ABA_PARAM.INC'
C
      CHARACTER*80 CMNAME
      DIMENSION STRESS(NTENS),STATEV(NSTATV),
     1 DDSDDE(NTENS,NTENS),DDSDDT(NTENS),DRPLDE(NTENS),
     2 STRAN(NTENS),DSTRAN(NTENS),TIME(2),PREDEF(1),DPRED(1),
     3 PROPS(NPROPS),COORDS(3),DROT(3,3),DFGRD0(3,3),DFGRD1(3,3),
     4 JSTEP(4)
...
      user coding to define DDSDDE, STRESS, STATEV, SSE, SPD, SCD
      and, if necessary, RPL, DDSDDT, DRPLDE, DRPLDT, PNEWDT
...
      RETURN
      END
```

12.5.2 State Variables and Constants

Abaqus allocates storage for the state variables (STATEV) for each Gauss point in the model. The number of STATEV is what you declare for DEPVAR in Module: Property, or inside the input file.

```
*Material, name=Material-1
*Depvar
     2,
*User Material, constants=3
230000.,      8.9, 3.92e-33
```

DEPVAR $= 2$ means that there are two state variables

– $\hat{\gamma}$ is the updated damage threshold (8.44).

– D is the accumulated damage (8.45).

CONSTANTS $= 3$ means that there are three user material constants

– The fiber modulus E_f

– The Weibull shape parameter $m = 8.9$.

– The product $\delta\alpha = 3.92 \times 10^{-33}$

12.5.3 User's Code

In nonlinear analysis, we have to recover the "state" of the material at the last iteration or increment from the STATEV array. Then retrieve the model parameters from the PROPS array.

```
c *** Recover state variables
     gamma_hat = statev(1)
     D = statev(2)
c *** get Properties, and Damage model parameters
     E = props(1) ! Elastic modulus
     m       = props(2) ! Weibul shape parameter
     delta_alpha = props(3) !parameter
```

A damaging material is nonlinear. Thus, the analysis is incremental. An increment of strain DSTRAN is calculated by Abaqus for each increment of time (Figure 12.5, p. 455), and when the current increment is converged, DSTRAN is added (in the UMAT) to the accumulated strain STRAN.

In this formulation, the damage threshold $\hat{\gamma}$ is an explicit function of stress, so we calculate it. Fortunately, the value of damage D is also an explicit function, so we calculate it explicitly in the code below. Finally, we calculate the STRESS as the product of Cauchy stress $(1 - D)E$ times the updated strain.

```
c *** compute effective stress
     sigma_b = E * (STRAN(1) + DSTRAN(1))
c *** update gamma_hat
     gamma_hat = max (gamma_hat, sigma_b)
c *** calculate D and secant stiffness 1D
     D = 1 - dexp(-delta_alpha*gamma_hat**m)
     C = dexp(-delta_alpha*gamma_hat**m)*E
c *** calculate the apparent stress *** output: S11
STRESS(1)= (1-D) * E * (STRAN(1) + DSTRAN(1))   ! see alternative below
```

It is worth noting that the stress that UMAT must calculate and return to Abaqus is Cauchy stress σ, to be used by Abaqus for the *structural* analysis side of the calculations, while the UMAT code takes care of the *material's* side of the calculations.

Furthermore, the mathematical formulation in this example uses *effective* stress $\tilde{\sigma}$, related to Cauchy stress by the damage state variable D as in (8.6). Therefore, the reduced modulus E needed in the last equation of the code above is given by (8.7). Note that in (8.7), $\tilde{E}$ is initial (undamaged) modulus of the unidirectional lamina, which in the code is labeled with plain E.

Before exiting the subroutine, we must store the updated values of the state variables, as follows

```
C *** Update state variables
      statev(1) = gamma_hat !effective stress, output: SDV1
      statev(2) = D         !damage          , output: SDV2
```

12.5.4   UMAT1D83.FOR

UMAT1D83.FOR is used in Example 8.3. It is called from the `Job, Tab General`, as shown in Figure 12.7.

To check the results, run the `Job`.

When completed, click `Results`.

In module: `Visualization, Toolbar: Field Output`, select S, S11, like this:

It will show the Cauchy stress at the end of the analysis (Step = 30).

Now with the Play controls, like these:

back up until you get the maximum stress = 3059 MPa.

Now use `Menu: Result, Step/Frame`, to discover from the table that the maximum stress occurs at Step = 30, Time = 30.

Click Cancel to close the table. Now with `Toolbar: Field Output`, select E, E11, you can see that the maximum stress occurs for Strain E11 = 1.5%.

Now compare to Figure 8.6, p. 310. Perfect!

The effective stress $\tilde{\sigma}$ is higher than the Cauchy stress σ because on account of (8.6), the effective stress is $\tilde{\sigma} = \sigma/(1 - D)$, and the damage grows from zero to 1 during the simulation, i.e., $0 < D < 1$.

At Step = 30, use the `Toolbar: Field Output`, and select SDV2, to learn that at the maximum stress (when the material breaks) the value of damage (the critical damage) is $D_{cr} = 0.1332$ (Figure 12.8).

Selecting SDV1, the value of effective stress is $\tilde{\sigma} = 3450$ MPa (not shown in Figure 12.8).

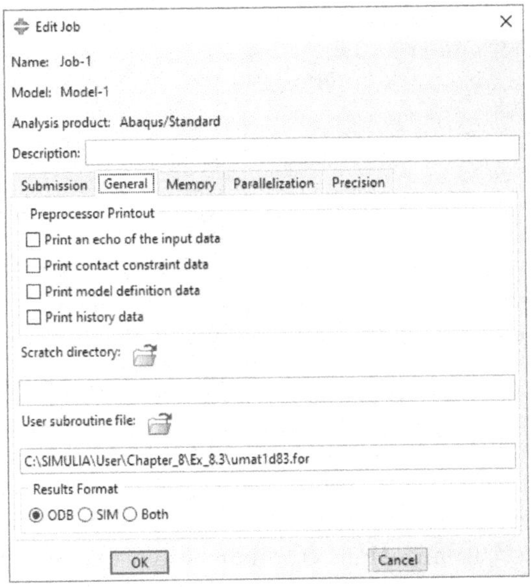

Fig. 12.7: Link to the user subroutine file.

The Fortran code is shown below and available in [2]:

```fortran
      SUBROUTINE UMAT(STRESS,STATEV,DDSDDE,SSE,SPD,SCD,              &
     & RPL,DDSDDT,DRPLDE,DRPLDT,STRAN,DSTRAN,                        &
     & TIME,DTIME,TEMP,DTEMP,PREDEF,DPRED,MATERL,NDI,NSHR,NTENS,    &
     & NSTATV,PROPS,NPROPS,COORDS,DROT,PNEWDT,CELENT,               &
     & DFGRD0,DFGRD1,NOEL,NPT,KSLAY,KSPT,KSTEP,KINC)
!     Copyright (2007,2012,2021) Ever J. Barbero, Abaqus Ex. 8.3
      INCLUDE 'ABA_PARAM.INC'
      PARAMETER (EPS=2.22D-16) !SMALLEST NUMBER REAL*8 CAN STORE
      CHARACTER*80 MATERL
      DIMENSION STRESS(NTENS),STATEV(NSTATV),                       &
     & DDSDDE(NTENS,NTENS),DDSDDT(NTENS),DRPLDE(NTENS),             &
     & STRAN(NTENS),DSTRAN(NTENS),TIME(2),PREDEF(1),DPRED(1),       &
     & PROPS(NPROPS),COORDS(3),DROT(3,3),                           &
     & DFGRD0(3,3),DFGRD1(3,3)
! -----------------------------------------------------------
!     UMAT FOR 1D, 2D, 3D SOLID ELEMENTS
!     F77 IMPLICIT NAME CONVENTION, F95 FREE FORMATTING
! -----------------------------------------------------------
!     NDI: # of direct components (11,...) of DDSDDE, DDSDDT, and DRPLDE
!     NSHR: # of engineering shear components (12,...)
!                                   of DDSDDE, DDSDDT, and DRPLDE
!     NTENS = NDI + NSHR: Size of the stress or strain component array
!     TIME(1): Value of step time at the beginning of current increment.
!     TIME(2): Value of total time at the beginning of current increment.
!     DTIME   : Time increment.
!     STRESS(NTENS):  passed in as the stress tensor at the beginning of
!                     the increment, must be updated.
! -----------------------------------------------------------
!23456
!     *********** User defined part ***********************************
      DOUBLE PRECISION E, m, delta_alpha
      DOUBLE PRECISION gamma_hat, D
      DOUBLE PRECISION sigma_b, C
      keycut   = 0
!     Recover state variables
      gamma_hat = statev(1)
      D = statev(2)
```

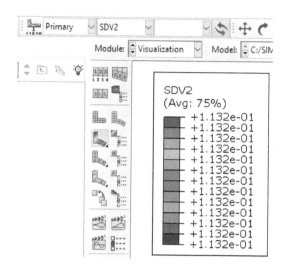

Fig. 12.8: In Module: Visualization. Toolbar: Field Output (top). SDV2 damage (bottom).

```
!       get Properties, and Damage model parameters
        E            = props(1)   ! Elastic modulus
        m            = props(2)   ! Weibul shape parameter
        delta_alpha = props(3)   !parameter
!       compute effective stress
        sigma_b = E * (STRAN(1) + DSTRAN(1))
!       update gamma_hat
        gamma_hat = max (gamma_hat, sigma_b)
!       calculate D and secant stiffness 1D
        D       = 1 - dexp(-delta_alpha*gamma_hat**m)
        C       = dexp(-delta_alpha*gamma_hat**m)*E
!       calculate the apparent stress *** output: S11
        STRESS(1)= (1-D) * E * (STRAN(1) + DSTRAN(1)) !see alternative below
!       INITIALIZE THE TANGENT STIFFNESS
        DO K1=1,NTENS
          DO K2=1,NTENS
             DDSDDE(K1,K2)=0.0D0
          ENDDO
        ENDDO
!       CALCULATE THE TANGENT STIFFNESS
        IF (NDI.EQ.1) THEN
          if (gamma_hat.eq.statev(1)) then   ! gamma_hat did not change
             DDSDDE(1,1)=C
          else   ! gamma_hat did change
             DDSDDE(1,1)=(1 - delta_alpha*m*gamma_hat**m)  &
     &                  *dexp(-delta_alpha*gamma_hat**m)*E
          endif
        ELSEIF (NDI.EQ.2) THEN
          write(*,*) "NDI=2 not implemented"
          call xit
        ELSEIF (NDI.EQ.3) THEN
          write(*,*) "NDI=3 not implemented"
          call xit
        ELSE
          WRITE (6,3)
        ENDIF
 3      FORMAT(1x,'NDI VALUE NOT IMPLEMENTED')
!       Update state variables
        statev(1) = gamma_hat !effective stress, output: SDV1
        statev(2) = D              !damage           , output: SDV2
```

```
       RETURN
       END
```

12.6   UMAT Plane Stress

A continuum damage mechanics (CDM) model for damage in the form of transverse matrix cracks is presented in Example 8.4. Damage is tracked by two variables, D_2 and D_6. The solution to Example 8.5 has all the equations that we need to implement it in a UMAT, which is available in [2, Ex_8.5].

We choose a `solid, plane stress` approximation because it will work for laminates subjected to in-plane load, which is the most efficient way to load a laminate, curved or not [3, p. 423].

12.6.1   User Subroutine Interface

Your UMAT code has to match the user subroutine interface described in the documentation for Release 2020, as follows

```
       SUBROUTINE UMAT(STRESS,STATEV,DDSDDE,SSE,SPD,SCD,
      1 RPL,DDSDDT,DRPLDE,DRPLDT,
      2 STRAN,DSTRAN,TIME,DTIME,TEMP,DTEMP,PREDEF,DPRED,CMNAME,
      3 NDI,NSHR,NTENS,NSTATV,PROPS,NPROPS,COORDS,DROT,PNEWDT,
      4 CELENT,DFGRD0,DFGRD1,NOEL,NPT,LAYER,KSPT,JSTEP,KINC)
C
       INCLUDE 'ABA_PARAM.INC'
C
       CHARACTER*80 CMNAME
       DIMENSION STRESS(NTENS),STATEV(NSTATV),
      1 DDSDDE(NTENS,NTENS),DDSDDT(NTENS),DRPLDE(NTENS),
      2 STRAN(NTENS),DSTRAN(NTENS),TIME(2),PREDEF(1),DPRED(1),
      3 PROPS(NPROPS),COORDS(3),DROT(3,3),DFGRD0(3,3),DFGRD1(3,3),
      4 JSTEP(4)
...
       user coding to define DDSDDE, STRESS, STATEV, SSE, SPD, SCD
       and, if necessary, RPL, DDSDDT, DRPLDE, DRPLDT, PNEWDT
...
       RETURN
       END
```

12.6.2   State Variables and Constants

The number of dependent (state) variables DEPVAR = 3, the number of constants (i.e., user's parameters) NPROPS = constants = 13, and those values, are provided in the input file, as follows

```
*Material, name=Material-1
*Depvar
     3,
*User Material, constants=13
171400., 9080.,  0.32,    0., 5290.,    0., 62.29, 92.34
  0.17,  0.23,   0.5,  -1.8,    1.
```

The state variables are

– SDV1 = δ is the hardening variable.

– SDV2 = D2 is the transverse damage variable.

– SVD3 = D6 is the shear damage variable.

At the beginning we recover the last increment's values of the state variables.

```
!       Recover state variables
        delta = statev(1)
        D2    = statev(2)
        D6    = statev(3)
```

The user's parameters are

```
!       Get the elastic properties and model parameters
        E1      = props(1)
        E2      = props(2)
        nu12    = props(3)
        nu23    = props(4)    ! Not used for plane stress
        G12     = props(5)
        G13     = G12
        G23     = props(6)    ! Not used for plane stress
        nu21    = nu12*E2/E1
        F2t     = props(7)    ! Transverse tensile strength
        F6      = props(8)    ! In-plane shear strength
        GIc     = props(9)    ! critical ERR mode I
        GIIc    = props(10)   ! critical ERR mode II
        c1      = props(11)   ! damage hardening property 1
        c2      = props(12)   ! damage hardening property 2
        gamma0  = props(13)   ! damage surface threshold
```

12.6.3   User's Code

The hardening function

$$\hat{\gamma} = \gamma + \gamma_0 = c_1 \left[\exp\left(\frac{\delta}{c_2}\right) - 1 \right] + \gamma_0 \quad ; \quad \gamma_0 - c_1 \leq \hat{\gamma} \leq \gamma_0 \qquad (12.3)$$

controls the growth of the damage surface (Figure 8.8) as a function of the *hardening variable* δ defined by (8.103). The damage surface has to grow so that damage stops at some value when the strain stops. More strain needs to be applied if one wants to see more damage.

The *rate of damage* $\dot{D}$ has magnitude $\dot{\lambda}$ and direction $\partial f/\partial Y$ (8.102). As long as damage cannot heal [226, 227], the *damage increment* $\dot{\lambda}$ satisfies $\dot{\lambda} \geq 0$.

The damage surface g (Figure 8.8) grows with the hardening δ according to (12.3) and $\dot{\delta} = -\dot{\gamma}$ in (8.112).

Integrating, we get $\delta_i = \delta_{i-1} - \lambda_i$. Finally, since $\dot{\lambda} \geq 0$, $\dot{\delta} \leq 0$, and starting with $\delta_0 = 0$ we get $\delta \leq 0$ for all i. The *hardening variable* can be tracked by SDV1 as shown in Section 12.6.4.

The solution to Example 8.5, complemented by Sections 8.4 and 8.4.1 have all the equations needed to implement the user's code below.

```
!       START the RETURN MAPPING ALGORITHM
        conv = 0
        k = 0
        kmax = 1000
        do while (conv.eq.0)
!           STEP 3. compute thermodynamic forces and damage hardening
```

```fortran
!              calculate secant stiffness 2D (plane stress)
!              as a function of current damage
               C       = 0
               C(1,1) = -E1 / (nu21 * nu12 - 1)
               C(1,2) = (-1 + D2) * E2 * nu12 / (nu21 * nu12 - 1)
               C(2,1) = E1 * (-1 + D2) * nu21 / (nu21 * nu12 - 1)
               C(2,2) = -(-1 + D2) ** 2 * E2 / (nu21 * nu12 - 1)
               C(3,3) = 2 * G12* (-1 + D6) ** 2 !incorporates ERRATA 2007
!!              calculate the Cauchy stress using the current secant stiffness
               STRESS = matmul(c,eps)
               sigma1 = STRESS(1)
               sigma2 = STRESS(2)
               sigma6 = STRESS(3)
!              calculate thermodynamic loads Y1,Y2,Y6 as f. of Cauchy stresses
               Y(1) = 0
               Y(2) = sigma2 ** 2 / (1 - D2) ** 3 / E2 - nu12 / E1 /        &
        &              (1 - D2)** 2 * sigma1 * sigma2
               Y(3) = sigma6 ** 2 / (1 - D6) ** 3 / 2 / G12 !ERRATA 2007
               Y2s = Y(2)
               Y6s = Y(3)
!              evalutate gamma
               gamma1 = c1 * (dexp(delta / c2) - 1.0D0)
!              STEP 4. compute damage surface
!              evaluate damage surface g
               if (Y2s+Y6s.ge.0.0) then
                   g_hat = dsqrt(((1 - GIc / GIIc) * Y2s * E2 / F2t ** 2 + &
        &              GIc / GIIc * Y2s ** 2 * E2 ** 2 / F2t ** 4 + Y6s ** &
        &              2 * G12 ** 2 / F6**4))
               else
                   g_hat = 0
               endif
               g =  g_hat - gamma1 - gamma0
!              elastic (load or unload) or inelastic load (new damage)
               if (g.le.tol) then
                   conv = 1
               else
                   k = k  + 1
!                  STEP 5. compute lambda multiplier
!                  compute dg_dY*dY_dD*df_dY
                   prod = dble(1 / ((1 - GIc / GIIc) * Y2s * E2 / F2t ** 2 &
        &              + GIc / GIIc * Y2s ** 2 * E2 ** 2 / F2t ** 4 + Y6s ** 2 &
        &              * G12 ** 2 / F6 ** 4) * ((1 - GIc / GIIc) * E2 / F2t ** &
        &              2 + 2 * GIc / GIIc * Y2s * E2 ** 2 / F2t ** 4) ** 2 *   &
        &              (-nu12 ** 2 * sigma2 / E1 / (-1 + D2) ** 2 * E2 / (nu21 &
        &              * nu12 - 1) * eps2 + (2 * sigma2 / (1 - D2) ** 3 / E2 - &
        &              nu12 / E1 / (1 - D2) ** 2 * sigma1) * (E1 * nu21 /       &
        &              (nu21 * nu12 - 1) * eps1 - 2 * (-1 + D2) * E2 / (nu21 * &
        &              nu12 - 1) * eps2))) / 0.4D1 + dble(32 / ((1 - GIc /      &
        &              GIIc) * Y2s * E2 / F2t ** 2 + GIc / GIIc * Y2s ** 2 *   &
        &              E2 ** 2 / F2t ** 4 + Y6s ** 2 * G12 ** 2 / F6 ** 4) *   &
        &              Y6s ** 2 * G12 ** 4 / F6 ** 8 * sigma6 / (1 - D6) ** 3  &
        &              * (-1 + D6) * eps6)
!                  compute dgamma_ddelta
                   dgamma_ddelta=c1/c2*dexp(delta/c2)
                   lambda = - g / (prod + dgamma_ddelta)
!                  STEP 6. update internal variables
!                  compute df_dY
                   df_dY(2) = (((1 - GIc / GIIc) * Y2s * E2 / F2t ** 2 +    &
        &              GIc / GIIc * Y2s ** 2 * E2 ** 2 / F2t ** 4 + Y6s ** 2 * &
        &              G12 ** 2 / F6 ** 4) ** (-0.1D1 / 0.2D1) * ((1 - GIc /  &
        &              GIIc) * E2 / F2t ** 2 + 2 * GIc / GIIc * Y2s * E2 ** 2 &
        &              / F2t ** 4)) / 0.2D1
                   df_dY(3) = ((1 - GIc / GIIc) * Y2s * E2 / F2t ** 2 +      &
        &              GIc / GIIc * Y2s ** 2 * E2 ** 2 / F2t ** 4 + Y6s ** 2 * &
        &              G12 ** 2 / F6 ** 4) ** (-0.1D1 / 0.2D1) * Y6s *         &
        &              G12 ** 2 / F6 ** 4
!                  update the state variables
```

```
            delta = delta - lambda        ! hardening variable
            D2 = D2 + lambda * df_dY(2)    ! transverse damage
            D6 = D6 + lambda * df_dY(3)    ! df_dY(3)=df/dY6
         endif
         if (k.ge.kmax) then  !RMA did not converge
            write(*,*) "*** ERROR: RMA in UMATPS85 did not converge &
     &                       ***"
            call XIT
         endif
!        STEP 7 . end linearized process
      enddo
```

Abaqus needs the tangent stiffness matrix DDSDDE but the secant is available from the code above, and it is easier to calculate, so we approximate the tangent by the secant.

```
!        STEP 8. compute tangent stiffness matrix.
!        For now, use the secant as tangent
!        Instead, we should use eq. (8.111) and step 8 in Sect. 8.4.1
         DDSDDE(1:3,1:3) = C
         DDSDDE(3,3) = C(3,3)/2   ! come back to engineering notation
         IF (NSHR.GT.1) THEN !shells
            DDSDDE(4,4) = G13 !out-of plane do not damage, Abaqus notation
            DDSDDE(5,5) = G23 !out-of plane do not damage, see Table 1.1 (2007)
            STRESS(4) = G13 * ( STRAN(4) + DSTRAN(4) )
            STRESS(5) = G23 * ( STRAN(5) + DSTRAN(5) )
         ENDIF
```

Next we update the state variables to their current value. In-plane stresses computed inside the return mapping algorithm (RMA) are current, and out-of-plane values are updated in the previous step.

```
!        STEP 9. Update the stress and state variables
!        In-plane stresses computed inside the RMA are current
!        and out-of-plane values are updated in the previous step
!        STRESS = STRESS + MATMUL(DDSDDE,DSTRAN) !Alternative method
!        Update state variables
         statev(1) = delta
         statev(2) = D2
         statev(3) = D6
```

Before exiting the subroutine, we update the state variables.

```
!        Update state variables
         statev(1) = delta
         statev(2) = D2
         statev(3) = D6
```

12.6.4   UMATPS85.FOR

UMATPS85.for is used in Examples 8.5 and 8.6. The Fortran code is available in [2, Ex_8-5].

In Example 8.5, SDV1 = δ (hardening variable), SDV2 = D2 (transverse damage), and SDV3 = D6 (shear damage) are all zero until damage D2 starts at step time = 15.

Using ex85.cae, in Module: Visualization, Toolbar: Field Output, select SDV1, then in Menu: Result, Step/frame, select increment = 6 (step time 15.0), then select increment = 7 (step time = 17.5) to see SDV1 and SDV2 grow but SDV3 remains at zero.

Remember that in CDM, damage variables are just a representation of the reduction of stiffness due to damage, not a direct representation of damage such as crack density. In Example 8.5, SDV3 remains zero because no shear is applied, and thus the simulation is unable to detect the reduction of shear stiffness.

In Example 8.6, SDV1 = δ (hardening variable), SDV2 = D2 (transverse damage), and SDV3 = D6 (shear damage) are all zero until damage D2 and D6 start to grow simultaneously at step time = 30. They grow simultaneously because the reductions of stiffness D2 and D6 are both a reflection of the underlying damage, which is in the form of transverse matrix cracks.

Using `ex86.cae`, in Module: Visualization, Toolbar: Field Output, select SDV1, then in Menu: Result, Step/frame, select increment = 10 (step time 30.0), then select increment = 11 (step time = 33.0) to see SDV1, SDV2, and SDV3 grow together.

12.7 UGENS. User General Section

Unlike UMAT, UGENS represents a non-local constitutive model. That is, by representing the material in the form of ABDH matrices, the constitutive model at a Gauss point in a conventional shell element takes into account what happens to all the laminas through the thickness of the shell.

A non-local constitutive model is a crucial necessity for fracture mechanics and discrete damage mechanics calculations because the Griffith-Irwin criterion compares energy release rate (ERR) to critical ERR (fracture toughness). Since the ERR is a non-local quantity that must be computed over the entire laminate, the computation of the Griffith-Irwin criterion at a conventional shell Gauss point requires knowledge of the state of the material in all laminas at once. If one were to use a local constitutive model for one lamina at a time, such as using a UMAT, the information regarding the state of damage on all other laminas would have to be iterated by Abaqus at the structural level, together with the equilibrium iterations, severely hampering convergence.

12.7.1 Alternative Solution for Example 9.1

To solve Example 9.1 we can use the scripts `Ex_9.1_pre.py` and `Ex_9.1_props.py`.

In `Ex_9.1_pre.py`, load the Python Modules:

```
from abaqusConstants import *
from mesh import *
from step import *
from regionToolset import Region
from multiprocessing import cpu_count
from visualization import openOdb
from abaqus import mdb        # mdb.model defined
import csv                    # utilities to write a .CSV file
```

`UgenKeyword` contains utilities to write UGENS parameters on the .inp file.

```
from UgenKeyword import *
```

The following statement simplifies parameterization of a script generated by using CAE.

```
session.journalOptions.setValues(
    replayGeometry=COORDINATE, recoverGeometry=COORDINATE)
```

GetKeywordPosition finds the block position in the input file where we want to insert additional lines that cannot be inserted by CAE.

```
# Define GetKeywordPosition after only 'mdb.model' is defined
import string # needed by GetKeywordPosition
def GetKeywordPosition(modelName, blockPrefix, occurrence=1):
    # Usage: set "position" so that
    #    *Initial Conditions is placed before *Step on the .inp file
    # position = GetKeywordPosition( 'Model-1', '*Step')-1
    # Model.keywordBlock.insert(
    #    position, """*Initial Conditions, type=SOLUTION, USER""")
    if blockPrefix == '':
        return len(mdb.models[modelName].keywordBlock.sieBlocks)-1
    pos = 0
    foundCount = 0
    for block in mdb.models[modelName].keywordBlock.sieBlocks:
        if string.lower(block[0:len(blockPrefix)])==\
            string.lower(blockPrefix):
            foundCount = foundCount + 1
            if foundCount >= occurrence:
                return pos
        pos=pos+1
    return -1 #next line empty
```

Model parameters are provided next:

```
NL = 3                                  # number of laminas (scalar)
ModelParameters = {}                    # define a dictionary data tye
ModelParameters['E1']      = '44700'    # add to the dictionary
ModelParameters['E2']      = '12700'
ModelParameters['G12']     =  '5800'
ModelParameters['Nu12']    = '.297'
ModelParameters['Nu23']    = '.41'
ModelParameters['Cte1']    = '3.7'
ModelParameters['Cte2']    = '30.'
ModelParameters['GIIc']    = '1e16'
ModelParameters['GIc']     = '0.254'
ModelParameters['tk']      = '.144'
ModelParameters['dummy']   = '0'
# A [list] of tuples (thickness angle)
ModelParameters['Lss']     = [(1,0), (8,90), (.5,0)]
ModelParameters['Strain'] = '1.8735'   # max. strain to run the model in %
ShellDimensionX = 10.                  # model dimensions
ShellDimensionY = 10.                  # model dimensions
```

Next, generate the Abaqus model. Ignore Warning "The following parts have some elements without any section assigned: Laminate".

```
mdb.close()
Model = mdb.models['Model-1']
# shell dimensions
Sketch = Model.ConstrainedSketch(
    name='LaminateSketch', sheetSize=10.*ShellDimensionX)
Sketch.rectangle(point1 = (0.0, 0.0),
                 point2 = (ShellDimensionX, ShellDimensionX)
                 )
Part = Model.Part(dimensionality = THREE_D,
                  name           = 'Laminate',
                  type           = DEFORMABLE_BODY
```

```
                    )
Part.BaseShell(sketch=Sketch)
Instance = Model.rootAssembly.Instance(dependent = ON,
                                       name='Laminate-1',
                                       part=Part
                                       )
Part.Set(faces=Part.faces,
         name='WholeLaminate'
         )
```

Next, define the Step and Field Output Requests:

```
# minInc, reduce it as much as needed
# maxNumInc, set to a large number
Step = Model.StaticStep(timePeriod=float(ModelParameters['Strain']),
                        initialInc=0.01,
                        minInc=0.0001,
                        maxInc=0.01,
                        maxNumInc=1000,
                        name='ApplyStrain',
                        previous='Initial'
                        )
Model.fieldOutputRequests['F-Output-1'].setValues(
    variables=('U', 'RF', 'SDV'))
```

Next, define the boundary conditions:

```
Model.XsymmBC(createStepName='Initial',
              name='SymmetryX',
              region=Region(edges=Instance.edges.findAt(((
              0.0, ShellDimensionY/2., 0.0), ), ))
              )
Model.YsymmBC(createStepName='Initial',
              name='SymmetryY',
              region=Region(edges=Instance.edges.findAt(((
              ShellDimensionX/2., 0.0, 0.0), ), ))
              )
Model.EncastreBC(createStepName='Initial',
                 name='NoRigidBodyMotion',
                 region=Region(vertices=Instance.vertices.findAt(((
                 0.0, 0.0, 0.0), ), ))
                 )
Model.DisplacementBC(createStepName='ApplyStrain',
                     name='Strain',
                     region=Region(edges=Instance.edges.findAt(((
                     ShellDimensionX, ShellDimensionY/2., 0.0), ), )),
                     u1=ShellDimensionX*float(ModelParameters['Strain'])
                     /100.
                     )
```

Next, apply a membrane strain:

```
mdb.models['Model-1'].DisplacementBC(
    amplitude=UNSET, createStepName='Initial',
    distributionType=UNIFORM, fieldName='', localCsys=None, name='BC-4',
    region=Region(
    faces=mdb.models['Model-1'].rootAssembly.instances['Laminate-1'].faces.\
    findAt(((ShellDimensionX/2., ShellDimensionY/2., 0.0), (0.0, 0.0, 0.0)),
    )), u1=UNSET, u2=UNSET, u3=SET, ur1=SET, ur2=SET, ur3=SET
    )
```

The controls may need to be adjusted for convergence. Abaqus default $R_n^\alpha =$ 0.005 (first value in the list) changed to 0.02. R_n^α is the convergence criterion for ratio of largest residual to corresponding average flux norm.

```
mdb.models['Model-1'].steps['ApplyStrain'].control.setValues(
    allowPropagation=OFF, resetDefaultValues=OFF, displacementField=(0.02,
    0.01, 0.0, 0.0, 0.02, 1e-05, 0.001, 1e-08, 1.0, 1e-05, 1e-08),
    electricalPotentialField=(0.005, 0.01, 0.0, 0.0, 0.02, 1e-05, 0.001,
    1e-08, 1.0, 1e-05), hydrostaticFluidPressureField=(0.005, 0.01, 0.0,
    0.0, 0.02, 1e-05, 0.001, 1e-08, 1.0, 1e-05), rotationField=(0.005,
    0.01, 0.0, 0.0, 0.02, 1e-05, 0.001, 1e-08, 1.0, 1e-05)
    )
```

Next, we mesh:

```
Part.setElementType(elemTypes=(ElemType(elemCode=S4R), ),
                    regions=Region(faces = Part.faces)
                    )
Part.seedPart(size=ShellDimensionX) # only one element
Part.generateMesh()
```

Next, define a temporary job before adding keyword blocks. Abaqus does not allow UGENS with multiple Cpus. Do not use numCpus=cpu_count(). Use numCpus=1.

```
Job = mdb.Job(name='Job-ddm-exe',
              model='Model-1',
              nodalOutputPrecision=FULL,
              numCpus=1,
              numDomains=cpu_count(),
              userSubroutine='abaqusddm-std.obj'
              )
```

Next, we use the method `GetKeywordPosition` to find the position in the input file so that *Transverse Shear Stiffness is placed before *End Part in the new input file. Then, insert the model parameters.

```
Model.keywordBlock.synchVersions(storeNodesAndElements=False)
position = GetKeywordPosition( 'Model-1', '*End Part')-1
Model.keywordBlock.insert(position, UgenKeyword(ModelParameters))
```

Next, we find the position so that *Initial Conditions is placed before *Step in the new input file. Then, insert *Initial Conditions, type=SOLUTION, USER.

```
position = GetKeywordPosition( 'Model-1', '*Step')-1
Model.keywordBlock.insert(position, """*Initial Conditions
    , type=SOLUTION, USER""")
```

The new input file will be written automatically when the Job is submitted for execution. Then, create a new Job "from input file" and submit it.

```
mdb.JobFromInputFile(activateLoadBalancing=False, atTime=None,
    explicitPrecision=SINGLE, getMemoryFromAnalysis=True, inputFileName=
    'Job-ddm-exe.inp', memory=90, memoryUnits=PERCENTAGE,
    multiprocessingMode=DEFAULT, name='Job-ddm-exe-1'
    , nodalOutputPrecision=SINGLE, numCpus=1, numDomains=1
    , parallelizationMethodExplicit=DOMAIN, queue=None
    , scratch='', type=ANALYSIS, userSubroutine=
    'abaqusddm-std.obj', waitHours=0, waitMinutes=0)
mdb.jobs['Job-ddm-exe-1'].submit(consistencyChecking=OFF)
Job.waitForCompletion()
```

Now in CAE, open the Job Manager, select Job-ddm-exe-1, and click 'Results'. This will switch the focus to Module: Visualization.

Table 12.1: Information stored by SDVs in `abaqusddm-std.obj`

SDV	Physical quantity
SDV1	CrackDensity λ_1 in lamina 1
SDV2	Transverse damage D_2 in lamina 1
SDV3	Transverse damage D_6 in lamina 1
... repeats for all laminas	

Using Menu: Result, Field Output or the Field Output Toolbar you should get this:

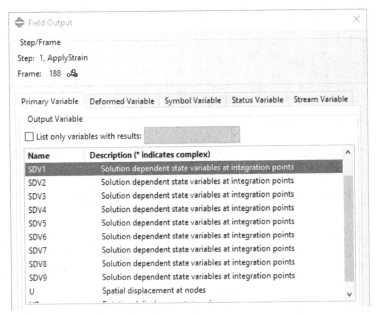

The plugin `abaqusddm-std.obj` uses 3N state variables, where N is the number of laminas in the symmetric one-half of the actual laminate. For each lamina starting with the bottom (external) lamina, the state variables are shown in Table 12.1. This example has three laminas and nine SDVs.

The crack density λ is the only true state variable. The damage variables D_2 and D_6 can be calculated from λ using (9.33) but they are stored for convenience because they are used many times in the code. In this example SDV1, 4, and 7 are crack densities for laminas $1, 2, 3$. Lamina $k = 2$ damages first because it is much thicker then the others. Its damage is tracked by SDV4 as a function of increment (frame) number. Damage starts at increment 47 and grows up to a value of 1.016 for increment 188. See Example 9.1 for additional details and results.

12.7.2 Alternative Solution for Example 11.1

To solve Example 11.1 and 11.2 we can use the Python scripts `Ex-11-1.py` and `Ex-11-2.py` that include everything needed for the simulation plus post-processing to examine the results.

For Example 11.1, that is, monotonic cooling during the first cycle (N = 1), we set ModelParameters['Cycles'] = 1, that is, only the first thermal cycle (N = 1) is performed (no fatigue). In this case, the defect nucleation rates β_I, β_{II} are not needed, so we set them to zero.

For Example 11.2 we set ModelParameters['Cycles'] = N with $N > 1$, and the analysis is performed for that cycle. In this case, values for the defect nucleation rates β_I, β_{II} are needed. To get the complete fatigue story one had to repeat the analysis for every cycle or modify the script to do that automatically.

The script, which is available in [2], can be explained as follows.

First, Load the Python Modules.

```
from abaqusConstants import *
from mesh import *
from step import *
from regionToolset import Region
from multiprocessing import cpu_count
from visualization import openOdb
from abaqus import mdb # defines object "mdb"
import csv # utilities to write a .CSV file
import os
import string # needed by GetKeywordPosition
```

The following simplifies the parameterization of models created during a CAE session.

```
session.journalOptions.setValues(
    replayGeometry=COORDINATE, recoverGeometry=COORDINATE)
```

Next we import utilities to write UGENS parameters on the Job.inp file. The utilities UgenKeywordTM_UGENS() and InitialConditionsKeyword() are defined in the script UgenKeywordTM_UGENS.py, available in [2]. They are imported as follows

```
# utilities to write UGENS parameters on the Job.inp file
from UgenKeywordTM_UGENS import *
```

Then, define `GetKeywordPosition` but not before the 'mdb' object is defined (see above). This finds the position in the .inp file where you want to insert something that CAE cannot insert in the `.inp` file.

```
def GetKeywordPosition(modelName, blockPrefix, occurrence=1):
    # Usage: set "position" so that *Initial Conditions
    #     is placed before *Step on the .inp file
    #     position = GetKeywordPosition( 'Model-1', '*Step')-1
    # Model.keywordBlock.insert(
    #     position, """*Initial Conditions, type=SOLUTION, USER""")
    if blockPrefix == '':
        return len(mdb.models[modelName].keywordBlock.sieBlocks)-1
    pos = 0
    foundCount = 0
    for block in mdb.models[modelName].keywordBlock.sieBlocks:
        if string.lower(block[0:len(blockPrefix)])==\
            string.lower(blockPrefix):
            foundCount = foundCount + 1
            if foundCount >= occurrence:
                return pos
        pos=pos+1
    return -1
```

Next, enter the model parameters. These are data that the script later inserts in the modified .inp file. Start with the elastic properties.

```
# Enter your model parameters here
ModelParameters = {}           # define a dictionary data type
# Elastic properties
# Quadratic temp-dependent properties: P = a + bT + cT^2
ModelParameters['E1a']         = '271270.586' # [MPa]
ModelParameters['E1b']         = '-8.10997'
ModelParameters['E1c']         = '1.18938E-02'
ModelParameters['E2a']         = '6554.2638'
ModelParameters['E2b']         = '-11.6689'
ModelParameters['E2c']         = '4.9329E-04'
ModelParameters['G12a']        = '3998.0213'
ModelParameters['G12b']        = '-8.84364'
ModelParameters['G12c']        = '6.1187E-03'
ModelParameters['Nu12a']       = '0.3147'
ModelParameters['Nu12b']       = '-6.9707E-05'
ModelParameters['Nu12c']       = '-4.0521E-07'
ModelParameters['Nu23a']       = '0.5557'
ModelParameters['Nu23b']       = '-1.0089E-04'
ModelParameters['Nu23c']       = '-1.1402E-06'
```

Now the thermal properties.

```
# Thermal properties
# Cubic temp-dependent properties: P = a + bT + cT^2 + dT^3
ModelParameters['CTE1a']       = '-0.9766711'    # [microstrain]
ModelParameters['CTE1b']       = '-9.0549E-05'
ModelParameters['CTE1c']       = '-7.6732E-06'
ModelParameters['CTE1d']       = '2.3038E-08'
ModelParameters['CTE2a']       = '38.4684'
ModelParameters['CTE2b']       = '8.9456E-02'
ModelParameters['CTE2c']       = '-3.6454E-04'
ModelParameters['CTE2d']       = '0.0'
```

Next the fracture properties.

```
# Fracture properties
# Quadratic temp-dependent properties: P = A + BT + CT^2
ModelParameters['GIcA']        = '0.18187' # [kJ/m^2]
ModelParameters['GIcB']        = '0.0'
ModelParameters['GIcC']        = '0.0'
ModelParameters['TGImax']      = '24'        # [Celsius]
ModelParameters['TGImin']      = '-184'      # [Celsius]
ModelParameters['GIIcA']       = '1'
ModelParameters['GIIcB']       = '0.0'
ModelParameters['GIIcC']       = '0.0'
ModelParameters['TGIImax']     = '24'        # [Celsius]
ModelParameters['TGIImin']     = '-184'      # [Celsius]
```

Next, the temperature range for which the elastic and thermal properties were measured. Since the properties are not know outside that range, they are not extrapolated but assumed to be constant outside of the range.

```
# Temperature range for elastic (e) and thermal (t) properties
ModelParameters['SFT']         = '177'        # Stress free temperature
ModelParameters['Temax']       = '120'        # Upper limit(e) [Celsius]
ModelParameters['Temin']       = '-156'       # Lower limit(e) [Celsius]
ModelParameters['Ttmax']       = '120'        # Upper limit(t) [Celsius]
ModelParameters['Ttmin']       = '-156'       # Lower limit(t) [Celsius]
```

For failure prediction, the fiber failure properties are needed.

```
# Fiber failure properties
ModelParameters['F1t']        = '1900.87' # [MPa]
ModelParameters['F1c']        = '441.2'
ModelParameters['F2c']        = '57.23'
ModelParameters['mWeibull']   = '8.9'      # []
```

The defect nucleation rates β_I, β_{II} are only needed for thermal fatigue, to define the defect nucleation function f(N) in mode I and II (see Section 11.4). If monotonic cooling is simulated, then cycles $N = 1$, no fatigue and β_I, β_{II} are now used, so we set them to zero. Otherwise, for fatigue, cycles $N > 1$, enter the number of cycles N. If $N > 1$, then β_I and β_{II} cannot be zero.

For N = 1 (thermal cycle only).

```
ModelParameters['Cycles']   = '1'     # number of cycles
ModelParameters['BetaGIc']  = '0.0'   # [1/m]
ModelParameters['BetaGIIc'] = '0.0'   # [1/m]
```

For $N > 1$ (fatigue cycle).

```
ModelParameters['Cycles']   = '10'      # number of cycles
ModelParameters['BetaGIc']  = '-0.204'  # [1/m], mode I
ModelParameters['BetaGIIc'] = '-0.204'  # [1/m], not needed
```

Note that β_{II} is not needed because for a cross-ply subjected to thermal load $G_{II} = 0$. After damage starts, the laminate is no longer a true cross-ply but still $G_{II} << G_I$ and $G_{IIC} > G_{IC}$ for most composite materials, so the effect of mode II could be ignored.

The rest of the data defines the model. The LSS is 1/2 of the symmetric laminate, starting at k = 1 (bottom surface) with split middle lamina.

```
# a [list] of tuples (angle,thickness)
ModelParameters['Lss']      = [(0,1), (90,1), (0,1), (90,.5)]
ModelParameters['tk']       = '0.127'    # Ply thickness
NL = len(ModelParameters['Lss'])
```

The geometry of the shell.

```
ShellDimensionX = 100.                      # model dimensions
ShellDimensionY = 100.                      # model dimensions
Arclength       = 30.
ModelParameters['IntegrationPoints']    = '3' # Simpson
```

The load is provided by a strain and a temperature range.

```
ModelParameters['Strain']   = '0'       # max. strain
ModelParameters['Temp']     = '-156'    # lowest temperature Tmin
ModelParameters['SFT']      = ' 177'    # stress free temperature SFT
```

The plugin, called DDM6TM, needs 45 material properties and 12 state variables per layer (3 true state variables + 3 damages + 3 stresses + 3 energies). Of the 12 state variables stored by Abaqus, only three are true state variables for the model. The other nine are simply stored as state variables to allow for visualization. The whole laminate thickness (Shell Thickness) and the number of Simpson integration points are necessary for the definition of the Section (Figure 12.9).

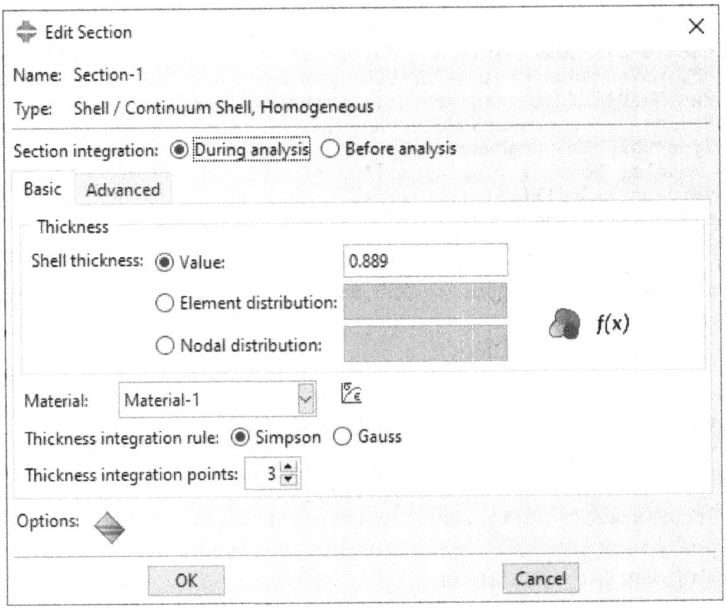

Fig. 12.9: Section Definition

```
PROPERTIES = 45 + 2*NL
VARIABLES = 12*NL
WholeLaminateThickness = 2.*sum(
    [x[1] for x in ModelParameters['Lss']])*float(ModelParameters['tk'])
IntegrationPoints = float(ModelParameters['IntegrationPoints'])
```

Start the generation of the model.

```
# Generate the Abaqus model
mdb.close()
Model = mdb.models['Model-1']
```

Part.

```
# Part
Sketch = Model.ConstrainedSketch(
    name='LaminateSketch', sheetSize=10.*ShellDimensionX)
Sketch.rectangle(point1 = (0.0, 0.0),
                 point2 = (ShellDimensionX, ShellDimensionX))
Part = Model.Part(dimensionality = THREE_D,
                  name           = 'Laminate',
                  type           = DEFORMABLE_BODY)
Part.BaseShell(sketch=Sketch)
Instance = Model.rootAssembly.Instance(dependent = OFF,
                                       name='Laminate-1',
                                       part=Part)
Part.Set(faces=Part.faces, name='WholeLaminate')
SetEdge=Model.rootAssembly.instances['Laminate-1'].edges
```

Define a dummy material used for later locating the KeywordBloc, that is the position where to put the actual material properties required by the fatigue damage model.

```
# Material properties
Material = mdb.models['Model-1'].Material(name='Material-1')
```

```
Model.HomogeneousShellSection(idealization=NO_IDEALIZATION,
    integrationRule=SIMPSON, material='Material-1', name='Section-1',
    numIntPts=3, poissonDefinition=DEFAULT, preIntegrate=OFF,
    temperature= GRADIENT, thickness=WholeLaminateThickness,
    thicknessField='', thicknessModulus=None,
    thicknessType=UNIFORM, useDensity=OFF)
Section = mdb.models['Model-1'].sections['Section-1']
Section.TransverseShearShell(k11=1.0, k12=0.0, k22=1.0)
Part.SectionAssignment(offset=0.0,
    offsetField='', offsetType=MIDDLE_SURFACE, region=
    Part.sets['WholeLaminate'], sectionName=
    'Section-1', thicknessAssignment=FROM_SECTION)
```

Instance the part as dependent.

```
# Assembly
Model.rootAssembly.DatumCsysByDefault(CARTESIAN)
Instance = Model.rootAssembly.Instance(dependent = OFF,
                                        name='Laminate-1',
                                        part=Part)
SetEdge=Model.rootAssembly.instances['Laminate-1'].edges
```

Create a step for the simulation.

```
# STEP
# minInc, reduce it as much as needed
# maxNumInc, set to a larger number to allow for reduced minInc
Step = Model.StaticStep(
    timePeriod=float(ModelParameters['SFT'])
    - float(ModelParameters['Temp']),
                        initialInc=1.,
                        minInc=0.0001,
                        maxInc=1.,
                        maxNumInc=1000,
                        name='ApplyThermalStrain',
                        previous='Initial')
Model.fieldOutputRequests['F-Output-1'].setValues(variables=(
'U', 'RF', 'SDV'))
```

Boundary conditions.

```
# BC
# symm X=0
Model.XsymmBC(createStepName='Initial', name='SymmetryX',
            region=Region(edges=Instance.edges.findAt(
            ((0.0, ShellDimensionY/2., 0.0), ), )))
# symm Y=0
Model.YsymmBC(createStepName='Initial', name='SymmetryY',
            region=Region(edges=Instance.edges.findAt(
            ((ShellDimensionX/2., 0.0, 0.0), ), )))
# fix (0,0,0)
Model.EncastreBC(createStepName='Initial', name='NoRigidBodyMotion',
                region=Region(vertices=Instance.vertices.findAt(
                ((0.0, 0.0, 0.0), ), )))
# Membrane only
mdb.models['Model-1'].DisplacementBC(amplitude=UNSET,
    createStepName='Initial', distributionType=UNIFORM,
    fieldName='', localCsys=None, name='BC-4', region=Region(
    faces=mdb.models['Model-1'].rootAssembly.instances['Laminate-1'].\
    faces.findAt(
    ((ShellDimensionX/2., ShellDimensionY/2., 0.0), (0.0, 0.0, 0.0)), ))
    , u1=UNSET, u2=UNSET, u3=
    SET, ur1=SET, ur2=SET, ur3=SET)
```

Predefined fields to define the initial and final temperature fields.

```
# Predefined Field
# apply initial temperature = SFT
Model.Temperature(createStepName='Initial',
    crossSectionDistribution=CONSTANT_THROUGH_THICKNESS, distributionType=
    UNIFORM, magnitudes=(float(ModelParameters['SFT']), ),
    name='Predefined Field-1', region=
    Model.rootAssembly.instances['Laminate-1'].sets['WholeLaminate'])
# apply final temperature = Temp
Model.Temperature(createStepName='ApplyThermalStrain',
    crossSectionDistribution=CONSTANT_THROUGH_THICKNESS, distributionType=
    UNIFORM, magnitudes=(float(ModelParameters['Temp']), ),
    name='Predefined Field-2', region=
    Model.rootAssembly.instances['Laminate-1'].sets['WholeLaminate'])
```

Job controls, adjust as needed to get convergence.

```
# Controls
Model.steps['ApplyThermalStrain'].control.setValues(
    allowPropagation=OFF, resetDefaultValues=OFF, displacementField=(0.02,
    0.01, 0.0, 0.0, 0.02, 1e-05, 0.001, 1e-08, 1.0, 1e-05, 1e-08),
    electricalPotentialField=(0.005,0.01,0.0,0.0,0.02,1e-05,0.001,1e-08,
    1.0, 1e-05), hydrostaticFluidPressureField=(0.005, 0.01, 0.0, 0.0, 0.02,
    1e-05, 0.001, 1e-08, 1.0, 1e-05), rotationField=(0.005, 0.01, 0.0, 0.0,
    0.02, 1e-05, 0.001, 1e-08, 1.0, 1e-05))
```

Now create a 4 × 4 mesh of S4R elements.

```
# Mesh
InstancePart = Model.rootAssembly
InstancePart.setElementType(elemTypes=(ElemType(elemCode=S4R), ),
                regions=Region(faces = Part.faces))
InstancePart.seedPartInstance(deviationFactor=0.1,
    minSizeFactor=0.1, regions=(
    InstancePart.instances['Laminate-1'], ), size=ShellDimensionX/4)
InstancePart.generateMesh(regions=(
    InstancePart.instances['Laminate-1'], ))
```

Abaqus does not allow UGENS to run with multiple Cpus. Do not use numC-pus=cpu_count(). Use numCpus = 1. If using a script, do not write the .inp file incrementally. Just define the Job and wait until everything is done to modify the .inp file in memory, then Submit it.

```
Job = mdb.Job(name='Job',
            model='Model-1',
            nodalOutputPrecision=FULL,
            numCpus=1,
            numDomains=cpu_count(),
            userSubroutine='ddm6tm-2017-std.obj')
```

Now replace *Shell Section by *Shell General Section.

```
Model.keywordBlock.synchVersions(storeNodesAndElements=False)
WholeLaminateThickness = 2.*sum([x[1] for x in ModelParameters['Lss']])
    * float(ModelParameters['tk'])
# set the position
position = GetKeywordPosition( 'Model-1', '*Shell Section')
Model.keywordBlock.replace(
    position, '*Shell General Section, elset=WholeLaminate, USER,
    PROPERTIES='+str(PROPERTIES)+', VARIABLES='+str(VARIABLES)
    +'\n'+str(WholeLaminateThickness)+', '+str(IntegrationPoints))
```

Set the position so that `*properties` are placed before `*Transverse shear` on the `.inp` file.

```
position = GetKeywordPosition( 'Model-1', '*Transverse Shear')-1
Model.keywordBlock.insert(position, UgenKeywordTM_UGENS(ModelParameters))
```

Remove the dummy material.

```
position = GetKeywordPosition( 'Model-1', '*Material')
Model.keywordBlock.replace(position, '**')
```

Set the position so that `*Initial Conditions` is placed before `*Step` on the `.inp` file. The initial conditions are defined in a subroutine.

```
position = GetKeywordPosition( 'Model-1', '*Step')-1
Model.keywordBlock.insert(position, """*Initial Conditions,
    type=SOLUTION, USER""")
```

The subroutine SDVINI defines the initial conditions. It is called by Abaqus if the following command line is present. In this case the initial crack density SDV1 is set to 0.02.

```
*Initial Conditions, type=SOLUTION, USER
```

Next, write the `.inp` file (only once), and Submit the Job.

```
# write the .inp file (only once)
mdb.jobs['Job'].writeInput(consistencyChecking=OFF)
# execution starts here
Job.submit()
Job.waitForCompletion()
```

Once the Job is completed, post-processing can take place. Basically, we write the crack density in all laminas for the last frame, i.e., at $T = T_{min}$ for whatever value of number of cycles N the analysis was run. The following is a basic script that can be easily modified to extract crack density for all the frames and so on.

```
# Post-processing
fo = open('Job.txt', 'wb')        # log file
ofile = open('Job.csv', 'wb')     # results file
Writer = csv.writer(ofile)
Header = ['Life']
for i in range(NL):
    Header.append('Crack density '+str(i+1))
Writer.writerow(Header)
Results = openOdb(path='Job.odb', readOnly=True)#avoid warning
print len(Results.steps['ApplyThermalStrain'].frames) # all frames
frame  = Results.steps['ApplyThermalStrain'].frames[-1] # at Tmin
cd = [] # crack density
cd.append(ModelParameters['Cycles'])
for i in range(NL):
    cd.append(frame.fieldOutputs['SDV'+str(12*i+1)]
    .values[0].dataDouble)#2017 use dataDouble
Writer.writerow(cd) # results file
fo.write(str(cd))    # log file
fo.close()
ofile.close()
Results.close()
# mdb.close()
```

See Example 11.1 for additional details and results.

12.8 Execute Abaqus from MATLAB

See Section 10.5 for a demonstration on how to use Abaqus CAE as a calculation engine for MATLAB. Specifically, a complete Python script is developed to perform a simulation and to compare the predicted results with experimental data. The script is called repeatedly by MATLAB's function `fminsearch()` to adjust the model parameters so that the error between experimental data and simulation results is minimized.

Appendix A

Tie Constraints

Examples 6.3 and 6.5 have been updated in this edition to account for a change in Abaqus 2016 and newer with respect to previous versions. See also Section 2.4.

In Abaqus/Standard prior to 2016, when an offset exists between the master and slave surfaces, a tie constraint prevents rigid body rotation even when no rotational DOF is present for either of the participation surfaces, e.g., when the elements used have no nodal rotation d.o.f. For example, in Example 6.3, the tie constraint in Abaqus 2015 and older enforces $U_2 = 0$ at $y = a_3$ implicitly. This behavior is maintained in Abaqus/Explicit even after 2016.

Since 2016, in Abaqus/Standard the *tie constraints* allow for rotation to occur. Thus, $U_2 = 0$ has to be enforced explicitly as it is now done in Example 6.3.

Furthermore, in order to achieve the same rotational behavior of the tie constraint between the surfaces, as it was prior to 2016, in newer versions one needs to use a node based tie constraint, i.e., instead of selecting master and slave *surfaces* one needs to select *node region* (i.e., node sets) for the master and slave surfaces in the tie constraint definition. This will provide the desired boundary condition $U_{2(slave)} = U_{2(master)}$ and $U_{3(slave)} = U_{3(master)}$. See [1, Example 6.3] at time 7:22.

Note that when using *node region*, the mesh on opposite surfaces participating in the tie constraint must be exactly the same, i.e., the nodes on the slave an master surfaces must pair at exactly the same in-plane position. If there is any deviation, one would see hot spots in the resulting contour plots. The simplest way to achieve this result is to use a *sweep* mesh. This is discussed in the documentation under the section "Mesh tie constraints" in the topic "Accounting for an offset between tied surfaces."

In Abaqus release 6.13, when neither surface has rotational degrees of freedom, we find:

"*If neither surface has rotational degrees of freedom, the global translational degrees of freedom of the slave node and the closest point on the master surface are constrained to be the same. When an offset exists, Abaqus will enforce the constraint through the fixed offset like a PIN-type MPC when the nodes in the MPC are not coincident. Because the fixed offset does not rotate, the surface-based constraint will not represent rigid body rotation correctly. The constraint will represent rigid body motion correctly when the offset is zero. This behavior can be ensured by specifying that all tied slave nodes should be moved onto the master surface."*

In release 2016 and later, when neither surface has rotational degrees of freedom, we find:

"If neither surface has rotational degrees of freedom, the global translational degrees of freedom of the slave node and the closest point on the master surface are constrained to be the same. When an offset exists, the behavior of Abaqus/Standard differs from that of Abaqus/Explicit.

Abaqus/Explicit enforces the constraint through the fixed offset like a PIN-type MPC when the nodes in the MPC are not coincident. Because the fixed offset does not rotate, the surface-based constraint will not represent rigid body rotation correctly. The constraint represents rigid body motion correctly when the offset is zero. This behavior can be ensured by specifying that all tied slave nodes should be moved onto the master surface. If an offset needs to be maintained, general contact with surface-based cohesive behavior should be used, which correctly accounts for rigid body rotation of the offset (see "Surface-based cohesive behavior," Section 37.1.10).

In general, Abaqus/Standard enforces the constraint such that the surface-based constraint represents rigid body rotation correctly; the enforcement of this constraint will introduce nonlinearity in the model. There are, however, two exceptions in which rigid body rotation between the tied surfaces cannot be enforced: (1) when node-based master surfaces are used and (2) when using tie constraints for cyclic symmetry."

Appendix B

Tensor Algebra

Tensor operations are needed for the derivation of some of the equations in this textbook. Since most of these operations are not easily found in textbooks, they are presented here for reference [228].

B.1 Principal Directions of Stress and Strain

Since stress and strain tensors are symmetric and of second order, they have three real principal values and three orthogonal principal directions. The principal values λ^q and directions n_i^q of the stress tensor σ_{ij} satisfy the following

$$[\sigma_{ij} - \lambda^q \delta_{ij}]n_i^q = 0 \tag{B.1}$$

$$n_i^q n_j^q = 1 \tag{B.2}$$

where δ_{ij} is the Kronecker delta ($\delta_{ij} = 1$ if $i = j$, zero otherwise). Each of the principal directions is described by its direction cosines with respect to the original coordinate system.

The principal directions are arranged by rows into a matrix $[A]$. Then, the diagonal matrix $[A^*]$ of the principal values is

$$[A^*] = [a][A][a]^T \tag{B.3}$$

It can be shown that $[a]^{-1} = [a]^T$, where $[a]$ is the transformation matrix given by (1.21).

B.2 Tensor Symmetry

Minor symmetry provides justification for using contracted notation (Section 1.5). Minor symmetry refers to identical values of tensor components when adjacent subscripts are swapped. For example, minor symmetry of the stiffness tensor **C** means

$$A_{ijkl} = A_{jilk} = A_{\alpha\beta} \tag{B.4}$$

493

Major symmetry refers to identical values when adjacent pairs of subscripts are swapped, or when contracted subscripts are swapped. For example,

$$A_{ijkl} = A_{klij}$$
$$A_{\alpha\beta} = A_{\beta\alpha} \qquad (B.5)$$

B.3   Matrix Representation of a Tensor

A tensor A_{ijkl} with a minor symmetry has only 36 independent constants. Therefore it can be represented in contracted notation by a 6×6 matrix. Let $[a]$ be the contracted form of the tensor A. Each element of $[a]$ corresponds to an element in the tensor A according to the following transformation

$$a_{\alpha\beta} = A_{ijkl} \qquad (B.6)$$

with

$$\alpha = i \text{ when } i = j$$
$$\alpha = 9 - (i + j) \text{ when } i \neq j \qquad (B.7)$$

The same transformations apply between β and k and l, or in matrix representation, as

$$[a] = \begin{bmatrix} A_{1111} & A_{1122} & A_{1133} & A_{1123} & A_{1113} & A_{1112} \\ A_{2211} & A_{2222} & A_{2233} & A_{2223} & A_{2213} & A_{2212} \\ A_{3311} & A_{3322} & A_{3333} & A_{3323} & A_{3313} & A_{3312} \\ A_{2311} & A_{2322} & A_{2333} & A_{2323} & A_{2313} & A_{2312} \\ A_{1311} & A_{1322} & A_{1333} & A_{1323} & A_{1313} & A_{1312} \\ A_{1211} & A_{1222} & A_{1233} & A_{1223} & A_{1213} & A_{1212} \end{bmatrix} \qquad (B.8)$$

It is convenient to perform tensor operations using contracted form, especially if the result can be represented also in contracted form. This saves memory and time since it is faster to operate on 36 elements than on 81 elements. Examples of these operations are the inner product of two fourth-order tensors and the inverse of a fourth-order tensor. However, tensor operations in index notation do not translate directly into matrix operations in contracted form. For example, the double contraction of two fourth-order tensors is

$$\mathbf{C} = \mathbf{A} : \mathbf{B}$$
$$C_{ijkl} = A_{ijmn} B_{mnkl} \qquad (B.9)$$

Let $[a]$, $[b]$, and $[c]$ be the 6×6 matrix representations of the above tensors. Then, it can be shown that

$$[a][b] \neq [c] \text{ or}$$
$$a_{\alpha\beta} b_{\beta\gamma} \neq c_{\alpha\gamma} \text{ (matrix multiplication)} \qquad (B.10)$$

The rest of this appendix presents formulas for adequate representation of tensor operations in their contracted form.

B.4 Double Contraction

In (B.9), an element like C_{1211} can be expanded as

$$
\begin{aligned}
C_{1211} = {} & A_{1211}B_{1111} + A_{1222}B_{2211} + A_{1233}B_{3311} \\
& + 2A_{1212}B_{1211} + 2A_{1213}B_{1311} + 2A_{1223}B_{2311}
\end{aligned}
\tag{B.11}
$$

In order to achieve the same result by matrix multiplication, multiply the last three columns of the matrix $[a]$ by 2, and then perform the multiplication

$$
[c] =
\begin{bmatrix}
A_{1111} & A_{1122} & A_{1133} & 2A_{1123} & 2A_{1113} & 2A_{1112} \\
A_{2211} & A_{2222} & A_{2233} & 2A_{2223} & 2A_{2213} & 2A_{2212} \\
A_{3311} & A_{3322} & A_{3333} & 2A_{3323} & 2A_{3313} & 2A_{3312} \\
A_{2311} & A_{2322} & A_{2333} & 2A_{2323} & 2A_{2313} & 2A_{2312} \\
A_{1311} & A_{1322} & A_{1333} & 2A_{1323} & 2A_{1313} & 2A_{1312} \\
A_{1211} & A_{1222} & A_{1233} & 2A_{1223} & 2A_{1213} & 2A_{1212}
\end{bmatrix}
$$

$$
\begin{bmatrix}
B_{1111} & B_{1122} & B_{1133} & B_{1123} & B_{1113} & B_{1112} \\
B_{2211} & B_{2222} & B_{2233} & B_{2223} & B_{2213} & B_{2212} \\
B_{3311} & B_{3322} & B_{3333} & B_{3323} & B_{3313} & B_{3312} \\
B_{2311} & B_{2322} & B_{2333} & B_{2323} & B_{2313} & B_{2312} \\
B_{1311} & B_{1322} & B_{1333} & B_{1323} & B_{1313} & B_{1312} \\
B_{1211} & B_{1222} & B_{1233} & B_{1223} & B_{1213} & B_{1212}
\end{bmatrix}
\tag{B.12}
$$

This transformation can be produced by using the Reuter matrix $[R]$

$$
[R] =
\begin{bmatrix}
1 & 0 & 0 & 0 & 0 & 0 \\
0 & 1 & 0 & 0 & 0 & 0 \\
0 & 0 & 1 & 0 & 0 & 0 \\
0 & 0 & 0 & 2 & 0 & 0 \\
0 & 0 & 0 & 0 & 2 & 0 \\
0 & 0 & 0 & 0 & 0 & 2
\end{bmatrix}
\tag{B.13}
$$

Substituting in (B.12) we have

$$
[c] = [a]\,[R]\,[b]
\tag{B.14}
$$

B.5 Tensor Inversion

First, it is convenient to define the fourth-order identity tensor I_{ijkl} as a tensor that inner-multiplied by another fourth-order tensor yields this same tensor, or

$$
I_{ijmn}A_{mnkl} = A_{ijkl}
\tag{B.15}
$$

If A_{ijkl} has a minor symmetry, the following tensor achieves (B.15)

$$I_{ijkl} = \frac{1}{2} \left(\delta_{ik}\delta_{jl} + \delta_{il}\delta_{jk} \right) \tag{B.16}$$

where δ_{ij} is the Kronecker delta, defined as

$$\begin{aligned} \delta_{ij} &= 1 \quad \text{if } i = j \\ \delta_{ij} &= 0 \quad \text{if } i \neq j \end{aligned} \tag{B.17}$$

In Voigt contracted notation, the fourth-order identity tensor is denoted as $[i]$, which is equal to the inverse of the Reuter matrix (1.38)

$$[i] = \begin{bmatrix} 1 & 0 & 0 & 0 & 0 & 0 \\ 0 & 1 & 0 & 0 & 0 & 0 \\ 0 & 0 & 1 & 0 & 0 & 0 \\ 0 & 0 & 0 & 1/2 & 0 & 0 \\ 0 & 0 & 0 & 0 & 1/2 & 0 \\ 0 & 0 & 0 & 0 & 0 & 1/2 \end{bmatrix} = [R]^{-1} \tag{B.18}$$

Now, the inverse of a tensor is a tensor that multiplied by the original tensor yields the identity tensor, as follows

$$A_{ijmn}A_{mnkl}^{-1} = I_{ijkl} \tag{B.19}$$

Let us introduce the following notation:

$[a]^{-1}$ = inverse of the contracted form of A_{ijkl}

$[a^{-1}]$ = contracted form of the inverse of A_{ijkl}

If A_{ijkl} has a minor symmetry, the components of $a_{\alpha\beta}^{-1}$ are:

i. Multiply the last three columns of $[a]$ by 2 by using the matrix $[R]$.

ii. Invert the obtained matrix.

iii. Multiply the matrix by $[i]$.

In orther words, the matrix $[a^{-1}]$ is computed as

$$[a^{-1}] = [[a]\,[R]]^{-1}\,[i] = [i]\,[a]^{-1}\,[i] \tag{B.20}$$

B.6   Tensor Differentiation

B.6.1   Derivative of a Tensor with Respect to Itself

Any symmetric second-order tensor Φ_{ij} satisfies the following:

$$d\Phi_{ij} = d\Phi_{ji} \tag{B.21}$$

Therefore, differentiating a second-order symmetric tensor with respect to itself is accomplished as follows

$$\frac{\partial \Phi_{ij}}{\partial \Phi_{kl}} = J_{ijkl} \tag{B.22}$$

where J_{ijkl} is a fourth-order tensor defined as

$$\begin{array}{ll}
J_{ijkl} = 1 & \text{if } i = k, \text{ and } j = l \\
J_{ijkl} = 1 & \text{if } i = l, \text{ and } j = k \\
J_{ijkl} = 0 & \text{otherwise}
\end{array} \tag{B.23}$$

In contracted notation, the tensor J_{ijkl} is represented by

$$[j] = \begin{bmatrix}
1 & 0 & 0 & 0 & 0 & 0 \\
0 & 1 & 0 & 0 & 0 & 0 \\
0 & 0 & 1 & 0 & 0 & 0 \\
0 & 0 & 0 & 1 & 0 & 0 \\
0 & 0 & 0 & 0 & 1 & 0 \\
0 & 0 & 0 & 0 & 0 & 1
\end{bmatrix} \tag{B.24}$$

B.6.2 Derivative of the Inverse of a Tensor Respect to the Tensor

A second-order tensor contracted with its inverse yields the second-order identity tensor, or Kronecker delta

$$A_{ij} A_{jk}^{-1} = \delta_{ik} \tag{B.25}$$

Differentiating (B.25) with respect to A_{mn} and rearranging terms yields

$$A_{ij} \frac{\partial A_{jk}^{-1}}{\partial A_{mn}} = -\frac{\partial A_{ij}}{\partial A_{mn}} A_{jk}^{-1} \tag{B.26}$$

Pre-multiplying both sides by A_{li}^{-1} and rearranging yields

$$\frac{\partial A_{ij}^{-1}}{\partial A_{mn}} = -A_{ik}^{-1} \frac{\partial A_{kl}}{\partial A_{mn}} A_{lj}^{-1} \tag{B.27}$$

Finally, using (B.22) yields

$$\frac{\partial A_{ij}^{-1}}{\partial A_{mn}} = -A_{ik}^{-1} J_{klmn} A_{lj}^{-1} \tag{B.28}$$

Appendix C

Second-Order Diagonal Damage Models

Explicit expressions associated to second-order diagonal damage models are presented here for completeness.

C.1 Effective and Damaged Spaces

A *second-order damage tensor* can be represented as a diagonal tensor (see (8.61))

$$D_{ij} = d_i \, \delta_{ij} \quad ; \quad \text{no sum on } i \tag{C.1}$$

in a coordinate system coinciding with the principal directions of **D**, which may coincide with the fiber, transverse, and thickness directions, and d_i are the eigenvalues of the damage tensor, which represent the damage ratio along these directions. The dual variable of the damage tensor is the *integrity tensor*, $\mathbf{\Omega} = \sqrt{\mathbf{I} - \mathbf{D}}$, which represents the undamaged ratio.

The *second-order damage tensor* **D** and the *integrity tensor* $\mathbf{\Omega}$ are diagonal and have the following explicit forms

$$D_{ij} = \begin{bmatrix} d_1 & 0 & 0 \\ 0 & d_2 & 0 \\ 0 & 0 & d_3 \end{bmatrix} \tag{C.2}$$

$$\Omega_{ij} = \begin{bmatrix} \sqrt{1-d_1} & 0 & 0 \\ 0 & \sqrt{1-d_2} & 0 \\ 0 & 0 & \sqrt{1-d_3} \end{bmatrix} = \begin{bmatrix} \Omega_1 & 0 & 0 \\ 0 & \Omega_2 & 0 \\ 0 & 0 & \Omega_3 \end{bmatrix} \tag{C.3}$$

A symmetric fourth-order tensor, **M**, called the *damage effect tensor*, is defined (see (8.63)) as

$$M_{ijkl} = \frac{1}{2} \left(\Omega_{ik} \Omega_{jl} + \Omega_{il} \Omega_{jk} \right) \tag{C.4}$$

The *damage effect tensor* in contracted form multiplied by the Reuter matrix takes the form of a 6×6 array, as follows

$$
\mathbf{M} = \mathbf{M}_{\alpha\beta} =
\begin{bmatrix}
\Omega_1^2 & 0 & 0 & 0 & 0 & 0 \\
0 & \Omega_2^2 & 0 & 0 & 0 & 0 \\
0 & 0 & \Omega_3^2 & 0 & 0 & 0 \\
0 & 0 & 0 & \Omega_2\Omega_3 & 0 & 0 \\
0 & 0 & 0 & 0 & \Omega_1\Omega_3 & 0 \\
0 & 0 & 0 & 0 & 0 & \Omega_1\Omega_2
\end{bmatrix}
\tag{C.5}
$$

The damaged stiffness tensor $\mathbf{C}$ multiplied by the Reuter matrix can be written in explicit contracted notation for an orthotropic material by a 6×6 array as a function of the undamaged stiffness tensor $\overline{\mathbf{C}}$, as follows

$$
\mathbf{C}_{\alpha\beta} =
\begin{bmatrix}
\overline{C}_{11}\Omega_1^4 & \overline{C}_{12}\Omega_1^2\Omega_2^2 & \overline{C}_{13}\Omega_1^2\Omega_3^2 & 0 & 0 & 0 \\
\overline{C}_{12}\Omega_1^2\Omega_2^2 & \overline{C}_{22}\Omega_2^4 & \overline{C}_{23}\Omega_2^2\Omega_3^2 & 0 & 0 & 0 \\
\overline{C}_{13}\Omega_1^2\Omega_3^2 & \overline{C}_{23}\Omega_2^2\Omega_3^2 & \overline{C}_{33}\Omega_3^4 & 0 & 0 & 0 \\
0 & 0 & 0 & 2\overline{C}_{44}\Omega_2^2\Omega_3^2 & 0 & 0 \\
0 & 0 & 0 & 0 & 2\overline{C}_{55}\Omega_1^2\Omega_3^2 & 0 \\
0 & 0 & 0 & 0 & 0 & 2\overline{C}_{66}\Omega_1^2\Omega_2^2
\end{bmatrix}
\tag{C.6}
$$

where $\overline{C}_{44} = \overline{G}_{23}$, $\overline{C}_{55} = \overline{G}_{13}$, and $\overline{C}_{66} = \overline{G}_{12}$. The Voigt contracted notation for fourth-order elasticity tensors is used here: $C_{\alpha\beta}$ replaces C_{ijkl} where α, β take the values 1, 2, 3, 4, 5, 6, corresponding to the index pairs 11, 22, 33, 23, 13 and 12, respectively.

The relations between the effective and actual stress components assume the following expressions

$$
\begin{aligned}
\overline{\sigma}_1 &= \sigma_1\,\Omega_1^{-2}; & \overline{\sigma}_4 &= \sigma_4\,\Omega_2^{-1}\Omega_3^{-1}; \\
\overline{\sigma}_2 &= \sigma_2\,\Omega_2^{-2}; & \overline{\sigma}_5 &= \sigma_5\,\Omega_1^{-1}\Omega_3^{-1}; \\
\overline{\sigma}_3 &= \sigma_3\,\Omega_3^{-2}; & \overline{\sigma}_6 &= \sigma_6\,\Omega_1^{-1}\Omega_2^{-1};
\end{aligned}
\tag{C.7}
$$

and the strain components

$$
\begin{aligned}
\overline{\varepsilon}_1 &= \varepsilon_1\,\Omega_1^2; & \overline{\varepsilon}_4 &= \varepsilon_4\,\Omega_2\Omega_3; \\
\overline{\varepsilon}_2 &= \varepsilon_2\,\Omega_2^2; & \overline{\varepsilon}_5 &= \varepsilon_5\,\Omega_1\Omega_3; \\
\overline{\varepsilon}_3 &= \varepsilon_3\,\Omega_3^2; & \overline{\varepsilon}_6 &= \varepsilon_6\,\Omega_1\Omega_2;
\end{aligned}
\tag{C.8}
$$

where the over-line indicates an effective property.

C.2 Thermodynamic Force Y

By satisfying the Clausius-Duhem inequality, thus assuring non-negative dissipation, the following thermodynamic forces (see (8.129)) are defined

$$
Y_{ij} = -\frac{\partial\psi}{\partial D_{ij}} = -\frac{1}{2}\left(\varepsilon_{kl}-\varepsilon_{kl}^p\right)\frac{\partial C_{klpq}}{\partial D_{ij}}\left(\varepsilon_{pq}-\varepsilon_{pq}^p\right) = -\frac{1}{2}\varepsilon_{kl}^e\frac{\partial C_{klpq}}{\partial D_{ij}}\varepsilon_{pq}^e
\tag{C.9}
$$

The second-order tensor of the conjugate thermodynamic forces associated to the damage variables takes the following form

$$\mathbf{Y} = Y_{ij} = \begin{bmatrix} Y_{11} & 0 & 0 \\ 0 & Y_{22} & 0 \\ 0 & 0 & Y_{33} \end{bmatrix} \tag{C.10}$$

or in Voigt contracted notation as

$$\mathbf{Y} = Y_\alpha = \{Y_{11}, Y_{22}, Y_{33}, 0, 0, 0\}^T \tag{C.11}$$

Using (C.9), the explicit expressions for the thermodynamic forces written in terms of effective strain are found as

$$
\begin{aligned}
Y_{11} &= \frac{1}{\Omega_1^2} \left(\overline{C}_{11} \, \overline{\varepsilon}_1^{e\,2} + \overline{C}_{12} \, \overline{\varepsilon}_2^e \, \overline{\varepsilon}_1^e + \overline{C}_{13} \, \overline{\varepsilon}_3^e \, \overline{\varepsilon}_1^e + 2\overline{C}_{55} \, \overline{\varepsilon}_5^{e\,2} + 2\overline{C}_{66} \, \overline{\varepsilon}_6^{e\,2} \right) \\
Y_{22} &= \frac{1}{\Omega_2^2} \left(\overline{C}_{22} \, \overline{\varepsilon}_2^{e\,2} + \overline{C}_{12} \, \overline{\varepsilon}_2^e \, \overline{\varepsilon}_1^e + \overline{C}_{23} \, \overline{\varepsilon}_3^e \, \overline{\varepsilon}_2^e + 2\overline{C}_{44} \, \overline{\varepsilon}_4^{e\,2} + 2\overline{C}_{66} \, \overline{\varepsilon}_6^{e\,2} \right) \qquad \text{(C.12)} \\
Y_{33} &= \frac{1}{\Omega_3^2} \left(\overline{C}_{33} \, \overline{\varepsilon}_3^{e\,2} + \overline{C}_{13} \, \overline{\varepsilon}_3^e \, \overline{\varepsilon}_1^e + \overline{C}_{23} \, \overline{\varepsilon}_3^e \, \overline{\varepsilon}_2^e + 2\overline{C}_{44} \, \overline{\varepsilon}_4^{e\,2} + 2\overline{C}_{55} \, \overline{\varepsilon}_5^{e\,2} \right)
\end{aligned}
$$

The thermodynamic forces written in terms of actual stress are

$$
\begin{aligned}
Y_{11} &= \frac{1}{\Omega_1^2} \left(\frac{\overline{S}_{11}}{\Omega_1^4} \sigma_1^2 + \frac{\overline{S}_{12}}{\Omega_1^2 \Omega_2^2} \sigma_2 \sigma_1 + \frac{\overline{S}_{13}}{\Omega_1^2 \Omega_3^2} \sigma_3 \sigma_1 + \frac{2\overline{S}_{55}}{\Omega_1^2 \Omega_3^2} \sigma_5^2 + \frac{2\overline{S}_{66}}{\Omega_1^2 \Omega_2^2} \sigma_6^2 \right) \\
Y_{22} &= \frac{1}{\Omega_2^2} \left(\frac{\overline{S}_{22}}{\Omega_2^4} \sigma_2^2 + \frac{\overline{S}_{12}}{\Omega_2^2 \Omega_1^2} \sigma_2 \sigma_1 + \frac{\overline{S}_{23}}{\Omega_2^2 \Omega_3^2} \sigma_3 \sigma_2 + \frac{2\overline{S}_{44}}{\Omega_2^2 \Omega_3^2} \sigma_4^2 + \frac{2\overline{S}_{66}}{\Omega_2^2 \Omega_1^2} \sigma_6^2 \right) \qquad \text{(C.13)} \\
Y_{33} &= \frac{1}{\Omega_3^2} \left(\frac{\overline{S}_{33}}{\Omega_3^4} \sigma_3^2 + \frac{\overline{S}_{13}}{\Omega_3^2 \Omega_1^2} \sigma_3 \sigma_1 + \frac{\overline{S}_{23}}{\Omega_3^2 \Omega_2^2} \sigma_3 \sigma_2 + \frac{2\overline{S}_{44}}{\Omega_3^2 \Omega_2^2} \sigma_4^2 + \frac{2\overline{S}_{55}}{\Omega_3^2 \Omega_1^2} \sigma_5^2 \right)
\end{aligned}
$$

The derivative of the thermodynamic forces with respect to the damage $(\partial \mathbf{Y}/\partial \mathbf{D})$ is given by

$$
\frac{\partial \mathbf{Y}}{\partial \mathbf{D}} = \begin{bmatrix}
\dfrac{Y_{11}}{\Omega_1^4} & 0 & 0 & 0 & 0 & 0 \\
0 & \dfrac{Y_{22}}{\Omega_2^4} & 0 & 0 & 0 & 0 \\
0 & 0 & \dfrac{Y_{33}}{\Omega_3^4} & 0 & 0 & 0 \\
0 & 0 & 0 & 0 & 0 & 0 \\
0 & 0 & 0 & 0 & 0 & 0 \\
0 & 0 & 0 & 0 & 0 & 0
\end{bmatrix} \tag{C.14}
$$

The derivative of the thermodynamic forces with respect to the actual strain is given by

$$
\frac{\partial \mathbf{Y}}{\partial \varepsilon^e} =
\begin{bmatrix}
-\dfrac{P_{11}}{\Omega_1^2} & -\dfrac{\overline{C}_{12}\,\bar{\varepsilon}_1^e}{\Omega_1^2} & -\dfrac{\overline{C}_{13}\,\bar{\varepsilon}_1^e}{\Omega_1^2} & 0 & -2\dfrac{\overline{C}_{55}\,\bar{\varepsilon}_5^e}{\Omega_1^2} & -2\dfrac{\overline{C}_{66}\,\bar{\varepsilon}_6^e}{\Omega_1^2} \\[2mm]
-\dfrac{\overline{C}_{12}\,\bar{\varepsilon}_2^e}{\Omega_2^2} & -\dfrac{P_{22}}{\Omega_2^2} & -\dfrac{\overline{C}_{23}\,\bar{\varepsilon}_2^e}{\Omega_2^2} & -2\dfrac{\overline{C}_{44}\,\bar{\varepsilon}_4^e}{\Omega_2^2} & 0 & -2\dfrac{\overline{C}_{66}\,\bar{\varepsilon}_6^e}{\Omega_2^2} \\[2mm]
-\dfrac{\overline{C}_{13}\,\bar{\varepsilon}_3^e}{\Omega_3^2} & -\dfrac{\overline{C}_{23}\,\bar{\varepsilon}_3^e}{\Omega_3^2} & -\dfrac{P_{33}}{\Omega_3^2} & -2\dfrac{\overline{C}_{44}\,\bar{\varepsilon}_4^e}{\Omega_3^2} & -2\dfrac{\overline{C}_{55}\,\bar{\varepsilon}_5^e}{\Omega_3^2} & 0 \\[2mm]
0 & 0 & 0 & 0 & 0 & 0 \\[1mm]
0 & 0 & 0 & 0 & 0 & 0 \\[1mm]
0 & 0 & 0 & 0 & 0 & 0
\end{bmatrix}
\tag{C.15}
$$

where

$$
\begin{aligned}
P_{11} &= 2\overline{C}_{11}\,\bar{\varepsilon}_1^e + \overline{C}_{12}\,\bar{\varepsilon}_2^e + \overline{C}_{13}\,\bar{\varepsilon}_3^e \\
P_{22} &= \overline{C}_{12}\,\bar{\varepsilon}_1^e + 2\overline{C}_{22}\,\bar{\varepsilon}_2^e + \overline{C}_{23}\,\bar{\varepsilon}_3^e \\
P_{33} &= \overline{C}_{13}\,\bar{\varepsilon}_1^e + \overline{C}_{23}\,\bar{\varepsilon}_2^e + 2\overline{C}_{33}\,\bar{\varepsilon}_3^e
\end{aligned}
\tag{C.16}
$$

The derivative of the thermodynamic forces with respect to the actual unrecoverable strain is given by

$$
\frac{\partial \mathbf{Y}}{\partial \varepsilon^p} = -\frac{\partial \mathbf{Y}}{\partial \varepsilon^e}
\tag{C.17}
$$

The derivative of the actual stress with respect to damage is given by

$$
\frac{\partial \boldsymbol{\sigma}}{\partial \mathbf{D}} =
\begin{bmatrix}
P'_{11} & 0 & 0 & 0 & 0 & 0 \\[2mm]
0 & P'_{22} & 0 & 0 & 0 & 0 \\[2mm]
0 & 0 & P'_{33} & 0 & 0 & 0 \\[2mm]
0 & -\dfrac{1}{2}\dfrac{\Omega_3\overline{C}_{44}\,\bar{\varepsilon}_4^e}{\Omega_2} & -\dfrac{1}{2}\dfrac{\Omega_2\overline{C}_{44}\,\bar{\varepsilon}_4^e}{\Omega_3} & 0 & 0 & 0 \\[2mm]
-\dfrac{1}{2}\dfrac{\Omega_3\overline{C}_{55}\,\bar{\varepsilon}_5^e}{\Omega_1} & 0 & -\dfrac{1}{2}\dfrac{\Omega_1\overline{C}_{55}\,\bar{\varepsilon}_5^e}{\Omega_3} & 0 & 0 & 0 \\[2mm]
-\dfrac{1}{2}\dfrac{\Omega_2\overline{C}_{66}\,\bar{\varepsilon}_6^e}{\Omega_1} & -\dfrac{1}{2}\dfrac{\Omega_1\overline{C}_{66}\,\bar{\varepsilon}_6^e}{\Omega_2} & 0 & 0 & 0 & 0
\end{bmatrix}
\tag{C.18}
$$

where

$$
\begin{aligned}
P'_{11} &= -\overline{C}_{11}\,\bar{\varepsilon}_1^e - \overline{C}_{12}\,\bar{\varepsilon}_2^e - \overline{C}_{13}\,\bar{\varepsilon}_3^e \\
P'_{22} &= -\overline{C}_{12}\,\bar{\varepsilon}_1^e - \overline{C}_{22}\,\bar{\varepsilon}_2^e - \overline{C}_{23}\,\bar{\varepsilon}_3^e \\
P'_{33} &= -\overline{C}_{13}\,\bar{\varepsilon}_1^e - \overline{C}_{23}\,\bar{\varepsilon}_2^e - \overline{C}_{33}\,\bar{\varepsilon}_3^e
\end{aligned}
\tag{C.19}
$$

C.3   Damage Surface

An anisotropic damage criterion expressed in tensor form, introducing two fourth-order tensors, $\mathbf{B}$ and $\mathbf{J}$, defines a multiaxial limit surface in the thermodynamic force space, $\mathbf{Y}$, that bounds the damage domain. The damage evolution is defined by a damage potential associate to the damage surface and by an isotropic hardening function. The proposed damage surface g^d is given by

$$
g^d = \left(\hat{Y}_{ij}^N J_{ijhk} \hat{Y}_{hk}^N\right)^{1/2} + \left(Y_{ij}^S B_{ijhk} Y_{hk}^S\right)^{1/2} - \left(\gamma(\delta) + \gamma_0\right)
\tag{C.20}
$$

where γ_0 is the initial damage threshold value and $\gamma(\delta)$ defines the hardening.

The derivative of the damage surface with respect to thermodynamic forces is given by

$$\frac{\partial g^d}{\partial \mathbf{Y}} = \begin{bmatrix} \dfrac{J_{11} Y_{11}^N}{\Phi^N} + \dfrac{B_{11} Y_{11}^S}{\Phi^S} \\ \dfrac{J_{22} Y_{22}^N}{\Phi^N} + \dfrac{B_{22} Y_{22}^S}{\Phi^S} \\ \dfrac{J_{33} Y_{33}^N}{\Phi^N} + \dfrac{B_{33} Y_{33}^S}{\Phi^S} \\ 0 \\ 0 \\ 0 \end{bmatrix} \tag{C.21}$$

where

$$\begin{aligned} \Phi^N &= \sqrt{J_{11}(Y_{11}^N)^2 + J_{22}(Y_{22}^N)^2 + J_{33}(Y_{33}^N)^2} \\ \Phi^S &= \sqrt{B_{11}(Y_{11}^S)^2 + B_{22}(Y_{22}^S)^2 + B_{33}(Y_{33}^S)^2} \end{aligned} \tag{C.22}$$

The derivative of the damage surface with respect to damage hardening is

$$\frac{\partial g^d}{\partial \gamma} = -1 \tag{C.23}$$

C.4   Unrecoverable-Strain Surface

The unrecoverable-strain (yield) surface g^p is a function of the thermodynamic forces in the effective configuration $(\overline{\sigma}, R)$. Therefore, the unrecoverable-strain surface is

$$g^p = \sqrt{f_{ij}\overline{\sigma}_i\overline{\sigma}_j + f_i\overline{\sigma}_i} - (R(p) + R_0) \tag{C.24}$$

where $(i = 1, 2...6)$, R_0 is the initial unrecoverable-strain threshold, and R is the hardening function.

The derivative of the unrecoverable-strain surface with respect to effective stress is given by

$$\frac{\partial \mathbf{g^p}}{\partial \overline{\sigma}} = \begin{bmatrix} \dfrac{1}{2} \dfrac{f_1 + 2 f_{11}\overline{\sigma}_1 + 2 f_{12}\overline{\sigma}_2 + 2 f_{13}\overline{\sigma}_3}{\Phi^p} \\ \dfrac{1}{2} \dfrac{f_2 + 2 f_{22}\overline{\sigma}_2 + 2 f_{12}\overline{\sigma}_1 + 2 f_{23}\overline{\sigma}_3}{\Phi^p} \\ \dfrac{1}{2} \dfrac{f_3 + 2 f_{33}\overline{\sigma}_3 + 2 f_{13}\overline{\sigma}_1 + 2 f_{23}\overline{\sigma}_2}{\Phi^p} \\ \dfrac{f_4 \overline{\sigma}_4}{\Phi^p} \\ \dfrac{f_5 \overline{\sigma}_5}{\Phi^p} \\ \dfrac{f_6 \overline{\sigma}_6}{\Phi^p} \end{bmatrix} \tag{C.25}$$

where

$$
\begin{aligned}
\Phi^p \; = \; &(f_1\,\overline{\sigma}_1 + f_2\,\overline{\sigma}_2 + f_3\,\overline{\sigma}_3 + \\
&+ f_{11}\,\overline{\sigma}_1{}^2 + f_{22}\,\overline{\sigma}_2{}^2 + f_{33}\,\overline{\sigma}_3{}^2 + \\
&+ 2\,f_{12}\,\overline{\sigma}_1\,\overline{\sigma}_2 + 2\,f_{13}\,\overline{\sigma}_1\,\overline{\sigma}_3 + 2\,f_{23}\,\overline{\sigma}_2\,\overline{\sigma}_3 + \\
&+ f_6\,\overline{\sigma}_6{}^2 + f_5\,\overline{\sigma}_5{}^2 + f_4\,\overline{\sigma}_4{}^2)^{1/2}
\end{aligned}
\tag{C.26}
$$

The derivative of the yield surface with respect to unrecoverable-strain hardening is

$$
\frac{\partial g^p}{\partial R} = -1
\tag{C.27}
$$

Appendix D

Micromechanics

D.1 Periodic Microstructure Micromechanics

Source code for periodic microstructure micromechanics [38, PMM] is available in [2, sourceCode.html] and reproduced below, coded in PMM.sci using Scilab.

```
function [E1,E2,G12,v12,v23] = PMM(EA,ET,GA,vA,vT,EM,vM,Vf)
// Periodic Microstructure Micromechanics transversely isotropic fiber.
// (c) Ever J. Barbero (1994-2017)
// Equations taken from (please cite in your work):
// Barbero, E.J. and Luciano, R., "Micromechanical formulas for the
// relaxation tensor of linear viscoelastic composites with transversely
// isotropic fibers", International Journal of Solids and Structures,
// 32(13):1859--1872, 1995. http://barbero.cadec-online.com/papers/1995/
// 95BarberoLucianoMicromechanicalFormulas.pdf
  //
  // fiber coefficients
  f =ET/EA
  Delta = (1-2*vA^2*f-vT^2-2*vA^2*vT*f)/(EA*ET^2)
  C_11 = (1-vT^2)/(ET^2*Delta)
  C_22 = (1-vA^2*f)/(EA*ET*Delta)
  C_33 = C_22
  C_12 = (vA*f+vA*vT*f)/(ET^2*Delta)
  C_13 = C_12
  C_23 = (vT + vA^2*f)/(EA*ET*Delta)
  C_44 = ET/(2*(1+vT))
  C_55 = GA
  C_66 = C_55
  // matrix coefficients
  lam_m = (EM*vM)/((1+vM)*(1-2*vM))
  mu_m = EM/2/(1+vM)//isotropic matrix
  // geometry
  S3 = 0.49247 - 0.47603*Vf - 0.02748*Vf^2
  S6 = 0.36844 - 0.14944*Vf - 0.27152*Vf^2
  S7 = 0.12346 - 0.32035*Vf + 0.23517*Vf^2
  // a_values
  a1 = 4*mu_m^2 - 2*mu_m*C_33 + 6*lam_m*mu_m - 2*C_11*mu_m - ...
  2*mu_m*C_23 + C_23*C_11 + 4*lam_m*C_12 - 2*C_12^2 - lam_m*C_33 ...
  -2*C_11*lam_m + C_11*C_33 - lam_m*C_23
  a2 = 8*mu_m^3- 8*mu_m^2*C_33 + 12*mu_m^2*lam_m -4*mu_m^2*C_11 ...
  - 2*mu_m*C_23^2 + 4*mu_m*lam_m*C_23 + 4*mu_m*C_11*C_33 - ...
  8*mu_m*lam_m*C_33 - 4*mu_m*C_12^2 + 2*mu_m*C_33^2 - ...
  4*mu_m*C_11*lam_m + 8*mu_m*lam_m*C_12 + 2*lam_m*C_11*C_33 + ...
  4*C_12*C_23*lam_m - 4*C_12*C_33*lam_m - 2*lam_m*C_11*C_23 - ...
  2*C_23*C_12^2 + C_23^2*C_11 + 2*C_33*C_12^2 - C_11*C_33^2 + ...
  lam_m*C_33^2 - lam_m*C_23^2
  a3 = ((4*mu_m^2 + 4*lam_m*mu_m - 2*C_11*mu_m - 2*mu_m*C_33 - ...
```

```
C_11*lam_m - lam_m*C_33 - C_12^2)/a2)  + ...
((C_11*C_33 + 2*lam_m*C_12)/a2) - ((S3-((S6)/(2-2*vM)))/mu_m)
a4 = -1*((-2*mu_m*C_23 + 2*lam_m*mu_m - lam_m*C_23 - ...
C_11*lam_m - C_12^2 + 2*lam_m*C_12 + C_11*C_23)/a2) + ...
(S7)/(mu_m*(2-2*vM))
// C_values
C11t = lam_m + 2*mu_m - Vf*(-a4^2 + a3^2)*inv(-1*(((2*mu_m + ...
2*lam_m - C_33 - C_23)*(a4^2-a3^2))/a1) + ...
((2*(a4-a3)*(lam_m-C_12)^2)/a1^2))
C12t = lam_m + Vf*(((lam_m-C_12)*(a4-a3))/a1)*inv(1*(((2*mu_m + ...
2*lam_m - C_33 - C_23)*(a3^2-a4^2))/a1) + ...
((2*(a4-a3)*(lam_m-C_12)^2)/a1^2))
C22t = lam_m + 2*mu_m - Vf*(((2*mu_m + 2*lam_m - C_33 - ...
C_23)*a3/a1) -((lam_m-C_12)^2/a1^2))*inv(1*(((2*mu_m + 2*lam_m ...
- C_33 - C_23)*(a3^2-a4^2))/a1) + ((2*(a4-a3)*(lam_m-C_12)^2)/a1^2))
C23t = lam_m + Vf*(((2*mu_m + 2*lam_m - C_33 - C_23)*a4/a1) - ...
((lam_m-C_12)^2/a1^2))*inv(1*(((2*mu_m + 2*lam_m - C_33 - ...
C_23)*(a3^2-a4^2))/a1) + ((2*(a4-a3)*(lam_m-C_12)^2)/a1^2))
C44t = mu_m - Vf*inv((2/(2*mu_m - C_22 + C_23)) - ...
inv(mu_m)*(2*S3 - (4*S7/(2-2*vM))))
C66t = mu_m -Vf*inv(inv(mu_m-C_66) - S3/mu_m)
// C_total
C11 = C11t
C12 = C12t
C13 = C12t
C22 = (3/4)*C22t + (1/4)*C23t +(1/4)*C44t //corrected
C33 = C22
C23 = (1/4)*C22t + (3/4)*C23t -(1/4)*C44t //corrected
C55 = C66t
C66 = C66t
C44 = (C22-C23)/2 // TI enforced, do not use C44t here
// stiffness
C = [C11 C12 C13 0 0 0; C12 C22 C23 0 0 0; C13 C23 C33 0 0 0; ...
0 0 0 C44 0 0; 0 0 0 0 C55 0; 0 0 0 0 0 C66]
// compliance
S = inv(C);
// Elastic properties
E1 = 1/S(1,1)
E2 = 1/S(2,2)
v12 = -S(2,1)/S(1,1)
v23 = -S(3,2)/S(2,2)
G12 = 1/S(6,6)
G23 = 1/S(4,4)  //redundant
endfunction
```

D.2   Coefficients of Thermal Expansion

The tangent CTE is defined as

$$\alpha_i = \frac{\partial \epsilon_i}{\partial T} \quad \text{with} \quad i = 1, 2 \tag{D.1}$$

where ϵ_i are the components of strain along the longitudinal (fiber) and transverse directions, respectively, and T is the temperature.

The secant CTE is defined as

$$\bar{\alpha}_i = \frac{1}{T - SFT} \int_{SFT}^{T} \alpha_i \, dT \tag{D.2}$$

where SFT is the stress free temperature. The secant CTE relates the thermal strain of a unidirectional lamina with its CTE at any temperature without the need for specifying a reference temperature.

Source code for Coefficients of Thermal Expansion [3, Sect. 4.4] is included in `CLT.sci`, available in [2, sourceCode.html], coded using Scilab.

```
function [a1,a2] = Levin(EA,ET,GA,vA,vT,EM,vM,aA,aT,aM,Vf,E1,E2,v12,v23)
    // (c) Ever J. Barbero (2017) ISBN 978-1-138-19680-3 Sect. 4.4
    // computes lamina tangent CTE's (a1,a2) using Levin CTE equations
    // EA, ET, GA, vA, vT, aA, aT: T.I. elastic properties and CTEs of fiber
    // EM, vM, aM: isotropic matrix values at temp. T
    // E1, E2, v12, v23, must be previously calculated using PMM formulas
    // effective compliance 3x3 tensor
    S11_ = 1/E1 // E1, E2, v12, v23, previously calculated by PMM
    S22_ = 1/E2
    S12_ = -v12/E1
    S23_ = -v23/E2
    // volume average
    Vm = 1-Vf// matrix volume fraction
    a1h = aA*Vf + aM*Vm// axial. alpha tensors are diagonal
    a2h = aT*Vf + aM*Vm// transverse. alpha tensors are diagonal
    S11h = Vf/EA + Vm/EM
    S22h = Vf/ET + Vm/EM
    S12h = - vA/EA*Vf - vM/EM*Vm
    S23h = - vT/ET*Vf - vM/EM*Vm
    // A = Sf - Sm
    A11 = 1/EA - 1/EM
    A22 = 1/ET - 1/EM
    A12 = -vA/EA + vM/EM
    A23 = -vT/ET + vM/EM
    detA = A11*(A22^2-A23^2)+2*A12*(A12*A23-A22*A12)
    P11 = (A22^2-A23^2)/detA
    P22 = (A11*A22-A12^2)/detA
    P12 = (A12*A23-A22*A12)/detA
    P23 = (A12^2-A11*A23)/detA
    // effective tangent CTEs
    a1 = a1h + (S11_-S11h)*( (aA-aM)*P11 + (aT-aM)*2*P12 ) +..
        2*(S12_-S12h)*( (aA-aM)*P12 + (aT-aM)*(P22+P23) )
    a2 = a2h + (S12_-S12h)*( (aA-aM)*P11 + (aT-aM)*2*P12 ) +..
        (S22_-S22h + S23_-S23h)*( (aA-aM)*P12 + (aT-aM)*(P22+P23) )
endfunction
```

Appendix E

Software Used

Only four software applications are used throughout this textbook. Abaqus is by far the most used. BMI3 is used only in Chapter 4. MATLAB is used for symbolic as well as numerical computations. Finally, *Intel Fortran* must be available to compile and link Abaqus with user-programmed material subroutines, but its usage is transparent to the user because it is called by a batch file requiring no user intervention. Of course, some knowledge of *Fortran* is required to program new user-material subroutines, but programming is made easier by several example subroutines, which are provided and used in the examples.

The aim of this section is to present an introduction to the software used in this textbook, namely Abaqus and BMI3, as well as how to use *Intel Fortran* to compile and link user subroutines with Abaqus. It is assumed that the reader can use MATLAB without help besides that provided by the self-explanatory code included with the examples, either printed in this textbook or downloadable from the Web site [2].

Operation of the software is illustrated for a *Windows 10* platform but operation in a *Linux* environment is similar. For the sake of space, this section is very brief. The vendors of these applications have a wealth of information, training sessions, user groups, and so on, that the reader can use to get familiar with the software interface. One such source of information is the website for this textbook [2].

Abaqus is a commercial finite element analysis (FEA) application. It has a friendly graphical user interface (GUI) called *CAE* and an extensive documentation system. Once started, the user should have no difficulty navigating menus and so on.

All the mouse clicks in the GUI generate CAE script, which are saved in several files, but the script is very difficult to understand without significant time investment in learning CAE scripting [229].

The examples in this textbook were produced using Abaqus version 2020. They work identically in *Windows 10* and *Linux* platforms. On *Windows 10*, Abaqus is accessible from the Windows START menu, through "Dassault Systemes SIMULIA Established Products 2020", `Abaqus CAE`.

In this textbook, it is assumed that the user has created a folder `C:\Simulia\ User`, where the model files will reside. This folder is called `Work directory`,

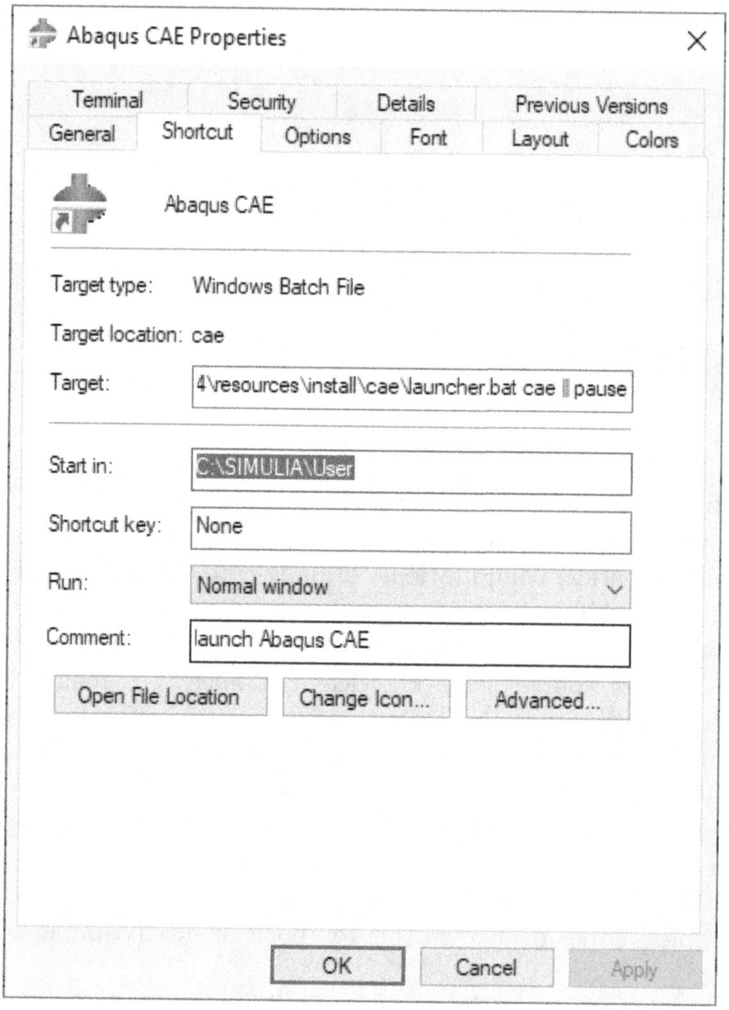

Fig. E.1: How to permanently set the default work directory.

and the location of it was set during installation of the Abaqus software, ideally at `C:\Simulia\User` but it might have been set at `C:\temp`. The default work directory can be changed by editing `Start in:` in Figure E.1. First, from Windows Start, `Dassault Systemes SIMULIA Established Products 2020`, find `Abaqus CAE`, right-click on it, then `More`, and `Open file location`. This points to the real shortcut file for Abaqus CAE. Right-click on it to get a popup and click `Properties` at the bottom. Then, change `Start in` as in Figure E.1.

Otherwise, the Work Directory for the current session can be changed from within Abaqus CAE by doing the following:

Menu: File, Set Work Directory

The default directory is used by Abaqus as a default location to store temporary files, including:

.dat The data file contains a printout of requested results.

.fil The result file contains results such as mode shapes and so on (see Example 4.3, p. 193). This file is unformatted.

.inp The input file contains the input to the Abaqus processor.

.jnl The journal file contains changes to the .mdb since the last `Save`.

.log The log file contains a brief log of the last execution of the Abaqus processor.

.odb The output database file contains all the results from an execution of the Abaqus processor.

.rec The recovery file contains a record of the operations performed by the user on the GUI since the last `Save`.

.rpy The replay file contains a replay of the whole procedure that you used in the GUI to generate the model.

In this textbook, we prefer to store and overwrite temporary files in the default `Work directory` to maintain clean user folders, but sometimes we will purposely store temporary files in a user folder. For example, if we want to read some output from the `.dat` file, we will store it in a user folder.

This section attempts to describe the installation and systems aspects of Abaqus software. On the other hand, a description of the Abaqus CAE GUI is integrated within the examples in the textbook, starting with Chapter 3. Therefore, it is highly recommended that the user installs Abaqus, then proceeds to study Chapter 3 and personally do all the examples in Chapter 3. By the end of Chapter 3, you should be quite skilled in Abaqus CAE, at least as far is concerned with composites modeling. Even higher proficiency will be achieved by studying and practicing the examples in the entire textbook. It has been my experience that students successfully teach themselves Abaqus CAE by working the examples in this textbook, along with the help system and the documentation included with Abaqus. Another source of help is the video recordings illustrating the execution of the examples [1].

E.1 How to Install Abaqus 2020 with Intel oneAPI

Intel oneAPI provides Fortran functionality for user-programmable features (UMAT, VUMAt, UGENS, etc.) Here is the way to install it.

i. Uninstall Intel Parallel Studio if it is installed.

ii. Install Abaqus 2020. When asked were to place the default work directory, you are offered `C:\temp`. Please don't use that, instead use `C:\SIMULIA\user`

iii. Before installing Intel oneAPI, download VS 2019 Community edition here: `https://tinyurl.com/2fekxrpb`

iv. Install VS with option: Desktop development with C++. I installed version 16.10.2

v. Download Intel oneAPI Base Toolkit: `https://tinyurl.com/bd9hm6ar`

You can choose online or local. I chose local to keep a copy of the installation package in case Intel discontinues the product.

vi. Install with these options (to save disk space):

Intel oneAPI DPC++/C++ Compiler
Threading building blocks
DPC++ Library
Math Kernel Library
DPC++ compatibility tool
Distribution for GDB
Integrate with MS VS 2019
Quit the installer before proceeding to the next step.
If you encounter problems:
`https://tinyurl.com/bdc3ssud`

vii. Download the HPC toolkit here: `https://tinyurl.com/2p9xhmnk`

You can choose online or local. I chose local to keep a copy of the installation package in case Intel discontinues the product.

viii. Install with these options (to save disk space):

Intel oneAPI DPC++/C++ Compiler and Intel C++ Compiler Classic
Intel MPI Library
Intel Trace Analyzer and Collector
Intel Fortran Compiler (Beta) and Intel Fortran Compiler Classic

ix. Find vcvarsall.bat. It will be in a location similar to this:

```
C:\Program Files (x86)\Microsoft Visual Studio\2019\
    Community\VC\Auxiliary\Build\
```

x. Find var.bat It will be in a location similar to this:

```
C:\Program Files (x86)\Intel\oneAPI\compiler\2021.2.0\env
```

xi. Add both paths to the PATH:

```
Start > Control Panel > System > Advanced System Settings
    > Environment Variables > System Variables > Path > Edit
    > New > OK
```

xii. In `C:/SIMULIA/Commands`

Make a backup copy of `abq2020.bat > abq2020_original.bat`

The original looks like this:

```
@echo off
setlocal
set ABA_COMMAND=%~nx0
set ABA_COMMAND_FULL=%~f0
"C:\SIMULIA\EstProducts\2020\win_b64\code\bin\ABQLauncher.exe" %*
endlocal
```

Change it to this:

```
call "C:\Program Files (x86)\Intel\oneAPI\setvars.bat" -arch intel64
call "C:\Program Files (x86)\Intel\oneAPI\setvars-vcvarsall.bat"
    -arch intel64
@echo off
rem setlocal
rem set ABA_COMMAND=%~nx0
rem set ABA_COMMAND_FULL=%~f0
@call
"C:\SIMULIA\EstProducts\2020\win_b64\code\bin\ABQLauncher.exe" %*
rem endlocal
```

xiii. Copy `C:\SIMULIA\EstProducts\2020\win_b64\SMA\site\abaqus_v6.env` to your default working directory. For me it is at `C:\SIMULIA\User\`

Abaqus will read the file if it exists in your default working directory. Do not modify the file in `C:\SIMULIA\EstProducts\2020\win_b64\SMA\site\`.

xiv. Add the following Python code at the end of `abaqus_v6.env` and make sure the file is in your default working directory, i.e. `c:\simulai\user`

```
# Solves problem with naming convention
compile_fortran += ['/names:lowercase',]
# EJB modification to simplify scripting
# Read and execute custom initialization commands
def onCaeStartup():
    ## recoverGeometry for easy parameterization
    session.journalOptions.setValues(recoverGeometry=COORDINATE)
    ## session colors
    session.viewports[
        'Viewport: 1'].viewportAnnotationOptions.setValues(title=OFF)
    session.graphicsOptions.setValues(backgroundStyle=SOLID,
        backgroundColor='#FFFFFF', translucencyMode=2)
# do not execute onCaeStartup() here, CAE does it!
# end modification
```

In this way, `recoverGeometry=COORDINATE` is ready to help you parameterize the script that you can capture from a CAE session (see Chapter 12) and the CAE window comes up with the color scheme that you like (background-Color='#FFFFFF').

xv. Make a backup copy of

`C:\SIMULIA\EstProducts\2020\win_b64\SMA\site\win86_64.env`

Then, modify it by adding 2 lines after `compile_fortran=['ifort',`,, as follows. The last line will not work for VUMAT.

```
compile_fortran=['ifort',
      '/Qmkl:sequential',    #EJB <-- enable MKL (math library)
      '/free', #EJB <-- free format Fortran95, do not use with VUMAT
```

An Errata for the textbook is maintained in [2]. Known installation issues will be reported there.

E.2 BMI3

Native BMI3 code accepts an input file in Abaqus `.inp` format, as long as the input file is filtered by the program `inp2bmi3.exe`. Therefore, one can create the model in Abaqus CAE, write the `.inp` file from the Job Manager, filter it with `inp2bmi3.exe`, and execute it with `bmi3.exe`. The executable files are available in [2]. The input/output file sequence is described in Table E.1. The sequence of execution is:

- Create a model in Abaqus CAE and write the `Job-1.inp` file.

- Execute `inp2bmi3.exe` to get the filtered files. If needed, modify the *mode, node, component* in the `BMI3.dat` file.

- Execute `bmi3.exe`. Record the results from the display or from `BMI3.out`

Not all of the Abaqus `.inp` commands are understood by the filter `inp2bmi3.exe`. Most of the limitations could be removed by modifying the filter, for which the source code `inp2bmi3.f90` is available in [2]. The current limitations are:

- The mesh must refer to S8R or S8R5 elements (see Example 4.2, p. 189).

- All loads must be concentrated loads.

- The material properties must be given using Shell General Stiffness.

- Only one material can be used for the entire model.

- Constrained boundary conditions cannot be used.

- Materials Orientation (*orientation) cannot be used.

Table E.1: Input/Output files to execute BMI3.

Program	Read	Write	Comments
inp2bmi3.exe	Job-1.inp	BMI3.inp	filtered model
		BMI3.dat	material properties
		ABQ.inp	filtered Abaqus
BMI3.exe	BMI3.inp	BMI3.out	results
	BMI3.dat	MODES.out	mode shapes

Abaqus CAE writes an .inp file every time a job is submitted, or **Write Input** is executed from the **Job Manager** window. Then, run **inp2bmi3.exe** to generate BMI3.inp, ABQ.inp, and BMI3.dat. If ABQ.inp is executed using Abaqus in command mode (see Example 2.1, p. 45), or imported into Abaqus CAE and executed, it will give the bifurcation loads $\Lambda^{(cr)}$ and the mode shapes.

The material properties and perturbation parameters are provided via BMI3.dat, which is automatically created by the filter. The last line in BMI3.dat contains *modenum, nodenum, component*. These are the *mode, node,* and *component* used as perturbation parameter δ. If all three values are zeros (the default), BMI3 picks the lowest mode and the node-component combination that yields the largest mode amplitude. The results of BMI3 are printed in BMI3.out and the mode shapes saved in the MODES.out file.

Note that the results (bifurcation loads, slopes, and curvatures) appear with negative sign. This is usual in stability analysis. Another peculiarity of the BMI3 software is that transverse deflections w (perpendicular to the plate) have opposite sign to Abaqus results. Since transverse deflections w are often used as perturbation parameters, the change in sign must be taken into account during interpretation of results (see Example 4.2, p. 189).

References

[1] Ever J Barbero. YouTube channel. `https://tinyurl.com/3nncz4ut` — or in `https://Youtube.com/c/EverBarbero` find the playlist "Abaqus Finite Element Analysis". Last accessed, 2023.

[2] E. J. Barbero. Web resource: `http://barbero.cadec-online.com/feacm-abaqus`.

[3] E. J. Barbero. *Introduction to Composite Materials Design* `http://barbero.cadec-online.com/icmd`. CRC Press, Boca Raton, FL, 3rd edition, 2018.

[4] D. Frederick and T.-S. Chang. *Continuum Mechanics*. Scientific Publishers, Cambridge, MA, 1972.

[5] F. P. Beer, E. R. Johnston Jr., and J. T. DeWolf. *Mechanics of Materials, 3rd Edition*. McGraw-Hill, Boston, MA, 2001.

[6] J. N. Reddy. *Energy and Variational Methods in Applied Mechanics*. Wiley, New York, 1984.

[7] S. S. Sonti, E. J. Barbero, and T. Winegardner. Mechanical properties of pultruded E-glass–vinyl ester composites. In *50th Annual Conference, Composites Institute, Society of the Plastics Industry (February) pp. 10-C/1-7*, 1995.

[8] Simulia. Abaqus documentation. Accessible from the windows start menu, through `Dassault Systemes Documentation SIMULIA 2020`.

[9] E. J. Barbero. *Finite Element Analysis of Composite Materials–First Edition*. CRC Press, Boca Raton, FL, 2007.

[10] ANSYS Inc. ANSYS mechanical APDL structural analysis guide, release 14.0, 2011.

[11] Simulia. Abaqus keywords reference manual. Accessible from the windows start menu, through `Dassault Systemes Documentation SIMULIA 2020`.

[12] SolidWorks. `http://www.solidworks.com/sw/engineering-education-software.htm`.

[13] E. J. Barbero. 3-D finite element for laminated composites with 2-d kinematic constraints. *Computers and Structures*, 45(2):263–271, 1992.

[14] R. J. Roark and W. C. Young. *Roark's Formulas for Stress and Strain, 6th Edition*. McGraw-Hill, New York, NY, 1989.

[15] J. N. Reddy. *Mechanics of Laminated Composite Plates and Shells, 2nd Edition*. CRC Press, Boca Raton, FL, 2003.

[16] E. Hinton and D. R. J. Owen. *An Introduction to Finite Element Computations*. Pineridge Press, Swansea, UK, 1979.

[17] NAFEMS. Test R0031/3. Technical report, National Agency for Finite Element Methods and Standards (UK), 1995.

[18] Simulia. Abaqus benchmarks manual. Accessible from the windows start menu, through Dassault Systemes Documentation SIMULIA 2020.

[19] E. J. Barbero and J. Trovillion. Prediction and measurement of post-critical behavior of fiber-reinforced composite columns. *Composites Science and Technology*, 58:1335–1341, 1998.

[20] E. J. Barbero. Prediction of compression strength of unidirectional polymer matrix composites. *Journal of Composite Materials*, 32(5)(5):483–502, 1998.

[21] A. Puck and H. Schurmann. Failure analysis of FRP laminates by means of physically based phenomenological models. *Composites Science and Technology*, 62:1633–1662, 2002.

[22] MIL17.org. The composite materials handbook, web resource, https://app.knovel.com/kn/resources/kpMHMILH43/toc.

[23] L. A. Godoy. *Theory of Stability-Analysis and Sensitivity*. Taylor and Francis, Philadelphia, PA, 2000.

[24] E. J. Barbero, L. A. Godoy, and I. G. Raftoyiannis. Finite elements for three-mode interaction in buckling analysis. *International Journal for Numerical Methods in Engineering*, 39(3):469–488, 1996.

[25] L. A. Godoy, E. J. Barbero, and I. G. Raftoyiannis. Finite elements for post-buckling analysis. I-The W-formulation. *Computers and Structures*, 56(6):1009–1017, 1995.

[26] E. J. Barbero, I. G. Raftoyiannis, and L. A. Godoy. Finite elements for post-buckling analysis. II-Application to composite plate assemblies. *Computers and Structures*, 56(6):1019–1028, 1995.

[27] I. G. Raftoyiannis, L. A. Godoy, and E. J. Barbero. Buckling mode interaction in composite plate assemblies. *Applied Mechanics Reviews*, 48(11/2):52–60, 1995.

[28] E. J. Barbero. Prediction of buckling-mode interaction in composite columns. *Mechanics of Composite Materials and Structures*, 7(3):269–284, 2000.

[29] S. Yamada and J. G. A. Croll. Buckling behavior pressure loaded cylindrical panels. *ASCE Journal of Engineering Mechanics*, 115(2):327–344, 1989.

[30] C. T. Herakovich. *Mechanics of Fibrous Composites*. Wiley, New York, 1998.

[31] J. D. Eshelby. The determination of the elastic field of an ellipsoidal inclusion and related problems. *Proceedings of the Royal Society*, A241:376–396, 1957.

[32] J. D. Eshelby. The elastic field outside an ellipsoidal inclusion. *Acta Metallurgica*, A252:561–569, 1959.

[33] T. Mori and K. Tanaka. The elastic field outside an ellipsoidal inclusion. *Acta Metallurgica*, 21:571–574, 1973.

[34] R. Hill. A self-consistent mechanics of composite materials. *Journal of Mechanics and Physics of Solids*, 13:213–222, 1965.

[35] S. Nemat-Nasser and M. Hori. *Micromechanics: Overall Properties of Heterogeneous Materials*. North-Holland, Amsterdam, 1993.

[36] R. Luciano and E. J. Barbero. Formulas for the stiffness of composites with periodic microstructure. *International Journal of Solids and Structures*, 31(21):2933–2944, 1995.

[37] R. Luciano and E. J. Barbero. Analytical expressions for the relaxation moduli of linear viscoelastic composites with periodic microstructure. *ASME J. Applied Mechanics*, 62(3):786–793, 1995.

[38] E. J. Barbero and R. Luciano. Micromechanical formulas for the relaxation tensor of linear viscoelastic composites with transversely isotropic fibers. *International Journal of Solids and Structures*, 32(13):1859–1872, 1995.

[39] J. Aboudi. *Mechanics of Composite Materials : A Unified Micromechanical Approach. Vol. 29 of Studies in Applied Mechanics*. Elsevier, New York, NY, 1991.

[40] Z. Hashin and S. Shtrikman. A variational approach to the elastic behavior of multiphase materials. *Journal of Mechanics and Physics of Solids*, 11:127–140, 1963.

[41] V. Tvergaard. Model studies of fibre breakage and debonding in a metal reinforced by short fibres. *Journal of Mechanics and Physics of Solids*, 41(8):1309–1326, 1993.

[42] J. L. Teply and G. J. Dvorak. Bound on overall instantaneous properties of elastic-plastic composites. *Journal of Mechanics and Physics of Solids*, 36(1):29–58, 1988.

[43] R. Luciano and E. Sacco. Variational methods for the homogenization of periodic media. *European Journal of Mechanics A/Solids*, 17:599–617, 1998.

[44] E. J. Barbero and D. H. Cortes. A mechanistic model for transverse damage initiation, evolution, and stiffness reduction in laminated composites. http://dx.doi.org/10.1016/j.compositesb.2009.10.001. *Composites Part B: Engineering*, 41(2):124–132, 2010.

[45] G. J. Creus. *Viscoelasticity: Basic Theory and Applications to Concrete Structures*. Springer-Verlag, Berlin, 1986.

[46] Eric W. Weisstein. Gamma function. MathWorld–A Wolfram Web Resource. http://mathworld.wolfram.com/GammaFunction.html.

[47] G. D. Dean, B. E. Read, and P. E. Tomlins. A model for long-term creep and the effects of physical ageing in poly (butylene terephthalate). *Plastics and Rubber Processing and Applications*, 13(1):37–46, 1990.

[48] B. F. Oliveira and G. J. Creus. An analytical-numerical framework for the study of ageing in fibre reinforced polymer composites. *Composites B*, 65(3-4):443–457, 2004.

[49] K. Ogatha. *Discrete-Time Control System*. Prentice Hall, Englewood Cliffs, NJ, 1987.

[50] K. J. Hollenbeck. Invlap.m: A MATLAB function for numerical inversion of Laplace transforms by the de Hoog algorithm (1998), http://www.mathworks.com/.

[51] R. S. Lakes. *Viscoelastic Solids*. CRC Press, Boca Raton, FL, 1998.

[52] J. D. Ferry. *Viscoelastic Properties of Polymers. 3rd. Edition*. Wiley, New York, 1980.

[53] R. M. Christensen. *Theory of Viscoelasticity: Second Edition*. Dover, Mineola, NY, 2010.

[54] P. Qiao, E. J. Barbero, and J. F. Davalos. On the linear viscoelasticity of thin-walled laminated composite beams. *Journal of Composite Materials*, 34(1):39–68, 2000.

[55] E. J. Barbero, F. A. Cosso, R. Roman, and T. L. Weadon. Determination of material parameters for Abaqus progressive damage analysis of E-Glass Epoxy laminates, http://dx.doi.org/10.1016/j.compositesb.2012.09.069. *Composites Part B:Engineering*, 46(3):211–220, 2012.

[56] J. Lemaitre and A. Plumtree. Application of damage concepts to predict creep-fatigue failures. In *American Society of Mechanical Engineers, 78-PVP-26*, pages 10–26, 1978.

[57] I. N. Rabotnov. *Rabotnov: Selected Works - Problems of the Mechanics of a Deformable Solid Body*. Moscow Izdatel Nauka, 1991.

[58] L. M. Kachanov. On the creep fracture time. *Izv. Akad. Nauk USSR*, 8:26–31, 1958.

[59] N. R. Hansen and H. L. Schreyer. A thermodynamically consistent framework for theories of elastoplasticity coupled with damage. *International Journal of Solids and Structures*, 31(3):359–389, 1994.

[60] J. Janson and J. Hult. Fracture mechanics and damage mechanics: A combined approach. *Journal of Theoretical and Applied Mechanics*, 1:S18–28, 1977.

[61] E. J. Barbero and K. W. Kelly. Predicting high temperature ultimate strength of continuous fiber metal matrix composites. *Journal of Composite Materials*, 27(12):1214–1235, 1993.

[62] K. W. Kelly and E. Barbero. Effect of fiber damage on the longitudinal creep of a CFMMC. *International Journal of Solids and Structures*, 30(24):3417–3429, 1993.

[63] J. R. Rice. *Continuum Mechanics and Thermodynamics of Plasticity in Relation to Microscale Deformation Mechanisms*, chapter 2, pages 23–79. Constitutive Equations in Plasticity. MIT Press, Cambridge, MA, 1975.

[64] J. C. Simo and T. J. R. Hughes. *Computational Inelasticity*. Springer, Berlin, 1998.

[65] D. Krajcinovic, J. Trafimow, and D. Sumarac. Simple constitutive model for a cortical bone. *Journal of Biomechanics*, 20(8):779–784, 1987.

[66] D. Krajcinovic and D. Fanella. Micromechanical damage model for concrete. *Engineering Fracture Mechanics*, 25(5-6):585–596, 1985.

[67] D. Krajcinovic. Damage mechanics. *Mechanics of Materials*, 8(2-3):117–197, 1989.

[68] A. M. Neville. *Properties of Concrete, 2nd Edition*. Wiley, New York, NY, 1973.

[69] ACI, American Concrete Institute.

[70] B. W. Rosen. The tensile failure of fibrous composites. *AIAA Journal*, 2(11):1985–1991, 1964.

[71] W. Weibull. A statistical distribution function of wide applicability. *Journal of Applied Mechanics*, 18:293–296, 1951.

[72] B. W. Rosen. *Fiber Composite Materials*, chapter 3. American Society for Metals, Metals Park, OH, 1965.

[73] A. S. D. Wang. A non-linear microbuckling model predicting the compressive strength of unidirectional composites. In *ASME Winter Annual Meeting*, volume WA/Aero-1, 1978.

[74] J. S. Tomblin, E. J. Barbero, and L. A. Godoy. Imperfection sensitivity of fiber micro-buckling in elastic-nonlinear polymer-matrix composites. *International Journal of Solids and Structures*, 34(13):1667–1679, 1997.

[75] D. C. Lagoudas and A. M. Saleh. Compressive failure due to kinking of fibrous composites. *Journal of Composite Materials*, 27(1):83–106, 1993.

[76] P. Steif. A model for kinking in fiber composites. I. Fiber breakage via micro-buckling. *International Journal of Solids and Structures*, 26 (5-6):549–561, 1990.

[77] E. Barbero and J. Tomblin. A damage mechanics model for compression strength of composites. *International Journal of Solids and Structures*, 33(29):4379–4393, 1996.

[78] S. W. Yurgartis and S. S. Sternstein. *Experiments to Reveal the Role of Matrix Properties and Composite Microstructure in Longitudinal Compression Strength*, volume 1185 of *ASTM Special Technical Publication*, pages 193–204. ASTM, Philadelphia, PA, 1994.

[79] C. Sun and A. W. Jun. Effect of matrix nonlinear behavior on the compressive strength of fiber composites. In *AMD*, volume 162, pages 91–101, New York, NY, 1993. ASME, American Society of Mechanical Engineers, Applied Mechanics Division.

[80] D. Adams and E. Lewis. Current status of composite material shear test methods. *SAMPE Journal*, 31(1):32–41, 1995.

[81] A. Maewal. Postbuckling behavior of a periodically laminated medium in compression. *International Journal of Solids and Structures*, 17(3):335–344, 1981.

[82] L. M. Kachanov. Rupture time under creep conditions problems of continuum mechanics. *SIAM*, pages 202–218, 1958.

[83] E. J. Barbero and P. Lonetti. Damage model for composites defined in terms of available data. *Mechanics of Composite Materials and Structures*, 8(4):299–315, 2001.

[84] E. J. Barbero and P. Lonetti. An inelastic damage model for fiber reinforced laminates. *Journal of Composite Materials*, 36(8):941–962, 2002.

[85] P. Lonetti, R. Zinno, F. Greco, and E. J. Barbero. Interlaminar damage model for polymer matrix composites. *Journal of Composite Materials*, 37(16):1485–1504, 2003.

[86] P. Lonetti, E. J. Barbero, R. Zinno, and F. Greco. Erratum: Interlaminar damage model for polymer matrix composites. *Journal of Composite Materials*, 38(9):799–800, 2004.

[87] E. J. Barbero, F. Greco, and P. Lonetti. Continuum damage-healing mechanics with application to self-healing composites. *International Journal of Damage Mechanics*, 14(1):51–81, 2005.

[88] S. Murakami. Mechanical modeling of material damage. *Journal of Applied Mechanics*, 55:281–286, 1988.

[89] J. M. Smith and H. C. Van Ness. *Introduction to Chemical Engineering Thermodynamics, 3rd Edition*. McGraw-Hill, New York, NY, 1975.

[90] E. R. Cohen, T. Cvitas, J. G. Frey, B. Holmstrom, K. Kuchitsu, R. Marquardt, I. Mills, F. Pavese, M. Quack, J. Stohner, H. L. Strauss, M. Takami, and A. J. Thor. *Quantities, Units and Symbols in Physical Chemistry : The IUPAC Green Book, 3rd Edition, http: // old. iupac. org/ publications/ books/ author/ cohen. html*. RSC Publishing, 2007.

[91] L. E. Malvern. *Introduction to the Mechanics of a Continuous Medium*. Prentice Hall, Upper Saddle River, NJ, 1969.

[92] Y. A. Cengel and M. A. Boles. *Thermodynamics: An Engineering Approach, 3rd Edition*. McGraw-Hill, New York, NY, 1998.

[93] J. Lubliner. *Plasticity Theory*. Collier Macmillan, New York, NY, 1990.

[94] J. A. Nairn. The strain energy release rate of composite microcracking: A variational approach. *Journal of Composite Materials*, 23(11):1106–1129, 11 1989.

[95] J. A. Nairn, S. Hu, S. Liu, and J. S. Bark. The initiation, propagation, and effect of matrix microcracks in cross-ply and related laminates. In *Proceedings of the 1st NASA Advanced Composite Technical Conference*, pages 497–512, Oct. 29–Nov. 1, 1990.

[96] S. Liu and J. A. Nairn. The formation and propagation of matrix microcracks in cross-ply laminates during static loading. *Journal of Reinforced Plastics and Composites*, 11(2):158–178, Feb. 1992.

[97] J. A. Nairn. Microcracking, microcrack-induced delamination, and longitudinal splitting of advanced composite structures. Technical Report NASA Contractor Report 4472, 1992.

[98] J. A. Nairn and S. Hu. *Damage Mechanics of Composite Materials*, volume 9 of *Composite Materials Series*, chapter Matrix Microcracking, pages 187–244. Elsevier, 1994.

[99] J. A. Nairn. Applications of finite fracture mechanics for predicting fracture events in composites. In *5th International Conference on Deformation and Fracture of Composites*, 18–19 March 1999.

[100] J. A. Nairn. *Polymer Matrix Composites*, volume 2 of *Comprehensive Composite Materials*, chapter Matrix Microcracking in Composites, pages 403–432. Elsevier Science, 2000.

[101] J. A. Nairn. Fracture mechanics of composites with residual stresses, imperfect interfaces, and traction-loaded cracks. In *Recent Advances in Continuum Damage Mechanics for Composites Workshop*, volume 61, pages 2159–2167, UK, 20–22 Sept. 2000. Elsevier.

[102] J. A. Nairn and D. A. Mendels. On the use of planar shear-lag methods for stress-transfer analysis of multilayered composites. *Mechanics of Materials*, 33(6):335–362, 2001.

[103] J. A. Nairn. *Finite Fracture Mechanics of Matrix Microcracking in Composites*, pages 207–212. Application of Fracture Mechanics to Polymers, Adhesives and Composites. Elsevier, 2004.

[104] Y. M. Han, H. T. Hahn, and R. B. Croman. A simplified analysis of transverse ply cracking in cross-ply laminates. *Composites Science and Technology*, 31(3):165–177, 1988.

[105] C. T. Herakovich, J. Aboudi, S. W. Lee, and E. A. Strauss. Damage in composite laminates: Effects of transverse cracks. *Mechanics of Materials*, 7(2):91–107, 11 1988.

[106] J. Aboudi, S. W. Lee, and C. T. Herakovich. Three-dimensional analysis of laminates with cross cracks. *Journal of Applied Mechanics, Transactions ASME*, 55(2):389–397, 1988.

[107] F. W. Crossman, W. J. Warren, A. S. D. Wang, and G. E. Law. Initiation and growth of transverse cracks and edge delamination in composite laminates. II-Experimental correlation. *Journal of Composite Materials Supplement*, 14(1):88–108, 1980.

[108] D. L. Flaggs and M. H. Kural. Experimental determination of the in situ transverse lamina strength in graphite/epoxy laminates. *Journal of Composite Materials*, 16:103–116, 03 1982.

[109] S. H. Lim and S. Li. Energy release rates for transverse cracking and delaminations induced by transverse cracks in laminated composites. *Composites Part A*, 36(11):1467–1476, 2005.

[110] S. Li and F. Hafeez. Variation-based cracked laminate analysis revisited and fundamentally extended. *International Journal of Solids and Structures*, 46(20):3505–3515, 2009.

[111] J. L Rebiere and D. Gamby. A decomposition of the strain energy release rate associated with the initiation of transverse cracking, longitudinal cracking and delamination in cross-ply laminates. *Composite Structures*, 84(2):186–197, 2008.

[112] S. C. Tan and R. J. Nuismer. A theory for progressive matrix cracking in composite laminates. *Journal of Composite Materials*, 23:1029–1047, 1989.

[113] R. J. Nuismer and S. C. Tan. Constitutive relations of a cracked composite lamina. *Journal of Composite Materials*, 22:306–321, 1988.

[114] T. Yokozeki and T. Aoki. Stress analysis of symmetric laminates with obliquely-crossed matrix cracks. *Advanced Composite Materials*, 13(2):121–140, 2004.

[115] T. Yokozeki and T. Aoki. Overall thermoelastic properties of symmetric laminates containing obliquely crossed matrix cracks. *Composites Science and Technology*, 65(11-12):1647–1654, 2005.

[116] Z. Hashin. Analysis of cracked laminates: A variational approach. *Mechanics of Materials*, 4:121:136, 1985.

[117] J. Zhang, J. Fan, and C. Soutis. Analysis of multiple matrix cracking in $[\pm\theta/90_n]_s$ composite laminates, part I, inplane stiffness properties. *Composites*, 23(5):291–304, 1992.

[118] P. Gudmundson and S. Ostlund. First order analysis of stiffness reduction due to matrix cracking. *Journal of Composite Materials*, 26(7):1009–1030, 1992.

[119] P. Gudmundson and S. Ostlund. Prediction of thermoelastic properties of composite laminates with matrix cracks. *Composites Science and Technology*, 44(2):95–105, 1992.

[120] P. Gudmundson and W. Zang. An analytic model for thermoelastic properties of composite laminates containing transverse matrix cracks. *International Journal of Solids and Structures*, 30(23):3211–3231, 1993.

[121] E. Adolfsson and P. Gudmundson. Thermoelastic properties in combined bending and extension of thin composite laminates with transverse matrix cracks. *International Journal of Solids and Structures*, 34(16):2035–2060, 06 1997.

[122] E. Adolfsson and P. Gudmundson. Matrix crack initiation and progression in composite laminates subjected to bending and extension. *International Journal of Solids and Structures*, 36(21):3131–3169, 1999.

[123] W. Zang and P. Gudmundson. Damage evolution and thermoelastic properties of composite laminates. *International Journal of Damage Mechanics*, 2(3):290–308, 07 1993.

[124] E. Adolfsson and P. Gudmundson. Matrix crack induced stiffness reductions in [(0m/90n/+p/-q)s]m composite laminates. *Composites Engineering*, 5(1):107–123, 1995.

[125] P. Lundmark and J. Varna. Modeling thermo-mechanical properties of damaged laminates. In *3rd International Conference on Fracture and Damage Mechanics, FDM 2003*, volume 251–252 of *Advances in Fracture and Damage Mechanics*, pages 381–387, Switzerland, 2–4 Sept. 2003. Trans Tech Publications.

[126] M. Kachanov. *Elastic Solids with Many Cracks and Related Problems*, volume 30 of *Advances in Applied Mechanics*, chapter X, pages 260–445. Academic Press, Inc., 1993.

[127] A. Adumitroaie and E. J. Barbero. Intralaminar damage model for laminates subjected to membrane and flexural deformations. *Mechanics of Advanced Materials and Structures*, 2013.

[128] T. Yokozeki, T. Aoki, and T. Ishikawa. Transverse crack propagation in the specimen width direction of cfrp laminates under static tensile loadings. *Journal of Composite Materials*, 36(17):2085–2099, 2002.

[129] T. Yokozeki, T. Aoki, and T. Ishikawa. Consecutive matrix cracking in contiguous plies of composite laminates. *International Journal of Solids and Structures*, 42(9-10):2785–2802, 05 2005.

[130] T. Yokozeki, T. Aoki, T. Ogasawara, and T. Ishikawa. Effects of layup angle and ply thickness on matrix crack interaction in contiguous plies of composite laminates. *Composites Part A (Applied Science and Manufacturing)*, 36(9):1229–1235, 2005.

[131] E. J. Barbero, F. A. Cosso, and F. A. Campo. Benchmark solution for degradation of elastic properties due to transverse matrix cracking in laminated composites, http://dx.doi.org/10.1016/j.compstruct.2012.11.009. *Composite Structures*, 98(4):242–252, 2013.

[132] S. Li, S. R. Reid, and P. D. Soden. A continuum damage model for transverse matrix cracking in laminated fibre-reinforced composites. *Philosophical Transactions of the Royal Society London, Series A (Mathematical, Physical and Engineering Sciences)*, 356(1746):2379–2412, 10/15 1998.

[133] Janis Varna, Roberts Joffe, and Ramesh Talreja. A synergistic damage-mechanics analysis of transverse cracking [$\pm\theta/90_4$]s laminates. *Composites Science and Technology*, 61(5):657–665, 2001.

[134] A. S. D. Wang, P. C. Chou, and S. C. Lei. A stochastic model for the growth of matrix cracks in composite laminates. *Journal of Composite Materials*, 18(3):239–254, 05 1984.

[135] S. Li, S. R. Reid, and P. D. Soden. Modelling the damage due to transverse matrix cracking in fiber-reinforced laminates. In *Proceedings 2nd International Conference on Nonlinear Mechanics (ICNP-2)*, pages 320–323. Peking University Press, 1993.

[136] D.H. Cortes and E.J. Barbero. Stiffness reduction and fracture evolution of oblique matrix cracks in composite laminates. http://dx.doi.org/10.1007/s12356-009-0001-5. *Annals of Solid and Structural Mechanics*, 1(1)(1):29–40, 2010.

[137] E. J. Barbero, G. Sgambitterra, A. Adumitroaie, and X. Martinez. A discrete constitutive model for transverse and shear damage of symmetric laminates with arbitrary stacking sequence. http://dx.doi.org/10.1016/j.compstruct.2010.06.011. *Composite Structures*, 93:1021–1030, 2011.

[138] G. Sgambitterra, A. Adumitroaie, E.J. Barbero, and A. Tessler. A robust three-node shell element for laminated composites with matrix damage. http://dx.doi.org/10.1016/j.compositesb.2010.09.016. *Composites Part B: Engineering*, 42(1):41–50, 2011.

[139] A. M. Abad Blazquez, M. Herraez Matesanz, Carlos Navarro Ugena, and E. J. Barbero. Acoustic emission characterization of intralaminar damage in composite laminates. In *Asociación Española de Materiales Compuestos MATCOMP 2013*, Algeciras, Spain, July 2–5, 2013.

[140] E. J. Barbero and J. Cabrera Barbero. Analytical solution for bending of laminated composites with matrix cracks. *Composite Structures*, 135:140 – 155, 2016.

[141] E. J. Barbero and F. A. Cosso. Determination of material parameters for discrete damage mechanics analysis of composite laminates. *Composites Part B:Engineering*, 2013.

[142] H. T. Hahn. A mixed-mode fracture criterion for composite materials. *Composites Technology Review*, 5:26–29, 1983.

[143] M. M. Moure, S. Sanchez-Saez, E. Barbero Pozuelo, and E. J. Barbero. Analysis of damage localization in composite laminates using a discrete damage model. *Composites Part B: Engineering*, 66:224 – 232, 2014.

[144] J. Varna, R. Joffe, N. V. Akshantala, and R. Talreja. Damage in composite laminates with off-axis plies. *Composites Science and Technology*, 59(14):2139–2147, 1999.

[145] H. Chai, C. D. Babcock, and W. G. Knauss. One dimensional modeling of failure in laminated plates by delamination buckling. *International Journal of Solids and Structures*, 17(11):1069–1083, 1981.

[146] M.-K. Yeh and L.-B. Fang. Contact analysis and experiment of delaminated cantilever composite beam. *Composites: Part B*, 30(4):407–414, 1999.

[147] W. L. Yin, S. N. Sallam, and G. J. Simitses. Ultimate axial load capacity of a delaminated beam-plate. *AIAA Journal*, 24(1):123–128, 1986.

[148] D. Bruno. Delamination buckling in composite laminates with interlaminar defects. *Theoretical and Applied Fracture Mechanics*, 9(2):145–159, 1988.

[149] D. Bruno and A. Grimaldi. Delamination failure of layered composite plates loaded in compression. *International Journal of Solids and Structures*, 26(3):313–330, 1990.

[150] G. A. Kardomateas. The initial post-buckling and growth behaviour of internal delaminations in composite plates. *Journal of Applied Mechanics*, 60(4):903–910, 1993.

[151] G. A. Kardomateas and A. A. Pelegri. The stability of delamination growth in compressively loaded composite plates. *International Journal of Fracture*, 65(3):261–276, 1994.

[152] I. Sheinman, G. A. Kardomateas, and A. A. Pelegri. Delamination growth during pre- and post-buckling phases of delaminated composite laminates. *International Journal of Solids and Structures*, 35(1-2):19–31, 1998.

[153] D. Bruno and F. Greco. An asymptotic analysis of delamination buckling and growth in layered plates. *International Journal of Solids and Structures*, 37(43):6239–6276, 2000.

[154] H.-J. Kim. Postbuckling analysis of composite laminates with a delamination. *Computer and Structures*, 62(6):975–983, 1997.

[155] W. G. Bottega and A. Maewal. Delamination buckling and growth in laminates–closure. *Journal of Applied Mechanics*, 50(14):184–189, 1983.

[156] B. Cochelin and M. Potier-Ferry. A numerical model for buckling and growth of delaminations in composite laminates. *Computer Methods in Applied Mechanics*, 89(1-3):361–380, 1991.

[157] P.-L. Larsson. On delamination buckling and growth in circular and annular orthotropic plates. *International Journal of Solids and Structures*, 27(1):15–28, 1991.

[158] W.-L. Yin. Axisymmetric buckling and growth of a circular delamination in a compressed laminate. *International Journal of Solids and Structures*, 21(5):503–514, 1985.

[159] K.-F. Nilsson, L. E. Asp, J. E. Alpman, and L. Nystedt. Delamination buckling and growth for delaminations at different depths in a slender composite panel. *International Journal of Solids and Structures*, 38(17):3039–3071, 2001.

[160] H. Chai and C. D. Babcock. Two-dimensional modeling of compressive failure in delaminated laminates. *Journal of Composite Materials*, 19(1):67–98, 1985.

[161] J. D. Withcomb and K. N. Shivakumar. Strain-energy release rate analysis of plates with postbuckled delaminations. *Journal of Composite Materials*, 23(7):714–734, 1989.

[162] W. J. Bottega. A growth law for propagation of arbitrary shaped delaminations in layered plates. *International Journal of Solids and Structures*, 19(11):1009–1017, 1983.

[163] B. Storåkers and B. Andersson. Nonlinear plate theory applied to delamination in composites. *Journal of Mechanics and Physics of Solids*, 36(6):689–718, 1988.

[164] P.-L. Larsson. On multiple delamination buckling and growth in composite plates. *International Journal of Solids and Structures*, 27(13):1623–1637, 1991.

[165] M. A. Kouchakzadeh and H. Sekine. Compressive buckling analysis of rectangular composite laminates containing multiple delaminations. *Composite Structures*, 50(3):249–255, 2000.

[166] J. M. Comiez, A. M. Waas, and K. W. Shahwan. Delamination buckling: Experiment and analysis. *International Journal of Solids and Structures*, 32(6/7):767–782, 1995.

[167] G. A. Kardomateas. Postbuckling characteristics in delaminated kevlar/epoxy laminates: An experimental study. *Journal of Composites Technology and Research*, 12(2):85–90, 1990.

[168] O. Allix and A. Corigliano. Modeling and simulation of crack propagation in mixed-modes interlaminar fracture specimens. *International Journal of Fracture*, 77(2):111–140, 1996.

[169] D. Bruno and F. Greco. Delamination in composite plates: Influence of shear deformability on interfacial debonding. *Cement and Concrete Composites*, 23(1):33–45, 2001.

[170] J. W. Hutchinson and Z. Suo. Mixed mode cracking in layered materials. *Advances in Applied Mechanics*, 29:63–191, 1992.

[171] N. Point and E. Sacco. Delamination of beams: An application to the dcb specimen. *International Journal of Fracture*, 79(3):225–247, 1996.

[172] J. G. Williams. On the calculation of energy release rate for cracked laminates. *International Journal of Fracture*, 36(2):101–119, 1988.

[173] A. Tylikowsky. Effects of piezoactuator delamination on the transfer functions of vibration control systems. *International Journal of Solids and Structures*, 38(10-13):2189–2202, 2001.

[174] C. E. Seeley and A. Chattopadhyay. Modeling of adaptive composites including debonding. *International Journal Solids and Structures*, 36(12):1823–1843, 1999.

[175] W. J. Bottega and A. Maewal. Dynamics of delamination buckling. *International Journal of Non-Linear Mechanics*, 18(6):449–463, 1983.

[176] G. Alfano and M. A. Crisfield. Finite element interface models for the delamination analysis of laminated composites: Mechanical and computational issues. *International Journal for Numerical Methods in Engineering*, 50(7):1701–1736, 2001.

[177] P. P. Camanho, C. G. Dávila, and M. F. Moura. Numerical simulation of mixed-mode progressive delamination in composite materials. *Journal of Composite Materials*, 37(16):1415–1438, 2003.

[178] Z. Zou, S. R. Reid, and S. Li. A continuum damage model for delaminations in laminated composites. *Journal of Mechanics and Physics of Solids*, 51(2):333–356, 2003.

[179] X-P Xu and A. Needleman. Numerical simulations of fast crack growth in brittle solids. *Journal of Mechanics and Physics of Solids*, 42:1397–1434, 1994.

[180] J. R. Reeder. A bilinear failure criterion for mixed-mode delamination in composite materials. *ASTM STP 1206*, pages 303–322, 1993.

[181] H. Chai. Three-dimensional fracture analysis of thin-film debonding. *International Journal of Fracture*, 46(4):237–256, 1990.

[182] J. D. Withcomb. Three dimensional analysis of a post-buckled embedded delamination. *Journal of Composite Materials*, 23(9):862–889, 1989.

[183] D. Bruno, F. Greco, and P. Lonetti. A 3d delamination modelling technique based on plate and interface theories for laminated structures. *European Journal of Mechanics A/Solids*, 24:127–149, 2005.

[184] J. G. Williams. The fracture mechanics of delamination tests. *Journal of Strain Analysis*, 24(4):207–214, 1989.

[185] W.-L. Yin and J. T. S. Wang. The energy-release rate in the growth of a one-dimensional delamination. *Journal of Applied Mechanics*, 51(4):939–941, 1984.

[186] J. R. Rice. A path independent integral and the approximate analysis of strain concentrations by notches and cracks. *Journal of Applied Mechanics*, 35:379–386, 1968.

[187] Simulia. Abaqus analysis user's manual, cohesive elements. Accessible from the windows start menu, through **Dassault Systemes Documentation SIMULIA 2020**.

[188] C. G. Dávila, P. P. Camanho, and M. F. De Moura. Mixed-mode decohesion elements for analyses of progressive delamination. In *42nd AIAA/ASME/ASCE/AHS/ASC Structures, Structural Dynamics, and Materials Conference, April 16 - 19, 2001*, volume 3, pages 2277–2288, Seattle, WA, 2001. AIAA.

[189] Simulia. Abaqus analysis user's manual, crack propagation analysis. Accessible from the windows start menu, through **Dassault Systemes Documentation SIMULIA 2020**.

[190] E.J. Barbero and J.N. Reddy. Jacobian derivative method for three-dimensional fracture mechanics. *Communications in Applied Numerical Methods*, 6(7):507–518, 1990.

[191] M. Benzeggagh and M. Kenane. Measurement of mixed-mode delamination fracture toughness of unidirectional glass/epoxy composites with mixed-mode bending apparatus. *Composite Science and Technology*, 56:439, 1996.

[192] J. Reeder, S. Kyongchan, P. B. Chunchu, and D. R.. Ambur. Postbuckling and growth of delaminations in composite plates subjected to axial compression. In *43rd AIAA/ASME/ASCE/AHS/ASC Structures, Structural Dynamics, and Materials Conference*, volume 1746, page 10, Denver, CO, 2002.

[193] D5528. Standard test method for mode i interlaminar fracture toughness of unidirectional fiber-reinforced polymer matrix composites. *ASTM*, pages 1–13, 2013.

[194] D7905. Standard test method for determination of the mode ii interlaminar fracture toughness of unidirectional fiber-reinforced polymer matrix composites. *ASTM*, pages 1–18, 2019.

[195] M. Cabello, A. Turon, J. Zurbitu, J. Renart, C. Sarrado, and F. Martínez. Progressive failure analysis of dcb bonded joints using a new elastic foundation coupled with a cohesive damage model. *European Journal of Mechanics A/Solids*, 63:22–35, 2017.

[196] SIMULIA. Defining the constitutive response of cohesive elements using a traction-separation description, 2020.

[197] D6671. Standard test method for Mixed Mode I-Mode II interlaminar fracture toughness of unidirectional fiber reinforced polymer matrix composites. *ASTM*, pages 1–15, 2019.

[198] SIMULIA. Simulia user assistance 2020. abaqus, elements, special-purpose elements, cohesive elements, defining the constitutive response of cohesive elements using a traction-separation description. 2020.

[199] Ever J Barbero and Javier Cabrera-Barbero. Damage initiation and evolution during monotonic cooling of laminated composites. *Journal of Composite Materials*, 52(30):4151–4170, 2018. DOI:10.1177/0021998318776721.

[200] D.F. Adams, R.S. Zimmerman, and E.M. Odom. Polymer matrix and graphite fiber interface study. *NASA Technical Report CR-165632*, 1985.

[201] T.R. King, D.M. Blackketter, D.E. Walrath, and D.F. Adams. Micromechanics prediction of the shear strength of carbon fiber/epoxy matrix composites: The influence of the matrix and interface strengths. *Journal of Composite Materials*, 26(4):558–573, 1992.

[202] D.F. Adams. A micromechanics analysis of the influence of the interface on the performance of polymer-matrix composites. *Journal of Reinforced Plastics and Composites*, 6(1):66–88, 1987.

[203] D.E. Bowles. Micromechanics analysis of space simulated thermal deformations and stresses in continuous fiber reinforced composites. *NASA Technical Report 90N21140*, 1990.

[204] D.S. Adams, D.E. Bowles, , and C.T. Herakovich. Characteristics of thermally-induced transverse cracks in graphite epoxy composite laminates. *NASA Technical Report TM-85429*, 1983.

[205] H. L. McManus. Prediction of thermal cycling induced matrix cracking. *Journal of Reinforced Plastics and Composites, 15:124-140*, 1996.

[206] J. R. Maddocks. Microcracking in composite laminates under thermal and mechanical loading. *NASA Technical Report N96-16101*, 1995.

[207] Adi Adumitroaie and E. J. Barbero. Intralaminar damage model for laminates subjected to membrane and flexural deformations. *Mechanics of Advanced Materials and Structures*, 22(9):705 – 716, 2015.

[208] Ever J Barbero and J. Cabrera Barbero. Determination of material properties for progressive damage analysis of carbon/epoxy laminates. *Mechanics of Advanced Materials and Structures, 26(11): 938-947*, 2019.

[209] E. J. Barbero. *Finite Element Analysis of Composite Materials Using Abaqus*. CRC Press, 2013.

[210] E. J. Barbero. *Finite Element Analysis of Composite Materials Using ANSYS*. CRC Press, 2014.

[211] E. J. Barbero and D. C. Cortes. A mechanistic model for transverse damage initiation, evolution, and stiffness reduction in laminated composites. *Composites Part B, 41:124-132*, 2010.

[212] Mohammadhossein Ghayour, H. Hosseini-Toudeshky, M. Jalalvand, and E. J. Barbero. Micro/macro approach for prediction of matrix cracking evolution in laminated composites. *Journal of Composite Materials*, 50(19):2647–2659, 2016.

[213] M. M. Moure, F. Otero, S. K. Garcia-Castillo, S. Sanchez-Saez, E. Barbero, and E. J. Barbero. Damage evolution in open-hole laminated composite plates subjected to in-plane loads. *Composite Structures*, 133:1048 – 1057, 2015.

[214] M. M. Moure, S. K. Garcia-Castillo, S. Sanchez-Saez, E. Barbero, and E. J. Barbero. Influence of ply cluster thickness and location on matrix cracking evolution in open-hole composite laminates. *Composites Part B: Engineering*, 95:40 – 47, 2016.

[215] M.M. Moure, S.K. Garcia-Castillo, S. Sanchez-Saez, E. Barbero, and E.J. Barbero. Matrix cracking evolution in open-hole laminates subjected to thermo-mechanical loads. *Composite Structures*, 183:510–520, 2018.

[216] Javier Cabrera Barbero. *Thermal-Fatigue and Thermo-Mechanical Equivalence for Transverse Cracking Evolution in Laminated Composites*. PhD thesis, West Virginia University, 2018.

[217] R. G. Budinas and J. K. Nisbet. *Shigley's Mechanical Engineering Design, 10th ed.* McGraw Hill, 2015.

[218] S. S. Sternstein. Yielding in glassy polymers. *Polymeric Materials, 369-410*, 1975.

[219] H. H. Kausch. *Polymer Fracture, Polymers, Properties and Applications*. Springer Verlag, Berlin, 1978.

[220] T. L. Brown. *The effect of long-term thermal cycling on the microcracking behavior and dimensional stability of composite materials.* PhD thesis, Virginia Polytechnic Institute and State University, 1997.

[221] S. S. Tompkins, J. Y. Shen, and A. J. Lavoie. Thermal cycling of thin and thick ply composites. *NASA TR NAG-1-1912,* 1(15):326-335, 1994.

[222] J. A. Nairn and S. Hu. *Damage Mechanics of Composite Materials,* chapter 6. Matrix Microcracking, pages 1–46. Elsevier, 1994.

[223] E. J. Barbero. *Time-temperature superposition principle for predicting long-term response of linear viscoelastic materials,* chapter 2: Time-temperature-age superposition principle for predicting long-term response of linear viscoelastic materials, pages 48–69. Woodhead, Cambridge (ISBN978-1-84569-525-5), 2011.

[224] Hamed Pakdel and Bijan Mohammadi. Characteristic damage state of symmetric laminates subject to uniaxial monotonic-fatigue loading. *Engineering Fracture Mechanics 199:86-100,* 2018.

[225] John A. Nelder and R. Mead. A simplex method for function minimization. *Computer Journal. 7 (4): 308-313,* 1965.

[226] E. J. Barbero and Kevin J. Ford. Characterization of self-healing fiber-reinforced polymer-matrix composite with distributed damage. *Journal of Advanced Materials,* 39(4):20–27, 10 2007. http://barbero.cadec-online.com/papers/2007/07BarberoFordCharacterizationOfSelf-Healing.pdf.

[227] E. J. Barbero, F. Greco, and P. Lonetti. Continuum damage-healing mechanics with application to self-healing composites. *International Journal of Damage Mechanics,* 14(1):51–81, 2005. http://barbero.cadec-online.com/papers/2005/10.1177/1056789505045928.pdf.

[228] A. Caceres. *Local Damage Analysis of Fiber Reinforced Polymer Matrix Composites, Ph.D. dissertation.* PhD thesis, West Virginia University, Morgantown, WV, 1998.

[229] G. Puri. *Python Scripts for Abaqus.* abaquspython.org, Atlanta, GA, 2011.

Index

Printed in the United States
by Baker & Taylor Publisher Services